Literature and Cartography

Theories, Histories, Genres

Literature and Cartography

Theories, Histories, Genres

edited by
Anders Engberg-Pedersen

The MIT Press
Cambridge, Massachusetts
London, England

© 2017 Massachusetts Institute of Technology

All rights reserved. No part of this book may be reproduced in any form by any electronic or mechanical means (including photocopying, recording, or information storage and retrieval) without permission in writing from the publisher.

This book was set in Neue Haas Grotesk Text Pro and Deca Serif by Toppan Best-set Premedia Limited. Printed and bound in the United States of America.

Library of Congress Cataloging-in-Publication Data is available.

ISBN: 978-0-262-03674-0

10 9 8 7 6 5 4 3 2 1

Contents

Acknowledgments   vii

Introduction – Estranging the Map: On Literature and Cartography   1
Anders Engberg-Pedersen

I   THEORIES AND METHODOLOGIES   19

1. Cartographic Fiction   21
   Jean-Marc Besse

2. Literary Cartography: Mapping as Method   45
   Barbara Piatti

3. The (Un)Mappability of Literature   73
   Robert Stockhammer

4. Cartographic Tropes: From Kant's Maps to Foucault's Topology   99
   Oliver Simons

5. The Language of Cartography: Borges as Mapmaker   119
   Bruno Bosteels

II   HISTORIES AND CONTEXTS   141

6. Muses of Cartography: Charting Odysseus from Homer to Joyce   143
   Burkhardt Wolf

7. Diagrammatic Thought in Medieval Literature   173
   Simone Pinet

8. Hybrid Maps: Cartography and Literature in Spanish Imperial Expansion, Sixteenth Century   199
   Ricardo Padrón

9. Bend of the Baroque: Toward a Literary Hydrography in France   219
   Tom Conley

10 Goethe and the Cartographic Representation of Nature around 1800   253
    John K. Noyes

11 Conceptualizing the Novel Map: Nineteenth-Century French Literary Cartography   279
    Patrick M. Bray

12 African Cartographies in Motion   299
    Dominic Thomas

III GENRES AND THEMES   323

13 Popular Map Genres in American Literature   325
    Martin Brückner

14 Map Line Narratives   361
    Jörg Dünne

15 Material Cartography: João Guimarães Rosa's Paratexts   391
    Clara Rowland

16 Cartographies of War: Star Charts, Topographic Maps, War Games   411
    Anders Engberg-Pedersen

Conclusion   443

Contributors   453
Index   459

Acknowledgments

My gratitude extends first and foremost to the contributors to this book—Jean-Marc Besse, Barbara Piatti, Robert Stockhammer, Oliver Simons, Bruno Bosteels, Burkhardt Wolf, Simone Pinet, Ricardo Padrón, Tom Conley, John K. Noyes, Patrick M. Bray, Dominic Thomas, Martin Brückner, Jörg Dünne, and Clara Rowland. Without their willingness to take time from their own projects there would be no book. Among them, however, one person must be singled out for his pivotal role in the field as well as for his efforts in the production of this volume. With singular charm and enthusiasm, Tom Conley has for decades been a *force motrice* in the establishment and development of literature and cartography. Among the contributors are several of his students, myself included, and during my work on the book he has on numerous occasions displayed his unfailing generosity. A remarkable exploratory seminar—"Cartography and Spatial Thinking from Humanism to the Humanities"—that he and Katharina Piechocki organized at the Radcliffe Institute for Advanced Study at Harvard University convened scholars from a range of disciplines, some of them represented in this volume, and gave me useful feedback on the introduction and a wealth of new ideas. At MIT Press I am most thankful to my editors, Gita Manaktala and Marc Lowenthal, for taking on the project, to Jesús J. Hernández for his help in guiding the manuscript through the publication process, and to the design team. Two outside readers offered detailed and useful reports whose insights and suggestions I have tried to adopt. In the form of a Sapere Aude grant The Danish Council for Independent Research generously supported the production of the book. Andrew Patten translated the chapter by Jörg Dünne, and Emily Finer provided me with a scan of the wonderful doubly inverted map in Viktor Shklovsky's essay on *Tristram Shandy*. The unsung heroes of any book that deals with cartography are the librarians around the world who patiently indulge our requests for ever more information, material, and scans. There are far too many to name, but you know who you are and so do we.

Introduction – Estranging the Map: On Literature and Cartography

Anders Engberg-Pedersen

I am told that there are people who do not care for maps, and find it hard to believe.
— Robert Louis Stevenson[1]

It is not down in any map; true places never are.
— Herman Melville[2]

Readers of *Tristram Shandy* know well the lure of maps. Early in Laurence Sterne's novel, the narrator relates the difficulties that beset Uncle Toby, as he tries to describe the sequence of events at the siege of Namur in a well-ordered string of words. Stumbling across the linguistic obstacles of the unwieldy terms of fortification, however, Uncle Toby quickly loses his way in the story and only manages to get back on track when he thinks of a cartographic solution to his narrative problem: "If he could purchase such a thing, and have it pasted down upon a board, as a large map of the fortifications of the town and citadel of *Namur*, with its environs, it might be a means of giving him ease."[3] And indeed, bent over the map, Uncle Toby masters his subject and eventually manages to narrate the story of the siege with true eloquence.

In literary history, Uncle Toby is not alone in seizing on maps to clarify the events of narrative. To ground fictional space, to visualize sites and movements, to help readers get their bearings in the imaginative world conjured by the text, authors have frequently included cartographic material either as paratexts or as inserts in the narratives themselves. You can easily understand why. Maps entice and fascinate, and they whisper implicit promises of comprehension, of clarity, of transparency, and of order. In accordance with Gaston Bachelard's dictum that "the cleverer I am at miniaturizing the world, the better I possess it,"[4] a map would seem the ideal supplement and guide to literature.[5] In this sense, Sterne's scene is emblematic of the larger concerns of literature and cartography.

Attentive readers of *Tristram Shandy*, however, will also recall that Uncle Toby's engagement with the map is significantly more laborious and complex than what appears at first glance. For Toby's map, the narrator informs us, does not simply illustrate. It serves, rather, as an alternative symbolic medium that takes the place of language, freeing him from the unsteady slipperiness of discourse—a slipperiness that is particularly difficult to manage when it comes to questions "about space." Yet, the map itself is not immediately comprehensible. The lure of a transparent medium, of a self-explanatory representation, is thwarted by the cumbersome presence of an opaque object. At first, Toby has great difficulty making any sense of the map, and only after "a fortnight's close and painful application" does he manage to "form his discourse with passable perspicuity."[6] To read the map properly, to decipher its codes and symbols, he has to rely on the text in the maps' cartouche as well as on a book on military architecture. Following a zigzagging path to comprehensible narration and even eloquence a few months later, Toby meanders from text to map to text again and then back to the map before he eventually manages to give an orderly account of events. But this does not spell the end of his cartographic travails. "Be'pictur'd" by the appealing image of the map, his fascination evolves into an obsession. Toby procures ever more maps of other fortified towns and along with them a veritable library of military treatises on the principles of fortification and siege warfare. But now the complexity of the treatises outstrips the visualizing capacities of his maps and leaves him, once again, deeply perplexed. In the end, Toby abandons both of the media that proved insufficient tools for his narrative: texts and maps.

This more detailed account begins to suggest some of the complexities of the entwinement of literature and cartography. Not merely an illustrative guide, the literary map establishes a varied series of productive tensions, of dependencies and complementarities, but also of exclusions and frictions—a set of relations that all raise fundamental theoretical questions about representation, fiction, and space. Can literature be mapped? How do literary and cartographic topographies differ? Which map effects arise in literary texts? How have historical contexts brought literature and cartography into dialogue and with what consequences? In which ways did the development of new mapping techniques impact the writing and reading of texts? Are there distinct genres of literary cartographies or of cartographic writing? And which possibilities have emerged with the rise of the digital humanities? In recent years literary scholars in increasing numbers have begun probing these questions. Charting the myriad relations between literature and cartography, they have drawn the outlines of one of the most vibrant fields in contemporary scholarship. This book brings some of these scholars together to present a synoptic overview of the field spanning basic theoretical questions, key historical moments, and the essential themes and debates.

The Cartographic Present

The current scholarly interest in literary cartography no doubt stems from the ubiquity of maps in contemporary everyday culture. From the confluence of globalization and digitization, an unprecedented culture of maps has emerged. We live in a cartographic present in which maps of all sorts surround us: maps of our own street and local environment, of distant conflicts, of crime, of disease and epidemics, weather maps, polling maps, and so on, most of them digital and provided by archives such as Google Earth or generated live by the satellites of GPS. Reaching into our pocket, we have an immense atlas readily available at our fingertips. Without any formal training to develop a critical cartographic literacy we have become the skilled consumers and users of maps in an age of convenience cartography. Often unknowingly we also generate maps and even become maps, as the data of our travels, our purchases, our work habits is conglomerated, processed, and visualized with the aid of GIS on the very maps that are then presented for us to read or, precisely, are hidden from view behind secure firewalls.[7] The map has become such a part of the everyday that it often no longer appears as an external object or even, in Marshall McLuhan's terms, an extension of ourselves, but instead as a completely integrated part of our lives. In an unprecedented way, we live the map.

So, it seems, do widely read contemporary authors. The general cartographic turn of culture at large has also manifested itself in works of literature. Throughout the twentieth century more literary works engaged with maps than ever before.[8] But in the early twenty-first century, the literary fascination with maps has only intensified and made it onto international bestseller lists.[9] In 2005, for example, Daniel Kehlmann achieved widespread acclaim for *Measuring the World* (*Die Vermessung der Welt*), a novel that aligns the lives and scientific careers of Carl Friedrich Gauss and Alexander Humboldt and compares their complementary but diverging attempts to explore, map, and measure the phenomena of the world. Or consider the book Judith Schalansky issued four years later, *Atlas of Remote Islands: Fifty Islands I Have Never Set Foot on and Never Will* (*Atlas der abgelegenen Inseln: Fünfzig Inseln, auf denen ich nie war und niemals sein werde*), a smartly designed volume that juxtaposes fifty maps of distant islands with fictional narratives of each place. She thereby revives and develops the genre of the island fiction, or *isolario*–a genre that emerged in the fifteenth century and continued in works such as Thomas More's *Utopia*, Daniel Defoe's *Robinson Crusoe*, Robert Louis Stevenson's *Treasure Island*, and Jules Verne's *The Mysterious Island* (*L'Île mysterieuse*).[10] The following year, Michel Houellebecq won the Goncourt Prize for his novel *The Map and the Territory* (*La carte et le territoire*), in which the protagonist has his artistic breakthrough with the exhibition "The Map Is More Interesting Than the Territory"–at once in accordance with and an inversion of Alfred Korzybski's well-known dictum that "a map is *not* the territory."[11] Exhibiting photographs not of the territory, but of the Michelin regional maps representing the territory, the artist and Houellebecq's novel put cartography itself on display as a medium, as an artifact at once material and aesthetic, concrete and fictional.

With the rise of a cartographic culture pervaded by map users and map producers and the recent literary engagement with cartography, it is perhaps unsurprising that a large number of literary scholars have also directed their attention to the map. Among the many twists and turns within the humanities, the "cartographic turn" has already been diagnosed several years ago, and if we go by the number of publications on literature and cartography it seems by now to be not just a critical concept but a thriving reality. Already in 1998 Bruno Bosteels noted the onset of a "veritable cartographic turn" in the humanities and traced the notion back to Gilles Deleuze and Félix Guattari's cartographies of the unconscious in the late 1970s and early 1980s.[12] Recently Martin Brückner similarly suggested the emergence of a cartographic turn akin to a Kuhnian paradigm shift within American studies based on the steep rise in the number of new essays and books using and exploring concepts such as maps or mapping.[13] While this development no doubt owes much to the cartographic culture we live in, it also has an internal, theoretical genealogy that can be traced back through a number of related turns.

Spatial Turns

The critical discourse on space underwent a profound change in the last few decades of the twentieth century. As early as in 1967 Michel Foucault, in his text "Of Other Spaces" ("Des espaces autres"), surmised that the focus on history and temporality in the nineteenth century was giving way to a focus on space as the main experiential category of modernity. As he writes, "The present epoch will perhaps be above all the epoch of space. We are in the epoch of simultaneity: we are in the epoch of juxtaposition, the epoch of the near and far, of the side by side, of the dispersed."[14] While the text went largely unnoticed until its publication and translation in the mid-1980s, other thinkers in a variety of fields enlarged the critical vocabulary with new spatial concepts and metaphors. Philosophers such as Deleuze and Guattari and Michel de Certeau developed a set of new categories and figures (the rhizome, deterritorialization, space vs. place, etc.) with ontological and epistemological import; the cultural geographer David Harvey traced the impact of global capitalism on the ways we represent the world to ourselves; and Doreen Massey linked space as a form of cultural inquiry to questions of gender, labor, and wealth. The "spatial turn" as a critical term, however, can be traced to Edward Soja's *Postmodern Geographies: The Reassertion of Space in Critical Social Theory* from 1989. Soja diagnosed a shift from the "Zeitgeist" of the past to a contemporary "Raumgeist" and encouraged the development of a new "politicized spatial consciousness"—a "cognitive mapping" of the postmodern situation as also developed by Fredric Jameson in his analysis of late capitalism.[15] Countering Paul Virilio's claim that space as we know it has disappeared, these thinkers not only showed that space in its multiple instantiations and functions was fundamental for how we live, they also turned it into a gravitational center of critical discourse.[16]

Inflected differently by diverging interests, traditions, and fields, the spatial turn spun off both a topological and a topographic turn. Originating in mathematics and philosophy, topology concerns itself not with space itself, but with spatial relations. Focused on the geometric properties of objects and bodies in space, topologists examine the properties of spatial relations that remain unaffected by the mutations and deformations of such bodies when they are stretched, bent, or twisted without causing any rupture. As Stephan Günzel summarizes, a topological description concerns "relations that are not themselves spatial (in the sense of extension and materiality)."[17] Topography, on the other hand, has found great resonance in literary studies with its emphasis on the semiotic aspects of spatial representations.[18] Composed of the Greek words for space—*topos*—and writing—*graphein*—topography has been a favored critical category for the exploration of the semiotic constitution of spaces and places across different discourses. J. Hillis Miller, for example, in his book *Topographies*, examines the function of landscapes and cityscapes in literature as well as the philosophical import of topographic categories in critical thought.[19] Treating space as a text and texts as space or a way of producing spatial configurations, the topographic approach has served as the common ground for a comparison of literature and cartography.[20]

The multiple inflections of the spatial turn share basic assumptions. First and foremost, spatial theorists have brought to our attention that, as Denis Cosgrove writes, "position and context are centrally and inescapably implicated in all constructions of knowledge."[21] But they have also stressed the productive and performative aspects of space. Against Cartesian and Kantian notions of abstract space, against a conception of space as a container in which events and actions take place, of space as extension, space as natural, as the given, critics have unearthed the active, productive forces inherent in spatial practices and representations. As Henri Lefebvre noted in *The Production of Space (La Production de l'espace)*, space is "second nature, an effect of the action of societies on 'first nature.'"[22] In other words, social action does not simply take place in space, it constitutes space.[23] In like fashion, the demiurgic qualities of textual and visual representations have come to the fore. Representations of space are not simply a matter of registering the world, but ways of worldmaking.

As the latest in the series of spatial turns, the cartographic turn owes much to its predecessors, but it also derives from the impact of constructivism on historians of cartography. In 1989, J. B. Harley, in his insightful article "Deconstructing the Map," provocatively suggested that to make progress in the understanding of maps, the last people to consult were the cartographers themselves. Adopting theoretical claims from Derrida and Foucault to challenge the seductive transparency and alleged scientific objectivity of the map, he instead recast it as a complex of rhetorical structures, power, and hidden ideologies.[24] As Christian Jacob put it a few years later, "Maps [had] lost their 'innocence.'"[25] In this respect, the critical reception of maps in the late 1980s and early 1990s recapitulates the shift in Uncle Toby's stance toward the siege map. But instead of abandoning the medium once its seductive transparency and pretensions of objectivity, immediacy, and scientificity were revealed to be visual effects of complex and painstaking efforts of construction, selection, and projection, these very procedures became the subject of critical analysis. In illuminating and subtle ways, historians of cartography have since shown that only when we resist the visual seduction and notice the opacity that lies hidden in plain view, will we begin to understand what has been done to the map, what the map itself does, and thus what it is.

While many cartographers do have a much more nuanced notion of maps than Harley suggested, historians who approached the map with a hermeneutics of suspicion that questioned the most basic assumptions and definitions managed to wrest the map from the hands of cartographers and offered it to other disciplines as an object of inquiry. "Cartography Is Dead (Thank God)" wrote Denis Wood to celebrate the end of the exclusive rights on maps held by the official institutions of cartography.[26] And indeed, the success of the democratization has been remarkable. In literary studies, in particular, scholars in the past few decades, many of them represented in this volume, have latched onto the map and begun exploring the cartographic elements of literature. In the process they have not only brought traditional literary theory to bear on the map, thereby further detailing its intricate workings with deconstructive, semiotic, and psychoanalytic concepts, among many others. They have also enriched the vocabulary of literary studies with cartographic terms such as *maps*, *mappings*, *projected spaces*, and so on, thereby opening a range of new perspectives on literature itself.

The rise of cartography in the field of literature has also come at a cost. With the widespread appeal of cartographic terms as tools of analysis and basic definitions up in the air, literary studies is faced with a ubiquity problem. Not only are we surrounded by maps in our everyday lives; in critical discourse almost everything seems to be a map or a mapping. The very qualities that make terms from the cartographic lexicon highly useful and productive have resulted in an often inadvertent or at least unacknowledged slippage between concept and metaphor. The sharp edges that clearly marked the borders and limits of the material map have become increasingly fuzzy in their abstract iterations. As Martin Brückner rightly puts it, "the cartographic turn is poised to imbue maps and mappings with too much metaphor and thus empty them of meaning."[27] Stretched thin to cover all sorts of phenomena that might share a spatial or epistemological *something*, maps and mappings are in danger of obfuscating rather than enlightening. The problem is not restricted to the literary field. In a critique of the general overuse of the terms *mapping* and *remapping*, Soja warns against the false belief that "everyone thinks they take part in the *spatial turn* when they *map* something."[28] And Marion Picker echoed the critique when she diagnosed a crisis within cartography itself because all traditional definitions of the map have become unstable. Is the future of cartography, she muses, a future "without the map?"[29]

Estranging the Map

Aside from the appeal of fashionable terms, the underlying cause of the cartographic ubiquity problem in critical discourse may well be the expectation that map concepts and metaphors are as self-evident as the material objects themselves were once purported to be. As the material, topographic map entices the viewer with a supreme nonperspectival gaze that no human eye could ever have, so the discursive map or mapping promises a conceptual survey that ignores the distortions of projections and the silent suppression performed by selection and choice. All maps misrepresent, whether material, conceptual, or metaphorical. They are useful illusions that guide us and lead us astray at one and the same time.[30]

Literary scholars will do well to recall this double cartographic gesture and take to heart the (de)constructive efforts of historians of cartography since the 1980s. The solution is not to return to the exclusive focus on the material map and a naive mimetic theory of representation. Rather, as an antidote to the ubiquity and self-evidence of maps in society at large and in literary studies in particular, we need to once again deconstruct or estrange the map. To get a sense of what Jörg Dünne has called the "uncanniness of mapping,"[31] we need to tease out its codes and signs, its visual schemas and norms, the ideologies and epistemologies that are folded into its forms, into its making, and into its effects in concrete historical circumstances. For such a procedure literature has shown itself to be a useful and perhaps even a privileged medium. Estrangement has been a staple of literary theory ever since Viktor Shklovsky's essay "Art as Device" (Isskustvo kak priëm) from 1917. As you will recall, Shklovsky warned against the automation of perception, an automation that "eats away at things, at clothes, at furniture, at our wives"—and, we may add, at our maps.[32] "And so," he continues in a famous paragraph,

> in order to return sensation to our limbs, in order to make us feel objects, to make a stone feel stony, man has been given the tool of art. The purpose of art, then, is to lead us to a knowledge of a thing through the organ of sight instead of recognition. By "estranging" objects and complicating form, the device of art makes perception long and "laborious."[33]

One such object that literature has the ability to defamiliarize or estrange is the map. Dehabituating our familiar engagement with the maps that flood our everyday experience, literature frequently makes the map feel mappy. Once addressed by a different medium, the literary map accrues a density that unconceals its otherwise near-invisible operations. If we again take *Tristram Shandy* as our guide to literature and cartography, we find a telling example of cartographic estrangement in the rather unconventional plot map that Sterne inserts in the sixth volume of the novel (figure 0.1).

As indicated by the two inscriptions at the bottom right and left, the map was both designed (invenit) and carved (sculpsit) by Tristram Shandy himself, and it charts the

— To be sure, said my mother. So here ended the proposition, — the reply, — and the rejoinder, I told you of.

— It will be some amusement to him, too, said my father.

— A very great one, answered my mother, if he should have children.

— Lord have mercy upon me! said my father to himself —

* * * * * * * * * * *
* * * * * * * * * * *
* * * * * * * * * * *

CHAPTER CCI.

I AM now beginning to get fairly into my work; and by the help of a vegetable diet, with a few of the cold seeds, I make no doubt but I shall be able to go on with my uncle Toby's story, and my own, in a tolerable straight line. Now,

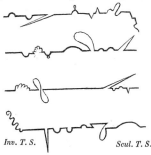

Inv. T. S. *Scul. T. S.*

These were the four lines I moved in through my first, second, third, and fourth volumes.* — In the fifth volume I have been very good, — the precise line I have described in it being this : —

* Alluding to the first edition.

By which it appears, that, except at the curve, marked A, where I took a trip to Navarre; — and the indented curve B, which is the short airing when I was there with the Lady Baussiere and her page, — I have not taken the least frisk of a digression, till John de la Casse's devils led me the round you see marked D; — for as for c c c c c, they are nothing but parentheses, and the common *ins* and *outs* incident to the lives of the greatest ministers of state; and when compared with what men have done, — or with my own transgressions at the letters A B D, — they vanish into nothing.

In this last volume I have done better still, — for from the end of Le Fever's episode, to the beginning of my uncle Toby's campaigns, — I have scarce stepped a yard out of my way.

If I mend at this rate, it is not impossible, — by the good leave of his Grace of Benevento's devils, — but I may arrive hereafter at the excellency of going on even, thus: —

which is a line drawn as straight as I could draw it by a writing-master's ruler (borrowed for that purpose) turning neither to the right hand nor to the left.

This *right line*, — the path-way for Christians to walk in! say divines, —

— The emblem of moral rectitude! says Cicero, —

— The *best line!* say cabbage-planters, — is the shortest line, says Archimedes, which can be drawn from one given point to another. —

I wish your ladyships would lay this matter to heart, in your next birth-day suits; —

— What a journey!

Pray can you tell me, — that is, without anger, before I write my chapter upon straight lines, — by what mistake, — who told them so, — or how it has come to pass, that your men of wit

Figure 0.1

The map of the meandering plot in Laurence Sterne's *Tristram Shandy* (Laurence Sterne, *The Life and Opinions of Tristram Shandy, Gentleman* [Leipzig: Bernh. Tauchnitz Jun., 1849], 368).

meandering course of the digressive plot in the first four volumes of the book.[34] A fifth plotline didn't make it onto the map but has been relegated to the following page. Just as Uncle Toby sought to grasp the science of fortification with a cartographic aid, this plot map is to serve as a "commentary, scholium, illustration, and key" to the novel itself.[35]

A cartographic satire, Sterne's metamap puts on display the incongruous relation between the contingent singularity of an unpredictable narrative and the generality and fixity of cartographic means of representation. Yet in its very incongruity, the plot map offers a dual perspective from which literary texts and literary maps can get a better view of themselves in the reflection of the other. For what makes this a map? The authority of the signature? The fact that the narrator claims to have produced one? The graphic representation of temporal progression from left to right in each volume and from top to bottom in the novel? Or, conversely, what makes it not a map? The complete absence of conventional cartographic elements and symbols such as a graticule, place names, or topographic details? The fact that the narrator's curious, erratic dance has been choreographed by Sterne in a smooth, abstract space rather than on the concrete territory described in the text? Do the axes indicate geographic movements, are they scales of suspense, or both? How, on the other hand, do the actual plot movements differ from their cartographic representation? What is the relation between the spatiality staked out by the narrative and that on the map? And which selection process made various narrative threads coalesce into a single line in each volume? Is plot, as Sterne seems to suggest, indeed unmappable?

Shklovsky, for one, did not think so. *Tristram Shandy* plays a central role in Shklovsky's theory of literature, and in an essay on the novel reprinted in his *Theory of Prose*, he included the plot map and offered his comments: even if Sterne fails to take into account the interruption of the motifs, the plotlines, he believed, "are more or less accurate." Indeed, they serve as "proof" that Sterne deliberately made use of the fundamental principle Shklovsky had come to believe governed the novel: estrangement. If any of Shklovsky's contemporary readers had attempted to establish a concordance between the map and the novel, however, they would certainly have been baffled, for in Shklovsky's article the plot map was accidentally printed doubly inverted: upside down and left to right (figure 0.2).[36]

The error is supremely ironic, for it undermines completely the oddly scientific and mimetic theory of visual representation that the father of literary estrangement entertained in his article on *Tristram Shandy*. Instead of taking one of the strangest maps in literary history (inverted or not) at face value, we might use it to extend Shklovsky's literary theory to the field of literary cartography. Taking apart both the conventional plot structure and the conventional map structure and then putting them together in a single representation, Sterne's plot map points to the mutual estrangement at work when literature and cartography are brought into contact. The illustrative function is

ГЛАВА CCI, стр. 433. «Я начинаю теперь совсем добросовестно приступать к делу, и я не сомневаюсь, что при помощи вегетарьянской диэты и, изредка прохладительного, мне удастся продолжать дяди Тобину повесть так же как и мою собственную, довольно таки прямолинейно.

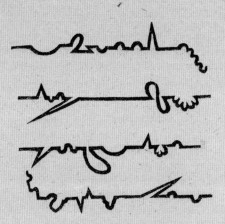

Вот те четыре линии, по которым я подвигался в моем первом, втором, третьем и четвертом томе.—В пятом я вел себя вполне благопристойно; точная, описанная мною линия, такова:

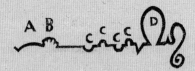

Из нее явствует, что кроме кривой, обозначенной А, где я завернул в Наварру, и зубчатой кривой В,

Figure 0.2

The doubly inverted plot map in Viktor Shklovsky's article on plot structure in *Tristram Shandy* (Viktor Shklovsky, "'Tristram Shendi Sterna' i teoriia romana," in *Sborniki po teorii poeticheskogo iazyka* [Petrograd: OPOIAZ, 1921], 38.

merely one, and perhaps the least interesting, of the multiform relations between maps and texts. Once we pause for a moment and look at the map with a long and laborious glance instead of simply recognizing it, we realize that the value of the field of literature and cartography lies not in charting their mimetic relation, but in the exploration of the various forms of their disjunctive overlaps and productive tensions.[37] These interactions, frictions, and interferences take place along a whole spectrum of relations: maps as a matrix for the generation of narratives; mapping as a method and a hermeneutic tool; the negotiation between literary immersion and cartographic distance; the development of hybrid genres, of cartographic models of language and forms of writing; the projection of map events; encounters with literary unmappability; overlapping and diverging presuppositions of the ideology of representation, of its power, ontological status, and epistemological virtues and vices, and so on. Literary engagements with cartography, however, might also exceed or transform the conventional map. Without resorting to vague metaphor, it is worth examining carefully how literary spaces contain cartographic elements that do not take the Ptolemaic map or even the standard geographic map as their model. Which innovative cartographies does literature build? And which novel definition of a map would we need to describe such a spatiality in cartographic terms? The productive tensions between literature and cartography are many and diverse, and sustained efforts to explore their astonishing multiplicity are only of recent date.

Literature and Cartography

This book presents an overview of the field of literature and cartography. Developed concurrently in different languages and countries following the *Denkstil* or thought style of national, disciplinary, or local traditions, the cartographic turn in literary studies is a truly global phenomenon. Yet, even though the contemporary republic of letters is more interconnected and multilingual than ever, there is no common point of reference. The field has taken form in monographs, book series, articles, new journals, special issues of already well-established journals, and at the ever-increasing number of conferences devoted to literature and cartography in English, German, French, Spanish, Portuguese, and other languages. When *Cartography in the European Renaissance*, the third volume of *The History of Cartography*, was published in 2007, it included a section devoted to cartographic literature.[38] The essays in that volume have been instrumental in establishing literature as an integral part of the study of cartography. Nevertheless, given the extraordinary interest in literary cartography also of earlier and later historical periods as well as in basic theoretical problems, there is a noticeable absence of an introduction to the field as such. Bringing together an international cohort of scholars who have all examined the productive tensions of texts and maps from a variety of perspectives, this volume presents a synoptic overview of the larger field of literature and cartography. Inherently interdisciplinary, the volume explores connections to science, philosophy, linguistics, and ethnography, among others, as well as important themes such as desire and subjectivity, exploration and navigation, empire and war, migration and settling. The chapters retain a clear focus on literary cartographies, but as part of this exploration they also engage literary geography, geocriticism, geopoetics, the spatial humanities, and other adjacent fields that have likewise been the subject of acute interest over the past few decades.

The book is organized into three main parts: *Theories and Methodologies* (part I), *Histories and Contexts* (part II), and *Genres and Themes* (part III). The chapters in the first section lay the theoretical foundations by addressing explicitly basic problems of fictionality, methodology, and language that run throughout the volume. This section conceives the relation between literature and cartography as a series of dialectics between fictionality and referentiality, mappability and unmappability, concept and metaphor, realism and nominalism. It thereby frames and analyzes the productive tensions that arise when we juxtapose literature and cartography—tensions that also vibrate between the approaches of some of the individual essays. In this way the section opens a space for a debate about the borders, the shortcomings, and the unrealized possibilities of the field.

If the conjunction of maps and texts both invites and necessitates fundamental theoretical reflections, the literary cartographies examined here often take on meaning through their embeddedness in the larger cultural sphere of their specific time and place. Any reckoning with literature and cartography must therefore take cognizance of their tremendous developments and shifts throughout history. Consider just how radically different are the functions and layers of meaning of Phoenician and Homeric *periploi*, of a medieval T-O map, or of a nineteenth-century Baedeker map. Plotting the trajectory of the literary map as it is passed on from one country and age to the next, the second, historical part traces key developments and important contexts in the history of literary cartography. Moving chronologically across a long arc from Homer to the twenty-first century, the essays showcase the variety of fields that literary cartography has engaged, the numerous purposes it has served, and the often unexpected objects it has depicted. Against or underneath the processes of modernization and rationalization that can be detected in the visual language of the maps themselves, the section uncovers the host of counterrepresentations that evade the graticules and ever more precise order of cartography proper. Thus desires, mnemonics, gendered topographies, itineraries, and postmigrant identities feature prominently. Setting these literary cartographies against the foil of the traditional image of the map, the essays tease out the nature and consequences of their overlaps and disjunctions. Not only do the literary cartographies reveal a wealth of nongeographic objects that can be mapped, the essays also show how textual fictions became the stuff on which mapmakers made their maps when cartographic aspirations outstripped factual knowledge and the practical limits of their craft. Merely samples of larger trends, the essays will need to be supplemented by additional histories and contextualizations, but they reveal at once the constancy of the literary fascination with cartography, as well as the mutability of its meanings and implications. The map becomes visible as a surface on which authors and cartographers alike have projected and inscribed their shifting ideas, worldviews, and dreams.

Literary and cartographic genres, however, allow us to notice the transhistorical connections and parallels when they link up distant times and disparate places. The essays in the final section establish typologies of genre in order to reveal the larger, generic links between texts and maps. But in the process they also challenge the traditional distinction between the stasis of the map and the mobility of narrative. Whether in the form of military event maps that project uncertain but possible futures, paratextual experiments that spatialize and reorder narrative linearity, or map line narratives in which events occur as the interruption of movement, the essays show how the genres of literary cartography often invert traditional assumptions about the spatial and temporal orders of both literature and cartography.

Evidently a book of this kind cannot be exhaustive. Other topics and questions are more than worthy of inclusion. While grounded in the European tradition the volume does try to cast a wide net and includes chapters on American, Latin American, and African literatures. For practical reasons, however, other important nations and, indeed, continents did not find room in the volume. Nor are the three main literary genres—poetry, prose, and drama—equally represented. A few chapters include discussions of lyric poetry, but the volume retains a clear focus on narrative literature as this has been particularly fecund for the exploration of literature and cartography. A volume like this could also be organized according to different principles, and some of the essays are enmeshed in both theoretical and historical as well as generic matters. The primary concern, however, has been to build an armature that offers an overview of major theoretical, historical, and generic questions and developments. Taken together, the essays expose the literary map as a surprisingly polyvalent and multifunctional phenomenon. As the contributors so ably show, cartography matters to literature in ways that far exceed the simple task of orienting the reader in the fictional world. As they also show, however, the precise valences and meanings of literary cartography become clear only when texts and maps have been framed in such a way as to make each other sufficiently strange for their complementarities, disjunctions, and tensions to appear.

Notes

1. Robert Louis Stevenson, *Treasure Island*, ed. John Sutherland (Peterborough, Ontario: Broadview Editions, 2012), 233.

2. Herman Melville, *Moby-Dick* (New York: Norton, 2002), 59.

3. Laurence Sterne, *The Life and Opinions of Tristram Shandy, Gentleman*, ed. Melwyn New and Joan New (London: Penguin Classics, 2003), 75.

4. Gaston Bachelard, *The Poetics of Space* (Boston: Beacon Press, 1994), 150.

5. This is also how the curators of an exhibition of literary maps from the Library of Congress view their function. See Martha Hopkins and Michael Buscher, eds. *Language of the Land: The Library of Congress Book of Literary Maps* (Washington, DC: Library of Congress, 1999).

6. Sterne, *Tristram Shandy*, 79.

7. See Laura Kurgan, *Close Up at a Distance: Mapping, Technology, and Politics* (Brooklyn, NY: Zone Books, 2013), 34.

8. See the survey by Adele J. Haft, "Literature and Cartography," in *The History of Cartography*, Vol. 6.1: *Cartography in the Twentieth Century*, ed. Mark Monmonier (Chicago: University of Chicago Press, 2015), 782–787. See also Eric Bulson, *Novels, Maps, Modernity: The Spatial Imagination, 1850–2000* (New York: Routledge, 2007).

9. Artists have also engaged with maps in great numbers. For a collection, introduced by the British novelist Tom McCarthy, of recent artistic work that reworks and reimagines more conventional cartographic representations see *Mapping It Out: An Alternative Atlas of Contemporary Cartographies*, ed. Hans Ulrich Obrist (New York: Thames and Hudson, 2014).

10. See Simone Pinet, *Archipelagoes—Insular Fictions from Chivalric Romance to the Novel* (Minneapolis: University of Minnesota Press, 2011). See also Ricardo Padrón, "Mapping Imaginary Worlds," in *Maps: Finding Our Place in the World*, ed. James Akerman and Robert Karrow, 255–287 (Chicago: University of Chicago Press, 2007).

11. Alfred Korzybski, *Science and Sanity: An Introduction to Non-Aristotelian Systems and General Semantics*, 5th ed. (Lakeville, CT: International Non-Aristotelian Library Pub. Co., distributed by Institute of General Semantics, 1994), 58. The book was originally published in 1933.

12. Bruno Bosteels, "From Text to Territory," in *Deleuze & Guattari: New Mappings in Politics, Philosophy, and Culture*, ed. Eleanor Kaufman and Kevin Jon Heller (Minneapolis: University of Minnesota Press, 1998), 146.

13. Martin Brückner, "The Cartographic Turn and American Literary Studies: Of Maps, Mappings, and the Limits of Metaphor," in *Turns of Event: American Literary Studies in Motion*, ed. Hester Blum (Philadelphia: University of Pennsylvania Press, 2016), 45.

14. Michel Foucault, "Des espaces autres," in *Dits et Écrits*, vol. 2, 1976–1988, (Paris: Gallimard, 2001), 1571. For an overview of the development of the spatial turn see also Jörg Döring and Tristan Thielmann, eds., *The Spatial Turn: Das Raumparadigma in den Kultur- und Sozialwissenschaften* (Bielefeld: transcript Verlag, 2008); Brückner, "The Cartographic Turn"; Barney Warf and Santa Arias, eds., *The Spatial Turn: Interdisciplinary Perspectives* (London: Routledge, 2008); Robert T. Tally Jr., *Spatiality* (London: Routledge, 2013); Ernest W. B. Hess-Lüttich, "*Spatial Turn*: On the Concept of Space in Cultural Geography and Literary Theory," *Meta-Carto-Semiotics / Journal for Theoretical Cartography* 5 (2012): 1–11.

15. Edward W. Soja, *Postmodern Geographies: The Reassertion of Space in Critical Social Theory* (New York: Verso, 1989), 75, 62; Edward W. Soja, "Vom 'Zeitgeist' zum 'Raumgeist': New Twists on the *Spatial Turn*," in *The Spatial Turn: Das Raumparadigma in den Kultur- und Sozialwissenschaften*, ed. Jörg Döring and Tristan Thielmann, 241–262 (Bielefeld: transcript Verlag, 2008); Fredric Jameson, *Postmodernism, or, The Cultural Logic of Late Capitalism* (Durham, NC: Duke University Press, 1991). See also Jameson's earlier essay, "Postmodernism, or the Cultural Logic of Late Capitalism," *New Left Review* 146 (1984): 53–92.

16. Paul Virilio, *Esthétique de la disparition* (Paris: Balland, 1980). See also Niels Werber, *Die Geopolitik der Literatur: Eine Vermessung der medialen Weltraumordnung* (Munich: Carl Hanser Verlag, 2007); *Raum, Wissen, Macht*, ed. Rudolf Maresch and Niels Werber (Frankfurt am Main: Suhrkamp, 2002).

17. Stephan Günzel, "*Spatial Turn–Topographical Turn–Topological Turn:* Über die Unterschiede zwischen Raumparadigmen," in *The Spatial Turn: Das Raumparadigma in den Kultur- und Sozialwissenschaften*, ed. Jörg Döring and Tristan Thielmann (Bielefeld: transcript Verlag, 2008), 222.

18. For a comparison of the different versions of the topographic turn in cultural studies and in the "Kulturwissenschaften" see Sigrid Weigel, "Zum 'Topographical Turn': Kartographie, Topographie und Raumkonzepte in den Kulturwissenschaften," *KulturPoetik* 2, no. 2 (2002): 151–165.

19. J. Hillis Miller, *Topographies* (Stanford, CA: Stanford University Press, 1995).

20. To mark the shared graphic foundation of various representations of space, Henry S. Turner coins the word *topographesis*, which he defines as "the representation of place in any graphic mode—writing, painting, drawing" (Henry S. Turner, "Literature and Mapping in Early Modern England, 1520–1688," in *The History of Cartography*, Vol. 3.1: *Cartography in the European Renaissance*, ed. David Woodward (Chicago: University of Chicago Press, 2007), 424).

21. Denis Cosgrove, "Introduction: Mapping Meaning," in *Mappings*, ed. Denis Cosgrove (London: Reaktion Books, 1999), 7.

22. Henri Lefebvre, *La Production de l'espace* (Paris: Anthropos, 2000), xix.

23. See also Jörg Dünne, "Die Karte als Operations- und Imaginationsmatrix: Zur Geschichte eines Raummediums," in *The Spatial Turn: Das Raumparadigma in den Kultur- und Sozialwissenschaften*, ed. Jörg Döring and Tristan Thielmann (Bielefeld: transcript Verlag, 2008), 51.

24. See J. B. Harley, "Deconstructing the Map," *Cartographica* 26, no. 2 (1989): 1–20.

25. Christian Jacob, *The Sovereign Map: Theoretical Approaches in Cartography throughout History*, trans. Tom Conley, ed. Edward H. Dahl (Chicago: University of Chicago Press, 2006), 6.

26. Denis Wood, "Cartography Is Dead (Thank God!)," *Cartographic Perspectives* 45 (2003): 4–7.

27. Brückner, "The Cartographic Turn," 60.

28. Soja, "Vom 'Zeitgeist' zum 'Raumgeist,'" 242.

29. Marion Picker, "Die Zukunft der Kartographie: Neue und nicht so neue epistemologische Krisen," in *Die Zukunft der Kartographie: Neue und nicht so neue epistemologische Krisen*, ed. Marion Picker, Véronique Maleval, and Florent Gabaude (Bielefeld: transcript Verlag, 2013), 15.

30. As Robert Stockhammer puts it, "Jede Ver*zeichnung verzeichnet*" (Robert Stockhammer, *Kartierung der Erde: Macht und Lust in Karten und Literatur* (Munich: Wilhelm Fink Verlag, 2007), 49).

31. Jörg Dünne, "Die Unheimlichkeit des Mapping," in *Die Zukunft der Kartographie: Neue und nicht so neue epistemologische Krisen*, ed. Marion Picker, Véronique Maleval, and Florent Gabaude, 211–240 (Bielefeld: transcript Verlag, 2013).

32. Viktor Shklovsky, "Art as Device," in *Theory of Prose*, trans. Benjamin Sher (Elmwood Park, IL: Dalkey Archive Press, 1990), 5. The essay is available in the original Russian as "Isskustvo kak priëm," in *Gamburgskii schët* (Moscow: Sovetskii pisatel', 1990), 58–72.

33. Shklovsky, "Art as Device," 6.

34. In the edition edited by Melwyn New and Joan New, the map is printed on p. 425. For an analysis of this map's military context see the first chapter in Anders Engberg-Pedersen, *Empire of Chance: The Napoleonic Wars and the Disorder of Things* (Cambridge, MA: Harvard University Press, 2015), 10–37.

35. Sterne, *Tristram Shandy*, 34.

36. This was first brought to my attention by Emily Finer, who was also kind enough to procure a copy of the map. See her own analysis in *Turning into Sterne: Viktor Shklovskii and Literary Reception* (London: Legenda, 2010), 60–99.

37. As early as 1996, Tom Conley developed a concept of "cartographic literature" based on the notion of "tension." As he puts it, "Writings can be called 'cartographic' insofar as tensions of space and of figuration inhere in fields of printed discourse" (Tom Conley, *The Self-Made Map: Cartographic Writing in Early Modern France* (Minneapolis: University of Minnesota Press, 1996), 3).

38. *The History of Cartography*, Vol. 3.1: *Cartography in the European Renaissance*, ed. David Woodward (Chicago: University of Chicago Press, 2007). See especially the essays by Tom Conley, Henry S. Turner, Nancy Bouzrara, Franz Reitinger, Theodore J. Cachey Jr., Neil Safier, Ilda Mendes dos Santos, and Simone Pinet.

Bibliography

Bachelard, Gaston. *The Poetics of Space*. Boston: Beacon Press, 1994.

Bosteels, Bruno. "From Text to Territory." In *Deleuze & Guattari: New Mappings in Politics, Philosophy, and Culture*, ed. Eleanor Kaufman and Kevin Jon Heller, 145–174. Minneapolis: University of Minnesota Press, 1998.

Brückner, Martin. "The Cartographic Turn and American Literary Studies: Of Maps, Mappings, and the Limits of Metaphor." In *Turns of Event: American Literary Studies in Motion*, ed. Hester Blum, 44–72. Philadelphia: University of Pennsylvania Press, 2016.

Bulson, Eric. *Novels, Maps, Modernity: The Spatial Imagination, 1850–2000*. New York: Routledge, 2007.

Conley, Tom. *The Self-Made Map: Cartographic Writing in Early Modern France*. Minneapolis: University of Minnesota Press, 1996.

Cosgrove, Denis. "Introduction: Mapping Meaning." In *Mappings*, ed. Denis Cosgrove. London: Reaktion Books, 1999.

Döring, Jörg, and Tristan Thielmann, eds. *The Spatial Turn: Das Raumparadigma in den Kultur- und Sozialwissenschaften*. Bielefeld: transcript Verlag, 2008.

Dünne, Jörg. "Die Karte als Operations- und Imaginationsmatrix: Zur Geschichte eines Raummediums." In *The Spatial Turn: Das Raumparadigma in den Kultur- und Sozialwissenschaften*, ed. Jörg Döring and Tristan Thielmann, 49–69. Bielefeld: transcript Verlag, 2008.

Dünne, Jörg. "Die Unheimlichkeit des Mapping." In *Die Zukunft der Kartographie: Neue und nicht so neue epistemologische Krisen*, ed. Marion Picker, Véronique Maleval, and Florent Gabaude, 211–240. Bielefeld: transcript Verlag, 2013.

Engberg-Pedersen, Anders. *Empire of Chance: The Napoleonic Wars and the Disorder of Things*. Cambridge, MA: Harvard University Press, 2015.

Finer, Emily. *Turning into Sterne: Viktor Shklovskii and Literary Reception*. London: Legenda, 2010.

Foucault, Michel. "Des espaces autres." In *Dits et Écrits*, vol. 2, 1976–1988, 1571–1581. Paris: Gallimard, 2001.

Günzel, Stephan. "*Spatial Turn–Topographical Turn–Topological Turn:* Über die Unterschiede zwischen Raumparadigmen." In *The Spatial Turn: Das Raumparadigma in den Kultur- und Sozialwissenschaften*, ed. Jörg Döring and Tristan Thielmann, 219–237. Bielefeld: transcript Verlag, 2008.

Haft, Adele J. "Literature and Cartography." In *The History of Cartography*, Vol. 6.1: *Cartography in the Twentieth Century*, ed. Mark Monmonier, 782–787. Chicago: University of Chicago Press, 2015.

Harley, J. B. "Deconstructing the Map." *Cartographica* 26, no. 2 (1989): 1–20.

Hess-Lüttich, Ernest W. B. "*Spatial Turn*: On the Concept of Space in Cultural Geography and Literary Theory." *Meta-Carto-Semiotics / Journal for Theoretical Cartography* 5 (2012): 1–11.

Hopkins, Martha, and Michael Buscher, eds. *Language of the Land: The Library of Congress Book of Literary Maps*. Washington, DC: Library of Congress, 1999.

Jacob, Christian. *The Sovereign Map: Theoretical Approaches in Cartography throughout History*. Trans. Tom Conley, ed. Edward H. Dahl. Chicago: University of Chicago Press, 2006.

Jameson, Fredric. *Postmodernism, or, The Cultural Logic of Late Capitalism*. Durham, NC: Duke University Press, 1991.

Jameson, Fredric. "Postmodernism, or the Cultural Logic of Late Capitalism." *New Left Review* 146 (1984): 53–92.

Korzybski, Alfred. *Science and Sanity: An Introduction to non-Aristotelian Systems and General Semantics*. 5th ed. Lakeville, CT: International Non-Aristotelian Library Pub. Co., distributed by Institute of General Semantics, 1994.

Kurgan, Laura. *Close Up at a Distance: Mapping, Technology, and Politics*. Brooklyn, NY: Zone Books, 2013.

Lefebvre, Henri. *La Production de l'espace*. Paris: Anthropos, 2000.

de Man, Paul. *Allegories of Reading: Figural Language in Rousseau, Nietzsche, Rilke, and Proust*. New Haven, CT: Yale University Press, 1979.

Maresch, Rudolf, and Niels Werber. *Raum, Wissen, Macht*. Frankfurt am Main: Suhrkamp, 2002.

Melville, Herman. *Moby-Dick*. New York: Norton, 2002.

Miller, J. Hillis. *Topographies*. Stanford, CA: Stanford University Press, 1995.

Obrist, Hans Ulrich, ed. *Mapping It Out: An Alternative Atlas of Contemporary Cartographies*. New York: Thames and Hudson, 2014.

Padrón, Ricardo. "Mapping Imaginary Worlds." In *Maps: Finding Our Place in the World*, ed. James Akerman and Robert Karrow, 255–287. Chicago: University of Chicago Press, 2007.

Picker, Marion. "Die Zukunft der Kartographie: Neue und nicht so neue epistemologische Krisen." In *Die Zukunft der Kartographie: Neue und nicht so neue epistemologische Krisen*, ed. Marion Picker, Véronique Maleval, and Florent Gabaude, 7–19. Bielefeld: transcript Verlag, 2013.

Pinet, Simone. *Archipelagoes—Insular Fictions from Chivalric Romance to the Novel.* Minneapolis: University of Minnesota Press, 2011.

Shklovsky, Viktor. "Art as Device." In *Theory of Prose*, trans. Benjamin Sher. Elmwood Park, IL: Dalkey Archive Press, 1990.

Shklovsky, Viktor. "Isskustvo kak priëm." In *Gamburgskii schët*, 58–72. Moscow: Sovetskii pisatel', 1990.

Shklovsky, Viktor. "'Tristram Shendi Sterna' i teoriia romana." In *Sborniki po teorii poeticheskogo iazyka*. Petrograd: OPOIAZ, 1921.

Soja, Edward W. *Postmodern Geographies: The Reassertion of Space in Critical Social Theory.* New York: Verso, 1989.

Soja, Edward W. "Vom 'Zeitgeist' zum 'Raumgeist': New Twists on the *Spatial Turn*." In *The Spatial Turn: Das Raumparadigma in den Kultur- und Sozialwissenschaften*, ed. Jörg Döring and Tristan Thielmann, 241–262. Bielefeld: transcript Verlag, 2008.

Sterne, Laurence. *The Life and Opinions of Tristram Shandy, Gentleman.* Ed. Melwyn New and Joan New. London: Penguin Classics, 2003.

Sterne, Laurence. *The Life and Opinions of Tristram Shandy, Gentleman.* Leipzig: Bernh. Tauchnitz Jun., 1849.

Stevenson, Robert Louis. *Treasure Island.* Ed. John Sutherland. Peterborough, Ontario: Broadview Editions, 2012.

Stockhammer, Robert. *Kartierung der Erde: Macht und Lust in Karten und Literatur.* Munich: Wilhelm Fink Verlag, 2007.

Tally, Robert T., Jr. *Spatiality.* London: Routledge, 2013.

Turner, Henry S. "Literature and Mapping in Early Modern England, 1520–1688." In *The History of Cartography*, Vol. 3.1: *Cartography in the European Renaissance*, ed. David Woodward, 412–426. Chicago: University of Chicago Press, 2007.

Virilio, Paul. *Esthétique de la disparition.* Paris: Balland, 1980.

Warf, Barney, and Santa Arias, eds. *The Spatial Turn: Interdisciplinary Perspectives.* London: Routledge, 2008.

Weigel, Sigrid. "Zum 'Topographical Turn': Kartographie, Topographie und Raumkonzepte in den Kulturwissenschaften." *KulturPoetik* 2, no. 2 (2002): 151–165.

Werber, Niels. *Die Geopolitik der Literatur: Eine Vermessung der medialen Weltraumordnung.* Munich: Carl Hanser Verlag, 2007.

Wood, Denis. "Cartography Is Dead (Thank God!)." *Cartographic Perspectives* 45 (2003): 4–7.

Woodward, David, ed. *The History of Cartography*, Vol. 3.1: *Cartography in the European Renaissance*. Chicago: University of Chicago Press, 2007.

I

THEORIES AND METHODOLOGIES

1
Cartographic Fiction

Jean-Marc Besse

> The wide-eyed child in love with maps and plans
> Finds the world equal to his appetite.
> How grand the universe by light of lamps,
> How petty in memory's clear sight.
> — Charles Baudelaire, "Voyaging"[1]

In a celebrated passage from a short article published in 1929, Walter Benjamin imagines the difficulties a tourist in Paris might encounter when looking for points of reference on a blustery and rainy day. The tourist attempts to read a map of the city posted at a windy corner before it is torn or blown away. Finally managing to read it, the tourist, he adds, learns "what a city a map can be. And what a city is." Thus we can imagine Benjamin, his hand clinging to a Taride map, going about the Quartier de l'Europe near the Gare Saint-Lazare. And we can recall what is unique (and, at the same time, utterly banal) for those who at least once in their lives try to find their way in an unknown city: it is in these very places, by virtue of the map representing the Parisian quarter, that they find their bearings. "Near the great square in front of the Gare Saint-Lazare half of France and half of Europe surround the viewer. Names like Havre, Anjou, Provence, Rouen (for Rome?), London, Amsterdam, run through the gray streets as iridescent ribbons through gray silk. That's the so-called Europe Quarter. Thus, bit by bit, you can traverse the streets on the map."[2] Yet we cannot tell what Benjamin's imaginary tourist was really looking at when he was penning these lines, whether it was the map or the street. In question are two simultaneous ways of learning that take place on the terrain itself, that of reading the map and that of reading the city.

A play of reflections and mirrors: the map is a reflection of Paris. With their own eyes tourists can follow the streets on the map that sparkle like silken ribbons in a mirror. But the city is also a reflection, the reflection of greater territories, of France and Europe, to which it refers in the scatter of streets near the Gare Saint-Lazare. Around the station we take a few steps to go from Liège to Milan, then to Athens and London, and we do an about-face to go to Budapest, Vienna, Edinburgh, Lisbon, Madrid, and Naples. Leaving Edinburgh, Lisbon, Madrid, and Naples aside, the Paris that can be mapped in turn becomes a map. Can't we really say that that Taride map is a map of a map, and the reflection of a reflection? And why wouldn't the map of the Europe Quarter help in situating me in Europe? Something imaginary circulates in the space of the city (figure 1.1).

However, the street names, and in this instance geographic names, are what drive the imagination, the element allowing passage from one place to another and, at the same time, access to the entire city, its plan, and all of Europe. Benjamin writes: "Entire districts reveal their secrets in the names of their streets."[3] But in Paris, this magical power of street names causing tourists to go from one reflection to another transforms the place into the reflection of another place, and more generally, the city into the mirror of other cities, of a country, and of an entire continent. Noting an *Illustrated Guide to Paris*, Benjamin remarks, "The *passage* is a city, a world in miniature."[4] More precisely, by virtue of the geographic names of streets, the city soon becomes a vast map, a reservoir of signs for countless travels and new fictions. And the maps that represent it are maps of this map.

Can a Map Contain Itself? The Powers and Limits of Mimesis

Certain maps also pose logical questions, or else they are fictions that pose philosophical problems. In his novella titled "The Zahir," Borges relates the story of this painter from the Sind who had drawn in a little cell "a species of an infinite tiger": "This tiger was composed of many tigers in a vertiginous fashion; it was traversed, striped by tigers, it contained the seas and the Himalayas and armies that resembled tigers."[5] The first thing this painter wanted to do, Borges added, "had been to draw a world map." Whence a first question: Can a map contain itself?

This question refers to a philosophical tradition that Borges knows well. Its origin is found in a lecture given by the American philosopher Josiah Royce in which the final analysis leads to the notion of the perfect map, in other words to the exactitude of representation in the context of a reflection on systems of self-representation.[6] Royce imagines the possibility of drawing a perfectly exact map of England on a part of the very soil of England. If this map is exact, it follows that it must represent in itself another map that represents it in representing England, and in turn this other map contains another, that represents the two preceding maps, and so on up to infinity. Borges thus translates Royce's remark: "Let's imagine that a portion of England has been perfectly flattened, and that on it a cartographer draws a map of England. The work is perfect; there is not a detail of English soil, however reduced it might be, that fails to be recorded on the map; everything can be found therein. In this case, the map has to contain a map of the map, which must contain a map of the map of the map, and so on until infinity."[7]

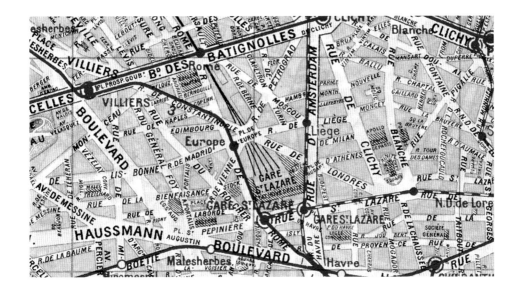

Figure 1.1
Nouveau plan de Paris avec toutes les lignes du Métropolitain et du Nord-Sud. A. Taride, Paris, 1915 (detail). Bibliothèque Nationale de France (code: GE D-7164).

Borges applies this philosophico-cartographic hypothesis in relation to other works of literature and painting: the *Ramayana,* the *Thousand and One Nights, Hamlet, Don Quixote, Las Meniñas.* In each instance it is matter of imagining a vertiginous experience—that of the infinite—through the intermediary of figurative or narrative means, of a *mise-en-abyme,* that can also be called a fiction within a fiction, a picture within a picture, a book within a book, a map within a map. What matters with the map of England drawn on the soil of England or with the map of the Empire is space within space, or rather, infinitely, space within space within space, and so on.

I will not dwell further on the logical dimension of the problem of the existence of this paradoxical space that contains and represents itself.[8] In a certain way Bertrand Russell solved the problem with his theory of types that Borges quotes elsewhere.[9] Royce's maps are illustrations of the notion of "reflexive class" (classes that contain themselves). Should this kind of paradox be avoided, according to Russell, the self-referential usage of language must be avoided and the levels of enunciation in discourse carefully distinguished. In other words, the map of England and England belong to two different types of reality between which an equivalence—but not an equality—can be grasped.

It might be worth pausing on the consequence that Borges draws from the matter: space that is opened by the map in the map is that of fiction, or better, a space in which the distinctions between reality and fiction are blurred. "Why," he asks, "are we upset if the map is included in the map and the thousand and one nights in the book of *A Thousand and One Nights*? I think I've found the cause: such inversions suggest that if the characters in a fiction can be readers or spectators, then we, their readers or their spectators, can be fictive characters."[10]

If in effect the map and the territory merge, who is going to tell me where I am, where I live? On what territory? Or on what map that represents it? Then who tells me that what I consider to be real territory in all truth is not a map? A representation? Who finally tells me that I am not in the middle of an infinite series of maps fitting into each other, much like two mirrors facing each other yield an infinite series of reflections? What I call real is perhaps nothing other than the map in which I happen to be at this moment. Perhaps, moreover, I am merely an object on a preceding map, just as I would be, unbeknownst to myself, a character whose story is being told at the moment I am living the story, but in a story told by another. At the spot where the map covers the territory (and a fortiori if it covers the entire territory), neither the real nor the image remains, no longer is there a distinction between reality and fiction: everything has become either fiction or reality. This space is indeed utopian, nowhere at all.

Yet surely these logically impossible, paradoxical, and infinite utopian spaces have existed concretely. In this case the logical contradiction is not a practical insufficiency. These maps that are territories that are maps can be traveled just as well in gardens and parks as at the edge of cities. For eons these spaces that are at once maps in themselves have been conceived, projected, and even realized. Projects of this kind have existed since the *Serapeum* of Hadrian's Villa at Tivoli, as Jean-Claude Grenier has shown, "put forward as if it were a veritable monumental map of Egypt" to the geographic gardens established in Paris up to the beginning of the twentieth century.[11]

Even in Europe today people can visit the *Swissminiatur* at Melide near Lugano, Holland in miniature in the Madurodam Park, *Italia in miniatura* in Rimini, *Catalunya en miniature* near Barcelona, *Portugal dos Pequenitos* in Coimbra, *France miniature* located in Élancourt near Paris, but also discover Alsace in the *Parc des miniatures* in Plombières-les-bains, the Val de Loire in Amboise inside the *Parc des Mini-Châteaux*, or even the landscapes of Saxon Switzerland in the *Miniaturpark die kleine Sächsische Schweiz* near Dresden. And there is also a *Miniatürk* near Istanbul, a miniature park of Silesian monuments near Wrocław in Poland, and no less a *Russie miniature* in Saint Petersburg. For a long time Europe has been visualized in Brussels in the *Mini-Europe* park situated near the Atomium. Whoever wants to see the world will go to the Beijing *World Park* or, not far away, near Shenzen, to the *Windows of the World*, or even to the *Tobu World Square* in the Tochigi Prefecture in Japan, but also to the *Minimundus* of Klagenfurt in Austria, even to the *Swiat Marzen* located in the province of Malopolska in Poland (where too Poland can be visited in miniature). Eleven countries are represented in the *World Showcase* at the Epcot Park in the heart of Orlando's Walt Disney World. Moreover, a few of these parks, like *France miniature* or *Italia in miniatura*, draw attention to the cartographic form of their plan. The analogical and mimetic drive seems to have been pushed to its limit. *France miniature* resembles France, or rather a map of France (figure 1.2).

Those who visit the park walk about in a map that, when all is said and done, is also a territory. They visit a map and they visit the territory that it represents. They discover a touristic geography in the walkways of the park that are structured around the major rivers (Loire, Rhône, Garonne, Seine), architectural schemes and exemplary technical achievements (the Eiffel Tower, the Versailles Castle, the Stade de France, the Garabit Viaduct, etc., among a total of 116), and large regional sections (the North and the Île-de-France, the East, Southwest, Center, West and Southwest). Yet must these disproportionate maps—whose projects endlessly follow one after another—also be considered vertiginous objects resembling the figure of an infinite tiger? To be sure, the existence of these spaces comprises a kind of counterhistory of cartography that continues to move alongside the scientific history of cartography. But what kind of epistemology and ontology would be needed to account for these spaces that are at once realized fictionalizations and fictional realizations? It goes without saying that in view of such "limit objects" the simple opposition between reality and fiction is inadequate, and as a result there needs to be recognized, at the very core of seemingly time-tested realities, the presence of zones of fiction. In the same way fiction will need to be seen as a dimension of our reality.

Figure 1.2
La France miniature. Google Earth.

The Reality Effect of Cartography

As Christina Ljungberg writes, "In fiction, maps have often been used as statements and as assertions of authenticity."[12] Maps (whether topographic or geographic) fill this function of authentification because, providing a reference to the work of fiction, they tie the work to a "reality" designated as foreign to the text; they anchor the fiction in a spatial and temporal reality to what elsewhere readers feel they can gain access. What fiction describes or reports having taken place took place because it is *there*, right where it is shown and designated by the map. It is as if Aristotle's assertion had to be verified: every being is somewhere (*Physics,* IV, 1), and to exist is to have a place. When the map shows the place told in the story it becomes in a certain way the very proof of the story.

Maps, however, not only attest to existence. They are also the operators of reading and understanding. We know well the famous formula that Flemish cartographer Abraham Ortelius set on the frontispiece of his *Parergon*: "Geography is the eye of history." Since the beginning of modernity, it is a commonplace in geography: to read histories, be they legendary, civil, or religious, a map is needed to visualize the places where these histories took place. Allowing the reader to see the places, the map also makes the action easier to follow and understand. Thus,

> in reading about Saint Paul's travels from the city of Jerusalem to Rome, should the reader be informed by means of the map in what place in the world Jerusalem is situated, and how the islands of Cyprus, Rhodes and Malta follow; that in sailing to the aforesaid Islands and in their presence, he finally arrived in Rome: not only will he better understand the voyage stated above, but also retain it in memory for a longer time and will be enabled to relate it in words more graciously, to explicate it, to have it understood by others.[13]

When placed before the eyes of the reader consulting the folio that Ortelius published in 1579, the map of Saint Paul's voyages thus acquires a kind of visual evidence. It would also be the case for the ancient history of Italy, Greece, Egypt, and Palestine, for the life of Abraham, for the peregrinations of Ulysses and Aeneas, for the adventures of the Argonauts, for the conquests of Alexander the Great, all of whose maps Ortelius progressively assembles in the *Parergon*.

The service that cartography renders to reading also holds for literature. It is impossible to count the number of maps drawn and printed to accompany the reading of novels. Thus for example, as Roger Chartier has noted, the editions of the *Orlando furioso* published by Valgrisi in 1556 and 1573 contain several geographic maps allowing readers to follow the travels of certain characters.[14] In the same way, inaugurating a long tradition, Spanish geographer Tomas Lopez proposed a "Map of a Part of the Kingdom of Spain Containing the Lands Traveled by Don Quixote" for the edition of the

Real Academia Española of 1780. Therein are drawn the itineraries of three "sorties" and then pinpointed thirty-five "adventures" on the part of the Hidalgo. A second map, titled "A Geographical Map of the Voyages of Don Quixote and the Sites of His Adventures," appears in the edition of 1797–1798. The insertion of maps into the narration lays stress on the latter's plausible character. As Chartier emphasizes, "The presence ... of maps that draw the imaginary wanderings of a fictional character into territories known to the reader enhances the reality effects of the writing."[15]

But if the presence of maps in literary (but also cinematographic) fiction fosters the attribution of a reality to the imaginary worlds their authors conceived, it nonetheless remains to be seen how this attribution of reality is made. In other words, how does cartography make possible this passage from fiction to reality? For the sake of argument we can distinguish two configurations in which cartography intervenes in a differentiated fashion:

1. On the one hand, there are maps representing invented places, places explicitly posited as fictive. Such is the case, from *Lord of the Rings* to *Game of Thrones,* in so-called *fantasy* literature whose developments are constructed around a detailed mapping of imaginary territories (*Middle Earth, Westeros*). Yet beyond this specific genre, many writers can be considered creators of territories. Thus, even if their topographies and their forms derive from real places, the maps of Faulkner's *Yoknapatawpha,* of Thomas Hardy's *Wessex,* of Juan Benet's *Región,* or of Michel Butor's *Bleston* represent spaces their authors have invented.

But what do these maps really display? What, it might be asked, is the referential value of these maps that represent fictive places? In this respect the map of the Island of Utopia found in the eponymous work of Thomas More (1516 and 1518) is paradigmatic of this operation that consists in representing nonexistent places as if they were real. From this perspective we can look at the later—and in fact quite mysterious—map drawn by Abraham Ortelius (circa 1595), ordered by one of his friends, Mattheus Wacker in Wackenfels (figure 1.3).

It matters little if this map corresponds to a desire on the part of Ortelius, who in fact did not include it in his *Parergon*. In my view it is more important to dwell on the type of paradox contained in this map, a paradox that at once constitutes and reveals the inner tension that, I believe, runs through most maps that designate invented or fictive places. The island represented by this map does not exist. It is a nonplace or perhaps a place-off, yet still within the map it is depicted as a geographic reality analogous to other territories, to other islands that can be found in Ortelius's *Theatrum orbis terrarum*. In other words, by dint of the map the Island of Utopia acquires a sort of effective territorial reality: yes, Utopia exists, but only on its map. Utopia does have a place, but this place is the map that represents it. Here Michel Foucault's famous expression is applicable: the map of Utopia is a *heterotopia*, a sort of "counteremplacement"

Figure 1.3

Abraham Ortelius, *Utopia Typus, ex Narratione Raphaelis Hythlodæi, Descriptione D. Thomas Mori.* Antwerp, 1595–1596.

inside of which Utopia acquires a reality that can be grasped—that is, a reality that can be thought.[16] Nonetheless the reading of the toponyms on Ortelius's map adds a supplementary dimension to this heterotopical affirmation: the cities are named "Horsdumonde," "Nulleville," "Keinstadt," the rivers are designated as "Nullipiscius," "Senzzaqua," "Sanspoisson," and so on, as if it were a question not only of displaying a nonexistent reality but furthermore a reality that in itself is contradictory, a reality without being of its own because it denies itself. Put otherwise, the map shows not only a reality that is *not* but also a reality *that cannot be.* Such is the thought that the map, a cartographic heterotopia, inspires: a condition of nonbeing, the reality of a nonexistent reality, of a nonplace that is in itself contradictory but nonetheless can be thought because it is represented, drawn in the map itself, which at once displays it while demonstrating its impossibility. But what does it mean to consider the map of a reality that does not exist? Simply, it is tantamount to affirming the reality of fiction and to positing fiction as a dimension of reality.

What is the ontological tenor of these places invented by literature (and no less, by philosophy, painting, and cinema) and, at the same time, represented in geographic maps? German philosopher Hermann Schmitz used the notion of "semi-thing" or "half-thing" (*Halbding*) to designate these real entities that are not things (because they lack substantiality, duration, material density, morphological permanence, etc.), but that act on us *as if* they were things, generating sensations, feelings, pain, and so on, or generally bodily affects.[17] I propose that we export the notion and apply our question to it: the fictive places that we see nowhere other than on the maps that represent them—that we follow so closely, that we trust so faithfully that our imagination gets carried away in the course of reading—become "half-places."

2. In literature there is a second category of maps in which the "reality effect" seems even more accentuated, namely, those that represent existing places, but inside of which original fictional worlds unfold. Thus, in preparing the fourth edition of *The Life and Surprising Adventures of Robinson Crusoe* (1719), Daniel Defoe uses as the frontispiece of the work a world map on which he draws the route and the place of the unfortunate hero's shipwreck. In the same way, Jules Verne inserts in *Around the World in 80 Days* (1873) a world map that allows the reader to follow Phileas Fogg's adventures. In another of Verne's novels, *Aventures de trois Russes et de trois Anglais dans l'Afrique Centrale* (1872) (*Meridiana: The Adventures of Three Englishmen and Three Russians in South Africa*), a map juxtaposes the itinerary of the expeditions Verne describes and the one made in the same places by David Livingstone. If, generally, the literature of voyage and adventure accords an important place to cartography, then the same holds for the whodunit or mystery novel that makes abundant use of maps at the very moment, for the first time in 1868, when Émile Gaboriau, the inventor of the genre, drew a map of the crime scenes in *Monsieur Lecoq.*[18]

How can we evaluate the way cartographic representation situates real places in these works of fiction?

a. In this respect a first observation seems fitting: whether it be in the adventure novel or mystery literature, space plays an essential role in the development of the narrative. It becomes the very element of the action. Characters go from one place to the next in search of things or of other characters. They traverse cities and countries such that the spatial displacement grounds the story and moves it ahead. The map on which this array of places and travels are inscribed in some way constitutes, Tom Conley argues, the narrative webbing of the work.[19] As if the time of the tale were structured as a function of the space represented on the map, it plays an organizing role in the development of the narrative. On this point more will follow.

b. A second observation: clearly the mobilization of the map anchors the narrated action or the scene being described in a spatial reality that, in an assemblage of familiar places, the reader knows or very quickly recognizes. Here the mimetic orientation of the map is exploited to its maximum, the map now becoming the equivalent of virtual travel in these places, and in this sense it confers on the fiction its power of authenticity. As I have noted above, the map is thus an operator that assures for both readers and viewers the credibility of the fiction.

c. Yet furthermore (a third observation), cartography is exploited here insofar as one of its essential virtues consists in making visible what without it would remain invisible or hidden. Ptolemy adduces the point in the first sentences of his *Geographia*: the map has the power of representing to human beings realities that they would otherwise be unable to see. On maps spatial relations can be discovered, and so also territorial forms; distance and mass can only be seen on maps, and not elsewhere because otherwise they are too remote or too immense to be seen. The miniature that is the map becomes the instrument that at the same time brings invisible realities into our field of vision. This is the resource that serves the ends of fiction. Because it attests to the reality of places, the map brings to light the secret relations that fiction recounts, the invisible stories that cut through the real space of territories and cities. It makes visible the unknown that nestles in the real.

d. In this sense (a fourth observation), it hardly suffices to say that the map draws fiction into a recognizable spatial reality. The situation needs to be envisaged in a symmetrical manner, such that a given fictive adventure can take place in an area that is generally taken to be real, that can be located and grasped, an area where the reader might eventually go, which *also* means that this area contains unsuspected, unexpected, imperceptible narrative powers in current evocations of the same setting. In other words, something else is taking place, other events are liable to take place in this seemingly stable space that the map represents, including other, less familiar, perhaps more disquieting, even uncanny stories. As if other spatialities—in other words, other uses of space—can be concealed inside the space represented by the map. This way, as it were, of causing space to be disquieting by inscribing it cartographically with fictional actions ultimately transforms real space (for example, that of the city) into a kind of playground, into a space open to the most diverse adventures.

e. And, in a fifth observation, we can go even further. A writing project can easily echo a reader's experience. When furnished with a book, a map, or a city view (or a tourist guide), in the fashion of a pilgrimage a reader can follow the steps of his or her authors of predilection or of fictional characters who have marked it. The tracks in Paris left by Proust's characters or Balzac's Lucien de Rubempré still resound under the steps of literary travelers for whom the city has become a great font of signs and indices. The Buttes-Chaumont, the Barrio Chino, Bloomsbury, Alexanderplatz count among many landmarks for literary guides promising, as it were, an experience of fiction. For their profit tourist speculators try to capture these fictional possibilities of urban spaces by proposing cultural tours following the tracks of given writers. But many of them, themselves included, have created travel agencies of sorts, and drawn itineraries of adventures and exploration in the heart of cities in the manner of Pessoa in *Lisbon*. Thus, and as it were, fiction is often built *on the places themselves*, in a sort of immediate transmutation of the real place onto the horizon of fictional spaces.

Nonetheless, the map is what makes possible this inscription of true urban reality into the territories of fiction. Cartography—that is, the possibility of relating the fictive universe to a real space via localization—allows the inverse, yet in the same way, of adding to real space the dimensions of fiction. The map operates transitions, passages, between the real and fiction.

By way of the map readers learn that an atmosphere of fiction floats like a cloud over the cities, and that fiction is written into the intersections where they can hope to see the characters of the novel whose steps they follow reappear and the fiction reflected in the shop fronts or street names. Basically, readers launch themselves in pursuit of a text that is distributed in space and materialized in buildings, streets, public places, but also in names. A real city is transformed into a space of fiction by virtue of the intrinsic power of displacement that the map transforms into a large-scale treasure hunt. Walter Benjamin compares Paris to the great reading room of a library through which flows the Seine: "With Notre-Dame we think of Victor Hugo's novel. The Eiffel Tower—Cocteau's *The Wedding Party on the Eiffel Tower*. ... The Opera: with Leroux's famous whodunit 'The Phantom of the Opera' we find ourselves in the lower depths of the edifice and, too, of literature."[20] The map reveals an imaginary, a floating, secondary, often invisible geography. Cartography awakens the immediate fictions that are asleep on the sidewalks and at the doorsteps of houses. If there is a dream in the living body of cities, cartography is what indicates the way to enter and how to find one's bearings.

Thus the operations of the map move simultaneously in two opposite but complementary directions: by means of localization the map makes it possible to anchor fiction in the real, but in the same way it makes possible the introduction of a fictionalizing dimension in the real. It is not only fiction that is bound to a spatial reality, but also geographic space itself that is opened from the inside onto the possibility of fiction, as if fiction were one of its dimensions and yet again one of its still-unexplored resources. The maps that transcribe the multiple trajectories, indeed the errant travels of fictional characters are at the same time operators that transform real space into a space of possible fictions.

Spatial Plotting

If space plays a decisive role in the development of fiction, then cartography occupies a strategic place. Maps have at least four functions in the general economy of a work, and especially in the process of writing and in the construction of the narration: (1) the map can function as an element setting the fiction into motion; (2) it allows a frame to be drawn around the plot, or it can place the intrigue within the perimeters of a theatrical stage; (3) it can become a rule or a principle for the ordering of narrative sequences; (4) finally, cartography constitutes a recourse or even a resource allowing the writer to be free of the linearity of a story or, in any case, to call the linearity into question.

1. Everyone knows Robert Louis Stevenson's famous account of how he began to write *Treasure Island* (1883). The map gave birth to the story, he writes in his essay "My First Book" (1924):

> On one of these occasions, I made the map of an island; it was elaborately and (I thought) beautifully coloured; the shape of it took my fancy beyond expression; it contained harbours that pleased me like sonnets; and with the unconsciousness of the predestined, I ticketed my performance 'Treasure Island.' ... Somewhat in this way, as I paused upon my map of 'Treasure Island,' the future characters of the book began to appear there visibly among imaginary woods; and their brown faces and bright weapons peeped out upon me from unexpected quarters, as they passed to and fro, fighting and hunting treasure, on these few square inches of a flat projection.[21]

In this case, the map is not only a reader's guide but also, containing a kind of creative power, it becomes an element that gets the story moving—in other words, it serves as a stimulus to the imagination of the writer who contemplates it. Here the image is not a simple illustration of the tale, but of its generative components.

For Émile Zola, in constructing the *Rougon-Macquart* series, what Stevenson presents as a fortuitous discovery becomes a methodical practice. His *Carnets d'enquête: Une ethnographie inédite de la France* (Notebooks of Inquiry: An Unedited Ethnography of France) reveal that Zola habitually drew sketches of the terrain in which he tried to characterize an architectural motif or an urban place, the characterizations reflected in his works.[22] In the same way, genetic analysis of the writings has often emphasized the role that visual—and notably spatial—images play in the development of Zola's novels.[23] Thus, in *L'Ébauche de Germinal* Olivier Lumbroso underscores the latent presence of geometric figures, of the square and the crossing of diagonals, in the structuring of the narrative topology. But Zola also draws a map of the region of Marchiennes (figure 1.4) that serves in a way as an experimental space in which he can put the story to work.

The map and the geometric figures, much like the sketches in the *Carnets d'enquête* (that furthermore interact with one another), belong to the narration's generative schemata. As Lumbroso has noted, they are a "springboard for the imagination: at once a miniature stage, a chessboard, a territory or a theater of operations, progressively inhabited and changed by the characters, by the scenarios and the values that the novelist invents in the course of preparatory work. The drawing becomes animated, and this animation comprises a fundamental process of the creation, indeed an initial drafting of the writing of the novel."[24]

2. The preparatory dossiers of the *Rougon-Macquart* contain many manuscript maps that allow Zola to establish the general geographic frame in which the plots of the novels will develop—for example, the map of the region of Plassans at the beginning of the detailed view of *La Fortune des Rougon* (1871), the map of Paradou in *La Faute de l'abbé Mouret* (1875), the summary map of Alsace in *La Débâcle* (1892), and so on. They allow the writer (the first reader of the work?) "to visualize and to get to know," as Stevenson noted, "his countryside, whether real or imaginary, like his hand."[25]

Yet, as Lumbroso notes, the novelistic maps chart a space that above all must respond to the inner strictures of the fiction. Thus, in *La Fortune des Rougon* Zola does not hesitate to create a route to Vidaubon and Aups that does not exist in reality.[26] In the same way, in the map of Alsace positions and distances are slightly altered to correspond to Zola's narrative intentions.[27] In *Germinal*, Zola inscribes the real geography of the region around Marchiennes (the city, the mineshaft) inside a social and symbolic geography (the miners and the placeholders, with the director between them). Put otherwise, if maps allow the author of the novel to construct and visualize space, they correspond more to narrative topologies than, in a strict sense, to topographic representations.

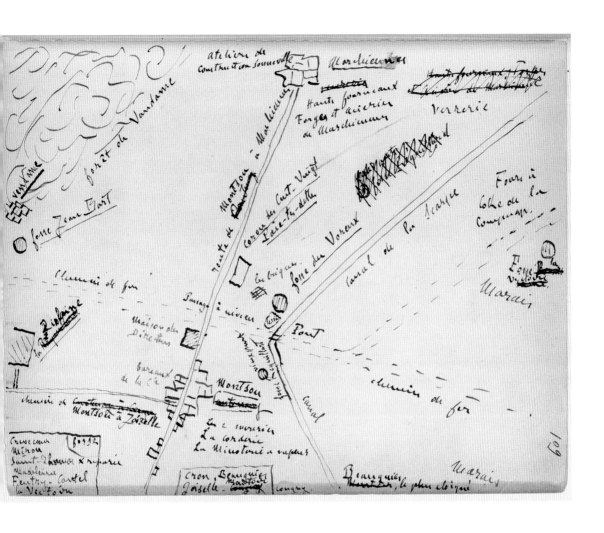

Figure 1.4
Émile Zola, manuscript draft of *Germinal*.
Bibliothèque Nationale de France, folio 109 (call
number: NAF 10308)

3. It is in this sense that maps can be considered with respect to the role they play in the organization of the story, in the structuring of the plot. The case of Joyce's *Ulysses* is exemplary. In no other novel is the role of topography and cartography as implicated in the conception and organization of the work. Joyce makes use of a map when he writes *Ulysses*—perhaps the Ordnance Survey Map of Dublin—and he uses it to construct the story.[28] "Joyce wrote the 'Wandering Rocks,'" writes Frank Budgen, "with a map of Dublin before him on which were traced in red ink the paths of the Earl of Dudley and Father Conmee. He calculated to a minute the time necessary for his characters to cover a given distance of the city."[29] In this example (but many others can be found) the action of the novel is not only deployed in but also according to the space of the map. There is a certain interaction between the spatial distribution of the places on the map and the succession of episodes that structure the story. The progression of the narration melds with the sum of the travels on the part of the characters between the places represented on the map, according to the paths drawn on it and the space of their measured time. The reader witnesses a double operation, at once one of a "spatialization of writing" and, as in Joyce, that of a "staging of the plot in space." The map is the site of this double operation.

4. But in certain cases cartography can become a mechanism of conflict, even of shattering, of the typical organization of narration. It is the principle (or secret model) of a spatialized writing, or in any case of a fiction that attempts to free the story of a linear temporality for the sake of what might be called a "dissemination" of meaning and of a multiplication of narrative units. The cartographic paradigm, and this whether a map is actually drawn or not, makes possible the simultaneity of several narrative trajectories, the dispersion and dislocation of the point of view, or the parallel montage of several stories—in other words, a spatialization of the story or at the least a spatial ordering of its components. When Thomas Mann described to Theodor Adorno the "principle of montage" that he used in his *Doctor Faustus*, on his own account he adopts a method analogous to that which had generally been used in literature written between the First and Second World Wars, in, for example, Döblin or Dos Passos.[30]

The analysis that Walter Benjamin proposes for Döblin's *Berlin Alexanderplatz* might be extended to certain forms of cartography: "The stylistic principle of this book is montage. In this piece of writing there is a cavalcade of petty-bourgeois writings, scandalous stories, news items of accidents, sensational events of 1928, popular songs, short announcements. The montage shatters the 'novel' both from its structural and stylistic points of view, thus creating new epic possibilities, notably in the formal design."[31] The map can be considered a writing mechanism of the same genre, allowing the novels to assemble, collate, bring forward, and set back stories and images of every kind. The map does not moreover lead inevitably to the dispersion and dislocation of the various meanings but rather to the consciousness of their complex tessellation within a work constructed spatially.

Therein cartography bears a form of literary rationality that can be called a mode of thinking with space and a thinking of writing with montage. In *Mobile* (1962) Michel Butor, like Georges Perec in *La vie mode d'emploi* (1978), made it manifest: the map and the atlas are modalities or possibilities of writing and of creation (figure 1.5).

The writer becomes a cartographer, both of which are taken up with the description and creation of a space that is fundamentally a space made of words and lines. Which Georges Perec underscores: "Space thus begins, only with words, signs traced on the white page. To describe space: to name it, to draw it, much as the artists of the portolan charts that saturated the coastlines with the names of ports, the names of capes, the names of creeks until a continuous ribbon of text separates the land from the sea."[32]

Conclusion

At the beginning of his "Berlin Chronicle," having imagined the possibility of a "biocartography" or a biographical cartography, Walter Benjamin writes:

> For a long time, for years I confess, I have been caressing the idea of graphically organizing on a map the space of life—*bios*. First off, I was vaguely dreaming of a Pharus plan but today I would be more inclined to return, if there existed one for inner cities, to the *état-major* map. ... I imagined a system of conventional signs and on the grey background of these maps there would be visible in every color the lodging of my friends, men and women, of the meeting places of diverse collectives ..., the hotel rooms and the whorehouses in which I stayed overnight, the decisive benches of the Tiergarten, the paths to the school and the tombs I saw being filled, the places where cafés held a place of honor whose names today have disappeared and that had always been on my lips, the tennis courts where today are empty rowhouses and rooms decorated with gilt and stucco, where the fears that came with dancing lessons were almost the equal of gymnasiums, if all that could be distinctly drawn.[33]

Extending Benjamin's intuition, we can wonder what a cartographic writing would be, or a cartographic moment of writing, indeed a carto-graphy. In other words, what does cartography teach and bring to writing? We can also ask if cartography itself, insofar as it would be conceived not only as a means of representing the territories to which it refers, but also as a form or mechanism of writing, has the power to shape information, ideas, and values. It is this organizational and configurative power of cartography that needs to be studied. As we have seen with Zola, when the map is deployed in the novel it becomes a formal operator that allows the articulation of a real topography with a

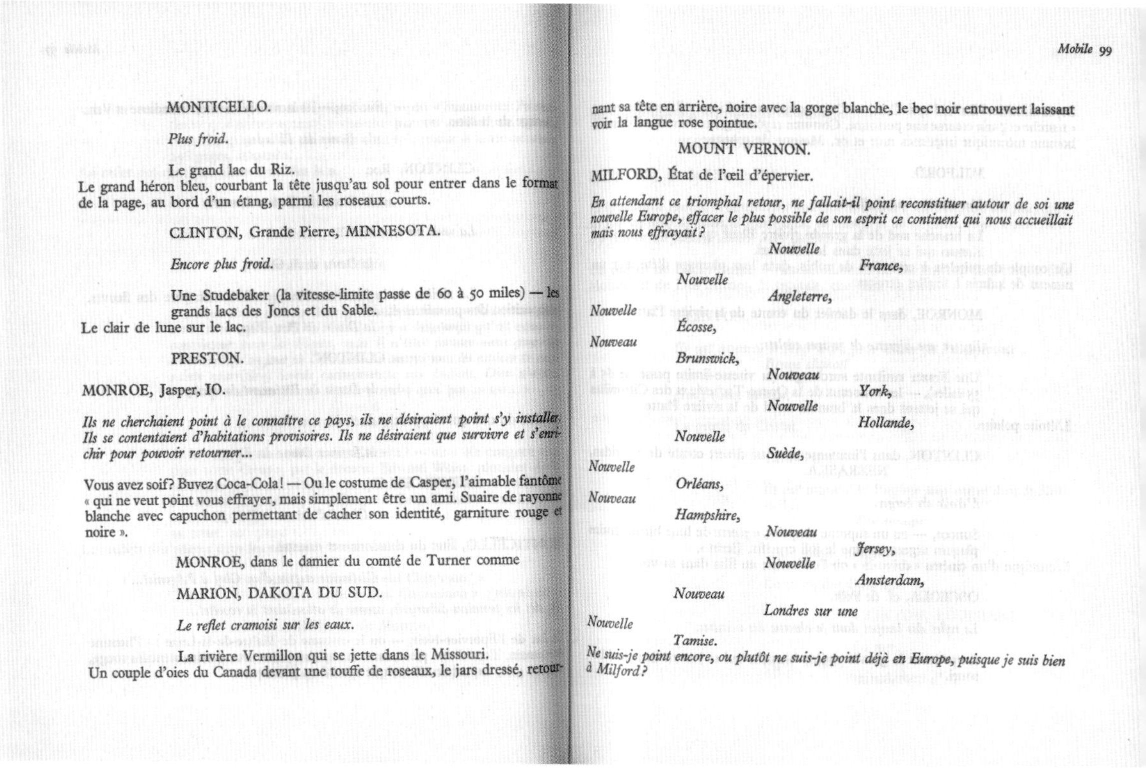

MONTICELLO.

Plus froid.

Le grand lac du Riz.
Le grand héron bleu, courbant la tête jusqu'au sol pour entrer dans le format de la page, au bord d'un étang, parmi les roseaux courts.

CLINTON, Grande Pierre, MINNESOTA.

Encore plus froid.

Une Studebaker (la vitesse-limite passe de 60 à 50 miles) — les grands lacs des Joncs et du Sable.
Le clair de lune sur le lac.

PRESTON.

MONROE, Jasper, IO.

Ils ne cherchaient point à le connaître ce pays, ils ne désiraient point s'y installer. Ils se contentaient d'habitations provisoires. Ils ne désiraient que survivre et s'enrichir pour pouvoir retourner...

Vous avez soif? Buvez Coca-Cola! — Ou le costume de Casper, l'aimable fantôme « qui ne veut point vous effrayer, mais simplement être un ami. Suaire de rayonne blanche avec capuchon permettant de cacher son identité, garniture rouge et noire ».

MONROE, dans le damier du comté de Turner comme

MARION, DAKOTA DU SUD.

Le reflet cramoisi sur les eaux.

La rivière Vermillon qui se jette dans le Missouri.
Un couple d'oies du Canada devant une touffe de roseaux, le jars dressé, retournant sa tête en arrière, noire avec la gorge blanche, le bec noir entrouvert laissant voir la langue rose pointue.

MOUNT VERNON.

MILFORD, État de l'œil d'épervier.

En attendant ce triomphal retour, ne fallait-il point reconstituer autour de soi une nouvelle Europe, effacer le plus possible de son esprit ce continent qui nous accueillait mais nous effrayait?

		Nouvelle	
			France,
		Nouvelle	
			Angleterre,
	Nouvelle		
		Écosse,	
	Nouveau		
		Brunswick,	
			Nouveau
			York,
		Nouvelle	
			Hollande,
		Nouvelle	
			Suède,
Nouvelle			
	Orléans,		
	Nouveau		
	Hampshire,		
			Nouveau
			Jersey,
		Nouvelle	
			Amsterdam,
		Nouveau	
Nouvelle		*Londres sur une*	
	Tamise.		

Ne suis-je point encore, ou plutôt ne suis-je point déjà en Europe, puisque je suis bien à Milford?

Figure 1.5
Michel Butor, *Mobile* (Paris: Éditions Gallimard, 1962)

sum of fictional intentions. It can surely present a documentary aspect that embellishes the construction of the novel, but in itself it is subordinate to the project of the novel. In the literary story, the map carries meaning only consequently as a blueprint of a space-for-the-story: it represents the space-of-the-story. But to say that the map represents the space of a narrative topology makes it possible to indicate the veritable stake of its presence in the novel: the question is not that of the opposition between the "real" and "fiction," but rather that of the project of the fiction itself and of the graphic means it mobilizes so as to be shown and developed. In other words, where cartography is in question, the configuration of a space of reference pertaining to the unfolding of the narrative scheme is the very crux of the fiction.

Translated by Tom Conley

Notes

1. Pour l'enfant, amoureux de cartes et d'estampes,
L'univers est égal à son vaste appétit.
Ah! que le monde est grand à la clarté des lampes!
Aux yeux du souvenir que le monde est petit!
(Charles Baudelaire, *The Flowers of Evil*, trans. James McGowan (Oxford: Oxford University Press, 2008), 282)

2. Walter Benjamin, "Paris, die Stadt im Spiegel," in *Gesammelte Schriften* 4:1 (Frankfurt am Main: Suhrkamp Verlag, 1972), 357.

3. Benjamin, "Paris, die Stadt," 357.

4. Walter Benjamin, "Paris, the Capital of the Nineteenth Century," in *Walter Benjamin: Selected Writings 3: 1935–1938*, ed. Howard Eiland and Michael W. Jennings, trans. Edmund Jephcott, Howard Eiland, and associates (Cambridge, MA: Belknap Press of Harvard University Press, 2004), 32.

5. Jorge Luis Borges, "The Zahir," in *Labyrinths: Selected Stories and Other Writings*, ed. Donald A. Yates and James E. Irby (New York: New Directions Books, 1964), 162.

6. The issue is fleshed out in Royce's *The World and the Individual* (New York: Macmillan, 1900), 501–507.

7. Jorge Luis Borges, "Partial Magic in the *Quixote*," in *Labyrinths: Selected Stories and Other Writings*, ed. Donald A. Yates and James E. Irby (New York: New Directions Books, 1964), 195–196.

8. On several occasions Charles Sanders Peirce, the logician and philosopher who also worked at the IS Coast and Geodesic Survey, encountered Royce's problem (see Peirce's *Collected Papers* (Cambridge, MA: Harvard University Press, 1932), 2.23, 3.609, 5.71, 8.122). We do not know whether Borges was familiar with these writings. Reference can surely be made to Umberto Eco in *How to Travel with a Salmon & Other Essays* with respect to what he calls the "normal map" in Borges's imagination (that is, the map of the Empire; clearly, in *Sylvie and Bruno* Lewis Carroll developed the same hypothesis of a 1:1 map covering the entirety of the territory it represented). Eco describes a variety of proposed technical solutions that are all doomed to failure. And so, from the moment it is placed on the territory and merges with it,

the map changes the whole (physically, with differences in luminosity, etc.). Which causes the map to become inexact (something it fails to take into account: it does not account for the fact that, as an image, it is itself an object that modifies the reality it seeks to represent). Also an image, the map must account for the fact that it represents a territory that the map entirely covers. There again, in the strict operation of representation, the problem of infinite regression exists: a second map has to be drawn on the first in order to represent the territory in addition to the map. We can fancy a "normal map" that puts an end to the process—in other words, "a final map ... that represents all the maps between itself and the territory, but does not represent itself" (Umberto Eco, *How to Travel with a Salmon & Other Essays*, trans. William Weaver (San Diego: Harcourt Brace, 1995), 105). But his normal map submits to the Russell-Frege paradox: of maps that have the property of not representing themselves, to which the ultimate map belongs (it is part of the series). Does the final map that represents all of them represent itself (if it represents all the maps, it must represent itself)? If the answer is affirmative (that is, it represents itself), then, since by definition all the maps of the series share the trait of not representing themselves, it does not belong to the series: a contradiction. If the answer is negative (that is, it does not represent itself), then it can neither complete the series nor be considered a final map: yet another contradiction. Which, in sum, means that a map of all maps, a map that would represent all maps, is a paradoxical idea. Ultimately, then: if the map merges with the territory, the territory cannot be represented. (For more discussion see Eco, *How to Travel with a Salmon*.)

9. Bertrand Russell, *Introduction to Mathematical Philosophy* (London: Allen & Unwin, 1919), 163–166, to which Borges refers in "Tlön, Uqbar, orbis tertius" in the *Obras completas 1: 1923–1949* (Buenos Aires: Emecé Editores, 1993), 437.

10. Borges, "Partial Magic in the *Quixote*," 192.

11. Jean-Claude Grenier, "La Décoration statuaire du 'Serapeum' du 'Canope' de la Villa Adriana: Essai de reconstitution et d'interprétation," *MEFRA* 101, nos. 101–102 (1989): 925–1019.

12. Christina Ljungberg, *Creative Dynamics: Diagrammatic Strategies in Narrative* (Amsterdam: John Benjamins, 2006), 41.

13. Abraham Ortelius, *Théâtre de l'univers* (Antwerp, 1587), preface.

14. *Annuaire Collège de France, Résumés 2013–2014*, 633–634.

15. *Annuaire*, 634.

16. Michel Foucault, "Des espaces autres," in *Dits et écrits: 1954–1988*, vol. 4 (Paris: Éditions Gallimard, 1994), 752–762.

17. Hermann Schmitz, *Was ist neue Phänomenologie?* (Rostock: Koch Verlag, 2003).

18. See Andrea Goulet, "Lecoq cartographe: Plan des lieux et terrains vagues dans le roman judiciaire," *Romantisme* 3 (2010): 39–52. Film noir and the spy film also use maps of real places, cities, or more extensive territories, to locate the action that will develop. Examples that come to mind include *Casablanca* (d. Michael Curtiz, 1942), *M* (d. Fritz Lang, 1931), *Detour* (d. Gregory Ulmer, 1945), and *La mani sulla città* (d. Francesco Rosi, 1963). See Tom Conley, *Cartographic Cinema* (Minneapolis: University of Minnesota Press, 2007), and Teresa Castro, "Les Cartes vues à travers le cinema," *Textimage* 2 (2008), https://www.revue-textimage.com/03_cartes_plans/castro2.htm.

19. Tom Conley, "Du cinéma à la carte," *Cinémas: Revue d'études cinématographiques / Cinemas: Journal of Film Studies* 10, nos. 2–3 (2000): 65–84.

20. Benjamin, "Paris, die Stadt," 356.

21. Robert Louis Stevenson, *Treasure Island* (Oxford: Oxford University Press, 1985), 194.

22. Olivier Lumbroso, "Espace et création: L'invention de l'espace dans la genèse de *Germinal*," ITEM, http://www.item.ens.fr/index.php?id=44574.

23. See Olivier Lumbroso, "De la palette à l'écritoire: Pratiques et usages du dessin chez Emile Zola," *Romantisme* 107 (2000): 71–95; Philippe Hamon, "Génétique du lieu Romanesque: Sur quelques dessins d'E. Zola," in *Création de l'espace et narration littéraire* (Nice: Université Nice Sophia Antipolis, 1997), 27–43.

24. Lumbroso, "De la palette à l'écritoire," 74.

25. Stevenson, *Treasure Island*, 200.

26. "In the novel Plassans is located on the road, going by Alboise and Plassans, that links La Palaud and Saint-Martin-de-Vaulx to Ochrères and Sainte-Roure. In reality Lorgnes is located halfway, as the crow flies, on the one hand, between La Garde-Freinet, Le Luc, and Vidauban while on the other between Salernes and Aups. No direct route links Lorgnes to each of its two groups of locality" (in Émile Zola, *Les Rougon-Macquart*, ed. Henri Mitterand, vol. 1 (Paris: Éditions Gallimard/Pléiade, 1960), 1544).

27. Lumbroso, "De la palette à l'écritoire," 80–81.

28. Eric Bulson, *Novels, Maps, Modernity: The Spatial Imagination, 1850–2000* (London: Routledge, 2007), 77–80.

29. Quoted in Bulson, *Novels, Maps, Modernity*, 68.

30. Theodor Adorno and Thomas Mann, *Correspondance 1943–1955* (Paris: Klincksieck, 2009), 21–27. See Jean-Pierre Morel, "Roman de montage et poétique historique," *Revue de Littérature Comparée* 51, no. 2 (1977): 241–248; Jean-Pierre Morel, "Montage, collage et discours romanesque dans les années vingt et trente," in *Collage et montage au théâtre et dans les autres arts dans les années vingt*, ed. Denis Bablet (Lausanne: l'Age d'Homme, 1978), 38–73.

31. Walter Benjamin, "The Crisis of the Novel," in *Walter Benjamin: Selected Writings 2: 1927–1934*, ed. Howard Eiland and Michael W. Jennings, trans. Edmund Jephcott, Howard Eiland, and associates (Cambridge, MA: Belknap Press of Harvard University Press, 2004), 303.

32. Georges Perec, *Espèces d'espaces* (Paris: Galilée, 1974), 26.

33. Walter Benjamin, "A Berlin Chronicle," in *Reflections: Essays, Aphorisms, Autobiographical Writings,* ed. Peter Demetz, trans. Edmund Jephcott (New York: Harcourt Brace Jovanovich, 1978), 5.

Bibliography

Adorno, Theodor, and Thomas Mann. *Correspondance 1943–1955*. Paris: Klincksieck, 2009.

Baudelaire, Charles. *The Flowers of Evil.* Trans. J. McGowan. Oxford: Oxford University Press, 2008.

Benjamin, Walter. "A Berlin Chronicle." Trans. Edmund Jephcott. In *Reflections: Aphorisms, Autobiographical Writings*, ed. Peter Demetz. New York: Harcourt Brace Jovanovich, 1978.

Benjamin, Walter. "The Crisis of the Novel." Trans. Edmund Jephcott, Howard Eiland, and associates. In *Walter Benjamin: Selected Writings 3: 1935–1938*, ed. Howard Eiland and Michael W. Jennings. Cambridge, MA: Belknap Press of Harvard University Press, 2004.

Benjamin, Walter. "Paris, the Capital of the Nineteenth Century." Trans. Edmund Jephcott, Howard Eiland, and associates. In *Walter Benjamin: Selected Writings 3: 1935–1938*, ed. Howard Eiland and Michael W. Jennings. Cambridge, MA: Belknap Press of Harvard University Press, 2004.

Benjamin, Walter. Paris, die Stadt im Spiegel. In *Gesammelte Schriften 4:1*. Frankfurt am Main: Suhrkamp Verlag, 1972.

Borges, Jorge Luis. Partial Magic in the *Quixote*. In *Labyrinths: Selected Stories and Other Writings*, ed. Donald A. Yates and James E. Irby. New York: New Directions Books, 1964.

Borges, Jorge Luis. Tlön, Uqbar, orbis tertius. In *Obras completas 1: 1923–1949*. Buenos Aires: Emecé Editores, 1993.

Borges, Jorge Luis. The Zahir. In *Labyrinths: Selected Stories and Other Writings*, ed. Donald A. Yates and James E. Irby. New York: New Directions Books, 1964.

Bulson, Eric. *Novels, Maps, Modernity: The Spatial Imagination, 1850–2000*. London: Routledge, 2007.

Castro, Teresa. "Les Cartes vues à travers le cinéma." In *Textimage* 2 (2008), https://www.revue-textimage.com/03_cartes_plans/castro2.htm.

Chartier, Roger. *Annuaire Collège de France, Résumés 2013–2014*. https://annuaire-cdf.revues.org/11942.

Conley, Tom. *Cartographic Cinema*. Minneapolis: University of Minnesota Press, 2007.

Conley, Tom. ""Du cinéma à la carte." In *Cinémas: Revue d'études cinématographiques / Cinemas*." Journal of Film Studies 10 (2–3) (2000): 65–84.

Eco, Umberto. *How to Travel with a Salmon & Other Essays*. Trans. W. Weaver. San Diego: Harcourt Brace, 1995.

Foucault, Michel. Des espaces autres. In *Dits et écrits: 1954–1988*. vol. 4. Paris: Éditions Gallimard, 1994.

Goulet, Andrea. "Lecoq cartographe: Plan des lieux et terrains vagues dans le roman judiciaire." *Romantisme* 3 (2010): 39–52.

Grenier, Jean-Claude. "La Décoration statuaire du 'Serapeum' du 'Canope' de la Villa Adriana: Essai de reconstitution et d'interprétation." *MEFRA* 101 (101–102) (1989): 925–1019.

Hamon, Philippe. Génétique du lieu Romanesque: Sur quelques dessins d'E. Zola. In *Création de l'espace et narration littéraire*. Nice: Université Nice Sophia Antipolis, 1997.

Ljungberg, Christina. *Creative Dynamics: Diagrammatic Strategies in Narrative*. Amsterdam: John Benjamins, 2006.

Lumbroso, Olivier. "De la palette à l'écritoire: Pratiques et usages du dessin chez Emile Zola." *Romantisme* 107 (2000): 71–95.

Lumbroso, Olivier. "Espace et création: L'invention de l'espace dans la genèse de *Germinal*." ITEM, www.item.ens.fr/index.php?id=44574.

Morel, Jean-Pierre. "Montage, collage et discours romanesque dans les années vingt et trente." In *Collage et montage au théâtre et dans les autres arts dans les années vingt*, ed. Denis Bablet. Lausanne: l'Age d'Homme, 1978.

Morel, Jean-Pierre. "Roman de montage et poétique historique." *Revue de Littérature Comparée* 51 (2) (1977): 241–248.

Ortelius, Abraham. *Théâtre de l'univers.* Antwerp: 1587.

Peirce, Charles Sanders. *Collected Papers.* Cambridge, MA: Harvard University Press, 1932.

Perec, Georges. *Espèces d'espaces.* Paris: Galilée, 1974.

Royce, Josiah. *The World and the Individual.* New York: Macmillan, 1900.

Russell, Bertrand. *Introduction to Mathematical Philosophy.* London: Allen & Unwin, 1919.

Schmitz, Hermann. *Was ist neue Phänomenologie?* Rostock: Koch Verlag, 2003.

Stevenson, Robert Louis. *Treasure Island.* Oxford: Oxford University Press, 1985.

Zola, Émile. *Les Rougon-Macquart.* vol. 1. Ed. Henri Mitterand. Paris: Éditions Gallimard/Pléiade, 1960.

2
Literary Cartography: Mapping as Method

Barbara Piatti

Literary Landscapes

"Regardless of the late hour and the setting sun, we will have an hour's rest up by the little cabin; for in front of us, a green, but not all too steep slope of vertiginous depth extends itself between the dark woods, lonely at the top and below covered with hundreds of little houses, and at the very bottom—a tip of Lake Lucerne, embedded in a veritable maze of defiant alpine mountains thrust into each other in a chaotic jumble. This is not a 'view,' it is more than that: it is a landscape, in fact a landscape that only the imagination of a Leonardo da Vinci could dream of."[1] This passage was penned by Carl Spitteler, the almost forgotten Swiss Nobel laureate in literature (1919). His novella "Xaver Z'Gilgen" is a telling example of a continuously growing anthology of literary accounts across several centuries, which is still a work in progress and mostly involves a specific area, the region of Lake Lucerne / Gotthard in the heart of Switzerland. It would be no exaggeration to say that the Lake Lucerne and Gotthard region deserves the designation of a gravitational center or crossroads on the European map of literature.[2]

From the very first legends and chronicles of the early modern era to contemporary novels, hundreds of fictional texts have featured the rocky and at the same time very charming landscape between Lucerne in the north and the Gotthard Heights in the south as their setting. What makes this collection of texts particularly striking is the mixture of authors that are regionally "embedded" (e.g., Heinrich Federer, Josef Maria Camenzind) or accomplished Swiss ones (Gotthelf, Keller, Inglin, Spitteler, Walser, Frisch, among others) as well as foreign writers of international standing (Goethe, Schiller, Hesse, Strindberg, Scott, Twain, Cooper, D. H. Lawrence, Tolstoy, Dumas (the elder), and Flaubert, just to name a few). They all describe the Lake Lucerne and Gotthard region in distinct literary genres, and some make it the backdrop of their work.

From the perspective of literary analysis, how do you approach such a "literary landscape"? Interest in this phenomenon is not new, as an almost unmanageable number of studies, anthologies, collections of essays, and even travel guides for such hauts-lieux of literature attests. In more recent times, however, a series of productive scholarly approaches have been devised or refined. These approaches seem highly promising and at the same time they exhibit a number of similarities. Geocriticism, geopoetics, literary geography—each of them deals, under a different label, with literary texts that refer to an existing landscape, region, or city.[3]

To this end, I have developed my own research under the already existing banner of *literary geography*, a field that, as I will show, has a long tradition and in some (but not all) instances entertains close relations with *literary cartography*. Put simply, a clear line between both expressions can be drawn by differentiating between object and method: what is of interest is the *geography of literature*, which in turn implies the use of cartographic instruments.

The majority of literary-cartographic approaches have as a common denominator their concern with the interactions between fictional spaces and real spaces, sometimes also called *geospace* or *first space*, the latter according to the terminology of Edward Soja. This has already been the object of extensive thinking in (human) geography, literary theory, and philosophy. After all, we are talking about an issue of ontological dimensions: Is it even admissible to assume that there are (spatial) references between fictional worlds and our real world(s)? A significant number of publications have dealt with exactly that question and hence with a major controversial point of literary cartography.[4] Within the limited space of this chapter, however, I will just note that it is quite inspiring and thought-provoking to deal with possible interactions and overlays of fictional and real worlds instead of completely denying such interfaces. Literary cartography aims at making these spaces visible in their fascinating intermediate status between reality and fiction.

The sublime landscape of Lake Lucerne and Gotthard mentioned above forms the focal point of a first illustrative study: it is about gauging opportunities to explore a section of first space in its entire literary richness. Figure 2.1 shows a selection of 150 fictional texts written between 1477 and 2005 with their settings drawn onto a topographic map of the region. Where the setting is rather clearly locatable, one can find a point symbol on the map. For settings with a vague reference toward geospace the ellipsoid symbol has been chosen to mark a "zonal location." Looking at the map one can quickly get an overview—about gravitational centers (Lucerne, Schwyz, Lake Uri, Gotthard) and unwritten areas. In other words, the literary layer of that region, previously invisible, suddenly becomes visible. Furthermore, the map deals with degrees of reference. How meticulously does a fictional text refer to the given section in real space—mimetically or rather in a defamiliarizing way, superimposing and thereby redefining and shifting fictive and existing elements? In this context, I have compiled one of the results under the designation "strong landscape." The bulk of these texts refer topographically and toponymically in a precise manner to Lake Lucerne and Gotthard (orange points and ellipses). Only a few exceptions, marked in violet, invent settings or transform parts of the landscape. A revealing example is Walter Scott's novel *Anne of*

Figure 2.1

The fictional saturation of a region. The map aggregates 150 fictional accounts, written between 1477 and 2005, that refer to the region between Lucerne and Gotthard. Point symbols: exact location (settings); ellipsoid symbols: vague location (zones of action); violet symbols: settings or zones of action that significantly alter the given geospatial toponymy or topography.
© Barbara Piatti, Switzerland

Geierstein, or The Maiden of the Mist (1829), which makes use of the area around Lake Lucerne in a fairly accurate way. But at one point a fictitious gothic castle is inserted in the landscape. This is a classic literary-geographic move: adding a nonexistent element to an existing territory to create a suitable setting for a fictional scene.

Further visualized aspects on other maps in that study include settings connected to famous plots or even myths. Some of these plots might be so dominant that they literally "block" other storylines or motifs. In other words, contemporary or later writers avoid those spots, already marked. The Wilhelm Tell material, the Swiss liberation saga, stands out as such an exclusive topography. It is an interesting fact that certain Tell settings around the lake and in the mountains only lend themselves to this particular literary plot and apart from that are not used (or cannot be used) for any other storyline. The *Rütli*, a meadow surrounded by trees, is a scene that is monosemanticized to the extent that it seems exclusively reserved for the theme of the Tell legend and the liberation saga, something that also applies to the *Hohle Gasse* (the narrow pass close to Mount Rigi, where, according to the legend, Wilhelm Tell killed his enemy Gessler with a single arrow). An "overwriting" of the Tell topography with a love story or thriller generally seems avoided at all costs. To sum up, literary landscapes, the way they should be understood here, are actual existing places and regions that have become objects of (narrative) literature: as a setting, a backdrop, sometimes carrying a story, in some cases limited to a specific historical moment, in other cases used diachronically throughout a succession of decades and centuries up to the present.

Needless to say, literary landscapes and urban topographies that are exceedingly rich in detail can never achieve bibliographical completeness. Hence the visualization will always remain partial. Accordingly, we must acknowledge an undeniable feature of literary cartography: a different choice of texts and/or different parameters of analysis would have yielded different maps. One of the major challenges for literary cartography is indeed to cope in a productive way with numerous uncertainties. These range from the unavoidable incompleteness of the primary sources (with respect to a certain landscape or region) to the individual reading of the text that clearly differs from expert to expert, to name just two seemingly problematic aspects. Against this background, literary cartography and its output, the maps, should not be expected to deliver precise, reproducible results. Moreover, while literary cartography joins forces with the technical science of cartography, it is common knowledge that maps can no longer be regarded as reliable, unambiguous, and true depictions of facts. Under the rubric of "critical cartography," scholars such as J. B. Harley, Denis Cosgrove, Denis Wood, and Jeremy Crampton have from various angles shown that maps by no means simply reproduce "reality," but that their representational strategies are pervaded by at times overt, at times hidden political agendas and ideologies.

In other words, literary maps, as presented here, do visualize spatial aspects of fiction, but they themselves contain fictional elements. They are the product of selections, omissions, and interpretations, but as an analytic method literary cartography can nevertheless reveal unknown aspects of texts or unexpected patterns of a literary landscape. But maps that enable such discoveries should rather be seen as sources of inspiration, as a starting point for further research, not as its endpoint.

A Literary Atlas of Europe: Development of a Prototype (2006–2014)

Literary geography and cartography are anything but new inventions. In fact, they can look back on a hundred-year history. But systematic attempts to chronicle this field have just begun to get underway.[5] The visualizations generated in literary cartography range from handcrafted collages with artistic pretensions to high-tech data models. In recent years the situation has changed fundamentally. Although there have been database-supported literary research projects employing digital maps since the 1990s (most of these projects have not been developed any further),[6] it is digital and animated cartography, the alliance of the humanities and data technology, that have opened endless possibilities, especially after Web 2.0. To name a few, "Mapping the Lakes,"[7] "Mapping St. Petersburg: Experiments in Literary Cartography,"[8] "The Digital Literary Atlas of Ireland, 1922–1949,"[9] "Mapping Emotions in Victorian London,"[10] as well as "A Cultural Atlas of Australia: Mediated Spaces in Theatre, Film, and Literature"[11] all demonstrate that the interest in mapping fiction is vivid and widespread. Last but not least, tools such as Google Lit Trips[12] and participatory Internet services such as "Pinbooks.de," (Dein Buch zur Stadt/Your Book to the City), which allow readers to locate books and their settings geographically, serve the wider popularization of the concept.[13] Another joint project is "A Literary Map of Manhattan," on which readers of the *New York Times* have entered their favorite books.[14] One thing is clear: literary cartography has become a truly interdisciplinary challenge. Real progress can only be made if cartography and IT experts are included in these endeavors. That was precisely the case with "A Literary Atlas of Europe."

The reflections presented here have grown out of several years' work on the research project "A Literary Atlas of Europe," which was anchored at the Institute of Cartography and Geoinformation at the ETH Zürich between 2006 and 2014 (with research partners at the Georg-August-University in Göttingen and the Charles University in Prague). "A Literary Atlas of Europe" was launched and designed by the author of this chapter.[15] The collaboration included experts in literary studies, cartography, graphic design, and IT solutions. We developed a prototype of an interactive database-supported literary atlas that incorporated the autonomy of literary space in an appropriate manner and that produced numerous visualizations.[16]

Three literarily imbued model regions of a highly divergent character formed the starting point of the "Literary Atlas of Europe": Prague as an urban space, North Friesland as a coastal and border region, and the region of Lake Lucerne / Gotthard, introduced above, as a mountainous landscape. In a first step, we collected fictions (using "fiction" in a narrow sense of the word, hence novels, short novels, short stories, legends, etc.) and referenced them bibliographically. The corpus consisted of narrative texts set in the named regions from around 1750 to the immediate present. To study the three model regions accurately, to portray them cartographically, and subsequently to compare them using literary-geography parameters, we developed a database that specifically targeted these objectives. At the same time, we developed an extensive online-input form that allows for intuitive data entry by literary scholars. The locations of the action are sketched onto the map and described thematically along with annotations of various attributes.

Obviously this presupposes an a priori close reading by the literary scholar applying every trick of the trade. Concerning the parameters like localization, for example, one might reasonably ask: Are the settings localizable, are they just zonally localizable, or is their position entirely indeterminable? Or about function: Does the setting fulfill only the background function or does it interfere with the story? Does it usurp the protagonist function, so to speak, as may occur with avalanches, landslides, earthquakes, floods? Is a direct reference to the geospace noticeable, or does the author employ the technique of "indirect referencing"? The latter signifies that an existing place in the text is not explicitly named but is made identifiable through description and surroundings. Once the spatial units and the corresponding attributes have been saved, the system can handle (almost random) inquiries. Out of this extracted data a geographic information system (GIS) for literature is created, which enables a thematically and spatially diverse analysis of data. Fig. 2.2 presents the–filtered–aspects of a coastal landscape in the model region Northern Frisia, which in various fictional accounts turns into a kind of agent itself, when floods and storms, sliding landmasses and other natural hazards turn into opponents of the characters.

The results of these inquiries are issued as prototype maps or GIS visualizations based on a newly developed set of symbols. Space created in fiction is always incomplete; it has only vague boundaries and often is not precisely localizable. Moreover, a number of literary techniques can emphasize the reference to a real and existent geographic space as well as being able to cloud it. This implies that these different levels of transformation have to be taken into consideration during the mapping process.[17] Blurry boundaries, for example, are symbolized as fuzzy shapes and a vague localization with newly developed symbols along with a color gradation (the richer the color, the closer the fictional stage comes to the real-world setting; the paler the color, the more the one is dissociated from the other). The system thereby enables the generation of both statistical queries and maps of individual texts from the same set of data.

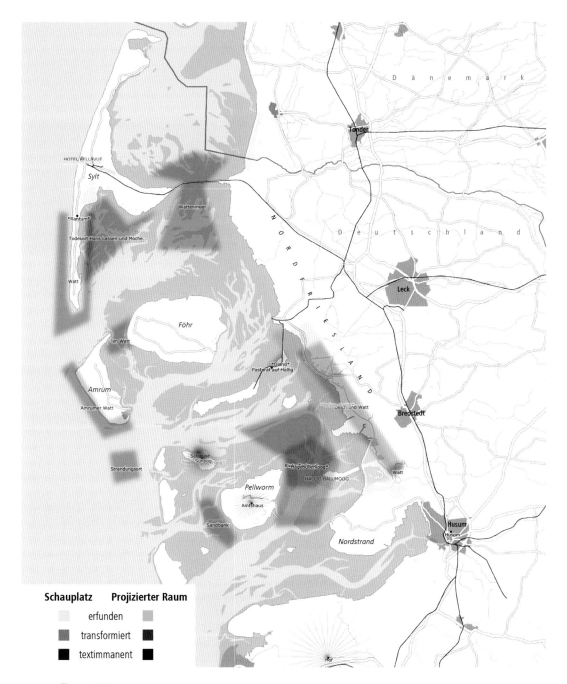

Figure 2.2

A literary "hazard map" of the shoreline of Northern Frisia. The map is based on analyses of 113 prose texts written between 1847 and 1938. Marked are actual settings (in red) and projected spaces (in blue), where natural disasters and accidents such as storms, floods, shipwrecks, and drownings happen or have happened. The attribute "dangerous" is one of the approximately fifty qualities of the settings/projected spaces listed in the large database of "A Literary Atlas of Europe." The content can be filtered via queries and selection commands, such as "Display all settings with the attribute 'dangerous' between 1900 and 1910." Each query results in a different map. © Institute of Cartography and Geoinformation, ETH Zurich (textual analysis: Kathrin Winkler, Kim Seifert; visualization: Anne-Kathrin Weber-Reuschel)

Test Cases: What Can Be Seen in Literary-Cartographic Visualizations?

In the words of Franco Moretti, the founding father of a new literary geography, the following question should be asked when developing a map: "What exactly do they do? What do they do that cannot be done with words, that is; because, if it can be done with words, then maps are superfluous."[18] Or in the words of Jörg Döring, who puts it to the test with "Berlin literature": "What do you see differently, what do you see more clearly, or what do you only see when the places in the storyline of 'Berlin literature' are mapped out?"[19] What can actually be discovered on literary maps?

One important category in our analysis are the so-called projected spaces. According to our definition, projected spaces are not "accessed" by actual characters, but rather inserted into the literary space by means of memories, dreams, and reveries. In Flaubert's *Madame Bovary*, set in the province of Normandy, the title hero dreams of Paris. Max Frisch's *Montauk* is another good example. While the melancholic main plot takes place in New York City and Long Island, over long stretches the narration consists of the first-person narrator's flashbacks and memory fragments that bring us through Swiss regions and places such as the Ticino and Zurich, locations that are quasi–"brought into" the main plot.[20]

Literature is full of projected places. Arguably they constitute one of the most important spatial categories because they are specific to literature. With the exception of film, no other art form has such a wide range of possibilities for creating projected spaces or for transitioning from scenes to projected spaces or vice versa. Even when it comes to individual texts, retrieving the "geography of projected places" can be highly illuminating. Bruce Chatwin's novel *Utz* (1988), about the eponymous quirky porcelain collector who is visited by an American expert (this is still during the Cold War), uses Prague almost exclusively as a backdrop. Yet, when the numerous projected places are added, a completely different picture emerges (figure 2.3). Although the main action does take place in Prague, Utz's stories and the story of the first-person narrator fabricate a dense network of places all over Europe and even globally, which within the framework of the fiction is only accessible via the imagination. One of the elaborately projected places is the French health resort Vichy. At first, Utz imagines the place and connects all sorts of mental pictures to it: Utz had "an idea, derived from *Russian novels* or his parents' love affair at Marienbad, that a spa-town was a place where the unexpected invariably happened."[21] Then he actually travels to Vichy for a stay at the health spa, which thereby changes its status and becomes a setting.

What does this mean? Especially with regard to projected spaces it is more than obvious that a literary landscape can never be viewed in isolation. The integration into a global network is sometimes stronger, sometimes weaker. The example of North Friesland is telling in the sense that the work of what appears to be an extremely locally anchored author such as Theodor Storm actually contains a wide variety of projected places (figure 2.4).

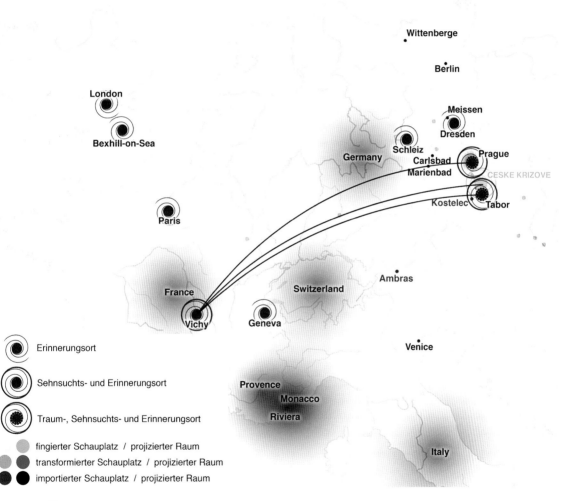

Figure 2.3

Projected spaces in Bruce Chatwin's novel *Utz* (1988). A so-called single-text map based on materials from "A Literary Atlas of Europe." The map gives an impression of how significant the layer of projected spaces (marked in violet) is for the structure of that narrative. Prague is the main setting, but almost every other place appears in the mode of a projected space. © Institute of Cartography and Geoinformation, ETH Zurich (textual analysis: Barbara Piatti; visualization: Anne-Kathrin Weber-Reuschel)

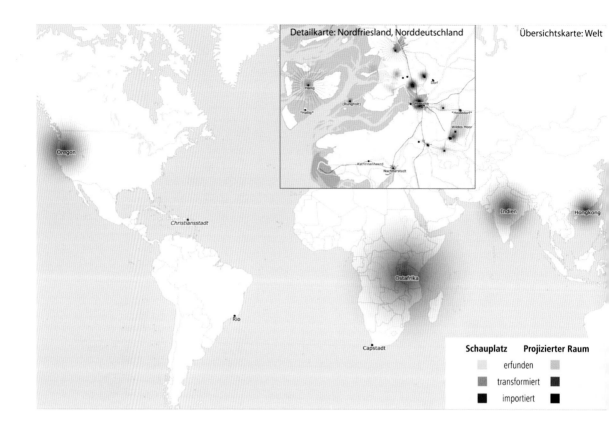

Figure 2.4

Theodor Storm and the global dimension. The map is based on the analysis of thirty novels and novellas by Storm. The main zone of action is Northern Frisia and the little town of Husum. But the layer of projected spaces (marked in violet) reveals that Storm's work has a much wider horizon that extends from Oregon to Hong Kong. © Institute of Cartography and Geoinformation, ETH Zurich (textual analysis: Kathrin Winkler, Kim Seifert; visualization: Anne-Kathrin Weber-Reuschel)

A world map of Storm's work features exotic places such as East Africa, India, Hong Kong, and Oregon—in Storm's novella "Bötjer Basch" (1886) the main character is back in Northern Frisia, but recounts his adventures during the gold rush in Oregon. The inclusion of such faraway destinations does not necessarily correspond to the tightly defined geographic horizon with which one would generally associate this poet and that is presented in the framed map extract, "Detailkarte: Nordfriesland, Norddeutschland." Storm and the tiny North Frisian fishing village Husum determine themselves mutually.

Figure 2.5 reveals the hidden spatial dynamic in Friedrich Schiller's famous play *Wilhelm Tell* (1804), set in the pre-alpine and alpine scenery around Lake Lucerne. The play is packed with descriptions of journeys made by the characters, be it journeys in the past (purple) or planned journeys (green) or real-time journeys witnessed by others (yellow). Looking at this map, it immediately becomes visible how crucial the flow of ideas, thoughts, and information is, as well as the transport of weapons and goods between the scattered settings, for the play is all about a fight for freedom and a conspiracy between oppressed tribes. Some of the heroes such as Wilhelm Tell and young Melchthal turn out to be veritable long-distance runners traversing the valleys, moving along the shores of the lakes and over passes and peaks. Only a map can reveal these undercurrents of movements, which are only very indirectly indicated in the original text, but that add greatly to the suspense of the play. Once you have seen the map, you will probably read the play in a different way. Indeed, the map may be considered from the point of view of Wolfgang Iser's literary theory. For Iser, a literary work consists of both written and unwritten portions of the text. According to his reader-response theory, a literary text is "full of unexpected twists and turns, and frustrations of expectations. … Thus whenever the flow is interrupted and we are led off in unexpected directions, the opportunity is given to us to bring into play our own faculty for establishing connections—for filling in the gaps left by the text itself."[22] In the present case, that is quite literally what the map provides. It visualizes the distances the characters travel, although the act of traveling itself is rarely described in a comprehensive way. Already in the very first scene, when "Baumgarten of Alzellen" arrives at the shores of Lake Uri, out of breath, hunted by the soldiers of Landvogt Wolfenschiessen, the entire dramatic downhill flight has been omitted in the description. The flight is marked on the map as are many other movements that appear only as "gaps" (in Iser's sense) in the play. Of course, the map does not allow a movielike complete immersion in the setting. But the shaded topographic background relief at least reveals that moving through that region must be connected with a great deal of physical strain. In other words, what is hardly obvious when *reading* the text—the long distances, the sheer number of journeys, the physical stress—becomes clear when *viewing* the text on the map.[23]

Figure 2.6 presents a map of F. Scott Fitzgerald's novel *Tender Is the Night* (1934). Individual text maps hold a whole string of legible information: the geographic radius

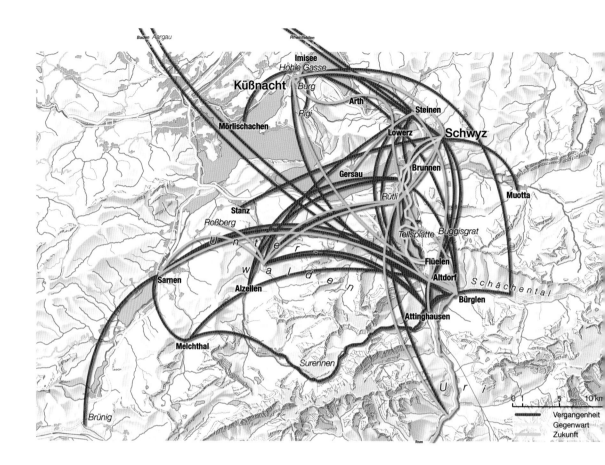

Figure 2.5

The network of all roads traveled by the characters in Friedrich Schiller's play *Wilhelm Tell* (1804). The colors indicate how the route is presented in the text. Purple = past, a character tells about a journey already made; yellow = present, one character acts as an eyewitness and reports a journey as it takes place; green = future, a character reveals his or her travel plans in the future. © Barbara Piatti / Anne-Kathrin Weber-Reuschel, Switzerland

Figure 2.6

The content-related weight of settings. A map of F. Scott Fitzgerald's novel *Tender Is the Night* (1934), including settings (in orange) and projected places (in violet). The bigger the label and symbol, the more important the setting or the projected space.
© Barbara Piatti / Anne-Kathrin Weber-Reuschel, Switzerland

of a text (does it take place entirely in the model region or is there another setting on the European or world map?), the reference level of settings and projected spaces, the function of the space (setting, projected space, marker, path), the labeling (are toponyms taken from the geo space or does it result in newly invented names?). Again, we find the settings (in orange) and projected places (in violet). But this time the map also reveals the so-called weight of the settings—some are more important than others, as indicated by the bigger circles. In other words, the map allows us to capture the spatial extension and structure along with the network of settings of the novel at a glance, including information about the main elements of the story and the projected dimensions.

The last example (figure 2.7) comes from a master's thesis written under my supervision. Andreas Bäumler maps two fictional texts, both set in the rocky landscape of Engadine in the South of Switzerland.[24] At first glance, one can see how differently the two authors make use of the given geospace. While Ulrich Becher in *Die Murmeljagd* (*The Woodchuck Hunt*, 1969), marked in red, narrates mainly along the roads, railways, and villages, Hans Boesch in *Der Kreis* (*The Circle*, 1998), marked in blue, by and large ignores the populated valleys (except for the part between Bever and La Punt) and lets his characters act almost exclusively within the alpine zones in the North and South.

The maps presented here illuminate different literary phenomena. Some deal with a single text, others with a whole group of texts. They visualize various themes and topics such as the paths of characters, the dimension of projected spaces, or the density of a literary landscape with regard to a specific topic. What they all have in common is, first, a (topographic) base map and hence the assumption of the mapmakers that the fictional texts refer to geospace. Second, one immediately notices the shortcoming of including much more on the map than what the text itself offers in terms of geographic and topographic information (you can find motorways, rivers, political borders, and other features on background maps that have nothing to do with the content of the works in question). In other words, the ideal type(s) of background maps for the purposes of literary studies are yet to be developed.

What becomes clear is that the functions of literary maps range from pure *illustration* (to present something that could also be explained in a text), to *inspiration* (the process of mapping or in some cases its impossibility may lead to a new train of thought), and finally to *instrument* (in the best-case scenario something will be visible on the maps that could not be seen without them). In this latter function, maps become hermeneutic tools. The visualization of literary cartography in no way replaces the classical hermeneutic methods. They are rather a sort of intermediate result, and starting from there the literary expert should return to the primary sources (the fictional texts) and continue analyzing them with the proven techniques of literary study: weigh

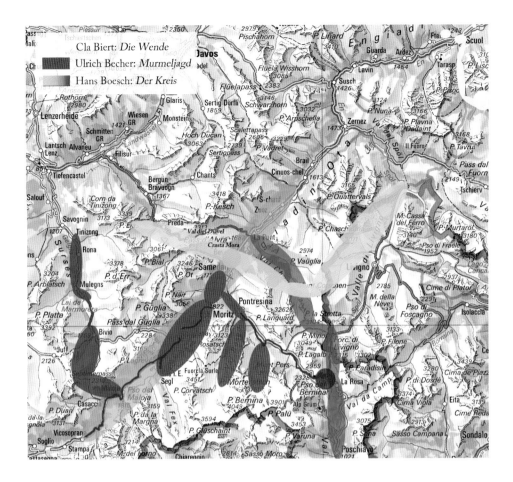

Figure 2.7

The same geographic area in novels by two different authors. Andreas Bäumler's map enables a comparison of the spatial arrangements of two novels. While Ulrich Becher's *Die Murmeljagd* (*The Woodchuck Hunt*, 1969), marked in red, narrates mainly along the roads, railways, and villages, Hans Boesch's *Der Kreis* (*The Circle*, 1998), marked in blue, places its characters almost exclusively in the alpine zones. © Andreas Bäumler, University of Lucerne, Switzerland

ambiguities, compare, contextualize, enlighten historical references, juxtapose several readings, combine, and include other methods and instruments.

Such an approach is even more necessary since every literary-cartographic study considerably reduces the complexity of a fictional text. For this reason, the maps should be seen as signposts toward further exploration rather than conclusive evidence. In some cases the map, once produced, might even become obsolete, since it was the actual process of mapping and maybe the impossibility of mapping some features of the text that stimulated new thoughts.

Of course, one could think of entirely different, alternative maps of fiction—ones that are more artistically oriented. These might include a kind of fluid, multidimensional map, which changes smoothly between scales and layers; maps that include "foggy zones" where the text opens up blank, unwritten spaces as opposed to richly elaborated settings; maps that play with distortions; or maps that could merge georeferential elements with invented places and spaces. But for the moment, such ideas still belong to the future.

Literary Cartography as a Tool for Conveying Culture

It is hardly surprising that literary geography is appealing to a broader audience. In 2014, the "Literary Maps of Switzerland" project drew quite a bit of media attention in both Germany and Switzerland. I developed this project together with Ann-Kathrin Weber (formerly Reuschel) on behalf of the Swiss Embassy in Berlin and on the occasion of the book fair in Leipzig (http://www.literatur-karten.ch). What is being presented is a comprehensive literary map of Switzerland. Many voices have contributed to this fictive and imaginative Switzerland. Swiss-born authors stand next to foreign ones—poets who temporarily or permanently have been living in Switzerland, some of their own free will, others as refugees from war, poverty, and fear. German literature takes center stage, complemented by treasures of the French, Italian, and Rhaeto-Romanic Swiss literature. Intriguing literary neighborhoods can be discovered on this map. Peter Stamm finds himself in the company of Carl Sternheim and Franz Hodjak in the Lake Constance area; Jean-Jacques Rousseau, Stefan Zweig, and Gustav Meyrink meet up with W. G. Sebald at Lake Geneva. Eras, cultures, nations, and destinies mingle on literary-geography maps, linked by only one commonality: the settings and inspirational locations are found in Switzerland. A strikingly large number of various origins and manifold backgrounds converge on this Swiss map: German, Austrian, Swiss, Danish-German, German-French, German-British, Hungarian-Swiss, and others (and there would be many more, if English, American, Italian, French, and other writers were included). The focus is not "national literature," but the complexity and many voices of a literary space. With all the arbitrariness of choice, these by-no-means-complete literature maps emphasize that between the Jura Heights and Lake Constance—covering centers like Zurich, Bern, Basel, and the lonely Ticino—stretches the territory of the jointly created cultural heritage that now becomes visible in this superimposition.

The records can be found on the overview map and can therefore be filtered according to different criteria. Moreover, special maps for large literary subjects can be retrieved: for love encounters, murders, and deaths; for utopian, dystopian, and counterfactual worlds; for journeys of the mind; as well as—under the title "Place Name Games"—for fictitious and renamed settings (figure 2.8). A feather quill indicates a famous place that has been given a pseudonym (e.g., Otto F. Walter's "Jammers"—a play on the meaning of the German verb *jammern* (moaning, lamenting)—disguises the existing provincial town Olten), and a tipped-over inkwell indicates a fictitious setting that can only roughly be localized such as Gottfried Keller's Seldwyla.

The map "Gedankenreisen" ("Journeys of the Mind") (figure 2.9) presents the projected spaces introduced above. Symbols that look like shooting stars indicate dreams, memories, and longings, and they visualize such journeys of the mind. In his novel *Das Jahr der Liebe* (*My Year of Love,* 1981) Paul Nizon writes that Wohlen Lake, close to Bern, "was my space of memory, desire and longing." The literary characters are not present in the marked places; they only summon them in the form of dreams, memories, and longings. Peter Bichsel's sailor sits in a bar in the U.S. state of Kentucky and seems to remember a deepwater port in Bern, the capital of a country that is widely known to be landlocked. … And Nicolas Bouvier experiences déjà-vu at a provincial train station in the extreme north of Japan, which "beams" him thousands of kilometers away to the Vaud train station Allaman. As already mentioned, the map slightly resembles the image of a shooting star: the symbol shows where the thought "lands," the tail where it was triggered.

Literary cartography is an approach that works on many levels—from theoretically highly demanding research endeavors to more popular, playful products like the "Literary Maps of Switzerland." One of the major advantages of that field is precisely its appeal to a wide range of audiences (scholars, teachers, students, tourists, interested readers in general and of all ages), as well as the mediation of literary-cartographic content in many formats such as research papers, scholarly books, popular online applications, educational material for schools, tourist guidebooks, and so on. In other words, besides many other things literary cartography can ease the way for reading experiences.

Limitations and Future Tasks

There are numerous possibilities for further studies in literary cartography. For example, it would be possible (and would surely make for a highly interesting map) to visualize the research of Immacolata Amodeo, who in her essay "Verortungen: Literatur und Litteraturwissenschaft" ("Locations: Literature and Literary Studies," 2010) addresses

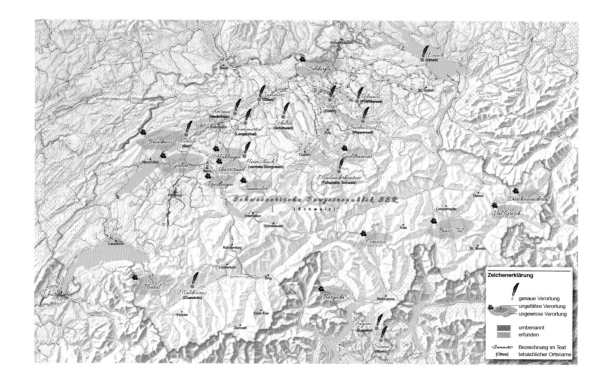

Figure 2.8

Playing with place names. From the project "Literaturlandkarten der Schweiz" ("Literary Maps of Switzerland") (2014), this thematic map visualizes renamed towns and villages as well as invented place names in fiction. A quill marks the renaming of an existing place for literary purposes; an inkwell symbolizes an invented place with only a vague location. The map is interactive and offers more information via mouseover: http://www.literatur-karten.ch. © Barbara Piatti / Anne-Kathrin Weber-Reuschel, Switzerland / Swiss Embassy in Berlin

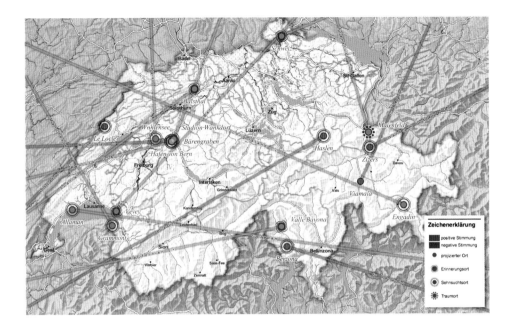

Figure 2.9

Imaginary trips. Another literary map from the series "Literaturlandkarten der Schweiz" ("Literary Maps of Switzerland"). The map focuses on the projected spaces–cities, towns, villages, and landscapes–that do not serve as actual settings in a work of fiction, but figure in the imagination of characters when they remember a place, long for it, or dream of it. These imaginary journeys are visualized as a sort of "shooting star." The tail is labeled with the actual setting where the imaginary trip starts, the actual falling star with the name of the imagined place. In blue: places connected with a positive atmosphere in the respective work of fiction. In red: places connected with a negative atmosphere. © Barbara Piatti / Anne-Kathrin Weber-Reuschel, Switzerland / Swiss Embassy in Berlin

the fictional geography of writers who have come to Germany as a result of the immigration waves since the 1960s. In smartly chosen categories Amodeo distinguishes between the "mythical elsewhere," which is usually described as the home left behind, the "intermediate/temporary place" such as train stations, trains, and transit hubs, and the "here-and-now-places of a new home country literature [Heimaten-Literatur]"— Berlin appears in numerous examples.[25] But for such complex content, appropriate symbols are yet to be developed.

In addition to continued experimentation and more case studies, an agenda for future research in the growing field of literary geography and literary cartography should include three major tasks:

1. *Developing a joint set of categories and terms*

There appears to be an unfortunate pattern within literary-geography and literary-cartography research, namely, that identical phenomena are being studied but with different terms. To give an example, what is called "projected spaces" in "A Literary Atlas of Europe" is called "imaginative-functional toponyms" in the study by Döring.[26] In the future it is less the individual maps (individual solutions for singular texts) that will be in demand, but rather systems of literary cartography that can manage large amounts of data and thereby enable the pooling of individual analyses. As Arnim von Ungern-Sternberg points out, "It is important and should be possible to develop cartographic standards that ensure the comparability of such undertakings."[27]

The same set of literary and spatial studies and theories is referenced time and again when new projects are developed (Bachelard, Bakhtin, Lotman, Lefebvre, Soja, Moretti, etc.), so there is reason to assume that a convergence of existing glossaries, methods, model studies, and references—in the sense of a literary-cartographic toolbox—would greatly facilitate future work in the field. One could start by studying and comparing the terminology in the works by some of the leading scholars in the field, such as Robert T. Tally Jr., David Cooper, Charles Travis, Michel Collot, and Tania Rossetto, to name just a few.[28] Based on those and other sources a quite useful *common* glossary might be extracted.

2. *Mapping imaginary spaces*

Since the very beginning of literary geography it has been postulated that imaginary spaces are cartographically difficult if not impossible to represent: "All the different attempts to create maps of literature always fail when literary texts—and this from the very beginning—depict spaces that evade mapping, such as fantastic spaces, fictional places, or utopias governed by spatiotemporal conditions that are neither physically, mathematically, nor geographically verifiable and presentable."[29] To cope with those spatial dimensions, one would need to leave the conventional base maps behind and move toward other modes of visualization—for instance, toward topologically structured images, where actual map scales and correct distances can be disregarded. As suggested above, such an unrestricted, artful way of mapping fiction has been sketched several times, but so far only in words. An early example was given by the Austrian writer and poet Ingeborg Bachmann in 1959–1960 in her "Frankfurter Poetik-Vorlesungen." There she speaks of an atlas of literature where the settings of fiction from James Joyce's Dublin to Marcel Proust's Combray build a new type of network, where the river Neva in Saint Petersburg, as described in Russian novels, is located next to the Seine in Paris, so extensively detailed by the narratives of Balzac, Flaubert, and several other writers; where we find not one Venice, but hundreds (one by Hugo von Hoffmannsthal, another by Thomas Mann, and so on); and where finally creations such as Atlantis and Orplid, fictitious countries, are seemlessly incorporated into that large continuous territory of literature.[30] In short, there are plenty of concepts around, but as far as I can see, we still lack convincing realizations as actual cartographic visualizations.

3. *Combining hermeneutic approaches with methods of distant reading*

As far as I can tell, both procedures that are currently used to produce literary maps are limited in their output as well as in their explanatory power: reading and analyzing individually, text by text, and with hermeneutic methods vs. not reading and analyzing enormous corpora with big data/macroanalysis tools.[31] If those two procedures could be more closely linked, new possibilities will undoubtedly open up, but for a variety of reasons this is easier said than done.

One of the most impressive and convincing examples of data mining in the humanities is "Mapping the Republic of Letters" at Stanford University. The project documents and animates the network of well-known academics in the eighteenth century: Who sent a letter to whom and when?[32] However, when it comes to literary cartography things are obviously somewhat more complicated, a fact that also makes it more contestable. Unlike clear criteria such as consignor and consignee location of letters or for methods of word counting, literary cartography as it is exercised today still relies on an interpreter or a team of interpreters to prepare a literary text for mapping. To return to our examples, the editor decides if a place is a "projected space" or just a simple marker (a brief mention in the text), if a character's path is carrying the story or not, how vaguely or precisely it can or should be entered, and so on. Although we have made an effort to capture the criteria of interpretation as intersubjectively as possible and to define them as precisely as possible in our own system, it is not unimaginable, but actually highly likely, that different interpreters would interpret one and the same text differently—just as it should be in literary studies.

One possible approach to merging hermeneutics and distant-reading procedures has been tested, again, by the Stanford Literary Lab. "Mapping Emotions in Victorian London" is designed as a crowdsourced research experiment: "The project has invited anonymous participants to annotate whether passages drawn from novels, published mainly in the Victorian era, represented London places in a fearful, happy, or unemotional manner. This data from the crowd allowed us to generate the maps you find here."[33]

The "Mapping Emotions in Victorian London" project focuses on text snippets and passages, not entire novels. Needless to say, the crowdsourcing approach is also a multiplication of individual readings. It might very well be that two readers would label the same text extract in different ways. And then we are again at the same point mentioned above: not only a different set of texts, but also different readers will generate different maps.

Another remarkable project is called "Reframing the Victorians" and is headed by Fred Gibbs and Dan Cohen, who have researched over 1.6 million English book titles between 1789 and 1914 using word-count and word-search programs to find out what really preoccupied the Victorian era. The two researchers "looked at what Victorians thought was sinful, and how those views changed over time."[34] The query led to a result set of nearly a hundred pages of "detailed descriptions of acts and behavior Victorian writers classified as sinful." In a next step, the researchers took a closer look at the full texts when necessary. "In other words, we can remain close to the primary sources and actively engage them *following* computational activity."[35] This is exactly the point: the fragile intersection of machine-driven data mining and the subsequent interpretations, based again on an engagement with the primary sources. In the case of the "Reframing the Victorians" project by Gibbs and Cohen, the team came up with some quite surprising observations. Regarding the concept of "sin," they were able to "trace a shift from biblically freighted terms to more secular language."[36] But they rightly conclude that any "robust digital research methodology must allow the scholar to move easily between distant and close reading, between the bird's eye view and the ground level of the texts themselves."[37]

The most interesting maps in the field of literary cartography are undoubtedly based on the analysis of such large text groups. But you either have to process the texts through a program or algorithm or you have to collaborate with the crowd as described above, since a limited group of researchers is obviously incapable of dealing with thousands of texts. Hence the challenge for literary scholars lies in the informed handling of technical procedures. The autonomy of our own area of expertise has to give way to a certain extent when forging an alliance with cartography and even more so when we work with data technology. But as long as this collaboration is accompanied by critical reflection on its methods and procedures, it is a path into the future.

Once the research community has come up with suggestions or solutions for those three desiderata—(1) a common toolbox and glossary, (2) extended cartographic means for charting invented fictional territories in a scientific way, and (3) a teaming up of macroanalysis and hermeneutic techniques—we might have an actual chance of approaching an ambitious vision of literary geography and cartography, something like a global "literary map." Such an overarching atlas system should ask (and maybe even answer) questions such as: Where are the areas of concentration in literature found? Are there entire unliterary landscapes? What is the density of fictional storylines in a given space? To what degree can one find international representation, or do native authors occupy it almost exclusively? When does the imaginary space of literature contract and when does it expand? In other words, literary geography and literary cartography could very well open up new horizons toward a spatially organized history of literature that combines big pictures with careful, detailed analyses of individual texts, and even more importantly, that allows us to read and study texts from different eras, areas, and authors side by side, in unprecedented and surprising combinations.

Concluding Remarks

Literary cartography as a method is about productive exchange and mutual complementarity. Jon Hegglund writes about *Ulysses*: "Cartography promises a surveying view, but this vantage is distant, abstract and ahistorical. Narrative, conversely, can project individual movements through time and space but ultimately must rely on partial views and situated knowledge. *Ulysses* frequently represents spaces that hover between these two perspectives but ultimately exist only in the negative space between literature and geography, narrative and historical space."[38] And cartographer Sébastien Caquard adds: "Neither cartography nor narrative on their own can capture the essence of place: both are required to get a better sense of it."[39] This insight may very well be extended to studies of literary landscapes and topographies. Together, narrative and visualization can undoubtedly open up new dimensions when a profound understanding of literary regions is sought.

Moreover, growing interest in sectors such as "story maps," "fictional cartography," "narrative atlases," and "geospatial storytelling" is noticeable. The common denominator in all these interests is the desire to understand more clearly how places and spaces function and how they interact with narratives of all types.[40] Literary maps, whether they are science- or audience-oriented, increase our access to literary layers of meaning in descriptive ways. Unlike architecture and natural attractions, "localized" literature is not visible, but it always leaves an imprint on the history of a city or landscape. Reflectively inserted literary cartography is an innovative, indeed eye-opening method, adding to existing approaches in the wider field of literature and geography or enhancing those approaches with new impulses and questions.

Despite the long tradition of mapping fiction and despite the numerous efforts made in the last few years, we might only be at the beginning of a literary cartography that meets the high standards of both cartography and literary studies—and one that, last but not least, does justice to the literary texts themselves. While I am a strong believer in the inspirational force provided by literary cartography, the ultimate, adequate maps of great literary works and of famous literary spaces are most likely still maps in our minds.

Notes

1. Carl Spitteler, "Xaver Z'Gilgen," in *Kleinere Erzählungen*, ed. Max Wehrli (Zürich: Artemis, 1945), 20. The translation is my own. The original text reads: "Aber wie sehr auch die Uhr und die sinkende Sonne mahnen mögen, oben bei dem Hüttchen werden wir eine Stunde ruhen; denn vor uns liegt zwischen dunklen Wäldern eine grüne, nicht allzu steile Halde von schwindelhafter Tiefe, oben einsam, unten mit hundert winzigen Häuschen besät, ganz zuunterst ein Zipfelchen Vierwaldstättersee, eingeschlossen in einem wahren Labyrinth von wirr durcheinander geschobenen trotzigen Alpenhäuptern. Das ist keine 'Aussicht,' es ist mehr als das: eine Landschaft, und zwar eine Landschaft, wie sie etwa die Phantasie eines Leonardo da Vinci hätte träumen mögen."

2. See Barbara Piatti, *Die Geographie der Literatur: Schauplätze, Handlungsräume, Raumphantasien* (Göttingen: Wallstein, 2008).

3. See Barbara Piatti, "Literaturgeographie und Literaturkartographie," in *Handbuch Literatur und Raum*, ed. Jörg Dünne and Andreas Mahler, 227–239 (Berlin: De Gruyter, 2015).

4. For an overview of the "contested field" of literary cartography and some critical remarks see Tania Rossetto, "Theorizing Maps with Literature," *Progress in Human Geography* 38, no. 4 (2014): 513–530.

5. The first historical sketches in the discipline and a number of map examples can be found in Piatti, *Die Geographie der Literatur*, 65–121. A very useful overview is given by Jörg Döring, "Zur Geschichte der Literaturkarte (1907–2008)," in *Mediengeographie: Theorie–Analyse–Diskussion*, ed. Jörg Döring and Tristan Thielmann (Bielefeld: transcript Verlag, 2008), 247–290.

6. See Piatti, *Die Geographie der Literatur*, 84–88.

7. "Mapping the Lakes," http://www.lancaster.ac.uk/mappingthelakes.

8. "Mapping St. Petersburg," http://www.mappingpetersburg.org/site.

9. "Digital Atlas of Ireland," http://www.tcd.ie/trinitylongroomhub/digital-atlas.

10. "Mapping Emotion in Victorian London," https://www.historypin.org/en/victorian-london.

11. "Australian Cultur Atlas," http://australian-cultural-atlas.info/CAA.

12. See http://www.googlelittrips.org and the extensive documentation by Terence W. Cavanaugh and Jerome Burg, *Bookmapping: Lit Trips and Beyond* (Eugene, OR: International Society for Technology in Education, 2011).

13. "Pinbooks: Dein Buch zur Stadt," http://www.pinbooks.de/index.html.

14. "A Literary Map of Manhattan," http://www.nytimes.com/packages/khtml/2005/06/05/books/20050605_BOOKMAP_GRAPHIC.html.

15. "A Literary Atlas of Europe," www.literaturatlas.eu. For another overview of the project see Barbara Piatti, "Mapping Fiction: The Theories, Tools and Potentials of Literary Cartography," in *Literary Mapping in the Digital Age*, ed. David Cooper et al., 88–102 (London: Routledge, 2016).

16. For more on the project design, on the categories used, and for examples of visualization in color, see Anne-Kathrin Reuschel and Lorenz Hurni, "Mapping Literature: Visualisation of Spatial Uncertainty in Fiction," in "Cartographies of Fictional Worlds," special issue, *Cartographic Journal* 48, no. 4 (2011): 293–308; Hans Rudolf Bär and Lorenz Hurni, "Improved Density Estimation for the Visualisation of Literary Spaces," in "Cartographies of Fictional Worlds," special issue, *Cartographic Journal* 48, no. 4 (2011): 309–316; and for an almost complete overview, see Anne-Kathrin Weber, "Mapping Literature: Spatial Data Modelling and Automated Cartographic Visualisation of Fictional Spaces," doctoral dissertation, ETH Zurich, 2014, http://e-collection.library.ethz.ch/view/eth:8350.

17. See Piatti, *Die Geographie der Literatur*, 123–147.

18. Franco Moretti, *Graphs, Maps, Trees: Abstract Models for a Literary Theory* (London: Verso, 2005), 35. His *Atlas of the European Novel* remains one of the main references for literary geography; see Franco Moretti, *Atlas of the European Novel 1800–1900* (London: Verso, 1999). It was first published in Italian two years earlier with the title *Atlante del romanzo europeo 1800–1900* by Einaudi in Turin, 1997.

19. Jörg Döring, "Distant Reading: Zur Geographie der Berlin-Toponyme nach 1989," *Zeitschrift für Germanistik* 18, no. 3 (2008): 597: "Was sieht man anders, was sieht man besser oder gar erst, wenn man die Handlungsorte von Berlin-Literatur kartiert?"

20. See Piatti, *Die Geographie der Literatur*, 128ff.

21. Bruce Chatwin, *Utz* (London: Vintage / Random House, 2005), 60.

22. Wolfgang Iser, "The Reading Process: A Phenomenological Approach," in "On Interpretation: I," *New Literary History* 3, no. 2 (Winter 1972): 285.

23. The map also shows that visualizing movements through fictional space is one of the most demanding tasks of literary cartography. Since it is impossible to follow a character's movements in great detail, we decided to work with a kind of flight path. In other words, only the starting point and the endpoint of a journey are registered in our database.

24. Andreas Bäumler, "Das Engadin als literarisierte Landschaft: Literaturgeographische Untersuchungen zum Verhältnis von Text und Raum," master's thesis, University of Basel, 2011.

25. Immacolata Amodeo, "Verortungen: Literatur und Literaturwissenschaft," in *Littérature(s) sans domicile fixe–Literatur(en) ohne festen Wohnsitz*, ed. Wolfgang Asholt et al., 1–12 (Tübingen: Narr, 2010). The original expressions are "Mythische[s] Anderswo" / "Dazwischen/Zwischenorten" / "Hier-und-Jetzt-Orte[n] einer neuen Heimaten-Literatur."

26. Jörg Döring, "Distant Reading," 597. The original reads "imaginativ-funktionale[n] Toponyme[n]."

27. Arnim von Ungern-Sternberg, "Dots, Lines, Areas and Words: Mapping Literature and Narration (With some remarks on Kate Chopin's 'The Awakening')," in *Cartography and Art* (Lecture Notes in Geoinformation and Cartography), ed. William Cartwright, Georg Gartner, and Antje Lehn (Heidelberg: Springer 2009), 244.

28. Robert T. Tally Jr., *Literary Cartographies: Spatiality, Representation, and Narrative* (Basingstoke: Palgrave Macmillan, 2014); David Cooper, "Critical Literary Cartography: Texts, Maps and a Coleridge Notebook," in *Mapping Culture(s): Place, Practice and Performance,* ed. Les Roberts, 29–52 (Basingstoke: Palgrave Macmillan, 2012); Charles Travis, *Abstract Machine: Humanities GIS* (Redlands, CA: Esri Press, 2015); Michel Collot, *Pour une géographie littéraire* (Paris: Edition José Corti, 2014); Rossetto, "Theorizing Maps with Literature."

29. Sylvia Sasse, "Poetischer Raum: Chronotopos und Geopoetik," in *Raum: Ein interdisziplinäres Handbuch,* ed. Stephan Günzel (Stuttgart: J. B. Metzler, 2010), 298: "All die unterschiedlichen Versuche, Literaturlandkarten herzustellen, scheitern allerdings immer dann, wenn literarische Texte—und dies von Beginn an—Räume entwerfen, die sich der Kartographierung entziehen: Phantastische Räume, fiktive Orte oder Utopien, in denen Raum- und Zeitverhältnisse herrschen, die weder physikalisch, mathematisch noch geographisch belegbar und darstellbar sind."

30. See Ingeborg Bachmann, "Frankfurter Poetikvorlesungen, IV (Der Umgang mit Namen)," in *Werke,* vol. 4, ed. Christine Koschel et al., 238–254 (Munich: Piper Taschenbuch Verlag, 1993). See also the suggestions by von Ungern-Sternberg in "Dots, Lines, Areas and Words."

31. See Matthew L. Jockers, *Macroanalysis: Digital Methods and Literary History* (Champaign: University of Illinois Press, 2013); Franco Moretti, *Distant Reading* (London: Verso, 2013). For a current overview see also Marko Juvan, "From Spatial Turn to GIS-Mapping of Literary Cultures," *European Review* 23, no. 1 (February 2015): 81–96.

32. "The Republic of Letters," https://republicofletters.stanford.edu.

33. "Mapping Emotions in Victorian London," https://www.historypin.org/en/victorian-london/geo/51.5128,-0.116085,12/bounds/51.477746,-0.195049,51.547827,-0.037121/project/about.

34. Frederick W. Gibbs and Daniel J. Cohen, "A Conversation with Data: Prospecting Victorian Words and Ideas," *Victorian Studies* 54, no. 1 (Autumn 2011): 77.

35. Gibbs and Cohen, "A Conversation with Data," 76.

36. Ibid., 74.

37. Ibid., 76.

38. Jon Hegglund, *World Views: Metageographies of Modernist Fiction* (Oxford: Oxford University Press, 2012), 87.

39. Sébastien Caquard, "Cartographies of Fictional Worlds: Conclusive Remarks," in "Cartographies of Fictional Worlds," special issue, *Cartographic Journal* 48, no. 4 (2011): 224.

40. See Sébastien Caquard, "Cartography I–Mapping Narrative Cartography," *Progress in Human Geography* 4 (2011): 135.

Bibliography

Amodeo, Immacolata. "Verortungen: Literatur und Literaturwissenschaft." In *Littérature(s) sans domicile fixe–Literatur(en) ohne festen Wohnsitz*, ed. Wolfgang Asholt et al., 1–12. Tübingen: Narr, 2010.

Bachmann, Ingeborg. "Frankfurter Poetikvorlesungen, IV (Der Umgang mit Namen)." In *Werke*, vol. 4, ed. Christine Koschel et al., 238–254. Munich: Piper Taschenbuch Verlag, 1993.

Bär, Hans Rudolf, and Lorenz Hurni. "Improved Density Estimation for the Visualisation of Literary Spaces." In "Cartographies of Fictional Worlds," special issue, *Cartographic Journal* 48, no. 4 (2011): 309–316.

Bäumler, Andreas. "Das Engadin als literarisierte Landschaft: Literaturgeographische Untersuchungen zum Verhältnis von Text und Raum." Master's thesis, University of Basel, 2011.

Caquard, Sébastien. "Cartographies of Fictional Worlds: Conclusive Remarks." In "Cartographies of Fictional Worlds," special issue, *Cartographic Journal* 48, no. 4 (2011): 224–225.

Caquard, Sébastien. "Cartography I–Mapping Narrative Cartography." *Progress in Human Geography* 4 (2011): 135–144.

Chatwin, Bruce. *Utz*. London: Vintage / Random House, 2005.

Collot, Michel. *Pour une géographie littéraire*. Paris: Edition José Corti, 2014.

Cooper, David. "Critical Literary Cartography: Texts, Maps and a Coleridge Notebook." In *Mapping Culture(s): Place, Practice and Performance*, ed. Les Roberts, 29–52. Basingstoke: Palgrave Macmillan, 2012.

Döring, Jörg. "Distant Reading: Zur Geographie der Berlin-Toponyme nach 1989." *Zeitschrift für Germanistik* 18, no. 3 (2008): 596–620.

Döring, Jörg. "Zur Geschichte der Literaturkarte (1907–2008)." In *Mediengeographie: Theorie–Analyse–Diskussion*, ed. Jörg Döring and Tristan Thielmann, 247–290. Bielefeld: transcript Verlag, 2008.

Gibbs, Frederick W., and Daniel J. Cohen. "A Conversation with Data: Prospecting Victorian Words and Ideas." *Victorian Studies* 54, no. 1 (Autumn 2011): 69–77.

Hegglund, Jon. *World Views: Metageographies of Modernist Fiction*. Oxford: Oxford University Press, 2012.

Iser, Wolfgang. "The Reading Process: A Phenomenological Approach." In "On Interpretation: 1," *New Literary History* 3, no. 2 (Winter 1972): 279–299.

Jockers, Matthew L. *Macroanalysis: Digital Methods and Literary History*. Champaign: University of Illinois Press, 2013.

Juvan, Marko. "From Spatial Turn to GIS-Mapping of Literary Cultures." *European Review (Chichester, England)* 23, no. 1 (February 2015): 81–96.

Moretti, Franco. *Atlas of the European Novel 1800–1900*. London: Verso, 1999.

Moretti, Franco. *Distant Reading*. London: Verso, 2013.

Moretti, Franco. *Graphs, Maps, Trees: Abstract Models for a Literary Theory*. London: Verso, 2005.

Piatti, Barbara. *Die Geographie der Literatur: Schauplätze, Handlungsräume, Raumphantasien*. Göttingen: Wallstein, 2008.

Piatti, Barbara. "Literaturgeographie und Literaturkartographie." In *Handbuch Literatur und Raum*, ed. Jörg Dünne and Andreas Mahler, 227–239. Berlin: De Gruyter, 2015.

Piatti, Barbara. "Mapping Fiction: The Theories, Tools and Potentials of Literary Cartography." In *Literary Mapping in the Digital Age*, ed. David Cooper et al., 88–102. London: Routledge, 2016.

Reuschel, Anne-Kathrin and Lorenz Hurni. "Mapping Literature. Visualisation of Spatial Uncertainty in Fiction." In "Cartographies of Fictional Worlds," special issue, *Cartographic Journal* 48, no. 4 (2011): 293–308.

Rossetto, Tania. "Theorizing Maps with Literature." *Progress in Human Geography* 38, no. 4 (2014): 513–530.

Sasse, Sylvia. "Poetischer Raum: Chronotopos und Geopoetik." In *Raum: Ein interdisziplinäres Handbuch*, ed. Stephan Günzel, 294–308. Stuttgart: J. B. Metzler, 2010.

Spitteler, Carl. "Xaver Z'Gilgen." In *Kleinere Erzählungen*, ed. Max Wehrli, 19–31. Zürich: Artemis, 1945.

Tally, Robert T. Jr. *Literary Cartographies: Spatiality, Representation, and Narrative*. Basingstoke: Palgrave Macmillan, 2014.

Travis, Charles. *Abstract Machine: Humanities GIS*. Redlands, CA: Esri Press, 2015.

von Ungern-Sternberg, Arnim. "Dots, Lines, Areas and Words: Mapping Literature and Narration (With some remarks on Kate Chopin's 'The Awakening')." In *Cartography and Art: Lecture Notes in Geoinformation and Cartography*, ed. William Cartwright, Georg Gartner, and Antje Lehn, 229–252. Heidelberg: Springer, 2009.

Weber, Anne-Kathrin. "Mapping Literature: Spatial Data Modelling and Automated Cartographic Visualisation of Fictional Spaces." Doctoral dissertation, ETH Zurich, 2014. http://e-collection.library.ethz.ch/view/eth:8350.

3
The (Un)Mappability of Literature

Robert Stockhammer

Two Types of Mappability and Literary Cartography

Mapping literature is fashionable. It offers opportunities for extensive cooperation between the field of literary studies and national or regional tourist agencies. The production of atlases or cellphone apps supporting walks through the Swiss or Slovenian "literary landscape" is likely to receive funding from the government. Certainly, the maps produced in Stanford and Zürich—to name only the most prominent examples of this popular trend—also serve other aims. One type of such maps focuses on the production, distribution, and reception of literature. These maps therefore situate authors, publishing houses, or other literary institutions. It is an enterprise that does not make any claims with regard to the setting of the literary texts themselves.[1] This chapter is concerned exclusively with another type of literary map: maps that claim to cartographically represent the setting of fictional narratives. For a more careful critique of the procedures involved in these mapping enterprises, it is necessary to make explicit some of their implicit assumptions.

First and foremost, these endeavors assume that literature is, in principle, mappable. Yet, Franco Moretti, for one, does not supply a definition of a map. All the graphic representations he produces do indeed correspond to the intuitive notions of maps most people share. These intuitive notions, however, would not be taken for granted by geographers. Ironically, it is precisely at the moment when mapmakers have developed important critical insights into their own practices that literary cartography is relying on an uncritical version of these practices.[2] It might therefore be instructive to begin with a working definition of a map, even if I will eventually challenge this definition in the third section of the chapter. In 1995, the International Cartographic Association defined a map as follows: "A map is a symbolised image of geographical reality, representing selected features or characteristics, resulting from the creative effort of its author's execution of choices, and is designed for use when spatial relationships are of primary relevance."[3] The definition of a map as a "symbolised image" has the advantage of being broad enough to cover outputs of all kinds of (digital or nondigital) geographic information systems. At the present time, when "the fabulous possibilities of a digital, interactive, animated cartography with database support have been discovered,"[4] it seems important to remind ourselves that maps have always been an output of geographic information systems: data first had to be collected in databases and then processed using specific techniques.[5] For example, it is possible to reconstruct Ptolemy's maps with a great degree of precision even though none of them have survived, because his *Geography* contains his database (the list of places and their coordinates in books II–V) as well as the processing methods (his projections, in the cartographic sense of methods for transforming a curbed surface into a plane, in book I).

The word *reality* in the ICA definition, however, causes several notorious problems. Even many geographers not dealing with fictional texts would refrain from using "geographic reality" as the simple subject matter of "symbolic images" and instead would emphasize the degree to which something called *reality* is itself *produced* by these very symbolic images.[6] Paradoxically, literary scholars—a group of people otherwise (in)famous for challenging notions of *reality*—have fewer reservations about it as soon as they produce a genre of "symbolic images" that they have learned to trust in their everyday experience. It may be hard to say whether we are in a real world when walking down 42nd Street, but it seems to go without saying that the toponym "42nd street" in a piece of fiction refers to the real 42nd Street.

Evidently, not all fictional texts are equally mappable. As a first step it seems convenient to distinguish between two different kinds of literary landscapes, some of which claim explicit relations to places out there (places to be reached by moving a human body, usually superimposed on some transportation vehicle), whereas others do not. Even this distinction is problematic, since it does not account for historical variations: changes in geographic reality itself or its generally accepted conception. Maps attached to the first edition of Swift's *Gulliver's Travels*, for example, depict, among other things, the peninsula of Brobdingnag, the islands of Balnibarbi, California, Laputa, and Luggnagg, and a country named Companys Land (somewhere near the Pacific Coast of Russia) (figures 3.1 and 3.2). On these maps, however, only Brobdingnag and *some* of the islands are augmentations of the base map Swift used for them: Herman Moll's *New & Correct Map of the Whole World* (1719) (figures 3.3 and 3.4). California-as-an-island and the existence of Companys Land were part of contemporary "reality" as depicted on Moll's widely accepted map itself. One might be tempted to distinguish between *fiction* and *error*—a distinction, however, that takes it for granted that California is not an island.

Taking these problems into account (and therefore avoiding the term *reality*), we may distinguish between two layers of mappability in literature:

• *Internal mappability* applies to cases in which all of the descriptions of geographic features rendered by a fictional text are consistent with each other and with the rules of Euclidean geometry, such that the world created by this text can be unequivocally depicted on a map.

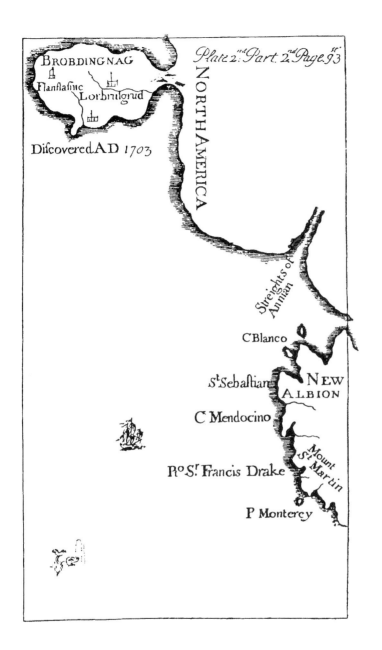

Figure 3.1

Map of Brobdingnag Jonathan Swift, *Gulliver's Travels*, ed. Christopher Fox (New York: St. Martin's Press, 1995), 90

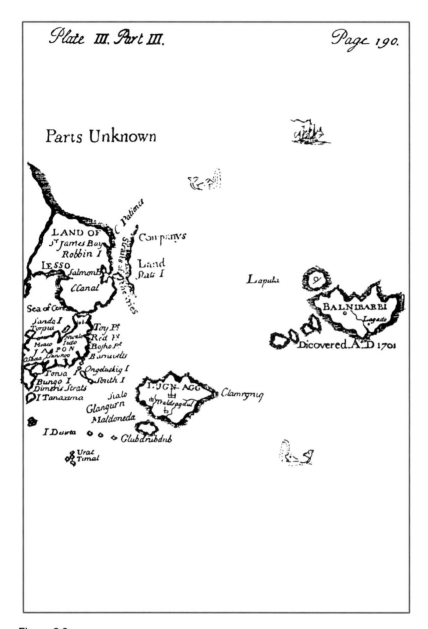

Figure 3.2

Map of Laputa etc. Jonathan Swift, *Gulliver's Travels*, ed. Christopher Fox (New York: St. Martin's Press, 1995), 148, according to Frederick Bracher, "The Maps in *Gulliver's Travels*," in *Huntington Library Quarterly* 8 (1944–1945): 59–74, here 60. These maps were included earlier in Lemuel Gulliver, *Travels into Several Remote Nations of the World* (London: Benjamin Motte, 1726).

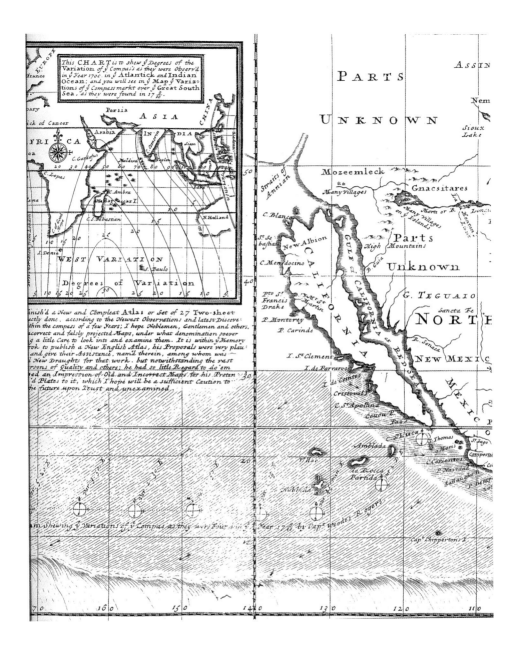

Figure 3.3

Herman Moll, *New & Correct Map of the Whole World*, 1719

Figure 3.4
Herman Moll, *New & Correct Map of the Whole World*, 1719

- *Referential mappability* applies to cases in which geographic features in a given fictional text correspond to features included on maps that are, at a given time, accepted as being useful for purposes other than mapping literature (e.g., world or cadastral maps).

In this distinction, internal mappability seems to be a necessary but not sufficient condition for referential mappability. A relatively unequivocal example of solely internal mappability is the map of Utopia produced by the renowned cartographer Abraham Ortelius in 1595. But obviously, the distinction between the two kinds of mappability is not clearcut. The map contained in Robert Louis Stevenson's *Treasure Island* (figure 3.5)–the germ of the novel itself according to its author–plays with the difference between the claims of internal and referential mappability. Its legend, "latitude and longitude struck out by J. Hawkins," suggests that the island could indeed be located within a system of global coordinates if only the protagonist had not veiled its location. In a different variation, the early maps in *Gulliver's Travels*, which follow and elaborate several indications contained in the text itself, mix the claims of internal and referential mappability at least insofar as they make statements about the position of the fictional countries, with Brobdingnag, for example, as a peninsula attached to today's Oregon. Thomas Hardy's Wessex and William Faulkner's Yoknapatawpha, which have both been mapped by their authors in peritexts for early editions of their novels, are striking examples of "fictionals worlds, where the real and the imaginary coexist in varying, often elusive proportions."[7] Or, if one still prefers to avoid the word *real*: these novels are particularly complex mixtures of idiosyncratic and generally acknowledged topological features. *Ulysses* might count as an extreme value of referential mappability, if one were to put faith in James Joyce's famous dictum that he wanted "to give a picture of Dublin so complete that if the city one day suddenly disappeared from the earth it could be reconstructed out of my book."[8]

By and large, all recent endeavors to map literature claim referential mappability to a very high degree. This claim underlies the methods of data collection and data processing silently used by Franco Moretti and partly revealed by Barbara Piatti, methods that can be described as a particular variety of thematic mapping. But even though it is broadly accepted, the distinction between a thematic map and a general reference map can be contested, since *any* map relies on "*selected* features or characteristics," to quote the ICA definition again, and any selection is guided by the theme or purpose of the given map. Why, for example, should political borders count as necessary features of "general reference maps" but not the location of drugstores?[9] Literary maps, however, clearly have such a particular theme that it seems safe to classify them as thematic maps.

Data acquisition, processing, and output in thematic mapping are usually characterized by their supplementary relation to other maps. They rely on an augmentation

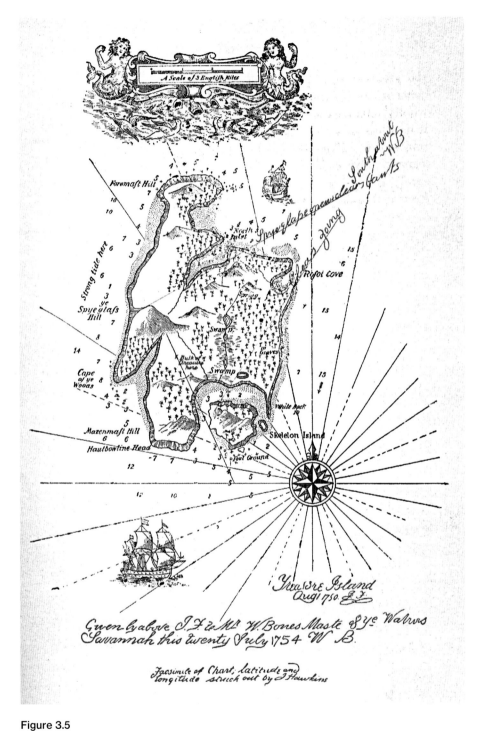

Figure 3.5

Map of Treasure Island. Robert Louis Stevenson, *Treasure Island* (London: Cassell & Company, 1883), frontispiece. For a facsimile, see https://en.wikisource.org/w/index.php?title=File%3AStevenson_-_Treasure_Island.djvu&page=8.

of already existing arrays of data by one or more dimensions. A helpful example for understanding these procedures is the table provided in Berghaus's *Physikalischer Atlas*, which presents the database used for the production of climate maps in the same atlas (figure 3.6).

The first four columns of this table contain latitudes, longitudes, altitudes, and names of locations where weather stations have been installed. All of this information had been collected beforehand. It is not the *theme* of the climate maps, but their basis which is taken for granted. Mean values for temperatures at the respective weather stations (for the year, for the four seasons, and for the coldest and warmest month) are only given in columns 5–11—while column 12 serves as a sort of metaepistemological control indicating the number of observations recorded and evaluated at the respective weather stations.

The procedures of "data acquisition" in the process of constructing a "Literary Atlas of Europe" are by no means identical, but they are comparable. Piatti underlines the novels she reads with a set of markers in many different colors. She elaborates a number of fine distinctions with regard to degrees of referentiality.[10] And she is remarkably honest in admitting the hermeneutic element in this transformation of *texts* into *data*.[11] Whereas the acquisition of weather data, or even the mere counting of letters or words, may to a large extent be automated, any evaluation of words that includes their semantic features still requires a human agent. Somebody must, for example, recognize that the word *Hamburger* in a novel does not necessarily refer to an inhabitant of Hamburg. Its classification as a geographic datum is debatable, and somebody has to resolve this debate, most often by an interpretation of the word's cotext.

With regard to the data output, the supplementary character of thematic maps is evident from the fact that most of them use base maps that had already been produced for other aims than mapping literature. The "features or characteristics" relating to literary texts are then simply *augmented* from those that already exist on the base maps.[12] Moretti's and Piatti's base maps differ strikingly with regard to the amount of detail they depict. But even Moretti's paper maps, which are relatively poor in detail, depict a large range of features obviously not derived from the literary texts themselves. Most gothic novels, for example, abound in Italian or German toponyms, but contain few individualizing descriptions of respective settings. Moretti's map of these novels, however, not only represents many towns and castles in "spatial relationships" that Ann Radcliffe or Charles Robert Maturin would probably not have been able to locate with such precision, but also elaborate coastlines of European countries. Even these maps display a much more worldlike coherence than one would get from reading these novels.

Neither Moretti nor Piatti discusses the geographic data themselves because they take them for granted in their thematic mapping: "The coordinates are directly

Figure 3.6

Climate Table. *Die Hauptmomente der Temperatur auf dem ganzen Erdboden gegründet auf die Beobachtungen an 307 Orten. Thermometer–Centigrade.* In Heinrich von Berghaus, *Physikalischer Atlas: Eine, unter der fördernden Anregung Alexander's von Humboldt verfasste, Sammlung von 93 Karten, auf denen die hauptsächlichen Erscheinungen der anorganischen und organischen Natur nach ihrer geographischen Verbreitung und Vertheilung bildlich dargestellt sind* (Gotha: Justus Perthes, 1845–1848, 2nd ed. (1852), pt. 1, Meteorologie, no. 4).

constructed through the geocoding tool."[13] Supplementary data are not expected to influence the primary ones; a change in the average temperatures of Paris cannot relocate the whole city—neither can a fictional description of Paris perform this feat. All this seems obvious. But is it true?

Innumerable Types of Unmappability

I admit that using the toponym *Paris* in a fictional narrative is highly likely to evoke various ideas about a city that a human being can reach by way of a plane, car, train, or bicycle. The assumption that any use of language, even in fictional texts, produces effects of referentiality, seems to me undeniable. In methodological side remarks, however, adherents of literary mapping often assume a somewhat crude oscillation between referentiality and nonreferentiality that has allegedly been developed in literary theory. In this narrative, so-called poststructuralist theories are said to have denied referentiality *tout court* in order to indulge in the endless self-referentiality of signs. They therefore feel it is time to turn the wheel again.[14] The strategy behind this line of argumentation repeats a situation after 1968 when close readings in the New Criticism tradition were attacked by more sociologically and historically interested scholars, a situation already ironized by Paul de Man: "Like the grandmother in Proust's novel ceaselessly driving the young Marcel out into the garden, away from the unhealthy inwardness of his closeted reading, critics cry out for the fresh air of referential meaning."[15]

De Man, however, did not deny the referential effects of language. Quite the contrary, he stressed the "irresistible motion that forces any text beyond its limits and projects it towards an exterior referent." He just doubted that referential effects simply bring in fresh air and therefore preferred the metaphor of "vertiginous possibilities of referential aberration."[16] Close reading does not negate referentiality. Rather, it demonstrates that it is much more complicated than it appears in the dim light of distant reading.

J. Hillis Miller's *Topographies*, a book sometimes mentioned but seldom really discussed in studies of literary cartography, closely investigates cases of referential aberration precisely with regard to literary topography. The book even includes a reading of Thomas Hardy's *The Return of the Native* and the map attached to its first edition, thereby discussing a typical example of the very tradition usually taken as a model for unproblematic mappability: the nineteenth-century realist novel. Miller, however, emphasizes: "Sooner or later, in a different way in each case, the effort of mapping is interrupted by an encounter with the unmappable."[17] It is important to stress that the distinction between the mappable and unmappable is not meant as a handy tool for classifying two different groups of entire texts.[18] It is rather a relation of conflicts, of interruptions, not easily classified. Hence Miller's insistence on the singularity of "each case."

The interruption of mappability may take the form of the peculiar spatiality of Kafka's *The Castle*. The setting of the novel does not seem to follow a Euclidean concept of space, at least not with regard to the location of the castle itself: a street that seemingly leads to the castle remains in constant distance from it.[19] This type of unmappability, by the way, transgresses the distinction of internal and referential mappability, as it is independent of the (im)possibility of identifying any model for the castle. The description would be unmappable even if the village and castle in the text shared a toponym with a village and castle in some Bohemia or Moravia once visited by Franz Kafka, the actual human being.

But unmappability may also take the form of a few but constitutive "points." That is the case with Herman Melville's *Moby-Dick*, a novel that in other respects seems so easily mappable that several of its editions include charts depicting "the cruise of the Pequod." Most parts of the *Pequod*'s route from Nantucket on the East Coast of the United States to the Pacific Ocean can indeed be traced by using geographic information gleaned from the text. At the climax of the novel, however, when the ship eventually meets *the* whale, the novel's eponymous hero, somewhere, but only *somewhere* close to Japan, mappability fails. This climax corroborates the enigmatic claim from a much earlier passage in the novel that Kokovoko, the birthplace of the harpooner Queequeg, "is not down in any map. True places never are."[20]

Other types of unmappability result from the incompatibility of various georeferential data given within one and the same text. For certain vague reasons, Marcel Proust's Combray, to quote a familiar example, seems identifiable with Illiers, a small town near Chartres. Once this identification was approved by the *Société des Amis de Marcel Proust et des Amis de Combray*, the district council renamed the town by hyphenating "reality" and "fiction": today, the town's official name is Illiers-Combray, and the visitor is invited to retrace Marcel's famous walks *du côté de chez Swann* (or Méséglise) and *du côté des Guermantes*. This visitor, however, will fail to understand the topography of the landscape within the *Recherche*, because the literary landscape consists of *two* spatial organizations that are entirely incompatible. Readers of the first volume have to choose between two alternative walks, since it is impossible to proceed from one side (*côté*) to the other. The literary geography of the first volume is organized in "two 'ways' ... so diametrically opposed that we would actually leave the house by a different door, according to the way we had chosen: ... And so to 'take the Guermantes way' in order to get to Méséglise, or vice versa, would have seemed to me as nonsensical a proceeding as to turn to the east in order to reach the west."[21] And it is only at the beginning of *Le temps retrouvé* that Gilberte, a Swann by birth and a de Guermantes by marriage, proposes to the first-person narrator that they take a walk that combines and reconciles the opposing sides (les "côtés ... si opposés")—a proposal that overturns the entire spatial conception of the narrator: "'If you like, we might go out one afternoon, and then we can go to Guermantes, *taking the road by Méséglise*, it is the nicest walk,' a sentence which upset all my childish ideas by informing me that the two 'ways' were not as irreconcilable as I had supposed [emphasis added]."[22] In other words, the hyphen between Combray and Illiers does not so much connect a fictional toponym with a real one, as it marks the difference between, on the one hand, a *readable* geography that contains two incompatible spatial conceptualizations within one and the same novel, and, on the other hand, a *walkable* geography supposed to be identical with itself. Even though the two opposing sides are eventually reconciled within the novel, its twofold conception of space cannot be reconciled with the self-identical conception of space somewhere near Chartres.

Alternative Mappings of Literature? Some Preliminary Reflections

Some of the literary features that are unmappable with regard to the ICA-map may nevertheless be mappable with regard to other kinds of maps.[23] On the one hand, I have my reservations about the widespread use of *mapping* as a metaphor for the most heterogeneous kinds of symbolic images, since this metaphor often merely insinuates the values of scientific exactness and a graphical as well as graphic appearance. In these cases, the metaphor veils the specific procedures and media used in such representations as well as their important epistemological differences from cartography in a narrower sense. On the other hand, I admit that there is no clear-cut distinction between creative work *on* concepts (not only *with* concepts) and metaphorical operations, concepts being neither more nor less than well-motivated metaphors.[24] In a similar vein, even the *History of Cartography*, an enterprise certainly not interested in blurring notions of *mapping*, starts out from a surprisingly broad definition: "Maps are graphic representations that facilitate a spatial understanding of things, processes, or events in the human world."[25] In contrast to the ICA definition, this definition carefully avoids any reference to "geographic reality," and even the adjective *spatial* has been moved from *relationships* (meaning something out there) to *understanding* (meaning something within the graphic representation itself, or within the brain of its beholder or reader).

Moreover, as the object of understanding according to this definition is "things, processes, or events," the definition is more suitable for including the mapping of such a thing, process, or event as a literary text itself, instead of simply its alleged setting, as is the aim of most conventional literary maps. A conventional map of the voyages of the protagonist of a novel, for example, would not be distinct from a conventional map of the voyages of a reader of this novel who traced the voyage of its protagonist. Alternative forms of mapping literature, however, can be expected not only to account for unmappable moments, but also to display specific features of literary texts *as* literary texts[26] (i.e., their *literariness*: the ways they differ from other texts or nontextual things, processes, or events).

Not being a cartographer myself, not having experimented with actual alternative forms of mapping literature, I am not able to suggest concrete examples. Instead, I propose three hypotheses as indispensable conditions for the development of alternative mappings. I will proceed *ex negativo* with an interrogation of a set of assumptions usually taken for granted in the production of literary maps. These hypotheses are therefore formulated as cautions, following the order of work steps in the production of those maps as already sketched in the first section of this chapter.

Caution against Conversion of Words into Data

Among the favorite readings of the young Marcel is a city map of Paris that allows him to locate the residence of his beloved Gilberte. In conversations with his parents he obsessively mentions a certain street simply because Gilberte lives there.[27] This habit is one of the numerous things he shares with Gilberte's father, Swann, who, while still young and desperately in love with Odette, preferred to have lunch at a restaurant named after the famous explorer Jean-François La Pérouse, and always seized the chance to mention him in conversation.[28] The reason for his obsession with this proper name, however, has nothing to do with its bearer, but only with Odette's habitation at Rue de La Pérouse. Swann uses the proper name as a signifier evoking his beloved in a doubly metonymical chain of association.

Toponyms, Proust teaches us, are signifiers whose georeferential function constitutes only *one* dimension of their potential for signification, a dimension that does not prevent its use as a signifier for various desires. The spatial relation between the restaurant *Lapérouse*, the Rue de La Pérouse, and the salons where Swann forces his interlocutors to talk about La Pérouse, follows quite another logic than the spatial relation between, say, the places La Pérouse had visited during his extensive journeys. Swann's use of the word in its double character as a proper name and a toponym is a striking example of the performative dimension of each speech act, of the importance of the very act of enunciating a word. This performative dimension tends to get lost during the process of data acquisition in literary cartography precisely because toponyms appear as words that are to be easily converted into data as mere constatives.[29]

I'm not suggesting that Swann's aberrant use of the toponym poses an insurmountable problem to conventional literary cartography. The difficulty is just that, from this point of view, it appears *as* a problem: "There [in written interpretations of literary texts], *ambiguity* of a reading is a sign of quality; here, in developing a system of signs and symbols for literary cartography, it becomes a problem."[30] In this case the ambiguity or the multifunctionality of the proper name / toponym (Rue de) La Pérouse is a quality not only for the readers of the *Recherche*, but even for the eponymous hero of its first volume *Du côté de chez Swann*. A map of the novel that "solved the problem" of the toponym would also rob Swann of his favorite way of enunciating his desire.

On the one hand, it does not seem altogether impossible to map a performative as maps are known to rely on performative acts. On the other hand, cartography is a medium particularly seductive in veiling its own performativity, more seductive than, for example, texts written in languages whose arbitrariness is much more noticeable. The more familiar maps appear, the more they hide their own performativity. Most likely, a map that would be able to translate (not just represent) literary performativity would have to be a map that defamiliarizes the reader's expectations of maps.

Caution against Using Traditional Projections

"In a novel it would have a good effect to present the concepts of its hero, for example of the earth, on a small map. The world would be represented as round, in the middle lies the village where he lives, represented as very large with all the mills etc., and then the other towns around it, Paris and London very small, in general everything becomes much smaller the further away it is located."[31] As proposed by Georg Christoph Lichtenberg, the map of the concept of the earth according to a novel's hero is reminiscent of Saul Steinberg's famous depiction of the New Yorker's worldview: centered on the home of the hero, its close neighborhood depicted in abundant detail, with everything else diminishing in proportion to its distance from the center, and Paris and London appearing miniscule (figure 3.7).

This "map" would probably not even be accepted *as* a map by an organization like the ICA. Since it is dependent on a particular point of view, it is rather a graphical hybrid, something like an extended version of a town view in cavalier perspective. As a mere "mental map" of a protagonist, it may seem of limited interest for literary cartography. But if the novel in question is told by a narrator with internal focalization, the hero's point of view is scarcely distinctive from that of the narrator. In such a case, and not an uncommon one according to narratologists, a Steinbergian graphical hybrid that evokes only some features of maps might be a more adequate representation of the spatial understanding supplied by the novel itself.

The metaphor of "focalizing,"[32] widely used by Gérard Genette, relies on the vehicle of optical instruments. This metaphor is somewhat plausible for a type of narration with a cavalier perspective like the one in Lichtenberg's proposal, but it is entirely incompatible with its own vehicle in the case of another of the most common types of focalizing: the *focalisation zéro* that usually characterizes the omniscient narrator. Neither a camera nor a telescope can be adjusted to this type of narration.

Since Genette explicitly admits that a narrator of a text does not, after all, possess a camera, a *focalisation zéro* might have more affinity for mapmaking. In this case, the metaphor of perspective could be replaced by the metaphor of narrative projection

Figure 3.7

Saul Steinberg's map of New York. In the *New Yorker*, March 29, 1976. © The Saul Steinberg Foundation / Artists Rights Society (ARS), New York/copydanbilleder.dk

in the technical sense of cartographic procedures. The advantage of this metaphor consists in describing points of "view" that are not views at all: maps constructed in accordance with almost all conventional projections *do not* depict territories from a bird's-eye view or an astronaut's view, as is usually assumed in everyday parlance. The astronaut's view of the earth would produce a distortion of its margins so extreme that it would not be acceptable to any cartographer. Rather, map projections mathematically eliminate *any* individual point of view. They are, of course, not "objective" in the sense of an undistorted representation—the curved surface of the earth will never be presented without distortions on a plane sheet of paper or a plane screen—but they are, after all, "objective" in the sense that these distortions depend on many factors, just not on the position of a viewer.

The shift from perspective to projection is enacted in a passage written by Adalbert Stifter, who was very familiar with surveying and cartography practices, as demonstrated in several of his fictional narratives. The protagonists of both "Kalkstein" ("Limestone") and *Nachsommer* (*Indian Summer*) work as surveyors, and a version of "Der beschriebene Tännling" ("The Inscribed Fir Tree") begins with the description of a map. In a passage from his story "Granite," which presents a topographic initiation of a boy by his grandfather, Stifter carefully transforms idiosyncratic points of view into a cartographic *non*view:

[Grandfather]   "That is the life of the forests. But now let us look at what is outside them. Can you tell me what those white buildings are that we can see through the double pine?"
[Boy]   "Yes Grandfather, those are the Prang farms."
"And farther left from the Prang farms?"
"Those are the buildings of the Front and Rear Seminary."
"And still farther left?"
"That is Glöckelberg."
"And farther toward us on the water?"
"That is the Hammer Mill and Farmer David."
"And the many buildings quite close to us, among which the church rises, and behind which there is a mountain on which there is another little church?"
"But Grandfather, that's our market town Oberplan, and the chapel on the mountain is the Chapel of the Good Water."
"And if the mountains weren't there and the heights that surround us, you would see many more buildings and villages: the Karl farms, Stuben, Schwarzbach, Langenbruk, Melm, Honnetschlag, and on the opposite side Pichlern, Pernek, Salnau, and several others."[33]

The narrative point of view of the protagonists relativizes itself in order to be transformed in two steps into a mere enumeration of places from a non–point of view.

Beyond the visible places (from the Prang farms to Oberplan), the grandfather insists there are places not visible from the point of view of the protagonists. And among these places, one group (from the Karl farms to Honnetschlag) is still, *ex negativo*, related to the beholders—they could be seen if mountains would not block "our" view— whereas the second group of places (from Pichlern to "several others") is no longer related to any beholder, but only to other places "on the opposite side" of the first group of places. In its abstraction from any concrete point of view, the passage is simulating a cartographic projection.[34] The increasing affinity of this passage for a map, its "cartographicity," is even visible on the level of the signifiers, since the percentage of toponyms in the total amount of words increases drastically during the passage.[35]

Most likely, this narrative projection is the exception rather than the rule within fictional narrative prose, and perhaps even within the nineteenth-century tradition of literary realism. Paradoxically, a conventional map of places in Stifter's version of the Böhmerwald would precisely fail to demonstrate that Stifter's texts themselves have a much greater affinity for these conventional cartographic techniques than other texts that take place in the same area. Its cartographicity could only be mapped in contrast to other mapping procedures, especially other projections.

Caution against Using Base Maps

In the last volume of the *Recherche*, only a few pages after the narrator's conception of the geography near Combray is "overturned," something similar happens with the reader's conception of this area. As a Proustian in Illiers-Combray, he was just enjoying the smell of the famous hawthorn (*aubépine*) as described in the first volume of the *Recherche*, a passage quoted in imitated handwriting on an aluminum plate close to a hedge that today forms one of the boundaries of an officially classified "Marcel Proust garden" (the *Pré Catalan / Jardin de Marcel Proust / classé par arrêté ministériel du 12 déc. 1945*). Now, however, a letter written by Gilberte and dating from 1916 informs the Proustien or Proustienne that the large grainfield adjacent to the hawthorn served as a famous battlefield during the "battle of Méséglise," which lasted for more than eight months and ought to be recorded "in the same way as Austerlitz or Valmy." The great wheatfield "is the famous slope '307,' the name you have so often seen recorded in the communiqués."[36] For many good reasons, however, the *Société des Amis de Marcel Proust et des Amis de Combray*, housed more than a hundred kilometers *west* of Paris, has to this day abstained from digging trenches for literary tourists—an act that would otherwise have made these events as traceable as the smell of the hawthorn.

It would be ridiculous to deny the high degree of referentiality in the passage about the battle of Méséglise by pointing out that such a battle never took place (but only a battle of, let's say, Verdun), and that no record of a slope 307 is found in official communiqués but only a slope 607. The kind of referentiality claimed in this passage even approaches types of what has recently become popular as "counterfactual narratives." But the passage does not simply add some individual details, as a typical historical novel is supposed to do. Rather, it challenges History as recorded and sanctioned in official communiqués.

I do not want to suggest that this passage makes the *Recherche* completely unmappable.[37] Even the location of Combray—west or east of Paris?—is less inconsistent than a tourist in Illiers-Combray might assume. During World War I, Proust was already eager to relocate Combray in the first volume. "Chartres," the name of a town close to Combray in the first edition of 1913, is replaced with "Reims" in the second edition of 1919.[38] Thus, Combray seems to be quite consistently located 250 km west of Illiers, whose present name, "Illiers-Combray," turns out to be nothing more than a fraud designed to mislead literary tourists. Even Cambrai, the seat of Hindenburg's headquarters in the first years of World War I, located 130 km north of Reims, is not just linguistically but geographically closer to Combray than Illiers.

But how could a map account for the "vertiginous possibilities of referential aberration," as explored in the *Recherche*? Should Combray simply be superimposed on a contemporary map of the Île-de-France and its adjacent regions? Should one depict the battle of Méséglise as yet another detailed map of the theater of war like the supplements to the multivolume official record *Schlachten des Weltkrieges* (*Battles of the World War*) (figure 3.8)?

Instead such a map would have to challenge the reliance on the base maps themselves. The authors of an illuminating article on thematic mapping in the age of digital navigation have distinguished a *navigational* use of maps from a *mimetic* one. The navigational use does not rely on any "*resemblance* between the map and the territory but on the detection of *relevant* cues allowing her [the navigator's] team to go through a heterogeneous set of datapoints from one *signpost* to the next."[39] The use of a base map that has been prepared for allegedly mimetic purposes may even mislead its reader. It suggests the existence of a whole "world" ("with all of its mills etc.") as a consistent unity of geographic elements. Most literary *texts*, on the contrary, have more affinity for a navigational than for a mimetic use of maps. They only supply some "*relevant* cues" in a discontinuous allocation,[40] and of very unequal ontological status: Méséglise (which only exists within the *Recherche*); a hawthorn (an existing species, but not unequivocally locatable as an individual plant); the Western front (which existed as a historical event); Cambrai (which exists on Google Maps); Illiers-Combray (which exists on Google Maps, but was spuriously invented *after* the *Recherche* had been published …).

Admittedly, few of these proposals will appear satisfactory to those currently involved in mapping literature. These proposals are derived from close readings of individual texts, so it would be even more difficult to realize these kinds of mappings

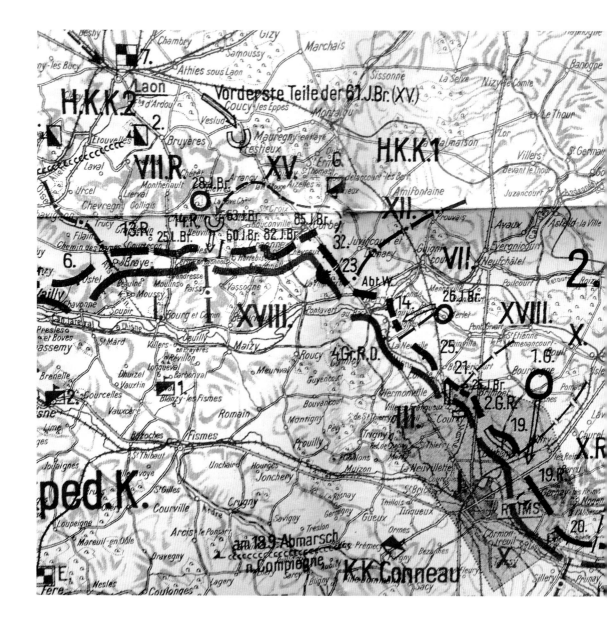

Figure 3.8

Early World War I map of the theater of war. *Der deutsche rechte Heeresflügel am 17. und 18. 9. 1914.* In *Der Weltkrieg 1914–1918*. Bearbeitet vom Reichsarchiv. *Die militärischen Operationen zu Lande*, vol. 5 (Berlin: Mittler & Sohn, 1929), Karte 2.

if the aim is to evaluate whole clusters of texts according to comparable data. But they might support an understanding of the specificity of literary texts, and they might lead to a more reflective usage of maps.

Notes

1. See Franco Moretti, *Atlante del romanzo europeo* (Turin: Einaudi, 1997), chap. 3, which follows the tradition of "literary atlases" as, for example, produced by Josef Nadler, but transforms this tradition into more sociologically oriented intentions. For a historical overview of these kinds of literary mappings see Jörg Döring, "Zur Geschichte der Literaturkarte (1907–2008)," in *Mediengeographie: Theorie Analyse Diskussion*, ed. Jörg Döring and Tristan Thielmann, 247–290 (Bielefeld: transcript Verlag, 2008); Barbara Piatti, *Die Geographie der Literatur: Schauplätze, Handlungsräume, Raumphantasien*, 65–122 (Göttingen: Wallstein, 2009).

2. See Döring, "Zur Geschichte der Literaturkarte," 285.

3. https://www.msu.edu/~olsonj/def.html.

4. Barbara Piatti and Lorenz Hurni, "Editorial: Cartographies of Fictional Worlds," *Cartographic Journal* 48, no. 4 (2011): 219.

5. In the context of a helpful summary of constants and variables in the development of geographic information systems "BC" (before computers) and "AC" (after computers), see Valéri November, Eduardo Camacho-Hübner, and Bruno Latour, "Entering a Risky Territory: Space in the Age of Digital Navigation," *Environment and Planning D: Society and Space* 28 (2010): 584.

6. See J. B. Harley, "Deconstructing the Map," *Cartographica* 26, no. 2 (1989): 1–20. For an overview of some of these discussions in the wake of Harley's work, see November, Camacho-Hübner, and Latour, "Entering a Risky Territory," 582.

7. Franco Moretti, *Graphs, Maps, Trees: Abstract Models for Literary History* (London: Verso, 2007), 63. For a critique of a theory of fiction based on notions of the "real" and the "imaginary" see Robert Stockhammer, "Exokeanismós: The (Un)Mappability of Literature," *Primerjalna književnost (Ljubljana)* 36, no. 2 (2013): 123–138, here 127–131.

8. James Joyce in a conversation with Frank Bugden, quoted in Piatti and Hurni, "Editorial," 218ff.

9. See Denis Wood and John Fels, *The Power of Maps* (New York: Guilford, 1992), 24.

10. See Barbara Piatti, "Mit Karten lesen: Plädoyer für eine visualisiere Geographie der Literatur," in *Textwelt-Lebenswelt*, ed. Brigitte Boothe et al. (Würzburg: Königshausen & Neumann, 2012), 275. See also the more detailed data submission form, which is better adapted to the practical procedures of data acquisition, at http://www.literaturatlas.eu/en/project/project-structure/data-acquisition.

11. See Piatti, "Mit Karten lesen," 275ff.

12. The use of digital GIS does not change the supplementary status of literary maps. It makes it possible to select features expressly for this purpose, or even to confine the output to the supplementary data themselves (see Piatti, "Mit Karten lesen," 274). But even in this case, their position relies on geographic data collected beforehand.

13. http://www.literaturatlas.eu/en/project/project-structure/data-acquisition.

14. See, for example, Piatti, "Mit Karten lesen," 267. The renewed desire for the fresh air of referential meaning is, remarkably enough, shared even by some scholars who know perfectly well that the alleged dichotomy between "referentiality" and "self-referentiality" is a hoax. Hans-Ulrich Gumbrecht will understand the allusion.

15. Paul de Man, *Allegories of Reading: Figural Language in Rousseau, Nietzsche, Rilke, and Proust* (New Haven, CT: Yale University Press, 1979), 4.

16. de Man, *Allegories of Reading*, 70, 10.

17. J. Hillis Miller, *Topographies* (Stanford, CA: Stanford University Press, 1995), 7.

18. See, for example, Piatti and Hurni, "Editorial," 220, for such a categorization. That, however, was not the purpose of the distinction.

19. See Franz Kafka, *Das Schloß*, ed. Malcolm Pasley (Frankfurt am Main: Fischer, 1982), 21.

20. For a more detailed account of mapping and unmappability in *Moby-Dick*, see Robert Stockhammer, *Kartierung der Erde: Macht und Lust in Karten und Literatur* (Munich: Fink, 2007), 187–209.

21. Marcel Proust, *Swann's Way*, trans. C. K. Scott Moncrieff, http://gutenberg.net.au/ebooks03/0300511.txt. The original French quotation reads: "Deux 'côtés' ... si opposés qu'on ne sortait pas en effet de chez nous par la même porte, quand on voulait aller d'un côté ou de l'autre. ... Alors, 'prendre par Guermantes' pour aller à Méséglise, ou le contraire, m'eût semblé une expression aussi dénuée de sens que prendre par l'est pour aller à l'ouest" (Marcel Proust, *A la recherche du temps perdu*, ed. Pierre Clarac and André Ferré, 3 vols. (Paris: Gallimard / Bibliothèque de la Pléiade, 1954), vol. 1, 134).

22. Marcel Proust *The Sweet Cheat Gone*, trans. C. K. Scott Moncrieff, http://gutenberg.net.au/ebooks03/0300541.txt. The original French quotation reads: "'Si vous voulez, nous pourrons ... aller à Guermantes, *en prenant par Méséglise*, c'est la plus jolie façon,' phrase qui en bouleversant toutes les idées de mon enfance m'apprit que les deux côtés n'étaient pas aussi inconciliables que j'avais cru" (Proust, *Recherche*, vol. 3, 693).

23. At an inspiring workshop in Copenhagen, Frederik Tygstrup justly criticized me for maintaining a conventional notion of *maps* or *mapping* even *ex negativo*, in my use of "unmappability." The third section of this chapter is an attempt to react to this critique.

24. For a highly thoughtful discussion of the term and/or metaphor of *mapping*, see Marion Picker, "Die Zukunft der Kartographie: Neue und nicht so neue epistemologische Krisen," in *Die Zukunft der Kartographie: Neue und nicht so neue epistemologische Krisen*, ed. Marion Picker, Véronique Maleval, and Florent Gabaude, 7–19 (Bielefeld: transcript Verlag, 2013).

25. J. B. Harley and David Woodward, eds., *The History of Cartography*, Vol. 1: *Cartography in Prehistoric, Ancient, and Medieval Europe and the Mediterranean* (Chicago: University of Chicago Press, 1987), 16. The multivolume work itself (still in progress) is hardly able to realize this definition, since it would have to include many forms of drawings or diagrams too remote from the realm of geography to be reasonably included. Which kind of understanding is not, after all, in some way "spatial"? See also the discussion of this definition in Jörg Dünne, "Die Unheimlichkeit des Mapping," in *Die Zukunft der Kartographie: Neue und nicht so neue epistemologische Krisen*, ed. Marion Picker, Véronique Maleval, and Florent Gabaude (Bielefeld: transcript Verlag, 2013), 222ff.

26. For a concise critique of this problem see Armin von Ungern-Sternberg, "Dots, Lines, Areas and Words: Mapping Literature and Narration (with Some Remarks on Kate Chopin's 'The Awakening')," in *Cartography and Art*, ed. W. Cartwright, G. Gartner, and A. Lehn, 229–252 (Heidelberg: Springer, 2009).

27. Proust, *Recherche*, vol. 1, 413.

28. Proust, *Recherche*, vol. 1, 296, 343.

29. Proust had his doubts about this conversion. He distinguished a chapter titled "Names of Countries: The Name," as part of the first volume of the *Recherche*, from another one, titled "Names of Countries: The Country," as part of the second volume.

30. Piatti and Hurni, "Editorial," 219.

31. The original German quotation reads: "In einem Roman müßte es sich gut ausnehmen, des Helden Begriffe z.B. von der Erde in einer kleinen Charte vorzustellen. Die Welt würde rund vorgestellt, in der Mitte liegt das Dorf wo er lebt, sehr groß mit allen Mühlen pp vorgestellt, und dann umher die andern Städte, Paris London sehr klein, überhaupt wird alles sehr viel kleiner, wie es weiter wegkömmt" (Georg C. Lichtenberg, "Sudelbücher," in *Schriften und Briefe*, vols. 1–3, ed. Wolfgang Promies (Munich: Hanser, 1967–1992), vol. 1, 772).

32. Translations of Genette's term *focalisation* (the established technical term for *focusing* in French) into English and German diminish its metaphoricity by introducing neologisms (*focalization* or *Fokalisierung*), producing a distinction not present in the original text.

33. Adalbert Stifter, "Granite," in *German Novellas of Realism*, ed. and trans. Jeffrey L. Sammons (New York: Continuum, 1989), 16. The original German quotation reads:

"Das ist das Leben der Wälder. Aber laß uns nun auch das außerhalb betrachten. Kannst du mir sagen, was das für weiße Gebäude sind, die wir da durch die Doppelföhre hin sehen?"
"Ja, Großvater, das sind die Pranghöfe."
"Und weiter von den Pranghöfen links?"
"Das sind die Häuser von Vorder- und Hinterstift."
"Und wieder weiter links?"
"Das ist Glökelberg."
"Und weiter gegen uns her am Wasser?"
"Das ist die Hammermühle und der Bauer David."
"Und die vielen Häuser ganz in unserer Nähe, aus denen die Kirche emporragt, und hinter denen ein Berg ist, auf welchem wieder ein Kirchlein steht?"
"Aber, Großvater, das ist ja unser Marktflecken Oberplan, und das Kirchlein auf dem Berge ist das Kirchlein zum guten Wasser."

"Und wenn die Berge nicht wären und die Anhöhen, die uns umgeben, so würdest du noch viel mehr Häuser und Ortschaften sehen: Die Karlshöfe, Stuben, Schwarzbach, Langenbruk, Melm, Honnetschlag, und auf der entgegengesetzten Seite Pichlern, Pernek, Salnau und mehrere andere."

(Adalbert Stifter, "Granit," in *Werke und Briefe: Historisch-Kritische Gesamtausgabe*, vol. 2.2, ed. Alfred Doppler and Wolfgang Frühwald (Stuttgart: Kohlhammer, 1978), 35)

34. See Albrecht Koschorke, "Das buchstabierte Panorama: Zu einer Passage in Stifters Erzählung 'Granit,'" *Vierteljahresschrift des Adalbert-Stifter-Institutes des Landes Oberösterreich* 38 (1989): 3–13; Stockhammer, *Kartierung der Erde*, 159–185 (also for discussions of other references to maps and mapping in Stifter's work).

35. It is true that in Stifter's text the list of toponyms still follows the arrangement of lines, whereas on a map they are distributed over the plane in some relation to their position in the territory. Poetic texts, however, experiment with analogous forms of "spacing" ("espacement"), to quote Mallarmé's own term in his introductory remarks to "A Throw of the Dice." For some tentative observations concerning this level of "cartographicity" see Stockhammer, *Kartierung der Erde*, 53, 78ff.

36. Marcel Proust, *Time Regained*, trans. Stephen Hudson [Sydney Schiff], http://gutenberg.net.au/ebooks03/0300691.txt. The original French quotation reads: "C'est la fameuse cote 307 dont vous avez dû voir le nom revenir si souvent dans les communiqués" (Proust, *Recherche*, vol. 3, 756).

37. See Barbara Piatti and Lorenz Hurni, "Mapping the Ontologically Unreal–Counterfactual Spaces in Literature and Cartography," *Cartographic Journal* 46, no. 4 (2009): 333–342. This article, however, does not contain any map depicting a counterfactual narrative, as drawn by the authors, but only maps already existing and reproduced from books or artworks.

38. Proust, *Recherche*, vol. 1, 136, 145, and the corresponding notes, 961.

39. November, Camacho-Hübner, and Latour, "Entering a Risky Territory," 585.

40. See von Ungern-Sternberg, "Dots, Lines, Areas, and Words," 238.

Bibliography

Döring, Jörg. "How Useful Is Thematic Cartography of Literature?" *Primerjalna književnost (Ljubljana)* 36, no. 2 (2013): 139–149, 291–292.

Döring, Jörg. "Zur Geschichte der Literaturkarte (1907–2008)." In *Mediengeographie: Theorie Analyse Diskussion*, ed. Jörg Döring and Tristan Thielmann, 247–290. Bielefeld: transcript Verlag, 2008.

Dünne, Jörg. "Die Unheimlichkeit des Mapping." In *Die Zukunft der Kartographie: Neue und nicht so neue epistemologische Krisen*, ed. Marion Picker, Véronique Maleval, and Florent Gabaude, 221–240. Bielefeld: transcript Verlag, 2013.

Harley, J. B. "Deconstructing the Map." *Cartographica* 26, no. 2 (1989): 1–20.

Harley, J. B., and David Woodward, eds. *The History of Cartography*, Vol. 1: *Cartography in Prehistoric, Ancient, and Medieval Europe and the Mediterranean*. Chicago: University of Chicago Press, 1987.

Kafka, Franz. *Das Schloß*. Ed. Malcolm Pasley. Frankfurt am Main: Fischer, 1982.

Koschorke, Albrecht. "Das buchstabierte Panorama: Zu einer Passage in Stifters Erzählung 'Granit.'" *Vierteljahresschrift des Adalbert-Stifter-Institutes des Landes Oberösterreich* 38 (1989): 3–13.

Lichtenberg, Georg C. "Sudelbücher." In *Schriften und Briefe*, vols. 1–3, ed. Wolfgang Promies. Munich: Hanser, 1967–1992.

de Man, Paul. *Allegories of Reading: Figural Language in Rousseau, Nietzsche, Rilke, and Proust*. New Haven, CT: Yale University Press, 1979.

Miller, J. Hillis. *Topographies*. Stanford, CA: Stanford University Press, 1995.

Moretti, Franco. *Atlante del romanzo europeo*. Turin: Enaudi, 1997.

Moretti, Franco. *Graphs, Maps, Trees: Abstract Models for Literary History*. London: Verso, 2007.

November, Valéri, Eduardo Camacho-Hübner, and Bruno Latour. "Entering a Risky Territory: Space in the Age of Digital Navigation." In *Environment and Planning D: Society and Space* 28 (2010): 581–599.

Piatti, Barbara. *Die Geographie der Literatur: Schauplätze, Handlungsräume, Raumphantasien*. Göttingen: Wallstein, 2009.

Piatti, Barbara. "Mit Karten lesen: Plädoyer für eine visualisiere Geographie der Literatur." In *Textwelt-Lebenswelt*, ed. Brigitte Boothe et al., 261–288. Würzburg: Königshausen & Neumann, 2012.

Piatti, Barbara, and Lorenz Hurni. "Editorial: Cartographies of Fictional Worlds." *Cartographic Journal* 48, no. 4 (2011): 218–223.

Piatti, Barbara, and Lorenz Hurni. "Mapping the Ontologically Unreal–Counterfactual Spaces in Literature and Cartography." *Cartographic Journal* 46, no. 4 (2009): 333–342.

Picker, Marion. "Die Zukunft der Kartographie: Neue und nicht so neue epistemologische Krisen." In *Die Zukunft der Kartographie: Neue und nicht so neue epistemologische Krisen*, ed. Marion Picker, Véronique Maleval, and Florent Gabaude, 7–19. Bielefeld: transcript Verlag, 2013.

Proust, Marcel. *A la recherche du temps perdu*. Ed. Pierre Clarac and André Ferré. 3 vols. Paris: Gallimard / Bibliothèque de la Pléiade, 1954.

Proust, Marcel. *Swann's Way*. Trans. C. K. Scott Moncrieff. http://gutenberg.net.au/ebooks03/0300511.txt.

Proust, Marcel. *The Sweet Cheat Gone*. Trans. C. K. Scott Moncrieff. http://gutenberg.net.au/ebooks03/0300541.txt.

Reuschel, Anne-Kathrin, and Lorenz Hurni. "Mapping Literature: Visualisation of Spatial Uncertainty in Fiction." *Cartographic Journal* 48, no. 4 (2011): 293–308.

Stifter, Adalbert. "Granit." In *Werke und Briefe: Historisch-Kritische Gesamtausgabe*, vol. 2.2, ed. Alfred Doppler and Wolfgang Frühwald, 21–60. Stuttgart: Kohlhammer, 1978.

Stifter, Adalbert. "Granite." In *German Novellas of Realism*, ed. and trans. Jeffrey L. Sammons. New York: Continuum, 1989.

Stockhammer, Robert. "'An dieser Stelle': Kartographie und die Literatur der Moderne." *Poetica* 33, nos. 3–4 (2001): 273–306.

Stockhammer, Robert. "Exokeanismós: The (Un)Mappability of Literature." *Primerjalna književnost (Ljubljana)* 36, no. 2 (2013):): 123–138.

Stockhammer, Robert. *Kartierung der Erde: Macht und Lust in Karten und Literatur*. Munich: Fink, 2007.

von Ungern-Sternberg, Armin. "Dots, Lines, Areas and Words: Mapping Literature and Narration (with Some Remarks on Kate Chopin's 'The Awakening')." In *Cartography and Art*, ed. W. Cartwright, G. Gartner, and A. Lehn, 229–252. Heidelberg: Springer, 2009.

Wood, Denis, and John Fels. *The Power of Maps*. New York: Guilford, 1992.

4
Cartographic Tropes: From Kant's Maps to Foucault's Topology

Oliver Simons

Introduction

In a sense, Kant and Foucault, both founders of a critical discourse, could not be further apart. The end of the eighteenth century and the second half of the twentieth century seem too disparate. After all, does Foucault not position himself at the end of the episteme that Kant had initiated? Is his thought not a critical response to the Enlightenment project?

The distance between both thinkers indeed seems insurmountable, particularly if one takes the term *distance* in a literal, spatial sense. The spatial theories of both thinkers are simply incongruous. And yet, if one considers how their concepts of space are intimately linked to their reflections on writing, could their distance not also be seen as proximity? Aren't Kant and Foucault both cartographers who chart spaces and map their philosophical discourses?

As this chapter argues, the distance and proximity of both thinkers can be best perceived if one takes not only their underlying spatial theories into account, but also their perspectives on literature. From Kant's maps to Foucault's topology, the parallels and divergences between both are most notable in their different understandings of literature. Kant's maps, I argue, are first and foremost attempts to exclude metaphorical language and arbitrary signs; Foucault's inventive cartography, on the other hand, is itself a poetic project.

Kant's Maps

In an anecdote handed down by Cicero and Vitruvius, the philosopher Aristippus recalls being shipwrecked on an unknown coast. Not knowing whether he finds himself in an inhabited country or on a lonely island, the castaway suddenly notices geometric figures in the sand. It is a comforting sight; only a human being could have drawn such images.[1] In the *Critique of the Power of Judgment*, Immanuel Kant recounts this famous anecdote:

> If someone were to perceive a geometrical figure, for instance a regular hexagon, drawn in the sand in an apparently uninhabited land, his reflection, working with a concept of it, would become aware of the unity of the principle of its

generation by means of reason, even if only obscurely, and thus, in accordance with this, would not be able to judge as a ground of the possibility of such a shape the sand, the nearby sea, the wind, the footprints of any known animals, or any other non-rational cause.[2]

This is but one example of the numerous geometric figures in Kant's critical philosophy and of a style that I will describe here as cartographic writing. Triangles, circles, or simple straight lines abound in his works as symbols of pure reason or rational humanity, particularly when the other element of Aristippus's anecdote is in sight: the raw, imponderable sea that denotes the opposite of reason; an image of the irrational, of the literary, it warns his readers not to stray too far from the safe shores of rational thinking. For Kant, "the straight line of truth" is a purely terrestrial phenomenon, and the philosopher is necessarily a cartographer.[3]

Philosophers need to control their metaphors to achieve the most exact form of language. The clear images of Euclidean geometry provide Kant with an archetype in this regard. Whereas metaphors can be ambiguous, geometric figures are reliable signs whose meaning can be directly conveyed at all times, even to the layperson. They are images of a controlled, rational faculty of imagination. Practical geometry is a model for philosophy, but not solely in a logical sense. Geometry is no less fundamental as a form of semiotics and a theory of signs in which every expression has a clear and comprehensible meaning. Geometry models an ideal language for philosophers, not least because their languages are fundamentally different from each other. Mathematics is synthetic, beginning with definitions and subsequently developing its geometric figures; philosophy is an analytic discipline that deals with abstract conceptions that it must resolve so as to conclude its deliberations, at best, with a definition. Consequently, the two disciplines have different forms of expression. The philosopher, in contrast to the geometer, thinks *in abstracto*; his medium is discursive speech and the certainty of this form of thought is always threatened by false terms or seductive metaphors. For Kant, geometry is therefore an unsuitable method of conducting philosophy. With respect to semiotics, however, it is exemplary. "I shall demonstrate this with the greatest possible clarity [auf das augenscheinlichste dartun],"[4] writes Kant, and this bringing-before-the-eyes is an oft-repeated formulation, a Kantian rhetorical flourish with which he seeks to approximate the ciphers of mathematics:

> For since signs in mathematics are sensible means to cognition, it follows that one can know that no concept has been overlooked, and that each particular comparison has been drawn in accordance with easily observed rules *etc*. And these things can be known with the degree of assurance characteristic of seeing something with one's own eyes. And in this, attention is considerably facilitated by the fact that it does not have to think things in their universal representation. It has rather to think the signs as they occur in their particular cognition.[5]

While geometric shapes and figures are self-evident, metaphysics must create intuitiveness by means of a discursive language; it must illustrate meaning and bring it "before the eyes." As I will argue in what follows, Kant's cartographic style is an attempt to reconcile geometric exactitude with philosophical rigor. I will unfold this thesis in three steps: first, I discuss Kant's theory of the relation between geometric figures and linguistic signs; second, I show how Kant associates the process of writing a philosophical text with the act of drawing images; and third, I conclude with passages in which Kant associates his project of a critical philosophy with a form of cartography. (1) Locating signs, (2) drawing lines, and (3) cartographic mappings are the three aspects of his writing that concern me here. Each aspect, I argue, is directly related to Kant's attempt to distinguish philosophy from literature.

Geometric Shapes as Signs

The realm of knowledge, Kant writes, is limited to visualizing and imaging, but is thereby also restricted to what can be schematized or reduced to an image. But how does it work for concepts of understanding or reason, which are foreign to the senses and for which no image is possible? Concepts of understanding, Kant writes in his *Critique of the Power of Judgment*, can still be demonstrated. Ideas of reason require a different kind of image that must vouch for the missing link, which Kant expounds as follows in his famous section 59. It can also be read as a reflection on the relation between philosophy and literature. Here, Kant distinguishes between two kinds of representation or "hypotyposis." It is either *schematic*, where to a concept grasped by the understanding the corresponding intuition is given a priori; or *symbolic*, where to a concept that only reason can think, and to which no sensible intuition can be adequate, an intuition is attributed with which the power of judgment proceeds in a way merely analogous to that which it observes in schematization. That is, it is merely the rule of this procedure, not of intuition itself, and thus merely the form of reflection, not the content, that corresponds to the concept.[6]

Kant's conception of hypotyposis has been much discussed, especially because he offers here one of his few comments on language.[7] Of particular interest was Kant's concept of symbolic hypotyposis, on which all philosophy depends because its signs cannot be illustrated *in concreto*, as in geometry. Symbols are used where concrete images cannot be produced; they function indirectly and by means of an analogy (for which empirical intuitions are also employed), in which the power of judgment performs a double task, first applying the concept to the object of sensible intuition, and then, second, applying the mere rule of reflection on that intuition to an entirely different object, of which the first is only the symbol.[8]

Symbolic hypotyposes, or metaphors, stem from analogies. They are terms that deploy an expression where no image is possible, attempting to fill this gap through an expressive resemblance to the concept of reason. In an extreme case, these symbolic representations are so arbitrary and disconnected from their meaning that they create a purely fantastic imagery. Literature, in Kant's understanding, would be an extreme example of this metaphorical mode of writing and the inverse of geometric exactitude. Philosophy is permanently threatened by this fictional mode of representation, for philosophy too depends on analogies. At the same time, however, and this is Kant's poetic project, it can be much more disciplined, namely as cartographic writing that controls its own images by drawing on geometry.

Drawing Lines and Writing Philosophy

Kant needs to use images and metaphors in a way that is (almost) as exact as the drawing of geometric figures. Establishing borders is therefore among his most important motifs, since drawing limits on knowledge seems to be the only way to avoid speculating about phenomena in purely fictional analogies. Philosophical texts follow the logic of a cartographic reason: "To that extent metaphysics is a science of the *limits of human reason*. A small country always has a long frontier; it is hence, in general, more important for it to be thoroughly acquainted with its possessions, and to secure its power over them, than blindly to launch on campaigns of conquest."[9] In his earlier writings, Kant himself had succumbed to this desire for conquest. After his critical turn, however, Kant eschews any transgression of the limits of reason, not only by restricting the space of knowledge to three dimensions, but also by striving for geometry's disciplined use of signs. Kant's geometry is thus at once a discipline of space and a poetological program; it is the definition of a space of perception and the field of metaphors for an exact but still visually perceptible science in which the author Kant looks for analogies. The repeatedly described "Ziehen einer Linie," drawing of a line, is therefore one of the most prominent examples of his geometric images; it shows how the hand of the philosophical subject appears to be guided by pure reason:

> But in order to cognize something in space, e.g., a line, I must draw it, and thus synthetically bring about a determinate combination of the given manifold, so that the unity of this action is at the same time the unity of consciousness (in the concept of a line), and thereby is an object (a determinate space) first cognized. The synthetic unity of consciousness is therefore an objective condition of all cognition, not merely something I myself need in order to cognize an object but rather something under which every intuition must stand in order to become an object for me, since in any other way, and without this synthesis, the manifold would not be united in one consciousness.[10]

The line is an image that humans have not copied from nature, but generated independently. It is not a mimetic imitation, but rather a product of the rational subject. As Kant illustrates in this passage, all perceptions exist only because of an "original synthetic unity of apperception." That is, in order to make them conscious—to add an "I think" to the perception—the various perceptions must be consolidated into an object, which means that the faculty of reason requires a synthetic effort. The "synthetic unity of apperception" is the "mere form of outer sensible intuition, space, [and] is not yet cognition at all; it only gives the manifold of intuition a priori for a possible cognition."[11] As a form, it also presupposes the possibility that diverse impressions can be joined together into a unified and meaningful figure of which the subject can be conscious. In drawing a line, this synthetic effort is illustrated quite literally. It is the expression of a controlled imagination and a countermodel to wild fantasy. For this reason, drawing a line is also never purely subjective; on the contrary, in drawing a line the hand of the subject is guided by reason, illustrating the subject's capabilities and a philosophical act of writing.

The line is thus not only an example of the rational subject and a specific space; it is also a metaphor for Kant's own writing. Here is another example:

> We cannot think of a line without drawing it in thought, we cannot think of a circle without describing it, we cannot represent the three dimensions of space at all without placing these lines perpendicular to each other at the same point, and we cannot even present time without, in drawing a straight line (which is to be the external figurative representation of time), attending merely to the action of the synthesis of the manifold through which we successively determine the inner sense, and thereby attending to the succession of this determination in inner sense.[12]

In terms of content, this passage resembles the one cited above, though with the addition of time—the time taken to draw the line, which makes succession visible as a form of synthesis. Yet it is also revealing that Kant's line connotes something more. It stands for space and for the time that elapses while producing a line. One could, therefore, read the line as an expression to which Kant assigns several meanings. In this specific instance he argues that the line should represent the conception of time. The line is not merely a mark; it "describes" the circle, portrays time, and thus comprises an abundance of meanings—or better, a *series* of meanings—that Kant links in their description. The movement he describes is not just the "I think," with which the subject accompanies his synthesized perceptions, nor is it merely the thinking subject. It is also the writing subject.[13] When a line is drawn, a chronology emerges, a linearity that unfolds like the text, which ends and is deferred by treating further connotations of the line. In this way, Kant's text enacts a movement that begins with the line as a physical stroke and an empirical image, and that extends into perceptions—that is, into meanings that can only be evoked indirectly. And yet at the same time, it is a concrete image, a drawing and a sign for the text and its temporality, a sign that portrays the literal unfolding of the very scene of writing. The line therefore represents the linearity of the text, the course of a certain path, and, insofar as it has two endpoints, also "stands" for a passage in a text, a figure. It represents the unity of an object that emerges from the path of the text: a philosophical text that, defined by lines of reason, is also a cartographic text.

The line can be found in multitudinous forms in Kant: as a border and an ideal path, or even as a kind of narrative thread that also traces a line.[14] Metaphysics must be a "science of the *limits of human reason*,"[15] he writes in the "Spirit-Seer," of limits against such "groundless sketches"[16] as Swedenborg's. If one transgresses the limits of reason, one might advance into the realm of fantasy, but one will not expand one's reason in doing so.[17] Plato's ideology serves Kant as an example of a philosophy that dares to advance past the material world into the "empty space of pure understanding" without having the groundwork for it.[18] For this reason, according to Kant, the first significant constraint on a critique of reason must be that all thought refers to sensory experience.[19] The line he draws signifies all of this; it is a symbol of reason, but it also serves Kant as a border between what is perceptible and the free space of speculation.

Kant as Cartographer

Kant's images of land surveying fit in with these spatial motifs, as in the *Critique of Pure Reason*, where he projects practical geometry onto the project of metaphysics, the very discipline that he had already illustrated geographically in his earlier texts:

> We have not only traveled through the land of pure understanding, and carefully inspected each part of it, but we have also surveyed it, and determined the place for each thing in it. This land, however, is an island, and enclosed in unalterable boundaries by nature itself. It is the land of truth (a charming name), surrounded by a broad and stormy ocean, the true sea of illusion, where many a fog bank and rapidly melting iceberg pretend to be new lands and, ceaselessly deceiving with empty hopes the voyager looking around for new discoveries, entwine him in adventures from which he can never escape and yet also never bring to an end. But before we venture out on this sea, to search through all its breadth and become certain of whether there is anything to hope for in it, it will be useful first to cast yet another glance at the map of the land that we would now leave, and to ask, first, whether we could not be satisfied with what it contains, or even must be satisfied with it out of necessity, if there is no other ground on which we could build; and, second, by what title we occupy even this land, and can hold it securely against all hostile claims.[20]

This imaginary journey can no longer be compared to Kant's fantasies about distant planets or the risky trip across an ocean, far from the shore, which one finds in his earliest, precritical writings. The royal path, the "Königsweg," of philosophy now runs as straight as possible, though always embedded in a homogeneous landscape and in continuous space. In a metaphorical sense, reason is territorialized: "In the progress of my labor I have been almost constantly undecided how to deal with this matter. Examples and illustrations always appeared necessary to me, and hence actually appeared in their proper place in my first draft."[21] Due to spatial constraints, Kant says he refrains from all too image-rich comparisons and analogies and avoids forms of popularization in order to focus all of his attention on the representation of reason.[22]

Yet even if the images in the *Critique of Pure Reason* are not nearly as eccentric as his early cosmological fantasies, his language still abounds in metaphor—in cartographic images, to be precise. In the "Transcendental Dialectic," for example, he describes how reason prepares the "field" of the intellect in a threefold manner. The images of concepts are comparable with the constant expansion of horizons in a level, homogeneous space: "One can regard every concept as a point, which, as the standpoint of an observer, has its horizon, i.e., a multiplicity of things that can be represented and surveyed, as it were, from it." Within this horizon, restricted visual fields are possible, but categories of meaning strive for a more universal horizon, "which one can survey collectively from its middle point, which is the higher genus, until finally the highest genus is the universal and true horizon, determined from the standpoint of the highest concept and comprehending all manifoldness, as genera, species, and subspecies, under itself."[23]

The horizon is not just a limit of knowledge, but—as a perpetually withdrawing line—also a sign of the progress of knowledge. This set of cartographic images describes the progress of reason. Kant envisions a level landscape as a uniform three-dimensional space in which the philosopher, always striding forward, attempts to approach a horizon. In Kant's Euclidean worldview, reason is controlled, and its ideal path runs alongside the horizon. Although the horizon arouses the desire for a final and universal horizon, it cannot be reached and has no beyond. Kant's didactic instructions for a proper philosophical method orient the reader in three-dimensional thought. Because metaphysics is henceforth a science about the limits of reason, one's loyalty to this science is measured according to one's specific attention to space: "The sum total of all possible objects seems to us to be a flat surface, which has its apparent horizon, namely that which comprehends its entire domain and which is called by us the rational concept of unconditioned totality."[24] According to Kant, reason is a discursive capacity, yet it needs analogies in order to illustrate its approach. The landscape image with which Kant paraphrases the faculty of reason is of course—as he himself notes—not quite as level a surface as the homogeneous space of a coordinate system. On the contrary, it resembles a curved surface, a sphere.[25] This very space of experience, however, not only corresponds to the space of human perception. It is also a domain in which Euclidean geometry—even if it describes more abstract, purer forms—is still applicable. In a small area of the sphere, in which the curvature can hardly be seen, Euclid's geometry promises a large measure of exactitude. One must only force oneself to proceed gradually, from one field of view to another, from one horizon to the next, to connect one's insights. This procedure of precise cartographic description creates evidence. And it also corresponds to land surveying in a very concrete way, even if Kant has no intention of transforming the spherical surface of the earth into a flat plane.[26] The description of this geography is called mapping; it is the projection of a geometric grid onto a sphere in the hopes of obtaining as unbroken a transfer as possible.[27] The space sketched out by Kant and the project of his mapping open up the dimension in which the line of the narrative can be developed as a continuous projection of Euclidean images and figures onto the course of the philosophical text.

Foucault's Inventive Cartography

Michel Foucault has been instrumental in the recent blossoming of interest in space, so much so that it is common to speak of a *spatial turn* in the humanities. But although Foucault is a key component of this trend's methodological canon, his writings are only seldom analyzed with reference to their own metaphors of space and architectonic structure. I suggest a reading that attempts to systematically understand Foucault's account of space. Well-known, substantive content in this regard is less important than the organization of his texts, their metaphors and figures of thinking. Foucault, too, is a cartographic writer; he resides in a different epistemic context than Kant, to be sure, but is likewise in search of a rigorous language. As the following pages intend to show, this search for exactitude does not sidetrack him from literature. Whereas Kant's cartographic lines are demarcations from the purely fantastical, in Foucault literature becomes a means of cartographic thought.

Perhaps Foucault's most remarkable spatial text is *Les mots et les choses* from 1966. To connect the knowledge of biology, economics, and language with philosophical reflections from the same era—to make visible each period's structures of knowledge—Foucault describes spaces of analogy and difference. As he notes in his introduction, borders must be redrawn, "things usually far apart are brought closer, and vice versa";[28] one must give history a new form. Foucault defines an epistemological space as the "totality of relations that can be discovered, for a given period, between the sciences when one analyses them at the level of discursive regularities."[29] Instead of a historical designation of an epoch, which showcases movements and trends in chronological order, Foucault outlines spaces of time that offer no possibility of transition, even though they border on one another. Every episteme is a spatial structure that differs fundamentally from other epistemes in its configuration. Although even widely disparate areas of knowledge harbor commonalities and analogies, these overlaps are only visible because they are distinct from other forms of regularity. There is no general space in which different epistemes can be classified; the transformations of knowledge from one epoch to the next are fundamental and radical. The only common basis for the varied epistemes is Foucault's search for analogies and differences and his own cartographic charting of these epochs.

The concrete formation of the three different epistemes he analyzes in his book is surprisingly simple, and interestingly enough Foucault's readers have consistently overlooked the schematic organization of his book. Whereas the Renaissance has a circular structure, the tableaus of the classical age resemble a square, and the nineteenth century reveals a triangular structure: "The criticism-positivism-metaphysics triangle of the object was constitutive of European thought from the beginning of the nineteenth century to Bergson."[30] But in contrast to the circle and square of the previous epochs, the triangle of the nineteenth century is not a planar figure. Foucault continues that one must imagine its edges as "diagonals" in different fields of knowledge. But in imagining a triangle, the edges of which are diagonals on two-dimensional planes, one must lend it depth. And if one sees in the basic triangular structure an opening onto a three-dimensional space, then the very image that Foucault ultimately describes as the figure of the nineteenth century appears: the so-called "trihedron of knowledge."[31] The triangle then enables a view onto a box enclosed by three planes. It is noteworthy that Foucault's description of this space is much more precise than the name he gives it. In accordance with its name, the trihedron belongs to the geometric group of polyhedrons, although these are bodies bordered by a certain number of planes. A trihedron would therefore have to be a body bordered by three planes, which is only possible if its sides are curved. Foucault is introducing a paradoxical figure. Three flat planes cannot create an enclosed space; a trihedron must therefore be open on one side, like a display case or a stage, and this is precisely the point of his description. The spatial structure of this trihedron is similar to the three-dimensional coordinate system, the same spatial concept that was crucial for Kant's philosophy. As discussed in the previous section, Kant's spatial a priori was not just any kind of space; it had a three-dimensional structure. Even more important is the dilemma that, according to Foucault, the human sciences that emerge with the nineteenth century cannot be located on the surface of the trihedron; they hover inside, which means, metaphorically speaking, they fail to give themselves a rigorous scientific status. In other words, the human sciences are trapped inside a spatial structure that they cannot see from the outside; theirs is consequently always a point of view akin to the cartographic images of the land surveyor we found in Kant. They view the world as a three-dimensional landscape unfolding before their eyes; they strive to reach a horizon that recedes with each added attempt to press deeper into this space. This is the epistemological dilemma of the humanities according to Foucault: they strive forward endlessly toward a constantly retreating horizon. The end of this three-dimensional space of knowledge is unattainable because space, as an a priori, is newly conceived with each step forward. Humans are living beings that see opening before them a space whose movable coordinates meet within them.[32] Being ensnared within a trihedron seems especially hopeless because a new horizon always appears and then retreats, no matter the progress. Every horizon awakens a desire in the subject that is simply unrealizable according to the logic of this structural space. To reach the exterior, the humanities would have to abandon the very perspective they occupy naturally.

To be sure, in Foucault's spatial narrative there are indeed sciences that are able to avoid the futile efforts of the humanities. The so-called countersciences—structural linguistics, psychoanalysis, and ethnology—decenter the subject in such a way that their knowledge cannot be attributed to the perspective of an advancing subject. These countersciences lead the way to the outside of the nineteenth century and define a new theoretical position, a position that Foucault adopts as well, and that is more clearly demonstrated in the book that followed *The Order of Things*, namely, *The Archaeology of Knowledge* from 1969.

Foucault's Topological Turn

The Archaeology of Knowledge, at least in its published form, is a kind of retrospective of Foucault's earlier works, a "discours de la méthode," as one reader conjectures,[33] or "less a discourse on his method than the poem of his previous work," as Deleuze writes.[34] As such, it is a book necessarily written from another perspective, demanding a space more general than that of *The Order of Things*: "I have tried to define this blank space from which I speak, and which is slowly taking shape in a discourse that I still feel to be so precarious and so unsure."[35] Without this blank space, without an exterior, the trihedron of knowledge would not have been thinkable.

The Archaeology of Knowledge is a text in which, and with which, Foucault attempts to cross this threshold. It is a book that must therefore also deal with the representation of a break. How does one write about discontinuities and those concepts that enable us "to conceive of discontinuity (threshold, rupture, break, mutation, transformation)?"[36] Will it be sufficient to choose another source of images besides Euclidean geometry, as Deleuze comments? But even if Foucault's new topology could not be understood without Bernhard Riemann or Albert Lautman, as Deleuze writes,[37] has Foucault simply transferred topology onto his discourse?

Foucault's own brief but verbose listing of the concepts of discontinuity—"threshold, rupture, break, mutation, transformation"—might suggest the difficulty in applying one's own language to an alternate discourse. Although topology describes the transformations and mutations of bodies, one of its rules is that such deformations must never lead to a "rupture" or "break" in the material, which is conceived of as a dough that one can freely knead and form but that should never be torn. Foucault's listing of "threshold, rupture, break, mutation, transformation" is not purely topological and is discontinuous insofar as these concepts cannot be clearly assigned to any particular science. This makes it all the more clear that his *Archaeology* needs a poetics and a theory of representation. Historical descriptions, he writes in the introduction, "are necessarily ordered by the present state of knowledge, they increase with every transformation and never cease, in turn, to break with themselves."[38] The text of history has its own historicity and must constantly be reinvented, for which reason it cannot rely on the same Euclidean thought as Kant's cartography. For Kant, Euclid was a timeless and ahistorical foundation of exactitude. As the following paragraphs show, Foucault draws on a different, topological model.

Foucault describes this search for a new language as a problem of narrative, as a search for form. In *The Archaeology of Knowledge* he writes:

> What, do you imagine that I would take so much trouble and so much pleasure in writing, do you think that I would keep so persistently to my task, if I were not preparing—with a rather shaky hand—a labyrinth into which I can venture, in which I can move my discourse, opening up underground passages, forcing it to go too far from itself, finding overhangs that reduce and deform its itinerary, in which I can lose myself and appear at last to eyes that I will never have to meet again. I am no doubt not the only one who writes in order to have no face. Do not ask who I am and do not ask me to remain the same: leave it to our bureaucrats and our police to see that our papers are in order. At least spare us their morality when we write.[39]

The investigation of the labyrinth, perhaps the necessity of getting lost in the blank spaces, to which Foucault attests repeatedly—"We soon lose our way, the path becomes confused"[40]—seems inextricably linked with a loss of identity. The space of history is difficult for the subject to leave, for the sole reason that its structure, just like perception for Kant, resembles the projection of images with a single perspective. Anthropology and humanism "are twins,"[41] according to Foucault. The narrator of *The Archaeology of Knowledge* enters the labyrinth to train a new set of eyes, to make possible a different kind of seeing, and to create a new form of evidence. After all, the *Archaeology* is also about images and seeing. *The Order of Things*, according to Foucault, was only an "imperfect sketch,"[42] in which he asks how it is possible to create other images, "or, in other words, what tables is it possible to draw up?"[43]

Foucault begins his attempt to make comprehensible the incremental formation of these images by defining the statement as the smallest unit of discourse. This is the raw material, the most elementary component of discourse, and for that very reason so difficult to grasp, since every view frames the past according to its own conceptions, integrating it into a space that may have a very different structure than the historical event itself. One would require the purest description possible, which would contain the actual, true facts. No subject can guarantee this purity, no primordial experience, no *cogito*, no pure consciousness.[44] The statement must not be confused with a logical proposition, Foucault says, since it cannot be reduced to the logic of what was said. Nor is it a sentence, since diagrams, tables, and graphics are also statements; it cannot be mistaken for a speech act, for a speech act often requires multiple statements. Foucault maintains that the linguistic analysis of sentences, speech act theory, and analytic philosophy, which concerns itself with logical propositions, are all unsuitable methodological frameworks. To designate statements as elementary units of discourse, a different method is required, a poetics of description. The statement is the basis of historical analysis, but in the history of theory it would seem that the "statement" never existed before Foucault. Foucault's considerations are therefore poetic in a very literal sense; he must first constitute the statement as an object within and together with a description that distinguishes itself from the representations of historians. And this is what he has in common with literature—literature not defined as representation but rather as a discourse that is poetic and literarily invents itself. Like Kant, Foucault uses mathematical images that are as exact and rigorous as possible, but that are at the same time—and in contrast to Kant's philosophical language—inventive: Foucault employs mathematics poetically and creates the spaces that he describes.

In his 1964 essay "The Language of Space," Foucault writes about the twentieth century as the period in which space had become "the most obsessive of metaphors,"[45] not on a thematic level, but rather as the very condition of its own discourse. In his readings of novels by Michel Butor, Roger Laporte, and Claude Ollier, he describes how literature itself has become a means of ordering spaces of knowledge, languages, and historical epochs. Butor's *Description of San Marco*, for instance, is not a mere geographic reproduction, "but more a deciphering"[46] of languages, legends, and books that constitute this particular location. Rather than describing visible spaces, Butor's text analyzes San Marco's archeological depths.

Foucault's own spatial poetics are similar to Butor's: rather than surveying visible surfaces, his cartography maps their preconditions, spaces that are not given entities, but that have to be brought into existence through their description. In the spatial poetics of *The Archaeology of Knowledge*, Foucault describes three aspects: every statement belongs to a collateral space, which is an associated or adjacent domain, a space formed by other statements; and then also to a correlative space, in which the statement is incorporated with subjects, objects, or concepts; and finally also to a complementary space of nondiscursive formations. What then are these collateral, correlative, and complementary spaces, and what is their purpose in Foucault?

Only when one carefully follows Foucault's description or sketch of these spaces, as Gilles Deleuze has done in his commentary, does it become evident that Foucault's dissociation from analytic philosophy, linguistics, and speech act theory also has a graphic equivalent. In a linguistic analysis of statements, for example, one would follow horizontal sentences and put them in corresponding order. According to the treatment of propositions in analytic philosophy, however, such statements would be ordered vertically. This schematization might seem strange, but if one reads Roland Barthes's "Introduction to the Structural Analysis of Narrative," for instance, it is impossible to overlook a similar categorization based on vertical and horizontal dimensions. Every sentence of a narrative marks a position in a chronology of events, but at the same time sentences can be given a logical order, which Barthes structures vertically. He reconstructs the structure of narrative in a two-dimensional coordinate system. Foucault's peculiar spatial metaphors must therefore be understood quite concretely, as references to processes of reading, as one might find in structuralism. It is of course all the more important that Foucault's own poetics disregards this kind of rigid geometrization, and that he find a language that can formally differentiate itself from structuralism.

To determine the rules for what can be stated, Foucault first describes collateral space, which specifies these rules based on vectors. One must try to visualize these descriptions, though they are initially less than intuitive. If the statement were a curve, then collateral space would determine the surroundings of this curve–that is, illustrate a space piece by piece. With vectors this could be grasped with greater precision and would also determine how the curve could be formed. It is as if one were to define the coordinate system in which the curve is inscribed. But for Foucault this space is not necessarily reducible to a flat plane with two dimensions. Collateral space is the concrete environment of the statement, an environment that can have various forms and would not be comparable to a flat coordinate system, like lines on a piece of paper for example.

The second form of space, correlative space, is different. It does not describe the surroundings of a statement, but rather its relationship to a subject or object. Statements thereby do not find their origin in a subject; rather, the subject is derived from the statement. The reading of a novel with an eye to its author is just one mode of reading, according to Foucault, and by no means a privileged mode. To use the metaphor of the curve, the correlative space is "derived" from the curve. It is a space that belongs to the curve, but that is not identical to it. Mathematically speaking, the derivative function would describe the course of the curve differently.

There is also a mathematical analogy for the third form, complementary space. It defines an extrinsic space that is occupied by institutions, which do not belong to the discursive order but to its conditions. Metaphorically speaking, this space concerns the limits of the functions that the graph approaches, with which it converges and from which it breaks away. One could render Foucault's description as a simple mathematical sketch, as a curve that has derivative functions and limits in a particular vector space.

One might compare these three aspects to Kant's geometric sign, the drawing of the line, and the cartographic space, which I discussed in the previous section. But whereas Kant's practical geometry was bound to the Euclidean space, Foucault's topological language creates a different kind of map. When he describes the historical dimension of his archeology, he delineates an "archive" that strings together the statement analyses, as it were, but thereby opens itself to a general space in which this layout is possible. In Kant, this space has the structure of a three-dimensional a priori, whereas in Foucault, one finds the following description:

> The domain of statements thus articulated in accordance with historical *a prioris*, thus characterized by different types of positivity, and divided up by distinct discursive formations, no longer has that appearance of a monotonous endless plain that I attributed to it at the outset when I spoke of "the surface of discourse"; it also ceases to appear as the inert, smooth, neutral element in which there arise, each according to its own movement, or driven by some obscure dynamic, themes, ideas, concept, knowledge. We are now dealing with a complex volume, in which heterogeneous regions are differentiated or deployed, in accordance with specific rules and practices that cannot be superposed. Instead of seeing, on the great mythical book of history, lines of words that translate in visible characters thoughts that were formed in some other time and place, we have in the density of discursive practices, systems that establish statements of events (with their own conditions and domain of appearance) and things (with their own possibility and field of use). They are all these systems of statements (whether events or things) that I propose to call *archive*.[47]

Foucault historicizes the a priori, or in other words, sketches a space that does not simply follow a third dimension in a straight line into its depth, but instead follows a space with a dimension of time that is multiplied into various vector spaces, each of which defines the field of statements in its own way. History can therefore take on different forms. In other words, "history" must be seen as just one of many spatial constellations. Like Kant, Foucault describes his archeology, a term borrowed from Kant, as a surveying of land, as mapping:[48] "its purpose is to map,"[49] as an analysis of "grids,"[50] the surveying of an "archaeological territory,"[51] and as in the history of cartography, the "blank spaces"[52] are particularly alluring. But whereas for Kant the progress of reason is distinctly oriented toward a horizon, given that perception is nothing other than the projection of a space modeled after the seeing subject, Foucault follows a different spatial logic in his *Archaeology*.

The never-completed, never-wholly-achieved uncovering of the archive forms the general horizon to which belong the description of discursive formations, the analysis of positivities, and the mapping of the enunciative field. The right word—which is not that of the philologists—authorizes, therefore, the use of the term *archeology* to describe all of these searches. This term does not imply the search for a beginning; it does not relate analysis to geological excavation. It designates the general theme of a description that questions the already-said at the level of its existence:

of the enunciative function that operates within it, of the discursive formation and the general archive system to which it belongs. *Archaeology* describes discourses as practices specified in the element of the archive.[53] In contrast to Kant's horizon, which was modeled on an actual form of visibility, Foucault's horizon is more a form of horizon likeness. It does not presuppose the unity of a certain space; rather, Foucault's archeological archive lacks linear spatial coordinates. The timeline of progress does not proceed one-dimensionally into a future that opens up before the subject; "archaeology maps the temporal vectors of derivation"[54] of temporalities in the plural. Foucault describes the move into this other "more general space"[55] as a striding into the unknown: "One is forced to advance beyond familiar territory, far from the certainties to which one is accustomed, towards an as yet uncharted land and unforeseeable conclusion."[56]

Conclusion

In a short book on Foucault, Gilles Deleuze describes in detail how Foucault develops a language of the functional, how he illustrates power not through a concrete figure such as a sovereign or state, but simply as "cardinality" in a more abstract sense. For Foucault, power becomes "simply operational,"[57] or in the logic of mathematical imagery, power is embodied by the might of a curve. Deleuze systematically reconstructs Foucault's poetics of topology and his purely functional understanding of space. His commentary also addresses the question posed multiple times in this chapter, namely, to what extent Foucault's spatial thought and his conception of an archeological description take up and continue Kant's search for evidence and his poetics of vivid illustration.

"Analysis and illustration go hand in hand," Deleuze writes about *Discipline and Punish*;[58] more concretely, Foucault's texts are blueprints of diagrams, insofar as he attempts to describe the conditions under which statements can be made. Simply put, Foucault attempts to show how curves can be drawn. His mapping in diagrams[59] is necessary for the sole reason that the curves of statements are spatiotemporal multiplicities, "a display of the relations between forces which constitute power."[60]

Kant's diagrams are entirely different. As described at length in the previous section, the triangle functions as such a forceful example of a schematic hypotyposis and diagram because its concrete form could not be separated from its construction. To understand a triangle, one must know how triangles are constructed. But as in the case of the line, the second example that Kant so often makes use of, a second requirement for these images is a homogeneous, level space: a sheet of paper or a two-dimensional coordinate system. In drawing a line, as we have seen, Kant not only demonstrates the laws of reason; the line also illustrates the space that is the condition for such vivid images.

Foucault draws images too, and according to Deleuze, his diagrams are also a metaphor for his own writing process, or more specifically, an expression of his creativity: "There is no diagram that does not also include, besides the points which it connects up, certain relatively free or unbound points, points of creativity, change and resistance, and it is perhaps with these that we ought to begin in order to understand the whole picture."[61] If Kant's triangle was a symbol of the rational subject, each line a demarcation, Foucault's curve shows resistance and creativity. Euclidean figures have become topological spaces. This is Foucault's transgression of the time period initiated by Kant. They are both cartographic writers, but whereas Kant seems to survey a land unfolding before his eyes that is always presupposed as a given entity, Foucault maps a different space. Whereas Kant's maps are illustrations and representations, Foucault's cartography is productive, a mapping that generates the space it describes. For this kind of project, then, literature is not a hazard or a risk; on the contrary, it is the model for the invention of new spaces.

If we return to Foucault's *Archaeology* once more, this book is not just about abstract figures and pure geometry, but also about a subject, a narrator who is heir to practical geometry and the art of surveying, who loses himself in the archive as if in a labyrinth—the labyrinth of his own narration. The "viewpoint" is one of Foucault's favorite metaphors in this text, just like the change in perspective, a discourse of seeing that is accompanied by the narrator's description of his own path:

> I am not proceeding by linear deduction, but rather by concentric circles, moving sometimes towards the outer and sometimes towards the inner ones: beginning with the problem of discontinuity in discourse and of the uniqueness of the statement (the central theme), I have tried to analyse, on the periphery, certain forms of enigmatic groupings; but the principles of unification with which I was then presented …, forced me to return to the center, to that problem of the statement; to try to elucidate what is meant by the term statement.[62]

Even in the most abstract topological images of curves and their spaces, the path of the narrator is always there. And this narrative path is very different from Kant's cartographer, who was always oriented toward the horizon. Foucault's space is not a given sphere; his *Archaeology* describes images that depict their own emergence, as well as a narrative thread that follows this development as an observer, an observer who nonetheless loses himself in the labyrinth of his own archive. This is as if the disappearance and decentering of human beings is what this narrative is all about. Whereas in Kant geometry is a model for a rigorous discourse that the philosophical narrator seeks to follow as closely and carefully as possible by mapping the space of knowledge against literature, Foucault describes a different scenario: geometry becomes itself inventive and a means of constructing spaces, in which his cartographer wanders about, following the literariness of his own discourse.

Notes

1. Gernot Böhme, *Kants "Kritik der Urteilskraft" in neuer Sicht* (Frankfurt am Main: Suhrkamp, 1999), 110–111.

2. Immanuel Kant, *Critique of the Power of Judgment*, ed. Paul Guyer and Allen W. Wood (Cambridge: Cambridge University Press, 2000), §64, 242.

3. Immanuel Kant, "Monadology," in *Theoretical Philosophy 1755–1770*, ed. and trans. David Walford (Cambridge: Cambridge University Press, 1992), 51.

4. Immanuel Kant, "Inquiry Concerning the Distinctness of the Principles of Natural Theology and Morality," in *Theoretical Philosophy 1755–1770*, ed. and trans. David Walford (Cambridge: Cambridge University Press, 1992), A87; 264.

5. Kant, "Inquiry Concerning the Distinctness," A88; 265.

6. Immanuel Kant, *Critique of the Power of Judgment*, 225.

7. On this, see Rodolphe Gasché, "Überlegungen zum Begriff der Hypotypose bei Kant," in *Was heisst "Darstellen"?*, ed. Christiaan Hart Nibbrig, 152–174 (Frankfurt/Main: Suhrkamp, 1994); Rüdiger Campe, "Vor Augen Stellen: Über den Rahmen rhetorischer Bildgebung," in *Poststrukturalismus: Herausforderung an die Literaturwissenschaft*, ed. Gerhard Neumann, 208–225 (Stuttgart: J. B. Metzler, 1997), here 210–212.

8. Kant, *Critique of the Power of Judgment*, 226.

9. Immanuel Kant, *Critique of Pure Reason*, trans. Paul Guyer and Allen W. Wood (Cambridge: Cambridge University Press, 1999), A116–117.

10. Kant, *Critique of Pure Reason*, B137–138.

11. Kant, *Critique of Pure Reason*, B137.

12. Kant, *Critique of Pure Reason*, B154–155.

13. Cf. Friedrich Kaulbach, "Schema, Bild und Modell nach den Voraussetzungen des Kantischen Denkens," in *Kant: Zur Deutung seiner Theorie von Erkennen und Handeln*, ed. Gerold Prauss, 105–129 (Cologne: Kiepenheuer & Witsch, 1973), here 105–106. Here Kaulbach compares reason's act of representing with the process of writing. See also Ludwig Jäger, "Das schreibende Bewusstsein," in *Philosophie der Schrift*, ed. Elisabeth Birk and Jan Georg Schneider, 97–122 (Tübingen: Max Niemeyer Verlag, 2009).

14. On bringing-before-the-eyes a syntagma of speech and the development of drawing, see also Campe, "Vor Augen Stellen," 217.

15. Immanuel Kant, "Dreams of a Spirit-Seer Illustrated by Dreams of Metaphysics," in *Theoretical Philosophy 1755–1770*, ed. and trans. David Walford (Cambridge: Cambridge University Press, 1992), 983.

16. Kant, "Dreams of a Spirit-Seer," 984.

17. Kant, *Critique of Pure Reason*, Bxxv.

18. Kant, *Critique of Pure Reason*, B9.

19. Kant, *Critique of Pure Reason*, B33.

20. Kant, *Critique of Pure Reason*, B294–295. In his commentary on this passage Goetschel contrasts the cartographer Kant with the geographer Hume. Cf. Willi Goetschel, *Kant als Schriftsteller* (Vienna: Passagen Verlag, 1990), 119.

21. Kant, *Critique of Pure Reason*, Axviii.

22. Cf. Kant, *Critique of Pure Reason*, Axix.

23. Kant, *Critique of Pure Reason*, B686–689.

24. Kant, *Critique of Pure Reason*, B787.

25. Kant, *Critique of Pure Reason*, B790.

26. Cf. Franco Farinelli, "Von der Natur der Moderne: Eine Kritik der kartographischen Vernunft," in *Räumliches Denken*, ed. Dagmar Reichert, 267–301 (Zurich: Zürcher Hochschulforum, 1996), here 280.

27. See also Jäger, "Das schreibende Bewusstsein," 98.

28. Michel Foucault, *The Order of Things: An Archaeology of the Human Sciences* (New York, Vintage Books, 1994), x.

29. Michel Foucault, *The Archaeology of Knowledge*, trans. Sheridan Smith (New York: Pantheon, 1972), 191.

30. Foucault, *The Order of Things*, 245.

31. Cf. Foucault, *The Order of Things*, 344–346.

32. Foucault, *The Order of Things*, 351.

33. Philipp Sarasin, *Michel Foucault zur Einführung* (Hamburg, Junius, 2005), 103.

34. Gilles Deleuze, *Foucault*, ed. and trans. Seán Hand (Minneapolis: University of Minnesota Press, 1986), 18.

35. Foucault, *Archaeology*, 17.

36. Foucault, *Archaeology*, 5.

37. Deleuze, *Foucault*, 13, 78.

38. Foucault, *Archaeology*, 5.

39. Foucault, *Archaeology*, 17.

40. Foucault, *Archaeology*, 40.

41. Foucault, *Archaeology*, 12.

42. Foucault, *Archaeology*, 15.

43. Foucault, *Archaeology*, 10.

44. Foucault, *Archaeology*, 54.

45. Michel Foucault, "The Language of Space." In *Space, Knowledge, and Power: Foucault and Geography*, ed. Jeremy W. Crampton and Stuart Elden (Burlington, VT: Ashgate, 2007), 163.

46. Foucault, "The Language of Space," 167.

47. Foucault, *Archaeology*, 128.

48. Foucault, *Archaeology*, 116.

49. Foucault, *Archaeology*, 155.

50. Foucault, *Archaeology*, 42.

51. Foucault, *Archaeology*, 184.

52. Foucault, *Archaeology*, 157.

53. Foucault, *Archaeology*, 131.

54. Foucault, *Archaeology*, 169.

55. Foucault, *Archaeology*, 26.

56. Foucault, *Archaeology*, 39.

57. Deleuze, *Foucault*, 27.

58. Deleuze, *Foucault*, 32.

59. Deleuze, *Foucault*, 34.

60. Deleuze, *Foucault*, 36.

61. Deleuze, *Foucault*, 44.

62. Foucault, *Archaeology*, 114.

Bibliography

Barthes, Roland. "An Introduction to the Structural Analysis of Narrative," *New Literary History*, vol. 6, no. 2, 1975, 237–272.

Böhme, Gernot. *Kants "Kritik der Urteilskraft" in neuer Sicht*. Frankfurt am Main: Suhrkamp, 1999.

Campe, Rüdiger. "Vor Augen Stellen: Über den Rahmen rhetorischer Bildgebung." In *Poststrukturalismus: Herausforderung an die Literaturwissenschaft*, ed. Gerhard Neumann, 208–225. Stuttgart: J. B. Metzler, 1997.

Deleuze, Gilles. *Foucault*. Ed. and trans. Seán Hand. Minneapolis: University of Minnesota Press, 1986.

Farinelli, Franco. "Von der Natur der Moderne: Eine Kritik der kartographischen Vernunft." In *Räumliches Denken*, ed. Dagmar Reichert, 267–301. Zurich: Zürcher Hochschulforum, 1996.

Foucault, Michel. *The Archaeology of Knowledge*. Trans. Sheridan Smith. New York: Pantheon, 1972.

Foucault, Michel. "The Language of Space." In *Space, Knowledge, and Power: Foucault and Geography*, ed. Jeremy W. Crampton and Stuart Elden. Burlington, VT: Ashgate, 2007.

Foucault, Michel. *The Order of Things: An Archaeology of the Human Sciences*. New York: Vintage Books, 1994.

Gasché, Rodolphe. "Überlegungen zum Begriff der Hypotypose bei Kant," in *Was heisst "Darstellen"?*, ed. Christiaan Hart Nibbrig, 152–174. Frankfurt/Main: Suhrkamp, 1994.

Goetschel, Willi. *Kant als Schriftsteller*. Vienna: Passagen Verlag, 1990.

Jäger, Ludwig. "Das schreibende Bewusstsein." In *Philosophie der Schrift*, ed. Elisabeth Birk and Jan Georg Schneider, 97–122. Tübingen: Max Niemeyer Verlag, 2009.

Kant, Immanuel. *Critique of the Power of Judgment*. Ed. Paul Guyer and Allen W. Wood. Cambridge: Cambridge University Press, 2000.

Kant, Immanuel. *Critique of Pure Reason*. Trans. Paul Guyer and Allen W. Wood. Cambridge: Cambridge University Press, 1999.

Kant, Immanuel. "Dreams of a Spirit-Seer Illustrated by Dreams of Metaphysics." In *Theoretical Philosophy 1755–1770*, ed. and trans. David Walford. Cambridge: Cambridge University Press, 1992.

Kant, Immanuel. "Inquiry Concerning the Distinctness of the Principles of Natural Theology and Morality." In *Theoretical Philosophy 1755–1770*, ed. and trans. David Walford. Cambridge: Cambridge University Press, 1992.

Kant, Immanuel. "Monadology." In *Theoretical Philosophy 1755–1770*, ed. and trans. David Walford. Cambridge: Cambridge University Press, 1992.

Kaulbach, Friedrich. "Schema, Bild und Modell nach den Voraussetzungen des Kantischen Denkens." In *Kant: Zur Deutung seiner Theorie von Erkennen und Handeln*, ed. Gerold Prauss, 105–129. Cologne: Kiepenheuer & Witsch, 1973.

Sarasin, Philipp. *Michel Foucault zur Einführung*. Hamburg: Junius, 2005.

5
The Language of Cartography: Borges as Mapmaker

Bruno Bosteels

A Taste for Maps

Maps and the art of mapmaking have always held a special interest for the Argentine writer Jorge Luis Borges. In "Borges and I," he defines his intimate sense of self—the "I" from the title of this brief prose piece—in terms of a series of aesthetic tastes or pleasures, neatly divided along the lines of time and space as the fundamental forms of intuition in a quasi-Kantian sense: "I like hourglasses, maps, eighteenth-century typography, etymologies, the taste of coffee, and Stevenson's prose."[1] According to one of Borges's earliest essays, moreover, there would be something inherently poetic about maps: "What's more, there are things that are poetic by merely implying a destiny: for example, the map of a city, a rosary, the names of two sisters."[2] But the language of mapmaking serves Borges above all to test the limits of language and representation in general. In this sense, the Argentine writer is no doubt best known for having pushed the ideal of cartographic representation to the absurd limit of an absolute map on a scale of one to one. Much less known is the fact that from an early age he was also fascinated by the enormous potential of language to keep on growing with the world, like a cartographer who adds whole new territories to the world as we know it. Thus, in his evaluation of language as cartography, Borges oscillates between a critical and a utopian perspective. In his own words, this debate is best understood in terms of the scholastic debate between realism and nominalism—a debate in which a possible mediating role is reserved for pragmatism as defined by the New England philosopher William James.

On several occasions in his essayistic production Borges indeed has recourse to an almost identical little vignette in which he summarizes the whole of Western philosophy as a set of millenarian variations on the quarrel between realists and nominalists, archetypically associated with the proper names of Plato and Aristotle. The image of "the map of the universe" invariably appears in this description as a metaphor representative of the Platonist or realist position with regard to the nature of language and representation, as opposed to the Aristotelian or nominalist view. For the latter, language would offer only "a system of arbitrary symbols," to quote the version of the vignette that Borges provides in his essay "From Allegories to Novels," included in *Other Inquisitions*:

Coleridge observes that all men are born Aristotelian or Platonist. The latter know by intuition that ideas are realities; the former, that they are generalizations; for the former, language is nothing but a system of arbitrary symbols; for the latter, it is the map of the universe. The Platonist knows that the universe is somehow a cosmos, an order, which, for the Aristotelian, may be an error or a figment of our partial knowledge. Across the latitudes and the ages, the two immortal antagonists change their name and language; one is Parmenides, Plato, Spinoza, Kant, Francis Bradley; the other is Heraclitus, Aristotle, Locke, Hume, William James. In the arduous schools of the Middle Ages they all invoke Aristotle, the master of human reason (*Convivio*, IV, 2), but the nominalists are Aristotle; the realists, Plato.[3]

In this summary the distinction between realism and nominalism gradually unfolds on three different levels, or in three distinct but intimately related fields of debate. I will enumerate these levels or fields in the order in which they appear in Borges's description, which is similar but not identical to the three questions asked at the origin of the scholastic dispute by Porphyry, in his commentary on Aristotle. First, at the level of epistemology, or in terms of the theory of knowledge, the distinction concerns the nature of ideas or universals as subsisting in reality or existing only as mental constructs. In the words of one of the few commentators quoted by Borges, "Hence there are two possible replies in terms of the Porphyrian formula: genera and species either exist as such, that is to say, humanity, life, rationality, etc., are subsistent realities (*subsistentia*), or else they are ultimately just mental facts (*nuda intellecta*)."[4] This first contrast, of course, is the one that will give rise to the labels for the two fundamental sides in this centuries-old debate, based on the scholastic opposition between *in re* and *in nomine* in Latin. Second, at the level of semiology, or in terms of the theory of language, the distinction between realism and nominalism concerns the nature of the sign as either motivated or arbitrary. In fact, using the modern terminology associated with Ferdinand de Saussure and already implicit in Borges's explanation, we can say that the Aristotelian position is based precisely on the famous principle of the arbitrariness of the sign, not in the sense of being random but in the sense of having been established by convention. The Platonist position sees the sign as essentially motivated, based on a relation of correspondence or mimetic analogy with the thing signified. The latter doctrine is also referred to as Cratylism, after the title character of the dialogue in which Plato offers some of his most illuminating insights into the nature of language.[5] Third and finally, at the level of cosmology, or in terms of the ontological theory of being, the distinction concerns the nature of all that is as either amounting to an ordered totality or being at best a figment of our partial knowledge of the world. Here *somehow* or *in some way* (*de algún modo*) is a technical term on which Borges relies quite systematically, both in his essays and in his short stories, to describe the realist position for which the particular in some way is, or participates in, the universal. His use of the terms *universe* and *world*, in fact, is also fairly systematic throughout his writings and in many instances can be shown to overlap with the scholastic distinction between realism and nominalism.

I might add that in other contexts, too, Borges's self-ascribed "basic skepticism"[6] takes the form of a questioning of the fundamental premises of our thought in these three domains that are the theories of knowledge, being, and the sign. In "The Analytical Language of John Wilkins," to mention only one key example, he relies on a radicalization of doubt that likewise moves from the epistemological to the ontological in order finally to account for the semiological principle according to which "there is no classification of the universe that is not arbitrary and conjectural."[7] To a first explanation of this principle ("The reason is simple: we do not know what the universe is"), Borges thus adds a second, even more radical expression of skepticism that concerns not just our limited knowledge of the world but the limit that belongs to the very being of all that is as an ordered universe or cosmos ("We must even go further; we must suspect that there is no universe in the organic, unifying sense inherent in that ambitious word"[8]).

Borges was clearly proud of his vignette on realism and nominalism, as witnessed by the fact that it appears on so many occasions in his work. To my knowledge, the first time he used the summary inspired by Coleridge was in his preface to the 1945 Spanish translation of William James's *Pragmatism*. But Borges also implicitly refers to the dispute between Platonists and Aristotelians in "Our Poor Individualism," first published in the magazine *Sur* in July 1946. In addition to the version in "From Allegories to Novels," first published in the newspaper *La Nación* in August 1949, he would repeat almost verbatim the same long paragraph in "The Nightingale of Keats," first published in *La Nación* in December 1951 and subsequently included in *Other Inquisitions*. What varies from one instance to the other in these cases does not concern the exact nature as much as the various applications of Coleridge's original statement. Borges thus continues and extends the paragraph quoted above with different versions of how and where the opposition between realism and nominalism can be said to be at work today, well beyond the narrow confines of scholastic philosophy alone. In "The Nightingale of Keats," for example, he ascribes a realist or nominalist character to certain national cultures. "Men, said Coleridge, are born Aristotelians or Platonists; one can state of the English mind that it was born Aristotelian. For that mind, not abstract concepts but individual ones are real; not the generic nightingale, but concrete nightingales," Borges remarks, not without adding a caveat: "Please do not read reprobation or disdain into the foregoing words. The Englishman rejects the generic because he feels that the individual is irreducible, unassimilable, and unique. An ethical scruple, not a speculative incapacity, prevents him from trafficking in abstractions like the German."[9] For Borges, the German mind would be predominantly realist or Platonist, and the English national character, more inclined toward nominalism or Aristotelianism. Similarly, in "Our Poor

Individualism," Borges contrasts the German, or even more broadly, the European and North American aptitude for abstraction with the attitude supposedly typical of the Argentine, who would not be able to identify with speculative notions such as the State. "Aphorisms like Hegel's–'The State is the reality of the moral idea'–seem like a vicious joke," Borges writes, before returning to some of his more familiar terms for defining the difference between realists and nominalists: "For the European the world is a cosmos where each person corresponds intimately to the function he performs; for the Argentine it is a chaos. The European and the North American believe that a book which has been awarded any sort of prize must be good; the Argentine acknowledges the possibility that it may not be bad, in spite of the prize."[10]

In addition to this first dividing line, drawn along national or geographic stereotypes, Borges also sees the distinction between realism and nominalism as marked by a profoundly historical or epochal break. This becomes particularly visible in the essay "From Allegories to Novels," where he makes the claim that the passage from realism to nominalism, which he identifies with the passage from allegories to novels or from genera and species to individuals, marks the transition of Western culture to modernity. About the fate of scholastic realism he observes: "A thesis that is inconceivable now seemed obvious in the ninth century, and it somehow endured until the fourteenth century. Nominalism, which was formerly the novelty of a few, encompasses everyone today; its victory is so vast and fundamental that its name is unnecessary. No one says that he is a nominalist, because nobody is anything else."[11] In fact, already in one of his first and most ambitious prose texts, the essay "History of Eternity," Borges had observed that nowadays the victory of nominalism has become so complete as to go without saying: "Now, like the spontaneous and bewildered prose-speaker of comedy, we all practice nominalism *sans le savoir*, as if it were a general premise of our thought, an acquired axiom. Useless, therefore, to comment on it."[12] Instead of opting for the usual years of 1789, 1776, or even 1492, Borges goes so far as to propose an "ideal date" for when this fateful beginning of modernity occurred: "That day in 1382 when Geoffrey Chaucer, who perhaps did not believe he was a nominalist, wished to translate a line from Boccaccio into English, *E con gli occulti ferri i Tradimenti* ('And Treachery with hidden weapons'), and he said it like this: 'The smyler with the knyf under the cloke.'"[13]

Third and finally, the great scholastic dispute about universals also has ethical and political applications, which Borges describes in terms of the relation between the individual and the State–similar to the way realists and nominalists in the Middle Ages struggled over the relation between individuation and the dogma of the Trinity under the watchful eye of the Christian Church. Just as according to "The Nightingale of Keats" the British mind stereotypically shies away from abstractions out of an ethical scruple, not because of a speculative incapacity, so too according to "Our Poor Individualism," Borges's deep skepticism with regard to the abstraction of the State may as yet have a political application, if we consider that our time according to him is marked by the twin evils of so-called totalitarianism, signaled by Hitler and Stalin: "The most urgent problem of our time (already proclaimed with prophetical clarity by the almost forgotten Spencer) is the gradual interference of the State in the acts of the individual; in the struggle against this evil—called communism and fascism—Argentine individualism, which has perhaps been useless or even harmful up to now, would find justification and a positive value."[14] Neither innate nor purely historical, the difference between realism and nominalism thus would be ethically and politically overdetermined. In no text is this made more poignant than in "Deutsches Requiem," a story first published in February 1946 in the magazine *Sur* and later included in *The Aleph*, in which Borges suggests that not just Nazism but all ideologies presuppose a realist or Platonist conception in which the individual in some way is the species of all those who identify with a supraindividual cause such as that of Nazi Germany. "It has been said that all men are born either Aristotelians or Platonists," the Nazi narrator, Otto Dietrich zur Linde, reflects in his would-be testament, going back to Coleridge's distinction but without mentioning his source by name, no doubt because this would ally him all too closely with the English. "Down through the centuries and latitudes, the names change, the dialects, the faces, but not the eternal antagonists. Likewise, the history of nations records a secret continuity," with the poet David Jerusalem—one of the Jewish victims in the camp—symbolizing the nominalist celebration of each thing in its strict singularity, and zur Linde in the manner of Platonic realism seeing himself as being *de algún modo* representative of all of Nazi Germany, albeit on the eve of the latter's defeat and his own death: "Tomorrow, by the time the prison clock strikes nine, I shall have entered the realms of death; it is natural that I should think of my elders, since I am come so near their shadow—since, somehow, I am they."[15]

There are, however, two important differences that set apart Borges's interpretation from the original idea borrowed from Coleridge. On the one hand, whereas the English romantic firmly identifies himself with the Platonic position and bestows only faint praise on Aristotle, most critics agree that Borges is a quintessential nominalist, albeit one haunted by occasional nostalgia for the mystical forces of Platonism, especially in the guise of Neoplatonism or secondary Cratylism. Thus, in his original vignette, dated July 1830, Coleridge had written: "Aristotle was, and still is, the sovereign lord of the understanding; the faculty judging by the senses. He was a conceptualist, and never could raise himself into that higher state, which was natural to Plato, and has been so to others, in which the understanding is distinctly contemplated, and, as it were, looked down upon from the throne of actual ideas, or living, inborn, essential truths."[16]

Borges, by contrast, always describes himself as part of the nominalist tradition that in his eyes ranges from Roscelinus to Fritz Mauthner and from David Hume to

Herbert Spencer. As the Argentine literary critic Jaime Rest concludes in an overview of the significance of nominalism for Borges's writing and thinking as a whole, "Thought is, for Borges, always language, always discourse; and language is always imperfect, artificial. Even though he occasionally denied it, this is what placed Borges on the opposite side of Platonism, which on the basis of our intellect pretended to erect a valid metaphysical order, constituted with pure ideas."[17] Borges, in other words, was never as confident as Coleridge that finite reason would have the capacity to overcome the fundamental maladjustment between thought and reality through the imperfections of human language.

On the other hand, a second difference that sets apart these two authors is the fact that for Coleridge there are no middle grounds between the positions of realism and nominalism, nor any possibility of changing from one to the other: "I do not think it possible that any one born an Aristotelian can become a Platonist; and I am sure no born Platonist can ever change into an Aristotelian. They are the two classes of men, beside which it is next to impossible to conceive a third. The one considers reason a quality, or attribute; the other considers it a power. I believe that Aristotle never could get to understand what Plato meant by an idea."[18] For Borges, by contrast, the history of philosophy is made up of an infinite number of third ways or intermediate positions between the theses of realism and nominalism, even though in the end he too must admit that we might be dealing with two fundamentally distinct ways of being in the world:

> As might be supposed, the passage of so many years multiplied the intermediate positions and the distinctions to the point of infinity. Nevertheless, for realism the universals (Plato would say the ideas, forms; we call them abstract concepts) were fundamental; and for nominalism, the individuals. The history of philosophy is not a vain museum of distractions and verbal games; the two theses probably correspond to two manners of intuitively perceiving reality.[19]

In any case what matters for us, which concerns the relation between language, literature, and cartography, is not changed by these discrepancies. In Borges's rewriting of Coleridge's fragment, cartography consistently describes a realist or Platonist view of language, based on a relation of correspondence or mimetic reflection between signs and things. Just as realists are said to take language to be the map of the universe, we could also put this the other way around by concluding that to rely on the cartographic metaphor seems in and of itself to be the mark of a deeply ingrained Platonism. In "Three Versions of Judas," for instance, Borges frames cartography precisely in terms of such a mimetic relation of reflection, or cosmic mirroring, between the earth and the heavens: "As below, so above; the forms of earth correspond to the forms of heaven; the blotches of the skin are a map of the incorruptible constellations; Judas is somehow a reflection of Jesus."[20]

This last example already gives us a hint of Borges's basic approach to the question of language and representation in light of the cartographic metaphor. Just as William of Ockham reproached his adversary, the realist Duns Scotus, for the blasphemy of conflating Judas and Christ under the same universal of humanity, so too will Borges draw the absurd conclusion that, if all individuals in some way are, or participate in, the universal, then according to the realist thesis Judas must in some way reflect Jesus and partake in the divine economy of redemption. Much like the nominalists of the late Middle Ages, Borges thus will proceed to interrogate the possibility of language as cartography, not by proving the validity of the Aristotelian thesis but by showing the absurdities that follow from a consistent application of the principles of Platonic realism.

The Aporias of Cartographic Mimetology

Borges, no doubt, is best known for this nominalist reductio ad absurdum of cartographic realism or mimeticism. Especially in the 1970s and 1980s, his quasi-deconstructive exercises in showing the inevitable aporias inherent in the art of mapmaking attracted the attention of a veritable *Who's Who* list of contemporary philosophers and theorists, most notably in France: from Louis Marin to Jean-François Lyotard and from Félix Guattari to Jean Baudrillard. In fact, these references became so commonplace that by the early 1990s Fredric Jameson in an uncharacteristic disparagement felt justified to write that "in the postmodern such essentially transcoded objects or symbiotic constructions as the famous Borges map (that always springs to mind on such occasions) or the images of Magritte cannot be used as figures or allegories for anything; and in the high theory of the postmodern they have all the vulgarity and lack of 'distinction' of Escher prints on the walls of middlebrow college students."[21] What Jameson seems to have in mind with "the famous Borges map" actually includes two very different treatments of the paradoxes of cartographic representation. The first and by far the most cited of these appears under the title "On Rigor in Science" in the final section, "Museum," of Borges's collection *El hacedor* (literally "The Maker," translated into English as *Dreamtigers*), later also repeated in "Naturalism Revived," one of the *Chronicles of Bustos Domecq* coauthored with Adolfo Bioy Casares. Here Borges reveals a principle of ruin and destruction at the heart of the ultimate map, drawn on the scale of one to one, of an unnamed Empire:

> ... In that Empire, the Art of Cartography reached such Perfection that the map of one Province alone took up the whole of a City, and the map of the Empire, the whole of a Province. In time, those Unconscionable Maps did not satisfy and the Colleges of Cartographers set up a Map of the Empire which had the size of the Empire itself and coincided with it point by point. Less Addicted to the Study of Cartography, Succeeding Generations understood that this Widespread Map was useless and not without Impiety they abandoned it to the inclemencies of the Sun and of the Winters. In the deserts of the West some mangled Ruins of the Map last on, inhabited by

Animals and Beggars; in the whole Country there are no other relics of the Disciplines of Geography.

Suárez Miranda: *Viajes de Varones Prudentes*, Book Four, Chapter XLV, Lérida, 1658.[22]

While the influence of Lewis Carroll's *Sylvie and Bruno Concluded* on this passage is well documented, much less remarked is the fact that Borges borrows a great deal of his vocabulary for this apocryphal quotation from a section in the second part of *Don Quixote* where Cervantes's character draws a long distinction between the true knight errant and those knights of the court who never leave their room but travel merely by looking at a map. The latter, in this sense, could very well be considered precursors of the "prudent men" alluded to in the title of the work attributed to the imaginary figure of Suárez Miranda:

> Not all knights can be courtiers, and not all courtiers can or should be knights errant: the world needs both kinds, and even if we are all knights there is a very great difference between one sort and the other; because courtiers, without leaving their chambers or the confines of the court, stroll all over the world just by perusing a map, which does not cost them a penny or cause them to suffer any heat or cold, or hunger or thirst; but we, the true knights errant, by night and by day, on foot and on horseback, measure the whole earth with our own paces; and we do not merely know our enemies from their portraits but in their own persons, and at every turn and on all occasions we attack them, without concerning ourselves over trifles like the laws of challenges: whether or not one lance or sword is longer than the other, whether the enemy has holy relics on him, or is practising some other form of covert trickery, whether or not the sun should be divided and carved up so that it shines into the eyes of neither, and other ceremonials of this sort that are involved in challenges between two individuals, which you do not know about and I do.[23]

Here as in the case of Coleridge, Borges's allegiances would appear to be the exact opposite of his favorite source: whereas Don Quixote warns his housekeeper that the true knight errant should not look at such childish trifles as maps or relics, Borges's parable appeals to the perfection of the cartographic drive, only to revel with an almost pataphysical delight in the moment of reversibility where its ideal operation turns into a monstrous, if also futile, aberration.

Often confused with "On Rigor in Science," there is a second instance alluded to under the vague heading of "the famous Borges map." This instance is better known from the essay "Partial Enchantments of the *Quixote*," in *Other Inquisitions*, which includes a passage adapted from Josiah Royce's *The World and the Individual*. Alfred Korzybski and Charles Sanders Peirce would also mention in their work and that Borges had already quoted in an earlier essay of his, "When Fiction Lives in Fiction," published in June 1939 in the magazine *El Hogar*. Here the project is not exactly a full-scale map but rather one that, falling just short of the ideal of one-to-one correspondence, endlessly includes itself in an abyssal relation of self-referentiality:

> But now suppose that this our resemblance is to be made absolutely exact, in the sense previously defined. A map of England, contained within England, is to represent, down to the minutest detail, every contour and marking, natural or artificial, that occurs upon the surface of England. ... For the map, in order to be complete, according to the rule given, will have to contain, as a part of itself, a representation of its own contour and contents. In order that this representation should be constructed, the representation itself will have to contain once more, as a part of itself, a representation of its own contour and contents; and this representation, in order to be exact, will have once more to contain an image of itself: and so on without limit.[24]

While they have given way to a slew of readings and misreadings, to the point of losing all connection to the original texts themselves, the upshot of these two formulations of the paradoxes of cartographic representation is not difficult to grasp. Cartography as Platonic mimesis not only collapses into an infinitely smaller map within the map in an endless series of self-embedded representations. A logic of supplementarity also impedes cartographers from stepping beyond the borders of the map, so to speak, in order to contain the territorial totality from an exterior Archimedean point of view. Even if a rigorous map of the entire territory were to be drawn on a scale other than the futile 1:1, this map would have to represent itself within its own structure, just as in the opposite direction England itself could be included in an ever larger or more encompassing map, and so on, virtually to no end. As a "layman's version" of Gödel's theorem, Lyotard writes in *The Postmodern Condition*, Borges's map shows us "the de facto impossibility of ever achieving a complete measure of any given state of a system," because "a complete definition of the initial state of a system (or of all the independent variables) would require an expenditure of energy at least equivalent to that consumed by the system to be defined."[25]

A Map of Pure Experience

And yet, a deconstruction *avant la lettre* of the aporias of Platonic mimeticism is not all there is to Borges's treatment of the language of cartography. The map metaphor also appears in some of his earliest essays, in which the view of language is far from leading only to the kind of critical or protodeconstructive conclusions for which, starting in the late 1960s, the Argentine writer was to become so famous. On the contrary, Borges's views in these instances exude an exhilarating sense of confidence in the powers of language not just to guide us through a world of appearances but also to add whole new provinces to being as such: "I am insisting on the inventive character of any language, and I do so intentionally," he writes: "Language constructs realities. The various disciplines of intelligence have engendered worlds of their own and possess an exclusive vocabulary to describe them."[26] Borges need not remain trapped in the scholastic alternative between Platonism and Aristotelianism, as though he could do no more than continue oscillating between a nominalist critique of the limits of language, on the one hand, and a mystical longing for the perfect word, on the other. For Borges, there exists another understanding of language, in which a privileged role is also reserved for the cartographic metaphor. Much of Borges's writing—both in his essays and in his short stories—seems determined to search for this intermediate position, one that strictly speaking would be neither realist nor nominalist but instead owes profound debts to the pragmatist philosophy of William James.

Here we may be confronted with a case of the synchronicity of the nonsynchronous. When French thinkers as diverse as Michel Foucault, Gilles Deleuze, and Jacques Derrida, in the late 1960s, all cited Borges as an authority on issues related to language, the history of philosophy, or the play of metaphor in key canonical works of theirs such as *The Order of Things*, *Difference and Repetition*, or *Writing and Difference*, they were certainly unaware of the enormous distance that separated their theories from the Argentine writer's original philosophical background.[27] By this I mean not just the large number of authors mentioned by Borges or exploited for aesthetic effect but the tradition of thought to which he remained loyal at a much more fundamental—ethical or even existential—level, which is the Anglo-American tradition that runs the gamut from British empiricism to New England pragmatism. This is the tradition that Borges in his "Preliminary Note" to the Argentine translation of *Pragmatism* sees as the prolongation of the nominalist position in philosophy. Basing himself on Coleridge's strict dichotomy between Platonism and Aristotelianism, Borges thus draws a straight line from Ockham, via Berkeley and Hume, to William James: "The English nominalism of the fourteenth century reemerges in the scrupulous English idealism of the eighteenth century; the economy of Ockham's formula, *entia non sunt multiplicanda praeter necessitatem*, permits or prefigures the no less taxative *esse est percipi*. From 1881 onward, William James enriches this lucid tradition."[28] Even more significantly, James's pragmatism would be ethically superior to all other philosophies, from Parmenides's thought of immobile being to Bradley's absolute idealism: "For an aesthetic criterion, the universes of other philosophies might be superior (James himself, in the fourth conference of this volume, speaks of 'the music of monism'); ethically, William James is superior. He is the only one, perhaps, for whom the human beings have something to do."[29] Coming from an author who tends to express his essential skepticism with the verb *descreer*, literally, "to disbelieve" or "to unbelieve" (whether in God, politics, or democracy), this is a rare and for this reason all the more astonishing confession of Borges's profound philosophical allegiance to the author of *A Pluralistic Universe* and *The Will to Believe*.

Borges's texts thus introduce something like a Trojan horse into the city of Paris as the capital of so-called French theory. When everyone was busy looking for anticipations of a structuralist, poststructuralist, or deconstructive discourse that would have been present in *Ficciones* or *Other Inquisitions*, few could have suspected that they were unwittingly being seduced by someone who from the 1920s onward had been an admirer of that same philosophy that most European thinkers still today tend to associate with the American way of life. Of all the French philosophers who turned to Borges in the late 1960s, only Deleuze—whose first book, *Empiricism and Subjectivity*, was dedicated to the philosophy of David Hume—may have been attuned to the Argentine's own mindset. In fact, aside from being among the few philosophers in France capable of spelling Charles S. Peirce's last name correctly, at the end of his life Deleuze was also not surprisingly the one who went so far as to place William James on the same level as Karl Marx, with the latter doing for the proletarian in Europe and the Soviet Union what the former would have done for the migrant in the United States. "The messianism of the nineteenth century has two heads and is expressed no less in American *pragmatism* than in the ultimately Russian form of socialism," Deleuze concludes in an essay on Herman Melville. And he continues: "Pragmatism is misunderstood when it is seen as a summary philosophical theory fabricated by Americans. On the other hand, we understand the novelty of American thought when we see pragmatism as an attempt to transform the world, to think a new world or new man insofar as they *create themselves*."[30]

We need not enter into detail about how Borges as a young boy was first introduced to the problems of philosophy by his father, Jorge Guillermo Borges, who was a fervent admirer of James's *Psychology*, the two-volume version as well as the abridged course.[31] Nor do we need to revisit the question of the almost mythical correspondence that another of Borges's mentors, the Socrates-like figure of Macedonio Fernández, briefly maintained with James.[32] Instead, what matters for our purposes concerns the way a firm commitment both to the pragmatic method for defining truth and to the philosophy of radical empiricism also profoundly changes the evaluation of language associated with the cartographic metaphor. In fact, this change in perspective presupposes a complete overhaul of the typical view, which defines language on the basis of a relation of correspondence between words and things, just as the process of thought would be defined by a relation of adequacy between concepts and things, or between a subject and an object.

In keeping with a form of radical empiricism, what exists is first of all reduced to a Jamesian "world of pure experience" in which the objective and subjective poles, or matter and spirit, have not yet been carved out. As Borges explains in his preface to *Varieties of Religious Experience*, a book chosen to be among the one hundred volumes of his *Personal Library*: "James affirmed that the elementary substance of what we call the universe is experience and the latter is anterior to the categories of subject and object, of knower and known, of spirit and matter. This curious solution of the problem of being is, obviously, closer to idealism than to materialism, to Berkeley's divinity than to the atoms of Lucretius."[33] With Hume doing for mind or spirit what Berkeley does for matter or reality, Borges early on in his essays or inquiries lets both of these polar opposites in the representational worldview collapse at the same time: "The object lapses together with the subject. Both of these ponderous nouns, matter and spirit, vanish at the same time as life changes into an intricate jumble of affective states, a dream without a dreamer."[34] The result is a world in which appearance is no longer merely the appearance of a hidden essence or the phenomenon of an ungraspable thing in itself: "Reality has no need of other realities to bolster it. There are no divinities hidden in the trees, nor any elusive thing-in-itself behind appearances, nor a mythological self that orders our actions. Life is true appearance."[35] Deleuze would call this the plane of life as pure immanence and, like Borges, he too would have found a friend and ally for this endeavor in William James's philosophy of radical empiricism.

There is no secret truth beyond the jumble of appearances, no hidden world of meaning behind the visible one, no *Hintenwelt* or ungraspable *Ding an sich*; only life as true appearance, or reality working in open mystery, as the young Borges also writes, this time borrowing a sentence from Macedonio Fernández: "La realidad trabaja en abierto misterio."[36] What language does to this world of pure experience cannot be described as an imitation that would copy the true nature of already existing things; instead, it provides the fundamental orientation that first makes things into what they are. Language, in other words, is above all a guide and an abridgment that leads us through a world of true appearance. This supposes a relation to the world measured in terms of efficacy rather than adequacy, a relation of hypothetical inferences rather than one based on the correspondence or copy theory of truth. Above all, it means that words, concepts, and things are constantly in motion, being made or produced on the go, literally thrown together in a series of open-ended conjectures, instead of remaining landlocked in a mimetic relation of analogy, imitation, or reflection: "The world stands really malleable, waiting to receive its final touches at our hands. Like the kingdom of heaven, it suffers human violence willingly. Man *engenders* truth upon it." James also concludes: "No one can deny that such a role would add both to our dignity and to our responsibility as thinkers. To some of us it proves a most inspiring notion."[37]

In keeping with the pragmatic definition of truth as made rather than found, engendered rather than discovered, words and concepts are maps that guide us through a world of pure appearance. Borges describes this process of abridgment and orientation in a free paraphrase of passages from William James's definition of truth:

> The world of appearances is a jumble of shifting perceptions. The vision of a rustic sky, that persistent aroma sweeping the fields, the bitter taste of tobacco burning one's throat, the long wind lashing the road, the submissive rectitude of the cane around which we wrap our fingers, all fit together in our consciousness, almost all at once. Language is an efficient ordering of the world's enigmatic abundance. Or, in other words, we invent nouns to fit reality. We touch a sphere, see a small heap of dawn-colored light, our mouths enjoy a tingling sensation, and we lie to ourselves that those three disparate things are only one thing called an orange.[38]

Or, to use the almost identical description in "A Study of Metaphors," in which Borges expressly returns to the metaphor of the map:

> Language is an efficacious ordering of this enigmatic abundance of the world. What we call a noun is nothing but an abbreviation of adjectives and, often, their fallacious probability. Instead of saying cold, hurting, unbreakable, shining, sharp-pointed, we state dagger; to substitute the absence of the sun and the progression of shadow, we say that it darkens. Nobody will deny that this nomenclature is a grandiose relief for our everyday life. Yet their aim is stubbornly practical: it is a prolix map that steers us through appearances, it is a most useful sign that our fantasy will at some point deserve to forget.[39]

Language, according to Borges, is not just an arbitrary system of symbols, as a strictly nominalist view would hold; it is also a practical map of relations. "Concepts not only guide us over the map of life, but we *revalue* life by their use," as James also posits in *Some Problems of Philosophy*. "They steer us practically every day, and provide an immense map of relations among the elements of things, which, though not now, yet on some possible future occasion, may help to steer us practically."[40]

Unlike Coleridge, Borges at this point no longer has to bow to the strict alternative between a nominalist observation of the limits and imperfections of human language and the mystical nostalgia for the perfect word that would fully express the Platonic Idea. Instead, we could conclude that James's pragmatic method for defining truth, based on the philosophy of radical empiricism, occupies an intermediate position in the age-old struggle between Aristotelians and Platonists. Borges himself seems to have been keenly aware of this possibility: "Middle solutions are one of the characteristics of pragmatism," he writes in his "Nota preliminar," before ending with a breathtaking summary in which William James no longer appears as the modern endpoint of the nominalist tradition, opposite Bradley or Royce. Instead James carves out a middle path between mechanical materialists and Hegelian idealists: "The universe of the materialists suggests an infinite, sleepless factory; that of the Hegelians, a circular labyrinth of vain mirrors, prison to one person who believes to be many, or to many who believe to be one; that of James, a river: the unending and unrecoverable river of Heraclitus.

Pragmatism does not seek to restrain or attenuate the richness of the world; it wants to keep on growing like the world."[41] To keep on growing like the world, instead of finding it ready-made before us, to add to the world, for example, by carving out groups of stars in the heavens and calling them constellations, such in the end is the heroic adventure to which Borges invites his readers and fellow travelers: "To add provinces to being, to envision cities and spaces of a hallucinatory reality, is a heroic adventure."[42] Heroic and, one might add, profoundly cartographic. Borges is less an author than a cartographer, less an imitator than a maker of maps. His view of the art of mapmaking is not limited to a demonstration of the critical aporias of language as the copying or imitation of an already existing territory but also envisions the utopian idea of cartographic language as the making or production of a whole new hallucinatory reality.

Notes

1. Jorge Luis Borges, "Borges and I," in *Dreamtigers*, trans. Mildred Boyer and Harold Morland (Austin: University of Texas Press, 1964), 51 (translation corrected). For Borges, the taste of coffee stands for leisure time, whereas Stevenson's *The Treasure Island* evokes the space of fiction.

2. Jorge Luis Borges, "A Profession of Literary Faith," in *Selected Non-fictions*, ed. Eliot Weinberger (New York: Penguin, 1999), 25.

3. Jorge Luis Borges, "From Allegories to Novels," in *Other Inquisitions, 1937–1952*, trans. Ruth L. C. Simms (Austin: University of Texas Press, 1964), 156 (translation modified). The reference is to Dante Alighieri's *Convivio*, trans. William Walrond Jackson (Oxford: Clarendon Press, 1909), 199. Let us note Borges's cleverness in using elements from both positions in this summary. To say that nominalists *are* Aristotle and realists, Plato, means to adopt a realist viewpoint, whereas to reduce this whole debate to nothing more than a change in *name* and *language* amounts to adopting the vantage point of nominalism.

4. Maurice de Wulf, *The History of Mediaeval Philosophy*, trans. E. C. Messenger (New York: Dover, 1952), vol. 1, 143. Another of Borges's named sources is George Henry Lewes; see, for example, "The Great Dispute," in *The History of Philosophy: From Thales to Comte*, vol. 2, 24–31 (London: Longmans, Green and Co., 1868). Also to be taken into account is the long entry on "Nominalismus," in one of Borges's favorite reference works, Fritz Mauthner's *Wörterbuch der Philosophie*, vol. 2, 416–432 (Leipzig: Felix Meiner, 1924). This entry already contains in a nutshell the whole argument of "From Allegories to Novels." For a more up-to-date overview of the scholastic dispute about universals, see Alain de Libéra, *La querelle des universaux: De Platon à la fin du Moyen Age* (Paris: Seuil, 1996).

5. For a massive study of Cratylism, see Gérard Genette, *Mimologics (Mimologiques: Voyages en Cratylie)*, trans. Thaïs E. Morgan (Lincoln: University of Nebraska Press, 1995). Genette begins his study of Cratylism, defined as the belief in "a relation of reflective analogy (imitation) between 'word' and 'thing' that *motivates*, or justifies, the existence and choice of the former," by quoting a passage from Borges's essay "The Analytical Language of John Wilkins": "At one time or another, we have all suffered through those unwinnable debates in which a lady, with copious interjections and anacoluthons, vows that the word *luna* is (or is not) more expressive

than the word *moon*" (*Mimologics* 7). This typically misogynist comment seems to suggest that if all *men* are born either Platonists or Aristotelians, *women* are born Cratylists. Genette's translator, on the other hand, concludes her introductory essay by referring to Borges's "Funes the Memorious" as the story not of an "extreme nominalist," which is how Funes is often presented, but of a "secondary Cratylist," so baptized by Genette after the model of Socrates in Plato's dialogue, who desires "to establish or reestablish in the language system, through some artifice, the state of nature that 'primary Cratylism'–that of Cratylus–naively believes to be still or already established there" (*Mimologics*, lvii, 27).

6. Borges describes his "basic skepticism" as the tendency "to evaluate religious or philosophical ideas on the basis of their aesthetic worth and even for what is singular and marvelous about them" (see Borges, *Other Inquisitions*, 189).

7. Borges, "The Analytical Language of John Wilkins," *Other Inquisitions*, 104.

8. Borges, "The Analytical Language of John Wilkins," 104.

9. Borges, "The Nightingale of Keats," *Other Inquisitions*, 123–124.

10. Borges, "Our Poor Individualism," *Other Inquisitions*, 33–34.

11. Borges, "From Allegories to Novels," 157. To support his argument regarding the passage from realism to nominalism as the epochal entrance into modernity, Borges cites Maurice de Wulf: "Ultrarealism gained the first adherents. The chronicler Heriman (eleventh century) speaks of those who teach dialectic *in re* as 'antiqui doctores'; Abelard calls the dialectic an 'antiqua doctrina,' and the name of 'moderni' is applied to its adversaries until the end of the twelfth century'" (quoted in Borges, "From Allegories to Novels," 156–157; cf. de Wulf, *History of Mediaeval Philosophy*, vol. 1, 142 (de Wulf's term in the original French edition is *réalisme outré*, translated in English as either "exaggerated" or "extreme" realism, whereas Borges uses *ultrarrealismo* in Spanish, which would seem to indicate that he consulted one of the French editions)). For a fascinating confirmation of Borges's claim that modernity begins not with the Enlightenment or the Renaissance but with the victory of nominalism and the breakup of the ontotheological cosmos in the late Middle Ages, see also Louis Dupré, *Passage to Modernity: An Essay in the Hermeneutics of Nature and Culture* (New Haven, CT: Yale University Press, 1993).

12. Jorge Luis Borges, "A History of Eternity," in *Selected Non-fictions*, ed. Eliot Weinberger (New York: Penguin, 1999), 135 (translation modified).

13. Borges, "From Allegories to Novels," 157.

14. Borges, "Our Poor Individualism," 35. Borges's typical source of authority in this context stems from his long-standing admiration for the libertarian doctrine of Herbert Spencer in his classic 1884 treatise *The Man versus the State*.

15. Borges, "Deutsches Requiem," 229, 233. For an analysis of what can be called Borges's political nominalism, see my "Manual de conjuradores: Borges o la colectividad imposible," in *Jorge Luis Borges: Políticas de la literatura*, ed. Juan Pablo Dabove, 251–270 (Pittsburgh: Instituto Internacional de Literatura Iberoamericana, 2008).

16. Samuel Taylor Coleridge, *Specimens of the Table Talk* (London: John Murray, 1835), vol. 1, 182. See also "Notes on Hooker": "Schools of real philosophy there are but two,–best named by the arch-philosopher of each, namely, Plato and Aristotle. Every man capable of philosophy at all (and there are not many such) is a born Platonist or a born Aristotelian" (Samuel Taylor

Coleridge, *The Literary Remains*, ed. Henry Nelson Coleridge (New York: Harper & Brothers, 1884), 37).

17. Jaime Rest, *El laberinto del universo: Borges y el pensamiento nominalista* (Buenos Aires: Librerías Fausto, 1976), 58.

18. Coleridge, *Specimens of the Table Talk*, 182. In his "Notes on Hooker," Coleridge also remarks: "There are, and can be, only two schools of philosophy, differing in kind and in source. Differences in degree and in accident, there may be many; but these constitute schools kept by different teachers with different degrees of genius, talent, and learning—auditories of philosophizers, not different philosophies" (see *The Literary Remains*, 37).

19. Borges, "From Allegories to Novels," 156.

20. Jorge Luis Borges, "Three Versions of Judas," in *Collected Fictions*, trans. Andrew Hurley (New York: Penguin, 1998), 164. See also a comparable passage in "The Aleph" in which the narrator explains the relevance of this letter: "In the Kabbala, that letter signifies the En Soph, the pure and unlimited godhead; it has also been said that its shape is that of a man pointing to the sky and the earth, to indicate that the lower world is the map and mirror of the higher" (Jorge Luis Borges, "The Aleph," in *Collected Fictions*, 285). Quickly glossing over the discrepancies in the scholastic debate over universals, Michel Foucault sees the similitude between heaven and earth, like the resemblance between God's "two books"—the Bible and the Book of Nature—as a dominant motive of the *episteme* of the West up to the end of the sixteenth century: "The universe was folded in upon itself: the earth echoing the sky, faces seeing themselves reflected in the stars, and plants holding within their stems the secrets that were of use to man" (see Foucault, *The Order of Things: An Archaeology of the Human Sciences* (New York: Random House: 1970), 17).

21. Fredric Jameson, *Postmodernism, or, the Cultural Logic of Late Capitalism* (Durham, NC: Duke University Press, 1991), 430n60. This is a puzzling judgment especially in view of the map's success in books by Jean Baudrillard, Jean-François Lyotard, Félix Guattari, Umberto Eco, and Jameson himself, elsewhere in his work! In part, the harshness of this disparagement is a reflection of the change of heart on the part of Baudrillard, who in *Symbolic Exchange and Death* had alluded to Borges's map as already being an example of postmodern simulation but in *The Beaubourg Effect* and then especially in the opening paragraphs of *Simulacra and Simulation* changed his mind and referred to it only as a second-order simulacrum, which maintains the minimal difference between medium and message. Compare Jean Baudrillard, *Symbolic Exchange and Death*, trans. Iain Hamilton Grant (London: Sage, 1993), 86n4; *L'Effet Beaubourg: Implosion et dissuasion* (Paris: Galilée, 1977), 41; and *Simulacra and Simulation*, trans. Sheila Faria Glase (Ann Arbor: University of Michigan Press, 1994), 1, 71. For a study of these different readings, see "Cartographies," in my dissertation *After Borges: Literary Criticism and Critical Theory* (Ann Arbor: ProQuest/UMI, 1995). More recently, see Ivan Almeida, "Borges à la carte (tres citas de Baudrillard)," *Variaciones Borges* 25 (2008): 25–51. Unfortunately, Almeida ignores Baudrillard's other references to "On Rigor in Science," prior to *Simulacra and Simulation*, allusions that are neither "inexistent" nor "unverifiable," and thereby loses sight of the possibility that the French author might actually be correcting his earlier misprisions.

22. Jorge Luis Borges, "On Rigor in Science," in *Dreamtigers*, trans. Mildred Boyer and Harold Morland (Austin: University of Texas Press, 1964), 90; "Naturalism Revived," in *Chronicles of Bustos Domecq*, trans. Norman Thomas di Giovanni (New York: Dutton, 1979), 43. For the French readings of this text, aside from Baudrillard, see also Louis Marin, "Utopia of the Map," in *Utopics: The Semiological Play of Textual Spaces,* trans. Robert A. Vollrath, 233–237 (Atlantic

Highlands, NJ: Humanities Press, 1984); Jean-François Lyotard, *The Postmodern Condition: A Report on Knowledge*, trans. Geoff Bennington and Brian Massumi (Minneapolis: University of Minnesota Press, 1984), 55; and Félix Guattari, *La Révolution moléculaire* (Fontenay-sous-Bois: Recherches, 1977), 52n1; *L'Inconscient machinique* (Fontenay-sous-Bois: Recherches, 1979), 234n28; and *Cartographies schizoanalytiques* (Paris: Galilée, 1989), 51n1. For a study of Guattari's case, see my "From Text to Territory: Félix Guattari's Cartographies of the Unconscious," in *Deleuze and Guattari: New Mappings in Politics, Philosophy, and Culture*, ed. Kevin Jon Heller and Eleanor Kaufman, 145–174 (Minneapolis: University of Minnesota Press, 1998). And, for the larger context behind this renewed fascination with cartographic metaphors in theory and philosophy, with special attention given to the case of Borges, see my "A Misreading of Maps: The Politics of Cartography in Marxism and Poststructuralism," in *Signs of Change: Premodern, Modern, Postmodern*, ed. Stephen Barker, 109–138 (Albany: State University of New York Press, 1996).

23. Miguel de Cervantes, *Don Quixote*, trans. John Rutherford (London: Penguin, 1999), 521. See also Lewis Carroll, *Sylvie and Bruno Concluded* (London: Macmillan, 1893), 169.

24. Borges, "Partial Enchantments of the *Quixote*," in *Other Inquisitions*, 46. The fragment is adapted from Josiah Royce, "The One, the Many, and the Infinite," in *The World and the Individual* (New York: Dover, 1959), vol. 1, 504–505. Alfred Korzybski, best known for his dictum "A map is *not* the territory," credits Royce with having pointed out the possibility of self-reflexiveness shared by cartography and language, in *Science and Insanity: An Introduction to Non-Aristotelian Systems and General Semantics* (Lancaster: Science Printing Press, 1933), 58, 751. Charles S. Peirce argues against Royce with the startling idea of a "continuous" map, in his *Collected Papers*, ed. Charles Hartshorne and Paul Weiss (Cambridge, MA: Harvard University Press, 1934), vol. 5, 71. In his text for *El Hogar*, "Cuando la ficción vive en la ficción," Borges mentions how he first read about Royce's text in a book by Bertrand Russell, *Introduction to Mathematical Philosophy*. See Jorge Luis Borges, *Textos cautivos: Ensayos y reseñas en El Hogar (1936–1939)*, ed. Enrique Sacerio-Garí and Emir Rodríguez Monegal (Barcelona: Tusquets, 1986), 325. For a study of this relation, see John Durham Peters, "Resemblance Made Absolutely Exact: Borges and Royce on Maps and Media," *Variaciones Borges* 25 (2008): 1–23.

25. Lyotard, *The Postmodern Condition*, 55.

26. Borges, "Verbiage for Poems," in *Selected Non-fictions*, 21.

27. See Foucault, *The Order of Things*, xv–xxiv; Gilles Deleuze, *Difference and Repetition*, trans. Paul Patton (New York: Columbia University Press, 1994), xxi–xxii; Jacques Derrida, *Writing and Difference*, trans. Alan Bass (London: Routledge, 1978), 114. For an analysis of Foucault's treatment of Borges, see my "Monstrosity and the Postmodern," in *Literature and Society: Centers and Margins*, ed. José García et al., 9–20 (New York: Department of Spanish and Portuguese, Columbia University, 1994). Deleuze also cites Borges's "Death and the Compass" in "On Four Poetic Formulae Which Might Summarize the Kantian Philosophy," a new preface written for his second monograph, *Kant's Critical Philosophy*, trans. Hugh Tomlinson and Barbara Habberjam (Minneapolis: University of Minnesota Press, 1984), vii. For a study of the relation between Borges and Derrida, see Lisa Block de Behar, *Al margen de Borges* (Buenos Aires: Siglo Vientiuno, 1987).

28. Jorge Luis Borges, "Nota preliminar," in William James, *Pragmatismo: Un nombre nuevo para algunos viejos modos de pensar*, trans. Vicente P. Quintero (Buenos Aires: Emecé, 1945), 9. James would have agreed with this assessment: "There is absolutely nothing new in the pragmatic method. Socrates was an adept at it. Aristotle used it methodically. Locke, Berkeley, and

Hume made momentous contributions to truth by its means," he says. "Being nothing essentially new, it harmonizes with many ancient philosophic tendencies. It agrees with nominalism for instance, in always appealing to particulars; with utilitarianism in emphasizing practical aspects; with positivism in its disdain for verbal solutions, useless questions and metaphysical abstractions" (see James, *Pragmatism* (New York: Meridian Books, 1974), 45, 47). On the relation between Borges and James, see my "The Truth Is in the Making: Borges and Pragmatism," *Romantic Review* 98, nos. 2–3 (2007): 135–151.

29. Borges, "Nota preliminar," 11.

30. Gilles Deleuze, "Bartleby; or, the Formula," in *Essays Critical and Clinical*, trans. Daniel W. Smith and Michael A. Greco (Minneapolis: University of Minnesota Press, 1997), 86. It also belongs to one of Deleuze's most original disciples to have studied William James from a point of view that would have pleased both his teacher and Borges. See David Lapoujade, *William James: Empirisme et pragmatisme* (Paris: PUF, 1997).

31. See Jorge Luis Borges, "An Autobiographical Essay," in *The Aleph and Other Stories, 1933–1969*, trans. Norman Thomas di Giovanni (New York: Bantam, 1971), 136.

32. See Jaime Nubiola, "William James and Borges Again: The Riddle of the Correspondence with Macedonio Fernández," *Streams of William James* 3, no. 2 (2001): 10–11; Jaime Nubiola, "Jorge Luis Borges y William James," in *Aproximaciones a la obra de William James: La formulación del pragmatismo*, ed. Jaime de Salas and Félix Martín, 201–218 (Madrid: Biblioteca Nueva, 2005).

33. Jorge Luis Borges, "William James, *Las variedades de la experiencia religiosa: Estudio sobre la naturaleza humana*," in *Biblioteca personal*, 126–127 (Madrid: Alianza, 1988).

34. Jorge Luis Borges, "La encrucijada de Berkeley," in *Inquisiciones* (Buenos Aires: Proa, 1925), 115.

35. Borges, "The Nothingness of Personality," in *Selected Non-fictions*, 8 (translation modified).

36. Borges, "The Nothingness of Personality," 8.

37. James, *Pragmatism*, 167.

38. Borges, "Verbiage for Poems," 21.

39. Borges, "Examen de metáforas," in *Inquisiciones*, 65–66.

40. See William James, *The Writings of William James*, ed. John McDermott (Chicago: University of Chicago Press, 1977), 243.

41. Borges, "Nota preliminar," 11–12. For further discussion of pragmatism as the middle solution in the scholastic debate between realism and nominalism, see my "The Truth Is in the Making."

42. Borges, "After Images," in *Selected Non-fictions*, 11. As James also writes, "The essential contrast is that *for rationalism reality is ready-made and complete from all eternity, while for pragmatism it is still in the making, and awaits part of its complexion from the future. On the one side the universe is absolutely secure, on the other it is still pursuing its adventures*" (*Pragmatism*, 167).

Bibliography

Alighieri, Dante. *Convivio*. Trans. William Walrond Jackson. Oxford: Clarendon Press, 1909.

Almeida, Ivan. "Borges à la carte (tres citas de Baudrillard)." *Variaciones Borges* 25 (2008): 25–51.

Baudrillard, Jean. *L'Effet Beaubourg: Implosion et dissuasion*. Paris: Galilée, 1977.

Baudrillard, Jean. *Simulacra and Simulation*. Trans. Sheila Faria Glase. Ann Arbor: University of Michigan Press, 1994.

Baudrillard, Jean. *Symbolic Exchange and Death*. Trans. Iain Hamilton Grant. London: Sage, 1993.

Block de Behar, Lisa. *Al margen de Borges*. Buenos Aires: Siglo Vientiuno, 1987.

Borges, Jorge Luis. *The Aleph and Other Stories, 1933–1969*. Trans. Norman Thomas di Giovanni. New York: Bantam, 1971.

Borges, Jorge Luis. *Biblioteca personal*. Madrid: Alianza, 1988.

Borges, Jorge Luis. *Collected Fictions*. Trans. Andrew Hurley. New York: Penguin, 1998.

Borges, Jorge Luis. *Dreamtigers*. Trans. Mildred Boyer and Harold Morland. Austin: University of Texas Press, 1964.

Borges, Jorge Luis. *Inquisiciones*. Buenos Aires: Proa, 1925.

Borges, Jorge Luis. *Other Inquisitions 1937–1952*. Trans. Ruth L. C. Simms. Austin: University of Texas Press, 1964.

Borges, Jorge Luis. *Selected Non-fictions*. Ed. Eliot Weinberger. New York: Penguin, 1999.

Borges, Jorge Luis. *Textos cautivos: Ensayos y reseñas en El Hogar (1936–1939)*. Ed. Enrique Sacerio-Garí and Emir Rodríguez Monegal. Barcelona: Tusquets, 1986.

Borges, Jorge Luis, and Adolfo Bioy Casares. *Chronicles of Bustos Domecq*. Trans. Norman Thomas di Giovanni. New York: Dutton, 1979.

Bosteels, Bruno. *After Borges: Literary Criticism and Critical Theory*. Ann Arbor: ProQuest/UMI, 1995.

Bosteels, Bruno. "Manual de conjuradores: Borges o la colectividad imposible." In *Jorge Luis Borges: Políticas de la literatura*, ed. Juan Pablo Dabove, 251–270. Pittsburgh: Instituto Internacional de Literatura Iberoamericana, 2008.

Bosteels, Bruno. "A Misreading of Maps: The Politics of Cartography in Marxism and Poststructuralism." In *Signs of Change: Premodern, Modern, Postmodern*, ed. Stephen Barker, 109–138. Albany: State University of New York Press, 1996.

Bosteels, Bruno. "Monstrosity and the Postmodern." In *Literature and Society: Centers and Margins*, ed. José García et al., 9–20. New York: Department of Spanish and Portuguese, Columbia University, 1994.

Bosteels, Bruno. "From Text to Territory: Félix Guattari's Cartographies of the Unconscious." In *Deleuze and Guattari: New Mappings in Politics, Philosophy, and Culture*, ed. Kevin Jon Heller and Eleanor Kaufman, 145–174. Minneapolis: University of Minnesota Press, 1998.

Bosteels, Bruno. "The Truth Is in the Making: Borges and Pragmatism." *Romantic Review* 98, nos. 2–3 (2007): 135–151.

Carroll, Lewis. *Sylvie and Bruno Concluded*. London: Macmillan, 1893.

de Cervantes, Miguel. *Don Quixote*. Trans. John Rutherford. London: Penguin, 1999.

Coleridge, Samuel Taylor. *The Literary Remains*. Ed. Henry Nelson Coleridge. New York: Harper & Brothers, 1884.

Coleridge, Samuel Taylor. *Specimens of the Table Talk*. London: John Murray, 1835.

Deleuze, Gilles. *Difference and Repetition*. Trans. Paul Patton. New York: Columbia University Press, 1994.

Deleuze, Gilles. *Essays Critical and Clinical*. Trans. Daniel W. Smith and Michael A. Greco. Minneapolis: University of Minnesota Press, 1997.

Deleuze, Gilles. *Kant's Critical Philosophy*. Trans. Hugh Tomlinson and Barbara Habberjam. Minneapolis: University of Minnesota Press, 1984.

de Libéra, Alain. *La querelle des universaux: De Platon à la fin du Moyen Age*. Paris: Seuil, 1996.

Derrida, Jacques. *Writing and Difference*. Trans. Alan Bass. London: Routledge, 1978.

de Wulf, Maurice. *The History of Mediaeval Philosophy*. Trans. E. C. Messenger. New York: Dover, 1952.

Dupré, Louis. *Passage to Modernity: An Essay in the Hermeneutics of Nature and Culture*. New Haven, CT: Yale University Press, 1993.

Foucault, Michel. *The Order of Things: An Archaeology of the Human Sciences*. New York: Random House, 1970.

Genette, Gérard. *Mimologics (Mimologiques: Voyages en Cratylie)*. Trans. Thaïs E. Morgan. Lincoln: University of Nebraska Press, 1995.

Guattari, Félix. *Cartographies schizoanalytiques*. Paris: Galilée, 1989.

Guattari, Félix. *La Révolution moléculaire*. Fontenay-sous-Bois: Recherches, 1977.

Guattari, Félix. *L'Inconscient machinique*. Fontenay-sous-Bois: Recherches, 1979.

James, William. *Pragmatism*. New York: Meridian Books, 1974.

James, William. *Pragmatismo: Un nombre nuevo para algunos viejos modos de pensar*. Trans. Vicente P. Quintero. Buenos Aires: Emecé, 1945.

James, William. *The Writings of William James*. Ed. John McDermott. Chicago: University of Chicago Press, 1977.

Jameson, Fredric. *Postmodernism, or, the Cultural Logic of Late Capitalism*. Durham, NC: Duke University Press, 1991.

Korzybski, Alfred. *Science and Insanity: An Introduction to Non-Aristotelian Systems and General Semantics*. Lancaster: Science Printing Press, 1933.

Lapoujade, David. *William James: Empirisme et pragmatisme*. Paris: PUF, 1997.

Lewes, George Henry. *The History of Philosophy: From Thales to Comte*. London: Longmans, Green and Co., 1868.

Lyotard, Jean-François. *The Postmodern Condition: A Report on Knowledge*. Trans. Geoff Bennington and Brian Massumi. Minneapolis: University of Minnesota Press, 1984.

Marin, Louis. "Utopia of the Map." In *Utopics: The Semiological Play of Textual Spaces*, trans. Robert A. Vollrath. Atlantic Highlands, NJ: Humanities Press, 1984.

Mauthner, Fritz. *Wörterbuch der Philosophie*. Leipzig: Felix Meiner, 1924.

Nubiola, Jaime. "Jorge Luis Borges y William James." In *Aproximaciones a la obra de William James: La formulación del pragmatismo*, ed. Jaime de Salas and Félix Martín, 201–218. Madrid: Biblioteca Nueva, 2005.

Nubiola, Jaime. "William James and Borges Again: The Riddle of the Correspondence with Macedonio Fernández." *Streams of William James* 3, no. 2 (2001): 10–11.

Peirce, Charles S. *Collected Papers*. Ed. Charles Hartshorne and Paul Weiss. Cambridge, MA: Harvard University Press, 1934.

Peters, John Durham. "Resemblance Made Absolutely Exact: Borges and Royce on Maps and Media." *Variaciones Borges* 25 (2008): 1–23.

Rest, Jaime. *El laberinto del universo: Borges y el pensamiento nominalista*. Buenos Aires: Librerías Fausto, 1976.

Royce, Josiah. *The World and the Individual*. New York: Dover, 1959.

II
HISTORIES AND CONTEXTS

6
Muses of Cartography: Charting Odysseus from Homer to Joyce

Burkhardt Wolf

Introduction

Poetry has executed a "spatial turn" from the very beginning. Unlike the regionally bound *Iliad*, the *Odyssey* turned toward open space, the sea, and its successive discovery. And both of Homer's epics, traditionally conceived as the "origin" of the Western poetic tradition, are "topographic," insofar as numerous poetic topoi, which had until then only been memorized and transmitted orally, first found their place in writing here. According to a contemporary scholar's thesis, the "Ur-text" of Odysseus's adventures at sea was first put to paper or papyrus around 800 BC. It thereby established and propagated the medium of Hellenic poetry and the Greek vocalized alphabet on both the poetic and actual seaways, by means of the landlocked reader and the increasingly literate sailor.[1] In the *Odyssey*, writing and seafaring come into contact in order to poetically "suture" the surface of writing and the surface of the sea. Against this background, in modernity–especially in the founding period of the history of cartography around 1900–Homer's maritime epic became the basis of the literary-historical question: What is the relationship between map and text? What is a cartographic text?

This chapter draws on the observation that the *Odyssey*'s plot and its respective "topography" is deterritorialized in a peculiar way. If, in antiquity, sailing manuals had to reckon with the nautical and existential disorientation experienced on the high seas, then it is highly probable that Homer fell back on them. In his maritime epic, technical instruction and poetic imagination, topos and tropes, actually seem to join together. One may therefore assume that the *Odyssey* is based on sailing manuals. But it is certainly more than their mere versification. It discloses the manual's poetic character by telling (on the plot level of Odysseus's travels) and by showing (on the level of its own toponymic creativity), how places within the placeless sea are generated, determined, and described at all. The *Odyssey* has recourse to sailing manuals, but only in order to carry on their "protocartographic" operations.

This principle of "recursion"–harking back to a putative origin in order to redetermine and not only to repeat it–also characterizes post-Homeric literary adaptations of the *Odyssey*: Roman epics, medieval romances, and modern novels all create "their" allegedly original *Odyssey* from which they emanate themselves. In other words, in the wake of the "polytropos" Odysseus, literature becomes self-referential, since it always has to grapple with placeless spaces, with their description and visualization. Particularly in modern approximations of the *Odyssey* that increasingly reference their own protocartographic makeup, this poetic generation and location is pushed to its limits: the charts are overstretched, the tools become dysfunctional, and explorations go methodically astray. The more complex the description of the modern world becomes, the more fatal its shortcomings. It is as if Odysseus, at the edge of modern cartography, has returned to the state of deterritorialization that spurred his departure to begin with.

Homer's Epic Topography

In its comprehensiveness, Homer's *Odyssey* remains authoritative among all preclassical sources on Western seafaring. Numerous nautical practices from shipbuilding to astronomical navigation are first documented or conceptually grasped in the text.[2] For the description of the seaman's experiences such as the wild sea and storms and for their allegorical or metaphorical interpretations, the *Odyssey* remained such an indispensable catalog of topoi into late antiquity that first rhapsodes, especially the Homeridae, then epic poets, lyric poets, and tragedians created veritable glossaries of its terms. These commonplaces were eventually taken up by the rhetors as the basis for exercises, before Alexandrian librarians compiled them as a poetic inheritance for posterity.

Compared to the *Iliad* and its pictoriality, what stands out in the *Odyssey* is its overlay of "real" places and poetic locations. For instance, the notorious Cape Maleas, whose promontory marked the outermost boundary of geographic describability, denotes here the node between the routes of Menelaus and Odysseus.[3] It became proverbial in Greek that whoever wanted to navigate this dangerous cape must forget about his home.[4] In the storms and currents of Cape Maleas, one was not just far from the Greek mainland but driven straight into the high seas, into the realm of no places. Already in Homer, "to lose one's way" means going astray from one's existence; thus the cape is not merely a place of delocalization but of a loss of self. Odysseus is "characterless" yet determined to survive, and it is only because forsakenness has become a "way of life" for him that the *Odyssey* can describe the "path of … flight" of its hero with exactly the semantic ambivalence that Adorno and Horkheimer formulated as "being cast up and being cunning."[5]

Homer's topoi are not derived from any neutral, underlying "topography." On the contrary, they make the description of elementary facts themselves into a poetic problem. In describing storms, for example, an existing topos is not simply applied to a geographic location. Rather, it is from a contingent event that a topical determinant is generated in the first place. As early as the turn of the twentieth century, Victor Bérard, a French Greek scholar, diplomat, and senator in the Département Jura, called attention to a sort of elementary realism in the *Odyssey*: "What we have here is not the storm of writers, rather a storm of seafarers, a storm in the Mediterranean."[6] This kind of topography is built on the inspiration of the muses, as well as on "prose fragments,"[7] which came to the poet by way of Phoenician sailors. The roots of the *Odyssey*, as purely philological findings also show,[8] are not to be found on dry land. Its locations and concepts of space are dynamized, its objects and actors are in motion. It is told from the sea with an eye on the coastline. In Bérard's words: "The interior of the mainland appears only indistinctly, is seen only from a great distance."[9]

With the ancient Hellenes in mind, Hegel speaks of a historical "state of turbulence, insecurity," of a positionally unmoored existence: "The physique of their country led them to this amphibious existence, and allowed them to skim freely over the waves."[10] In this abandonment of the terrane nomos with its ethical and political as well as topological and ontological liabilities; in the orientation toward a "becoming" that is "forgetful-of-being"; and in its turn to an amphibious existence that comes to fruition in the ventures of pirates, the sea trade, or the "thalassocracies," Plato had already seen the danger of a momentous uprooting and an unpredictable "being cast up" of his own polis.[11] But even in Homer's time before the classical epoch, a mentality of foreignness or even hostility toward the sea had manifested itself, in light of which the deterritorialized structure of the *Odyssey* seems all the more remarkable.[12] In the escape route of its polytropos—its hero "of many turns," or "of many twists"—who is at once widely traveled and cunning, trope and topos begin to overlap, to send him adrift and initiate his wanderings. This comes to pass on undescribed and even indescribable terrain: the sea of the unknown west, or simply the open sea far from any visible landmark. If one wants to understand it as a "topographic" text, the *Odyssey* seems to present a paradox: the endeavor of tracing and describing the places of a nonplace.

Mapping Odysseus's Routes

Evidently, this has been no great hindrance. Ever since antiquity many have attempted to reconstruct the journey of Odysseus in accordance with their own geographic knowledge and genealogical claims. Already Hesiod, and after him Thucydides and Plato, placed his route between western Italy, North Africa, and the mouth of the Atlantic. Roman founding myths reclaimed the sirens' homeland as the coast of Campania. Sometimes the journey of Odysseus is also transposed to the Black Sea or into the Atlantic.[13] The Alexandrian geographer Eratosthenes mockingly quipped that one would only be able to describe the original locations of the *Odyssey* after finding the cobbler who had stitched up Aeolus's sack of winds. The "true" route of "the cunning one" remained the subject of much speculation through the Middle Ages.[14]

Abraham Ortelius's Vlyssis Errores (1597) marks the beginning of the numerous attempts at cartographic representation that continue to the present—attempts whose proponents hoped to substantiate with all manner of cultural and linguistic theories. It was Johann Heinrich Voß, with his "Homeric world-plate" (a supplement to his translation of the *Odyssey*), who first went beyond projecting the poetic topoi onto a contemporary map. Instead, he interpreted them in the configuration that he himself saw laid out within the framework of the epic verse. (See figure 6.1.) Since then a tradition of an "imaginary" cartography has emerged in opposition to the proponents of the "realistic" one. As the future English prime minister William E. Gladstone claimed in 1858, Homeric cartographies should produce a "fictitious drawing" of Homer's "pictures of the imagination."[15] One could adduce good reasons for this, for in Homer's time no one had nautical charts. Sea travel was still undertaken without elaborate navigational instruments and mostly on the basis of experience and habit.

Victor Bérard, however, in pointing to ancient sailing handbooks, the so-called periploi, countered with a double argument, at once historical and aesthetic: "In my view, the *Odyssey* emerged as a Phoenician periplus (from Sidon, Carthage, or elsewhere) which was carried over into Greek verse and poetic myths," he wrote in 1902. "The anthropomorphic visualization of things, the humanization of the forces of nature, the Hellenization of the subject matter"[16] were to Bérard certain proof that Homer already fulfilled the basic requirement of all cartography: a kind of bird's-eye view shaping the perception of land formations, which, as the geographer Strabo also later claimed, imprints views of nature with a geometric pattern (such as the triangle of Sicily) or an organic form (such as the maple leaf of the Peloponnese).[17] At the same time Bérard disputes that the poet of the *Odyssey* had to fall back on mere imagination at sea due to a dearth of cartographic resources: "The amount of fantasy and imagination is limited here. The essential contribution of the poet consists in the organization and logic of the events"—for the epic had as its basis actual Phoenician periploi.[18]

Such handbooks are organized according to the principle of the list. They are one-dimensional maps, so to speak, and Bérard's handbook theory is substantiated up to the present by classical philological analyses. In accordance with the so-called Zielinski's law, narrative threads in Homer always unwind separately and are never interwoven with one another. Unlike the *Argonautika* of Apollonius of Rhodes (c. 250 BC), which appears to have been influenced by the first cartographic endeavors in the sixth century BC, the *Odyssey* eschews simultaneity and interconnectedness of narrative events entirely. Instead, as long as the setting and the action do not shift completely, it traces their course by constantly following successive points of orientation.[19] For Bérard it is simply beside the point whether Homer named real or fictitious locations. Instead, from a "scientific-poetological," an epistemological, and an aesthetic perspective, he follows the orientation process on the way to the map.

Figure 6.1

Johann Heinrich Voß's Plate of the Homeric World (1793) (Johann Heinrich Voß, *Homers Werke*, vol. 3 (Altona, 1793))

To place the *Odyssey* within a "realistic" topography, or to illustrate its "imaginary" map, is equally otiose. But it is crucial to consider to what degree Homer localizes Odysseus and topographizes his course. At the beginning of a chain of developments that stretches from floating toponyms in undescribed maritime space to a cartographic visualization of the world (if not yet of the sea itself), "two cultures" collide: in ethnic terms, the Greeks (in the character of Homer, according to Bérard) and the Phoenicians (in the character of Odysseus); and in terms of the order of things, the poetic places of the epic meet the nautical designations of the periplus. Periploi are, from their name, sailing handbooks meant to assist a "round trip" in the dangerous zone of the open sea. To this end, they referred—from the perspective of the sea—to clearing marks, specific maritime and meteorological phenomena, and eventually, to navigable coasts and anchor points.

Handbooks like these had found their way into Greek literature by the time the geographer Pseudo-Scylax composed one (itself titled *Periplus*) in the middle of the fourth century BC. But their first appearance in the Greek language, or more accurately, in Greek writing, are the sailing instructions given by Circe that make the adventurous nostos of Odysseus's epic possible in the first place. Handbooks of this kind had likely long since circulated among Egyptians and Phoenicians before the Greeks began their own tradition with the introduction of their vocalized alphabet. By the fourth century BC, the majority of seafarers were sufficiently literate to use written sailing instructions on their own. Along with the nautical markers and distance indications, which were translated into the duration of the journey when not designated by "stages," periploi also soon recorded descriptions of land and water bases, sacred sites, and defensive positions, and even advice on the morals, political relations, and commercial customs of the seigniories along the route. Thus, they developed into veritable descriptions of the lands themselves.

For Bérard, travel and adventure stories, regardless of what sort and era, are essentially poetic recursions of periploi.[20] To him they serve as media of narrative and fictionality that make it possible to see "with the eyes of another" in the first place.[21] As this kind of experience as transit (*Er-fahrung*) entails both the personal and the spatial, they are furthermore an exemplary medium of discovery. Periploi work their way into the process of generating locations in the first place. They stand at the beginnings of poetry and discovery alike. Without periploi, there would have been no Homer, nor the Greek journey into the blue. And this fact, as Bérard concedes, was already known to the ancients. His work is thus to be understood as recourse to "a sentence or two from Strabo."[22] For he, unlike Eratosthenes, recognized the poetic aspects of geography, and put forth the theory that Homer must have made use of Phoenician periploi.

Odysseus's Round Trip as Recursion

Against the background of Odyssean traditions, Homer's recourse to periploi is much more than a simple reference or application. It rather describes the putative origins of poetry, or more precisely, poetry's "recursive" approach in producing its supposed origins. Already Strabo problematized this approach with regard to the nautical and poetic experience of the *Odyssey*. One could define *recursion* as a circular motion that not only facilitates the return to the same, but also makes variation possible through self-referentiality.[23] In more technical terms, recursion means the reapplication of processing instructions to one variable, which is itself already output from these instructions. In the case of Homer, the processing instructions concern the generation and relation of locations, and the variables appear initially as descriptions of place, then as periploi, and finally as poetry. The variable value changes with every journey through this loop, and along with repetition the procedure produces difference. Recursions are processes of discovery, for as a rule, they lead to contingency. Here, especially, the alleged origin is always discovered anew. A clear and, at the same time, allegorical example of this recursive principle can be found in the siren episode. As Odysseus comes closer to the mythical origin of poetry itself, "the bewitching ones" (as *seirenes* has been translated) promise to strike up nothing other than songs from Homer's other—allegedly first—heroic epic.[24] The *Odyssey* therefore is a recursion of the *Iliad*, and the latter epic is said to have arisen from artistic inspiration allegorized here in the form of sirens, their singing, and their promise of supreme knowledge. Odysseus comes to this origin, at which song, poetry, and knowledge intersect, not by accident, but through the execution of the sailing instructions given by Circe.[25]

From Homer onward various series become recognizable within the history of literature and knowledge alike that not only involve transmission or adaptation, but that follow the principle of recursion. Apollonius of Rhodes's epic of the Argonauts, for instance, has always been understood as a supplement to the *Odyssey*. With mirrorlike symmetry it constructs the previously unknown area east of Greece and to this end has recourse to periploi like that of Scylax. Certainly the instructions from Homer's Circe already presuppose Jason's adventure, in which the Argo, the "first-ever ship," set sail with Odysseus's father Laertes on board.[26] The *Argonautika* becomes a recursion of the *Odyssey* in that it retroactively reestablishes its origin: the Black Sea instead of the Mediterranean, and Jason instead of Odysseus. Furthermore, the Argonaut material within the Greek tradition appears older than any preserved epic. Because they had no Greek alphabet to rescue them, these corresponding pre-Homeric songs were lost. Nevertheless the *Argonautika*, as a post-Homeric epic through which pre-Homeric material glimmers, indicates the narrative tradition that may underpin the *Odyssey*: "helper myths" that herald the rescue of the embattled hero gone astray,[27] as well as the ancient Egyptian "Tale of the Shipwrecked Sailor" and his meandering journey home, which has been identified as the source of the "Phaeacians."[28]

To summarize, we may conclude that the nostos, the heroic journey home, and the narrative of the sea journey point unmistakably to the use of periploi, whose existence in the Middle Kingdom of Egypt (2000–1700 BC) is attested to by numerous bas-relief inscriptions in Egyptian temples. From this perspective it becomes evident that the Phoenicians were already well-versed students of Egyptian seafaring long before the Greeks made Phoenician periploi the starting points for their own poetry.[29] According to Bérard, to read the *Odyssey* correctly, one must not only compare it to materially and thematically similar poetry, but also to the extant nautical writing. As he somewhat provocatively puts it, the *Odyssey* is ultimately just one piece in a "series of nautical instructions."[30]

This has consequences for the "Homeric question" concerning the originality of the *Odyssey*'s author. With every sentence, if not word, it seems that not only Greek aioidoi (songsters passing down their cantos without any written assistance), but also Phoenician or even Egyptian seafarers find their way into language. "The poet actually invents nothing: he brings to life, arranges, and disposes the material. Every periplus is initially just a wreath of proper names," writes Bérard. "Each adventure sets itself in motion through the use of toponymy."[31] The poet's imagination is therefore no free-floating faculty. It begins with the places and their naming; it vivifies the names in the form of actors, it arranges corresponding events, and it ultimately casts the whole of the narrative as nostos and "round trip."[32]

Even when toponymy, as Bérard concedes, can be decried as etymological "play," it still makes it possible to uncover the locations through a sort of lexical and grammatical cartography. And simple wordplay, which, especially in the ancient Greek context, implies hidden truths, indicates cunning linguistic and poetic routes of transmission.[33] Even a metric form like the hexameter could have served the purely mnemotechnical purpose of branding routes into the memory through individual descriptions of place and their succession. But the metric requirements of the Greeks also served to disperse the signifiers of real locations into their poetic counterparts: on purely poetic grounds, Cephalonia, for instance, seems to have been "cut up," so that another "long island" seems to have emerged from an isle called the "long Same." And the ethnic description of Homeric heroes goes adrift when the metric order of this or that passage requires it: "The Greeks are called Argeioi, Danaoi, or Achaioi, depending on whether the name appears at the beginning of a line, a medial caesura, or a line's end."[34] Every Greek in the *Odyssey* is thus a polytropos. But their topoi are in fact so exact that through Calypso's sailing instructions[35] and the constellations of Homer's time that have since been astronomically ascertained, the locations of "merely poetical" islands have been extrapolated.[36] In any case, Bérard finds the term "poetic description" insufficient to grasp the extent to which the verse of the *Odyssey* is entangled with "local details."[37]

Therefore, Bérard finally began to look at the stations of the *Odyssey* themselves as best as his research could localize them. Then, in 1933, he published an *Album Odysséen* along with numerous photographs. The result corroborated the hypothesis of the "equiprimordiality" of nautical and poetic experience: "All descriptions correspond to perceivable reality, to the scientific and experimental truth."[38]

The Tricky Polytropos

Precisely because they reference the real "delocalization" of the "placeless" ocean, the topoi of the *Odyssey* point as much to a firm origin as to an unstable route of many places and turns. This inevitable "being-off-course" concerns on the one hand the recursive déplacement of tradition and its crafty rescue on the other. Endowed with the gifts of song and supernatural knowledge, the sirens were once something like the daughters of a muse and a river god. But it is said that after their defeat in a contest with the muses themselves, they sank to the level of demonic harbingers of Hades. After that, sirens were regarded as "a kind of evil muse," as Jane Harrison writes, "rather of the barren sea than of the clear spring water," and were thought to personify the "perils of seafaring," or, more ambivalently, the fortuna di mare.[39] When they invoke the grammar, diction, and topics of the *Iliad* within the *Odyssey*,[40] the sirens aspire to occupy the place of the muses, previously the source of poetic inspiration as such. At the same time they promise Odysseus immortality by transforming him once more into the hero of the *Iliad* that he once was. Odysseus voluntarily chooses another way: the journey home, and along with it the peculiar cunning that turns him from a warrior into an adventurer and transmutes demonic songs into poetry.

Leaving the tradition of the old muses behind, the *Odyssey* opens up the possibility of a new, nonheroic epic: that of crafty survival in the depths of peril and of an adventurous route on a placeless ocean. This is evident in the fact that Odysseus is the first hero to be remembered without a heroic death. He is a hero who becomes immortal by deterritorializing the (traditionally landlocked) lieux de mémoire and by merging them (nautically) into "polytropy." The exact meaning of *polytropos* was debated in antiquity for this reason. The term could mean "cunning" and "dexterous" in the sense of spiritual and linguistic versatility, but it could also mean "of many turns" (i.e., "cast up" by fate and left wandering in open space). It has already been established that the *Odyssey* avoids any attempt at disambiguation. *Polytropos* appears only in a polyvalent context.[41] In a purely verbal sense, Odysseus is known as the most complex of all the heroes of antiquity precisely because of his boundless "adaptability" and the possibility for endless recursion that comes with it.[42] But already in antiquity that meant that there is no "true Odysseus" to grab hold of–just as there is no "true route"–beyond his tricks of speech and deceptive maneuvers.

In Latin poetry, however, whose tradition begins with Livius Andronicus's Homeric adaptation (titled *Odusia*), the poets fought this "polytropy" and its topographic consequences. In Virgil's *Aeneid*, the recursion on the *Odyssey* is transformed into a pattern of political self-assurance. The escape route of the expelled Trojans outpaces the development of the polytropos by the fact that six "Odyssean" books are followed by six "Iliadic" books that depict a protracted "conquest" in a "justified" war. The epic of the cast-up, who instead of heading home first becomes a refugee and then finally a conquerer and founder, confers the ideology of the principate with the legend of its birth. After all, Virgil set to work after the naval Battle of Actium in 31 BC with the purpose of writing the bloody origins of the Pax Augusta into oblivion.[43] That the epic became an instrument of naval supremacy for Virgil demonstrates perfectly how shipwrecks themselves can be transformed into a toponymy of power. Aeneas calls the promontory where his helmsman Palinurus went down "Cape Palinuro" and thereby marks a turning point in the future imperial topography of Rome.[44] Virgil likewise uses it to glorify the casualties that Octavian's fleet suffered here, before Rome's "thalassocracy" was to rasterize the Mediterranean after the rules of land surveying, and proclaim it mare nostrum.

Since the *Aeneid* connects the "polytropy" of the *Odyssey* with the empire's providential telos, Odysseus's cunning is simply regarded as deceitfulness.[45] His virtues are vices that go against the virtues of Roman legal certainty, namely the pietas as intergenerational contract and the fidelitas as sanctity of contract. From the Silver Age to the medieval Latin period, *Ulixice* therefore idiomatically means "fraudulent." In the Christian Middle Ages, as late as Dante's *Commedia*, a sort of "exit condition" is written into the endless cycle of pagan recursions. In canto XXVI of the *Inferno*, the circle of the nostos is broken. In a fiery speech Odysseus convinces his crew to sail beyond the Pillars of Hercules instead of continuing their journey home to Ithaca, which would include him in the "economy of salvation." Henceforth, he is to go down as hopeless and cast up. The trespass represented by Odysseus's "polytropy" in light of the Christian, namely Augustinian, catalog of sin is on the one hand curiositas, the quest for pure experientia oblivious to self and creation. On the other hand, it is the crafty or even sirenlike seduction of his listeners—the use of incendiary rhetoric to deliberately lead them astray, as described by Dante's teacher Brunetto Latini in his *Retorica*.

In the *Commedia*, Ulisse's infernal appearance is therefore one of a fork-tongued flame.[46] He ends up in Malebolge, the circle of Hell reserved for frauds, in the ditch for dishonest advisors. Around 1300, "adventures" that overstepped the bounds of approved itineraria irritated the established Christian order of words and places since they disregarded mortal being-within both verbally and geographically. Ulisses's violation of nec plus ultra and his alleged search for an earthly paradise in the uninhabited Southern Hemisphere drive him to Mount Purgatory, where he and his ship are thrown into a turbo,[47] or a cyclone, and devoured by the sea. The wreck comes to pass in this way, "com'altrui piacque," because the one Christian God wills it, and the pagan epic and its entire fictional cosmos go down along with Odysseus.[48] Conversely, this shipwreck corresponds to Dante's ascent. Only because of it can Dante's work become a "legno che cantando varca" (i.e., a piece of wood, or more accurately, a ship), which rather than going under, drifts upward in song.[49] From above, in hindsight and with a view of everything earthly, Dante can watch Ulisse's "mad flight," his "folle volo" once more.[50] Run aground in Christian Hell, Dante's Odysseus becomes an adventurer with no hope of return. But he is also the first Wi(e)dergänger,[51] or revenant, of that unholy discurrere, by which the Occident, since being (dis)oriented by empirical knowledge, will be endlessly haunted.

The Compasso of Navigation

Dante's Ulisse goes definitively adrift just when seafarers attempt to describe the Mediterranean in a new way: with ships equipped with new kinds of riggings and stern rudders, as well as sea charts and compasses. It is this multimedial navigation that first makes hydrography possible: in other words, the modern project of "writing the sea." And it is also this that is capable of producing the unity of order and place that those ashore had called nomos since antiquity. The compass is indispensable to this process as a means of orientation linked as much with navigation, astronomy, and cosmology, as with new literary techniques. The compass indicates the particular way it makes use of the earth's magnetic field. Historically, it was the first measuring device to visualize elemental forces through a pointer as well as their long-distance effects.

Since there was no proper concept of magnetism before the nineteenth century, this phenomenon prompted all sorts of different cosmological pictures: Pliny the Elder mentions the shepherd Magnes, who was said to have experienced a mysterious force pulling at his hobnailed shoes on Mt. Ida on the island of Crete. Ptolemy tells of ships getting stuck on magnetic islands in the Indian Ocean—probably a fabulous reversal of the account that local ship construction in those days made use of wooden nails.[52] After Ptolemy was translated into Arabic in the ninth century, magnetic mountains began to appear in regional myths such as "The Voyages of Sinbad," or in medieval verse novels such as Gottfried von Strassburg's *Tristan*, before establishing themselves as poetic and geographic myths. The *Gudrunlied*, for example, places the mountains in the northern waters, and they appear in Guido Guinizelli as relays between the North Star and the compass needle. The dangerous magnetic declinations, which seafarers of the Middle Ages were already fighting and that were imagined as magnetic mountains, stood for an aberration from the true faith. In this way the correct directions of the compass were inversely given a theological association with the right way to God: the needle as a sympathetic sensorium for the otherworldly force that appears in the heavens as the North Star and in the Aristotelian textbooks as the Prime Mover.

The oldest technical attestation of the compass needle, its magnetization and its practical use, comes from Alexander Neckham. In 1187, he mentions a needle in a bowl of water—in other words, a liquid compass. As unimportant as the differences between the magnetic, the geographic, and the celestial poles were to this system, the reassessment of the concept of "orientation" proved vital. While at sea, one no longer looked in the direction of the Orient or Jerusalem, as in overland travel or in the itineraria, but rather toward the polestar. Thus, "to lose one's bearings" came to be called *desnortear* in Portuguese, or in French, *perdre le nord*. Already by 1269, Petrus Peregrinus had described a dry compass with a compass rose, circular graduation, and sights. Only a precision suspension, the connection of needle and rose, and a closed case were needed to have the compass, the bussola in its classic form, ready at hand. (See figure 6.2.) The term *bussola* first appears in 1380 in Francesco da Buti's commentary on Dante as a corrupt variant of the Latin *buxida* (wooden box).[53] *Compasso*, on the other hand, is derived from the Latin *cum passare* ("to accompany") or *compassare* ("to gauge, measure"). It seems probable that from the meaning "circle," or "ornament in the shape of a circle," the nautical designation for the parhelic circle also came to be compasso.

The portolans, the medieval descriptions of harbors that have come about with the use of the compass, read like prosaically compiled, purely utilitarian sailing handbooks, when compared with the antique periploi. However, they contained a decisive new element: course instructions. Portolans give contextual information for the directions provided by the compass. Together with the bussola, they made possible the graphic system that first visualized the sea itself as seen on the portolan charts of the thirteenth and fourteenth centuries. These charts are the very first marine maps. They carry the lists of the handbooks over into the second dimension. Developed from the requirements of nautical praxis, applied as a means of information for dynamic, variable, or merely possible situations, and thus the visual counterpart of a discurso as "getting-there-narrative,"[54] the portolan charts have no basis in mathematics. They are neither projections nor do they have map grids. Unlike later sea charts, they are not organized from a transcendent perspective. Instead, they correspond to an empirical navigation, moving from point to point. But it is they, and not the portolan books, "that bring the relative locations of each place directly into view for others and designate them exactly."[55] (See figure 6.3.)

Although they only revealed the "unmarked space" of the sea in small "steps" or "measures," the portolan charts were the first to give a comprehensive view of what would later be called "Europe," whose continual expansion came from this self-reflection. Together with the portolan data, scales, and numerous compass rose coordinates, the charts made it possible to determine the correct rhumb lines on the surface of the map, and then, with the help of the compass, to choose the right course across the surface of the sea. This simultaneously symbolic, graphic, and operative system

Figure 6.2

Portuguese bearing compass (1780) (Wolfgang Köberer, ed., *Das rechte Fundament der Seefahrt: Deutsche Beiträge zur Geschichte der Navigation* (Berlin, 1982), 316)

Figure 6.3

Pisan chart (1290) (John Blake, *Die Vermessung der Meere: Historische Seekarten* (Stuttgart, 2007), 11)

circumscribes the ambiguity of the term *compasso* as it was used around 1300. It describes a handbook and a chart, a pointer, and a pair of compasses, employed to delimit distances. Even if your own place was uncertain, once you were locked into the correct course, you had a few days to find a coastline, at which point the last leg of the trip was manageable. In the confined space of the Mediterranean Sea this is a fairly easy task. The Mediterranean is thus a compasso that secures the empirical and vectorial navigation that promises a successful round trip and that guarantees it a specific "being-within."

Dante's Poetic *Navigatio Vitae*

It is against this hydrographic backdrop that the navigational compass found its way into poetry. In 1258, in an account of his stay in England, Brunetto Latini reported the discovery of the compass by the seafarers there—a discovery both wondrous and dangerous, possibly a tool of the devil. That the magnet could help reveal the secrets of creation without serving the devil's interests was shown by Latini's student, Dante Alighieri, by means of poetry. The locus classicus for the first poetic use of the compass is found in the *Commedia*, in canto XII of the *Paradiso*, in three lines that allude to the chants of the Franciscan Bonaventura, the pious author of *Itinerarium Mentis in Deum* ("Itinerary of the Mind to God"):

> Del cor dell'una delle luci nove
> si mosse voce, che l'ago alla stella
> parer mi fece in volgermi al suo dove.
> From the core of one of these new lights,
> as the north star makes a compass needle veer,
> rose a voice that made me turn to where it came from.[56]

The *Commedia* lays out a poetic topography that could be described as a scholastic-theological variant of Aristotelian cosmology and its concentric spheres. (See figure 6.4.) Hell, in which no star shines, in which no heavenly pull can be felt, and where everything is fixed in place, stretches all the way to the center of the earth—the furthest place from God. Mount Purgatory rises from the sea-covered, and thus deserted, Southern Hemisphere. From here the ascent into the earthly paradise may begin before the entry into the spheres of planets and stars, after which the crystalline sphere and Empyrean Heaven finally dissolve space, time, and all efforts into light and bliss. There is nothing in the crystalline sphere that is not raised up in its "being-within" and enclosed by the Primum Mobile—the highest unity in the world, which adjoins the one and highest being of the Empyrean. This does not affect the earthly sphere in the sense of a cause or according to the laws of nature, but rather immediately, in the manner of light or magnetism. And the border between this outermost

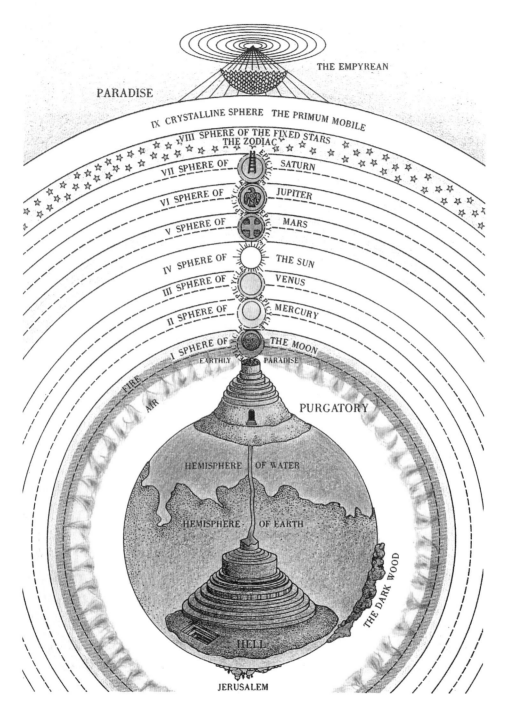

Figure 6.4

Dante's cosmography (Bruno Binggeli, *Primum mobile: Dantes Jenseitsreise und die moderne Kosmologie* (Zurich, 2006), 92

sphere and the Prime Mover is not spatial but an ontological border between the limited world and the limitless God.

Within this cosmographic setting, Dante introduces the compass on different levels. In the opening events of the *Commedia*, in which technical, theological, and philosophical knowledge translate into one another, it appears in persona: the cum passare, the escort to the highest destination, first takes the form of Virgil, then Beatrice, and finally Bernard of Clairvaux. On this side of the *Commedia*'s topography, the earthly region is located. It is spherically organized and already aligned "north-up" to the polestar, which of course may only be located in a clear sky. On the gran mar dell'essere, on the "grand sea of being," the normal occurrences of earthly obfuscation necessitate directions. For although everything and everyone are allocated their particular place, their propria essentie operatio (particular activity of an entity) remains decidedly in a sort of theological harbor finding. Just as the needle miraculously points to the polestar after having been coated with lodestone, the Prime Mover imbued torpid hearts with faith, hope, and love, and thus with an aspiration upward. Here on earth, we all find ourselves on a navigatio vitae. Hydrography therefore precedes the divine abrogation of all topography, and how we are guided here on earth determines our place in the hereafter—stuck forever in the place of suffering or on a vector of endless and blissful ascent.

Under these conditions, the journey described by Dante's Ulisse in canto XXVI of the *Inferno* must come to an infernal end, for Ulisse does not know the enigma of true belief. Without a compass, he sets sail for the Atlantic, breaks open the circuit of the old periploi, and goes beyond the Pillars of Hercules, which for Dante still marked the boundary of Christian "being-within." He follows the sun, heads first to the west, then turns to the south. In place of the polestar, a constellation of four stars appears. Instead of the hoped-for earthly paradise, Ulisse spots the highest of all mountains, and goes down in an abysmal undertow. What Dante already knows to be Mount Purgatory, Ulisse still believes to be a magnetic mountain. Certain of his fate in the approach of his own "lodestar" and helically ascending toward "the One," Dante himself transcends all that is merely conventional and symbolic.

Ulisse too traverses a winding path, but it remains bound by the terrestrial, and thus describes not ascent, but rather a sort of determined aberration from mortality. The failure of spatial sublation corresponds to a breakdown on the level of the sign. Because Ulisse relies not on the directions to the One, but on the "polytropy" of conventional signs—on mendacious words and deceitful rhetoric—he lacks the ontological magnetism that holds order together. Therefore he finally ends up in the pull of the turbo, in a vortex of signs, which, in an act of God, devours both Ulisse and his ship in the sea of garbled being and meaning. Ulisse exceeds the sublunar ordo, but he does not transcend it. Still, such transcendence is the only end of all existence and of all writing, as long as it deals with paradise and thus with our trasumanar, our "moving beyond mere humanity." To come closer to this "placeless" paradise and to provide it with poetic topoi, Dante relies on the media of hydrography.[57] He calls his work a legno, a wooden ship, which outstrips the segno, the definite sign. Thus, even against a theological backdrop, Dante's epic outlines a kind of hydrographic poetics—a mode of writing in response to the challenges of navigation and nautical cartography.

The *Commedia* is not only a topically organized allegory, but a dynamic frame of reference that begins in an uncertain place (such as the dark forest of the first canto), and navigates, from point to point, per luogo etterno, "through an eternal realm."[58] The narrated way becomes here the way of the narration; the cumpassare is at once metaphor and medium of transcendere; and art is no longer merely a handicraft, but something guided by genius or ingenio. It no longer merely follows a predetermined path. Instead, it continuously develops its own coordinates oriented toward a higher order. Yet this ingenio may merely be self-absorption and therefore pretension and hubris. Dante's fictive words, his "parole fittizie"[59] and their fiction of not being fiction, but theological discourse, may themselves be seen as sharing an affinity with Ulisse's deceitful speech. Thereby the signs of being may be read not as signs of the Creator, but as signs for their own sake and for the sake of their mortal author. In the end, Dante may well be the Ulisse of poetry, and poetry and seafaring just two aspects of one and the same transgression.

On the Edge of Modern Cartography

Like the instructions of the bussola- and portolan-guided navigation, the signs of Dante's autopoietic course are neither merely natural nor completely artificial, neither primal nor purely deduced. They unleash the *virtus fictiva* (the fictive force and force of fiction) that the *Commedia* seeks to legitimize, in order to simultaneously presuppose it. Dante's poetical compasso opens up or even builds a new world. And within a century of the *Commedia*, this nautical "construction" of a new world had moved into the realm of concentrated and institutionalized knowledge production. Unlike the Genoese brothers Guido and Ugolino Vivaldi, who disappeared on Africa's west coast in 1291 and served as models for Dante's Ulisse, the Portuguese prince Henry the Navigator organized a systematic circumnavigation of Africa based on his legendary sailing academy in Sagres. In the steady feedback loop between continuously updated sea charts and the latest information from sea expeditions, he wanted to realize his geopolitical vision of regular oceanic sea trade and the conversion of pagans and Arabs, but further, the world was from then on to be developed "empirically." And indeed, by the time of Bartolomeu Dias's doubling of the southern cape in 1488, the old Ptolemaic "worldview," according to which Africa could not be circumnavigated due to its connection to an unknown southern continent, had been overtaken. (See figure 6.5.) The future linkage of seafaring and sea trade based on the steady refinement of a general chart and general catalog can be described as the prototype of institutionalized occidental empirical

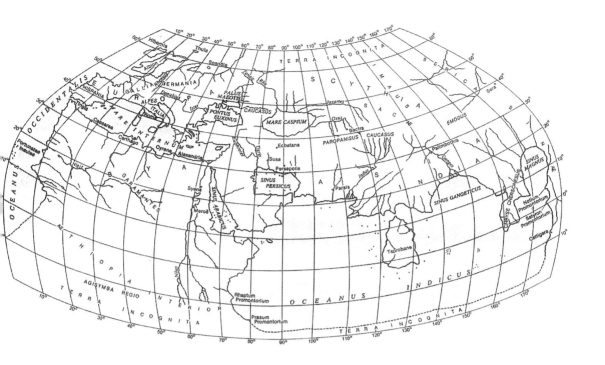

Figure 6.5

World map according to Ptolemy from Donnus Nicolaus Germanus, *Cosmographia* (1482) (http://cartanciennes.free.fr//maps/monde_ptolemee.jpg)

knowledge.[60] It was already practiced in 1503 in the Portuguese Casa da Índia, followed by the Spanish Casa de Contratatión, and eventually, based on the Iberian models, by the English Royal Society. That the fabrication of this global topography did not occur seamlessly, however—that this new "worldview" contained a number of blind spots—is evidenced by numerous cartographic complications, fabulous historical accounts, and the modern poetics of the cast-up.

Gaspar Correa's *Lenda da India* from the 1550s, for instance, adopts the style of imperial chronicles of seafaring and tells of Vasco da Gama's definitive opening of the seaway that made stable trade possible with the East Indies, and thereby of Portugal's rise as a maritime world power. Here the immense and decade-long endeavor to circumnavigate Africa's southern cape is compressed into a dramatic storm at sea, in which da Gama's convoy revolts and the failure of the whole venture looms. But the commander reacts promptly: he puts the mutineers in chains, gathers all the navigational instruments and charts, throws them overboard, and proclaims, "I do not require master nor pilot …, because God alone is the master and pilot."[61] As later comparisons with documents from da Gama's voyages show, Correa indulges in a fabulous historical narrative. In reality, there was neither a great storm nor a dramatic mutiny. In Correa's account, however, the historical figure of da Gama runs together with that of Bartolomeu Dias. Moreover, as the maritime sovereign by God's grace who can count on God's gubernatio even without charts or navigational instruments, da Gama appears as a pious and successful antitype to Dante's Ulisse. He acts as a prototype of the "captain," who would become indispensable to the command of the sea by Iberian, English, and Dutch naval powers. Nevertheless, Correa's foundation myth was soon perverted by the seaman's yarn spun all over the world: da Gama here seems to be the model of the accursed "Flying Dutchman," who, like a Widergänger of Ulisse, is damned to a "transcendental homelessness," an endless and directionless odyssey along the "Cape of Good Hope."

This yarn emerged at the very moment that the end of all real aberrances was promised by the scientification of seafaring, as it has been undertaken since the seventeenth century particularly in England. Francis Bacon conceived his Great Instauration in direct opposition to Dante's Ulisse. To him, the journey past the Pillars of Hercules was a duty rather than hubris. For in order to acquire "empirical knowledge," one must push beyond the "round trips" in the well-known mare nostrum. While it may lead into the unknown and the darkness, such a "great journey" is guided by method and science, and can thus reckon with the perpetual discovery of New Worlds. *Certi viæ nostræ sumus, certi sedis nostræ non sumus*—"We are certain of our road, but not of our position"—thus read Bacon's guiding maxim, which turned the direction of the compass, proven nautically and interpreted theologically by Dante, into the fundamental principle of "experimental science."[62] In this way, the compass came to serve as an emblem of the modern acquisition of power.

The more certain modernity became of its way, however, the more precarious its globalized system of orientation proved to be. From the outset, the great journeys confounded the old compasso. The portolan charts failed on longer ocean journeys because the rhumb lines could no longer be precisely located due to the curvature of the earth, so the distances led ships astray. The new "world picture" overwhelmed the old means of representation. The gridded world maps, made since the rediscovery of Ptolemy's *Geographia*, could not simply fall back on older imaging processes. Especially the gridlike positioning based on longitude and latitude conformed to an entirely different set of principles than the old vectorial localization. Despite Pedro Nunes's introduction of loxodromes and Gerhard Mercator's development of the conformal cylinder projection, usable nautical representation was limited to small sections of the world and faltered the closer one came to the upper latitudes where the scale became increasingly distorted. Furthermore, it had been known since Columbus's time that variation, the misleading angle between local magnetic and true geographic north, was stronger in some places and reversed in others. And the consequent revisions of geomagnetic hypotheses—from the assumption of a static axial or tilted dipole to that of a rotating or unconnected dipole[63]—could not thwart the further destabilization of the compasso. Alongside its declination, the compass needle's "inclination," its mutable tendency to point toward the center of the earth, and its temporal variability were discovered.[64] Around 1800, "deviation," the fact that the compass needle was thrown off by the ship's iron, was finally observed. To make matters worse, it was determined that the iron parts had a specific "magnetic signature" that could change anew with every route, according to its angle relative to the lines of force of the earth's magnetic field.

Even when the new maps with their grids and projections suggested completeness and precision, expanded to the global scale, the compasso system of location and order ultimately had to surrender. Understanding the dynamic field of reference between a ship and its parts, compass needle and geomagnetism, to say nothing of controlling or balancing it, appeared to be an endless task. Cartographically, for example, the difficulties could be seen in the attempt to depict the polar regions. Since they had hardly been explored, one could only wonder what tidal conditions reigned there and how they might affect the world's oceans. It was also unclear to what extent the magnetic North Pole coincided with the geographic North Pole, the location of which was in turn dependent on the compass, and, through the projective distortion in the polar regions, caused problems for cartographic representation. Even in 1861, Matthew Fontaine Maury, the father of modern oceanography, called the topo-, carto-, and hydrographically precarious polar region "a circle of mysteries," to which we are not driven by empty curiosity, but by the desire "to comprehend the economy of our planet."[65]

The *Odyssey* in Modern Literature

Against this backdrop, narrative texts in the tradition of the *Odyssey*—recursions on Homer's Odysseus, Dante's Ulisse, or the Flying Dutchman—proliferated in the nineteenth century, all highlighting the collapse of nautical and narrative orientation. In Edgar Allen Poe's "MS. Found in a Bottle" from 1833, for example, a cast-up protagonist finds himself once again aboard a Flying Dutchman, whose sea charts are long since out of date, but whose crew are determined to make their way to a polar whirlpool. The text ultimately opens itself to a topology of engulfment, which marks the limit of topographic representability. At long last, the narrative that is supposed to reach the reader as a message in a bottle from the "sirenlike" abyss of world knowledge is annotated by a short endnote: the text, it says, is only comprehensible with the help of Mercator's enigmatic polar map. For on this map "the Pole itself" is "represented by a black rock, towering to a prodigious height" and, at the same time, by "the bowels of the earth"—as if the indescribable finds its "analogous" depiction in the legendary magnetic mountain and its "negative" depiction in the maw of the earth.[66] A glance at Mercator's map makes the aporetic text comprehensible. (See figure 6.6.) Yet the map itself arose to depict the aporias of polar research. The ocean and its fatal pull, the pole and its enigmatic magnetism, and the map and its distorted projection all stand for one another in a kind of endless equation. And so, as though Poe wanted to plug one hole in knowledge with another, map and text stand for one another as well. Although the pole once more takes the form of a compass on the circular map, it has become an unreachable and fathomless region of the world and of knowledge as such.

Herman Melville's *Moby-Dick* (1851) represents a far-reaching recursion of Poe's story, as well as of Dante's canto XXVI and Homer's epic. On the one hand, the novel depicts contemporary maritime practices and navigational technology with great precision (as is obvious in Melville's explicit references to Maury); on the other hand, it pursues that very maelstrom that leads out of the topography and cartography of the modern disclosure of the world. Maury's sea maps, for which he continually analyzed countless logbooks of whalers around 1850, record global appearances of whales in grids he indexed numerically and topographically. In light of this, Ahab's "hunting instinct" is statistically underpinned—as the novel itself underscores in an extra footnote.[67] Yet he is absolutely possessed by The Whale, by an unprecedented example or unexampled precedent (*beispielloses Beispiel*) that at once incarnates and transcends its species and that therefore transgresses the field of statistical normality and deviation just as much as that of cartographic representation. The nearer the *Pequod* comes to him, the "less scientifically" it is piloted. Ahab destroys all navigational instruments, shirks the cartographic dispositif, and navigates vectorially—not to find Moby-Dick "down in any map," but where he has always been, in "limitless, uncharted seas."[68] If Moby-Dick exerts some kind of magnetic influence from outside,

Figure 6.6

Detail from Gerhard Mercator's world map (Gerhard Mercator, *Septentrionalium Terrarum descriptio*, VIII, Arktis, ([1595] 1634), taken from Roger Calcoen, *Le cartographe Gerard Mercator 1512–1594* (Brussels, 1994), 89) then, once again, external forces fundamentally transform the old Homeric nostos. But instead of Dante's highest of mountains, there appears the greatest of all living creatures; that which wills the descent into the whirlpool or turbo is no longer that big Other by the name of God, but a blind will behind a screen of pure, almost polar "Whiteness." If Ulisse's end was observed from God's perspective, then Ishmael now sees Ahab's disappearance from within an immersive situation; and if Ulisse found his place posthumously in the topography of the *Inferno*, then Ahab, on his entropic trajectory, vanishes completely.

The shift, which is not just genre-poetological, but also historico-philosophical, from the nostos in the rounded cosmos to the "transcendental homelessness" and eternal wandering that Georg Lukács characterized as the shift from epic to novel, becomes most explicit in the maritime recursions of Homer's *Odyssey* and in their dissolution of world pictures backed by cartography. In modern writing, this "chaodyssey," or "chao-errance," as Gilles Deleuze said, comes to pass apart from the sea. James Joyce understood *Ulysses* (1922) as his novelistic recursion of the *Odyssey*, and it picks up exactly where Odysseus's "last" journey, as prophesied by Teiresias, falls out of its epic framework, or where Odysseus simply left out one of the many forks in his path: the passage through the Symplegades (which Joyce calls "the Wandering Rocks"), for instance. *Ulysses* unfolds a space of possibility of different modalities of being, and in so doing it experiments with endless organizational principles—the linguistic and the list, the topographic and the topological—among which cartography is merely one possibility. Joyce recurses here on Bérard's toponymic work, and on his thesis that the Odyssean poet invents nothing, that everything narrated and all the methods of narration have a local character and are bound to a specific place.[69] While, for Bérard, the Phoenicians or the Greeks were a "polytropically" mixed race of traders and explorers with impure ethnic and linguistic origins, for Joyce, the same holds true of the places and words, the routes and roots, of the Irish. As Joyce knew from Giambattista Vico's teachings of historico-cultural ricorso and his early-Enlightenment study of Homer, the question of the "real" Odysseus and the "true" *Odyssey* is a question of cunning. Once they are taken to sea and thus to their own edges, maps guide the way between the Scylla of sheer realistic perception and the Charybdis of purely imaginary interpretation. But to escape the shipwreck of poetry, the map and the text alike must undergo a continual sea change. When it wants to open itself to that which is beyond representation without going astray, cartographic writing must again take up Odysseus's strategy of the cast-up and cunning hero. Or, as Ezra Pound tells us, it must assume Homer's perspective of the "periplum": "not as land looks on a map / but as sea bord seen by men sailing."[70]

Translated by E. A. Beeson

Notes

1. See Barry Powell, *Writing and the Origins of Greek Literature* (Cambridge: Cambridge University Press, 2002), 15, 195–196; Barry Powell, *Homer and the Origin of the Greek Alphabet* (Cambridge: Cambridge University Press, 1991), 66–67.

2. See *Odyssey*, vv. 243–261, 272–277; Samuel Mark, *Homeric Seafaring* (College Station: Texas A&M University Press, 2005), 70–96, 185–186.

3. See Robert Foulke, *The Sea Voyage Narrative* (New York: Twayne, 1997), 40–42.

4. See Fernand Braudel, *The Mediterranean and the Mediterranean World in the Age of Philip II*, vol. 1 (Berkeley: University of California Press, 1995), 109.

5. Max Horkheimer and Theodor W. Adorno, *Dialectic of Enlightenment: Philosophical Fragments*, ed. Gunzelin Schmid Noerr, trans. Edmund Jephcott (Stanford, CA: Stanford University Press, 2002), 37, 50. The original German wording is "verschlagen werden und verschlagen sein."

6. Victor Bérard, *Les Phéniciens et l'Odyssée*, vol. 1 (Paris: A. Colin, 1902), 482.

7. Victor Bérard, *Les navigations d'Ulysse* (Paris: A. Colin, [1927–1929] 1971), vol. 1, 312.

8. On the change of style compared to the *Iliad*, see Johannes Kahlmeyer, *Seesturm und Schiffbruch als Bild im antiken Schrifttum* (Hildesheim: Fikuart, 1934), IV, passim.

9. Bérard, *Les navigations*, vol. 3, 423.

10. Georg Wilhelm Friedrich Hegel, *Lectures on the Philosophy of History*, trans. J. Sibree (London: G. Bell and Sons, 1914), 237.

11. See Plato's *Critias*, 106a–121c, and *Timaeus*, 17a–27b.

12. See Albin Lesky, *Thalatta: Der Weg der Griechen zum Meer* (Vienna: Rohrer Verlag, 1947), 25–26, 33–35.

13. See Armin Wolf and Hans Helmut Wolf, *Die wirkliche Reise des Odysseus: Zur Rekonstruktion des Homerischen Weltbildes* (Munich: Langen & Müller, 1983), 145–147.

14. Strabo, *Geographica* I. 2, 15 (C. 22).

15. William Ewart Gladstone, *Studies on Homer and the Homeric Age* (Oxford: Oxford University Press, 1858), vol. 1, 219, 223.

16. Bérard, *Les Phéniciens*, vol. 1, 4.

17. See Strabo, *Geographica* II. 1, 30 (C. 83).

18. Bérard, *Les Phéniciens*, vol. 1, 295.

19. See Jonas Grethlein and Antonios Rengakos, *Narratology and Interpretation: The Content of Narrative Form in Ancient Literature* (Berlin: de Gruyter, 2009), 275–291.

20. See Bérard, *Les navigations*, vol. 3, 415.

21. See Bérard, *Les navigations*, vol. 1, 309.

22. Bérard, *Les Phéniciens*, vol. 1, 3. See also Bérard, *Les navigations*, vol. 4, 484.

23. See Odysseus's differentiation between recursive and simply repetitive narration in the *Odyssey*, XII. vv. 452–453.

24. See Homer, *Odyssey*, XII. 189–190.

25. See Homer, *Odyssey*, XII. 47sqq.

26. See Homer, *Odyssey*, XII. 69–70.

27. See Karl Meuli, *Odyssee und Argonautika* (Berlin: Weidmann, 1921), 112, 116–118.

28. See Uvo Hölscher, *Die Odyssee: Epos zwischen Märchen und Roman*, 3rd ed. (Munich: Beck, 1990), 110.

29. See Bérard, *Les navigations*, vol. 2, 438; vol. 3, 35; vol. 4, 498.

30. See Bérard, *Les Phéniciens*, 1902, 56–57.

31. See Bérard, *Les navigations*, vol. 3, 414.

32. See Bérard, *Les navigations*, vol. 4, 490; Bérard, *Les Phéniciens*, 18–19, 584.

33. See Bérard, *Les Phéniciens*, 5; Bérard, *Les navigations*, vol. 4, 500.

34. Walter Burkert, "Odysseen: Phantasien, Realitäten und Homer," in *Odysseen: Mosse-Lectures 2007*, ed. Elisabeth Wagner and Burkhardt Wolf (Berlin: Vorwerk 8, 2008), 15, 22.

35. See Homer, *Odyssey*, v. 271sqq.

36. See Richard Hennig, "Die Kenntnis des Sternenhimmels in ihrer Bedeutung für die Nautik vor der Einführung des Kompasses," *Marine-Rundschau* 35 (1930): 509, 546.

37. See Bérard, *Les navigations*, vol. 1, 335; vol. 4, 478.

38. See Bérard, *Les Phéniciens*, 582.

39. Jane Harrison, *Myths of the Odyssey in Art and Literature* (London: Rivingtons, 1882), 182; Ludwig Preller, *Griechische Mythologie*, vol. 1 (Berlin: Weidmann, 1860), 481.

40. See Pietro Pucci, "The Song of the Sirens," *Arethusa* 12, no. 2 (1979): 121, 124–125.

41. See Georg Danek, *Epos und Zitat: Studien zu den Quellen der Odyssee* (Vienna: Verlag der Österreichischen Akademie der Wissenschaften, 1998), 33–34.

42. See W. B. Stanford, *The Ulysses Theme: A Study in the Adaptability of a Traditional Hero* (Oxford: Blackwell, 1954), 6–7.

43. David Quint, *Epic and Empire: Politics and Generic Form from Virgil to Milton* (Princeton, NJ: Princeton University Press, 1993), 8, 53.

44. See Virgil, *Aeneid*, VI. 381.

45. See Virgil, *Aeneid*, VI. 529, IX. 602.

46. See Dante, *Inferno*, XXVI. 52–53. English translations verbatim from the Princeton Dante Project, http://etcweb.princeton.edu/dante/pdp.

47. See Dante, *Inferno*, XXVI. 137.

48. Dante, *Inferno*, 141.

49. Dante, *Purgatorio*, I. 3.

50. Dante, *Inferno*, XXVI. 125.

51. Translator's note: *Wiedergänger*, literally meaning "one who walks again," refers to a number of different folk superstitions about the undead who return to the world of the living, while *wider* means "against" or "contra." Roughly, yet quite neatly: "one who walks again(st)."

52. See Heinz Balmer, *Beiträge zur Geschichte der Erkenntnis des Erdmagnetismus* (Aarau: Sauerländer, 1956), 43, 526–530.

53. See Amir D. Aczel, *The Riddle of the Compass: The Invention That Changed the World* (New York: Harcourt, 2001), 36.

54. Isabel Capeloa Gil, "Hydrography, or the Anxiety of the Sea: The Nautical Chart as a Cultural Model," in *Fleeting, Floating, Flowing: Water Writing and Modernity* (Würzburg: Königshausen & Neumann, 2008), 109.

55. Konrad Kretschmer, *Die italienischen Portolane des Mittelalters: Ein Beitrag zur Geschichte der Kartographie und Nautik* (Berlin: Mittler, 1909), 100.

56. Dante, *Paradiso*, XII. 28–30.

57. See Karlheinz Stierle, *Das große Meer des Sinns: Hermenautische Erkundungen in Dantes "Commedia"* (Munich: Fink, 2007), 10–15, 193–195, 225–226.

58. Dante, *Inferno*, I. 114.

59. Dante, *Convivio*, II. 1, 3.

60. See David Watkin Waters, *Science and the Techniques of Navigation in the Renaissance* (London: National Maritime Museum, 1976), 28.

61. Gaspar Correa, *The Three Voyages of Vasco da Gama and His Viceroyalty*, ed. Henry E. J. Stanley (New York: Franklin, 1869), 62.

62. Francis Bacon, *Works*, vol. 9 (London: Rivington, 1826), 187.

63. See Art Roeland Theo Jonkers, *Earth's Magnetism in the Age of Sail* (Baltimore: Johns Hopkins University Press, 2003), 36.

64. See Jonkers, *Earth's Magnetism*, 63–75.

65. Matthew Fontaine Maury, *The Physical Geography of the Sea and Its Meteorology*, 10th rev. ed. (London: Sampson Low, 1861), 199.

66. Edgar Allan Poe, *Poetry and Tales*, ed. Patrick F. Quinn (New York: Literary Classics of the United States, 1984), 199.

67. See Melville's up-to-the-minute reference to Maury's new technique in Herman Melville, *Moby-Dick or The Whale*, ed. Harrison Hayford, Hershel Parker, and G. Thomas Tanselle, 4th ed. (Evanston, IL: Northwestern Newberry, 2000), 199.

68. Melville, *Moby-Dick*, 55, 183.

69. See Frank Budgen, *James Joyce and the Making of "Ulysses" and Other Writings* (London: Oxford University Press, 1972), 15, 174; Mary T. Reynolds, *Joyce and Dante: The Shaping Imagination* (Princeton, NJ: Princeton University Press, 1981), 35, 347.

70. Ezra Pound, *The Cantos* (New York: A New Directions Book, 1996), 324.

Bibliography

Aczel, Amir D. *The Riddle of the Compass: The Invention That Changed the World.* New York: Harcourt, 2001.

Bacon, Francis. *Works.* Vol. 9. London: Rivington, 1826.

Balmer, Heinz. *Beiträge zur Geschichte der Erkenntnis des Erdmagnetismus.* Aarau: Sauerländer, 1956.

Bérard, Victor. *Les navigations d'Ulysse.* Paris: A. Colin, [1927–1929] 1971.

Bérard, Victor. *Les Phéniciens et l'Odyssée.* Vol. 1. Paris: A. Colin, 1902.

Bittlestone, Robert, James Diggle, and John Underhill. *Odysseus Unbound: The Search for Homer's Ithaca.* Cambridge: Cambridge University Press, 2005.

Braudel, Fernand. *The Mediterranean and the Mediterranean World in the Age of Philip II.* Vol. 1. Berkeley: University of California Press, 1995.

Breusing, Arthur. "Flavio Gioja und der Schiffskompaß." In *Das rechte Fundament der Seefahrt: Deutsche Beiträge zur Geschichte der Navigation,* ed. Wolfgang Köberer. Hamburg: Hoffmann und Campe, 1982.

Budgen, Frank. *James Joyce and the Making of "Ulysses" and Other Writings.* London: Oxford University Press, 1972.

Burkert, Walter. "Odysseen: Phantasien, Realitäten und Homer." In *Odysseen: Mosse-Lectures 2007,* ed. Elisabeth Wagner and Burkhardt Wolf. Berlin: Vorwerk 8, 2008.

Correa, Gaspar. *The Three Voyages of Vasco da Gama and His Viceroyalty.* Ed. Henry E. J. Stanley. New York: Franklin, 1869.

Danek, Georg. *Epos und Zitat: Studien zu den Quellen der Odyssee.* Vienna: Verlag der Österreichischen Akademie der Wissenschaften, 1998.

Foulke, Robert. *The Sea Voyage Narrative.* New York: Twayne, 1997.

Gil, Isabel Capeloa. "Hydrography, or the Anxiety of the Sea: The Nautical Chart as a Cultural Model." In *Fleeting, Floating, Flowing: Water Writing and Modernity.* Würzburg: Königshausen & Neumann, 2008.

Gladstone, William Ewart. *Studies on Homer and the Homeric Age.* Oxford: Oxford University Press, 1858.

Grethlein, Jonas, and Antonios Rengakos. *Narratology and Interpretation: The Content of Narrative Form in Ancient Literature.* Berlin: de Gruyter, 2009.

Harrison, Jane. *Myths of the Odyssey in Art and Literature.* London: Rivingtons, 1882.

Hegel, Georg Wilhelm Friedrich. *Lectures on the Philosophy of History.* Trans. J. Sibree. London: G. Bell and Sons, 1914.

Hennig, Richard. "Die Kenntnis des Sternenhimmels in ihrer Bedeutung für die Nautik vor der Einführung des Kompasses." *Marine-Rundschau* 35 (1930).

Heubeck, Alfred, Joseph Louis Russo, and Manuel Fernández-Galiano. *A Commentary on Homer's Odyssey*. Vol. 1. Oxford: Clarendon Press, 1988.

Hölscher, Uvo. *Die Odyssee: Epos zwischen Märchen und Roman*. 3rd ed. Munich: Beck, 1990.

Horkheimer, Max, and Theodor W. Adorno. *Dialectic of Enlightenment: Philosophical Fragments*. Ed. Gunzelin Schmid Noerr, trans. Edmund Jephcott. Stanford, CA: Stanford University Press, 2002.

Jonkers, Art Roeland Theo. *Earth's Magnetism in the Age of Sail*. Baltimore: Johns Hopkins University Press, 2003.

Kahlmeyer, Johannes. *Seesturm und Schiffbruch als Bild im antiken Schrifttum*. Hildesheim: Fikuart, 1934.

Kretschmer, Konrad. *Die italienischen Portolane des Mittelalters: Ein Beitrag zur Geschichte der Kartographie und Nautik*. Berlin: Mittler, 1909.

Lesky, Albin. *Thalatta: Der Weg der Griechen zum Meer*. Vienna: Rohrer Verlag, 1947.

Mark, Samuel. *Homeric Seafaring*. College Station: Texas A&M University Press, 2005.

Maury, Matthew Fontaine. *The Physical Geography of the Sea and Its Meteorology*. 10th rev. ed. London: Sampson Low, 1861.

Melville, Herman. *Moby-Dick or The Whale*. Ed. Harrison Hayford, Hershel Parker, and G. Thomas Tanselle. 4th ed. Evanston, IL: Northwestern Newberry, 2000.

Meuli, Karl. *Odyssee und Argonautika*. Berlin: Weidmann, 1921.

Montanari, Franco. "Episches Meer, Epos des Meeres." In *Das Meer, der Tausch und die Grenzen der Repräsentation*, ed. Hannah Baader and Gerhard Wolf. Zurich: Diaphanes, 2010.

Münchberg, Katharina. *Dante: Die Möglichkeit der Kunst*. Heidelberg: Winter, 2005.

van Nes, Dirk. *Die maritime Bildersprache des Aischylos*. Groningen: J. B. Wolters, 1963.

Poe, Edgar Allan. *Poetry and Tales*. Ed. Patrick F. Quinn. New York: Literary Classics of the United States, 1984.

Pound, Ezra. *The Cantos*. New York: A New Directions Book, 1996.

Powell, Barry. *Homer and the Origin of the Greek Alphabet*. Cambridge: Cambridge University Press, 1991.

Powell, Barry. *Writing and the Origins of Greek Literature*. Cambridge: Cambridge University Press, 2002.

Preller, Ludwig. *Griechische Mythologie*. Vol. 1. Berlin: Weidmann, 1860.

Pucci, Pietro. "The Song of the Sirens." *Arethusa* 12, no. 2 (1979):

Quint, David. *Epic and Empire: Politics and Generic Form from Virgil to Milton*. Princeton, NJ: Princeton University Press, 1993.

Randles, William Graham L. *Geography, Cartography and Nautical Science in the Renaissance: The Impact of the Great Discoveries*. Aldershot: Ashgate/Variorum, 2000.

Reynolds, Mary T. *Joyce and Dante: The Shaping Imagination*. Princeton, NJ: Princeton University Press, 1981.

Stanford, W. B. *The Ulysses Theme: A Study in the Adaptability of a Traditional Hero.* Oxford: Blackwell, 1954.

Stierle, Karlheinz. *Das große Meer des Sinns: Hermenautische Erkundungen in Dantes "Commedia."* Munich: Fink, 2007.

Taylor, Eva G. R. *The Haven-Finding Art: A History of Navigation from Odysseus to Captain Cook.* Vol. 2. London: Hollis & Carter, 1971.

Waters, David Watkin. *Science and the Techniques of Navigation in the Renaissance.* London: National Maritime Museum, 1976.

Wolf, Armin, and Hans Helmut Wolf. *Die wirkliche Reise des Odysseus: Zur Rekonstruktion des Homerischen Weltbildes.* Munich: Langen & Müller, 1983.

7
Diagrammatic Thought in Medieval Literature

Simone Pinet

> Every medieval diagram is an open-ended one; in the manner of examples, it is an invitation to elaborate and recompose, not a prescriptive, "objective" schematic.
>
> — Mary Carruthers, *The Book of Memory*

In 1993, a group of archeologists working in Navarra discovered a 13,660-year-old etched stone, the size of a hand, which layered information about animal numbers and locations over geographic and strictly cartographic data.[1] Identifying this, the first known map of Western Europe, the archeologists describe their discovery as the depiction of a landscape, showing access routes, waterways, and animals, in short, as a sketch or map. But they also argue it could represent a "plan for a coming hunt, or perhaps a narrative story of one that had already happened." Already here, narrative and cartography seem to mirror each other, implying or suggesting common operations. Alexander Gerner uses the overlaps suggested by the Navarra discovery to argue for the cartographic as one of our oldest orienting tools, one that precedes the map. Such a statement turns the cartographic into a cognitive tool that, in some cases, might produce a map, linking the cartographic to the diagrammatic, but in others, as I will insist in these pages, this might turn out to be narration. As a cognitive tool for orientation not only in space, but also and simultaneously in time, as a tool for planning but also for preserving and reliving, Garner argues, "the diagrammatic imagination and the external epistemic diagrammatic tools of the cartographic are complementary in creating orientation."[2] A material result of this thinking toolkit might be just as likely a diagram, a proper map, or a narrative, which for my purposes and in tune with a much more medieval, capacious notion of discourse, I will call literature.

In recent years, diagrams have appeared recontextualized in different forms of discourse, analyzed because of their cognitive functions, other times, presented as actively serving as a stance from which to reframe discussions, particularly on the contemporary. Studies on the image, for instance, have revisited the idea of the diagram as a key concept, either as a subset of the image itself, or, as in the case of Frederik Stjernfelt, as a broader term than that of image or figure in order to reinvigorate theories of the visual. On the other hand, philosophers of science have also turned to diagrams in their role within cognitive science, and from this perspective often return to discussions on art. In visual and media studies, the strong presence of semiotic analyses has linked diagrams to grammar/rhetoric, if predictably in mostly a metaphorical way, which does not pursue the links with language further into the realm of literature or even poetics. In these very different fields, which do not always seem entirely aware of each other, from phenomenology to art history, and from cognitive science to anthropology, maps figure as core examples, sometimes prominently. Only in passing, however, do these arguments reference the long tradition of thinking and drawing diagrams—and maps—in medieval culture.

As a diagram, the map is above all an image, sharing with other diagrammatic expressions such as painting or drawing a certain "imageness," defined by Jacques Rancière as "a regime of relations between elements and between functions," between the visible and the sayable, between operations of reading and writing, seeing and interpreting.[3] James Elkins surveys the main theories of the image that condition modern thinking. Among these, some strike a particular resonance with what I wish to discuss about the medieval image. The first is what he calls the Susan Sontag position, or images as reminders: "Images don't tell us anything, they remind us what is important," an emphasis Elkins himself points out is similar to John of Damascus's, and to Christian doctrine in general, where images can be a mnemonics of divinity.[4] Elkins goes on to the role of images as models, "entailing a capacity for cognitive revelation," following Gottfried Böhm's idea that images are ways of thinking, similar to verbal thought in process though not in sign that might entail similar operations, such as *deixis* or *demonstratio*.[5] Both the idea of the image as a reminder and as model are crucial to how, in the following pages, I want to suggest medieval maps work within diagrammatic thought—that is, how they serve as indicators, as pretexts, as triggers for narration and, specifically, as tools for the invention of literature.

For a medieval scholar such as myself, the contrasting positions that Elkins finds in contemporary theories of the image highlight the simultaneity of function and the striking overlaps in the operations of the medieval image. Elkins's concerns with a contemporary "dissonance between fundamentally political understandings of the image and those that are not; between theological conceptualizations of the image and those that do not require theology; and between ideas of the image that take the visual to be nonrational, irrational, or nonlinguistic, and those that do not," are positions that the medieval scholar quite often finds herself in at once, not as positions distant from each other but available by dint of a mere change in posture, a tilt of the head.[6] Elkins anchors these perspectives in figures such as W. J. T. Mitchell for the political bent, Marie-José Mondzain for a theological and at times even a "new materialist" approach to images, along with a varied number of exponents of the third, from Jean-Luc Nancy and Georges Didi-Huberman to Rosalind Krauss, François Lyotard, and Gottfried Boehm. Such theoretical perspectives are problematized—or rendered more productive—when confronted with medieval visual culture. Here, I want to consider maps as part of a medieval visual culture that merges the theological, the materialist, and the political as it structures thought. As devices for the organization of subjects and materials, as structures that enable sequencing and elaboration, I want to explore how they not only act as archives of narrative material, but in fact provide operations for narration to be invented anew. Critical work on the overlaps and productive exchanges between the discourses of cartography and literature has enjoyed a continuous interest and rich investigation on all historical periods, often engaging with theoretical approaches. The modern period, from Gaston Bachelard's *The Poetics of Space* to Fredric Jameson's concept of cognitive mapping to Franco Moretti's *Atlas of the European Novel,* has put to work different disciplines—philosophy, anthropology, geography, urban planning, literature—with the entanglements of maps and writing, but surprisingly these have not included diagrams. Mary Carruthers has, for the medieval period, articulated most eloquently how diagrams are a vital part of archiving and inventing, and Suzanne Conklin Akbari's recent work looks precisely at the interactions between historiography, diagrams, and cartography, while Karl Whittington is investigating such a discursive interaction for medieval visual culture. But this work has just begun in many ways. To relate these three discursive frames, the diagrammatic, the cartographic, and the literary, and to dialogue with contemporary theories of the image will inform not only my study of medieval visual culture but also that of medieval literatures, while medieval culture, in its imbrication of diagrams, maps, and fiction, might shed light on contemporary theories of the image.

Figure, Interpretation

Like any diagram, a map is a particular sort of visual display, and as a working definition that will allow us to consider maps along with other devices specific to the medieval period, including writing, I will simply describe them as simultaneously graphic and iconic. For writing itself participates in this hybridity, what Sybille Krämer called "notational iconicity," which is both content and spectacle, knowledge and visuality.[7] This was, of course, a kinship between the visual and the verbal that from Horace to Gregory the Great, from Richard de Fournival to Geoffrey of Vinsauf, was constantly made use of in a variety of contexts—meditational, pedagogical, inventional—and that I will refer to with a word common to both, *figurae*, which Mary Carruthers has extensively and articulately studied.[8]

The visualization of *figurae* that many texts call for was not only a flourish, but an actual intention, the invoking of a mental operation that was sometimes put to parchment, sometimes not, but was nevertheless always to be present in the mind. Medieval diagrams, *formae* or *figurae*, offered different visual strategies of organization of material, including stemma, medallions, illumination, different forms of allegorical or metaphorical visualization ranging from trees, to ladders, to colonnaded buildings and systems of *rotae*. Melanie Holcomb argues that the pervasiveness of such a visual repertory, and the corresponding visual culture of medieval people, would have been such that "a thirteenth-century viewer ... would have been equipped with the interpretive skills required to decipher its contents."[9] But what were indeed the strategies shared or borrowed between the verbal and the painterly, between the scriptural and the diagrammatic? What shape in particular did such a hermeneutics take?

Michael Evans argues that one can already see different graphic strategies at work in a medieval diagram from verbal construction itself, going from the typographic, the stemmatic, or the geometric, to the emblematic, strategies that do not evolve in linear fashion but coexist with illumination, and mostly and most intimately, with prose.[10] The layout of the text itself—not only in relation to or in dialogue with or as explanation of the images depicted alongside it—was carefully considered with regard to its clarity and its usefulness as a hermeneutic aide. Whether one considers capitulars or rubrics, sizes of script to differentiate between text and commentary, or frontispieces and indices, these are all verbal and visual choices. As many scholars have noted, especially in the analysis of script in art objects such as caskets, textiles, or architectural elements, medieval diagrams and other displays participate equally in visuality and discursivity in such a way that it becomes possible for the discursive to be read as iconic and the iconic as discursive, resulting in overlays and junctures. The decorative potential of script, for instance, blurs these distinctions. In al-Istakhri's maps, script "not only serves to label certain areas, but also, through extension of the baseline of the script, to draw borders between adjacent regions of the map," fulfilling a denotative function while performing a limit through its materiality.[11] Here, the question of how to distinguish the visual from the verbal is rendered inoperative, as it fails to account for the multiple functions or interdependent tasks the visual and the verbal share. If this commonality of task at a basic level makes one rethink the fluidity between the verbal and the visual in medieval culture, complicating both theories of the image and those of literary form, the inquiry into other intersections, while less evident, results in more intimate interplays of operations.

"Medieval diagrams are often dense with data," underlines Holcomb: "frequently their aim was to show how one set of information coincides with another, often in closed systems that reinforce one another," where not only typological writing, but also intertextuality and allusion come to mind as parallel rhetorical tools triggered by the visual display.[12] The mindset that produces diagrams not only inserts these in works of science, such as Macrobius's *Commentary on the Dream of Scipio*, or Isidore of Seville's *Etymologies*. It also undergirds works of fiction such as Alexander romances and travel literature most obviously, from Mandeville to Marco Polo, courtly fiction at large, and long-winded genres such as chivalric fiction from *Zifar* to *Yvain* to *Amadís*, where spatiality calls for history and geography, and invites ethnography as well as cosmology[13] (figures 7.1 and 7.2). Like the mirrored spaces of the court and of the *studium* dialogue throughout the Middle Ages, the learned production of diagrams makes its way both into the most luxurious of manuscripts and into the humble versions drawn in ink that populate less spectacular specimens, but that frequently show up in monastic, university, and royal libraries. "In some [manuscripts], their utilitarian role overrides aesthetic considerations; stripped down to simple lines and circles, they are quite humble in their appearance," but remain nonetheless fully functional, operational, productive in their power of allusion and connectivity.[14] In the movements first to and from monasteries, then to and from cathedral schools, and even later from universities to courts, clerics would have become familiar with diagrammatic thought as an essential tool for composition in general, as a sort of rhetorical template flexible enough to move from the visual to the verbal and back, from *abbreviatio* to *amplificatio* and vice versa.[15] In the continuous and pervasive display of such diagrammatics, audiences would not only recognize the patterns, but would also follow the paths for thought inscribed in and through such diagrams to achieve the desired reading, bringing together aesthetics and didacticism from composition to interpretation. Such a process leading from didacticism to hermeneutics is precisely what makes the clerical context for diagrammatic thought most obvious, and what prompts the most basic connections between the visual arts and the arts of language in the hermeneutic process.[16]

Visualization, Thought

Visualization of abstract concepts is a distinct function of diagrams, and their use to explicate theoretical ideas was indeed characteristic of the medieval period, providing, as Evans has argued, "a valuable commentary on the way medieval thinkers approached some of the problems most important to them, furnishing an insight into thought-processes in the Middle Ages."[17] Medieval thinkers theorized this thought process. For example, al-Biruni reminded his younger colleague Ibn Sina that diagrams were an imperative necessity of scientific discourse and reasoning.[18] And Hugh of St. Victor would posit in the *Didascalicon* that geometry, as part of the curriculum, was both subject and method of medieval learning, in that it was "the fount of perceptions and the origin of utterances."[19]

However, visualization was not only useful for logic or astronomy; it was employed in works intended for a wider audience, often fulfilling simultaneous functions. A case in point is that of Beatus manuscripts, rich in a combination of manuscript painting techniques, diagrammatic organization of material, and a visual program that includes, most famously, *mappaemundi*. In the 1047 *Beatus of Fernando y Sancha*, folios 12v

Figure 7.1

Cosmological diagram: Eight concentric circles, *de positione septem stellarum errantium*. In *Codex Ovetensis*, fragment corresponding to Isidore's *De natura rerum*. Escorial r.II.18, fol. 14r. ©Patrimonio Nacional

Figure 7.2

Crónica Troyana de Alfonso XI. Escorial h.i.6, fol. 137v. ©Patrimonio Nacional

and 13r open to reveal a display that shows, on the left, a square T/O map and next to it, a genealogy of humankind from Adam and Eve onward organized as a sea of linked circles, a rosary of rotae[20] (figure 7.3). Map and genealogy are diagrams focused on the spatialization and the organization of prelapsarian and historical time. They also simultaneously trigger narration, invoking as hyperlinks the memory of biblical stories, the division of the continents among Noah's sons, or the scientific classification of continents according to climates. Such a narration might forge a new path between geography and history, it might invite the spectator, the reader, the orator to elaborate or digress on any given element, or to locate herself on the continents, in the history of humankind. Specifically used as a didactic tool, a second, apostolic mappamundi that accompanies all Beatus manuscripts is perhaps one of the best-known images in the entire visual program of the *Commentary of the Apocalypse.* This mappamundi has been extensively studied, both separated from the codex as part of a genealogy of medieval maps, and as part of the visual program that accompanies the apocalyptic commentary. The diagrams of the program, however, have usually been left to the side or referred to as decoration, while attention has focused on the interpretation of the rich illuminations that supplement the text. This visualization of geography, the mappamundi, was selected to decorate one of the walls of the monastery of San Pedro de Rocas, now almost invisible in its deterioration. In the region known as the Ribeira Sacra in Galicia, Spain, this monastery, founded at the end of the sixth century, was granted donations and privileges by Alfonso III and was later abandoned.[21] The mural map, painted in the last quarter of the twelfth century, is an apostolic diaspora map that can be contextually linked to a larger tradition of commissioned mural maps beginning with that made under Agrippa that hung in the Porticus Vipsania under Augustan rule, according to Pliny's account in his *Natural History* (book III).[22] The medieval period also used large-format maps as didactic aids for lecturing: the orator Eumenius used one at the academy at Autun; a map similar to our Galician one was painted on the walls of a monastery accessible to the laity, in the Church of Saint-Silvain in Chalivoi-Milon; and Pope Zacharias commissioned another in the eighth century for a dining hall at the Lateran Palace (also an apostolic map). There are also others painted on walls or as floor mosaics, known today by their literary descriptions, such as the painted tablets of Charlemagne, Bishop Theodulf's in the Palace of Orléans, and so on.[23]

Beyond their didactic function, in a way a form of narration, scholars have also looked to cartography for the visualization of different and simultaneous forms of telling stories, and of incorporating texts and discourses. Ernst Kitzinger, for instance, names mappaemundi as sources for the composition of the late twelfth-century floor mosaic in Turin.[24] There, the cartographic depiction of the world, which houses an image of Fortuna at its center, draws information from Isidore. Its exploitation of spatiality—a viewer of this map also "walks" the world—further emphasizes the interpellation of the

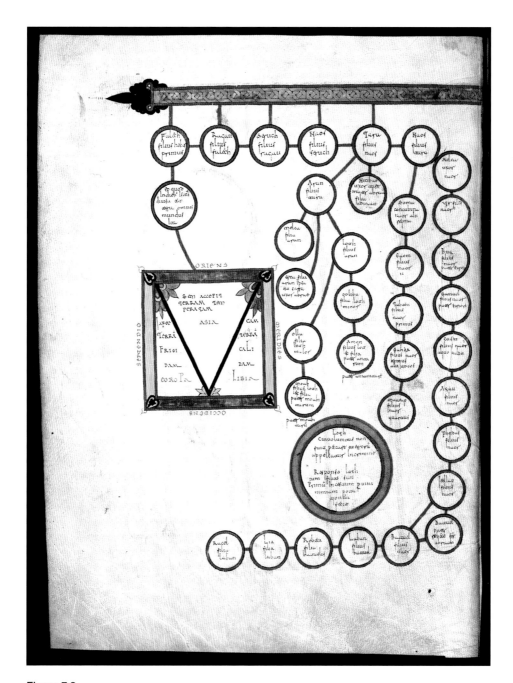

Figure 7.3

Commentarius in Apocalypsin or Beato *de Fernando I y doña Sancha*. c. 1047. Biblioteca Nacional de España, vit. 14–2, fol. 12v. ©Biblioteca Nacional de España.

allegory at its center, while not excluding the abstract idea of the worldly rise and fall of empires within the frame of mortality, surrounded by the eternal kingdom of God.[25] And finally, there are those maps we know only from confabulation, such as Baudri de Bourgueil's imagined floor mosaic as an emblem of power for Adela, countess of Blois.[26] Within this last category, but linked in their possible materiality to those on walls or on parchment, one would include the verbal mappamundi Hugh of St. Victor provides instructions for drawing, or those described and used in the *Libro de Alexandre* and other Alexander romances—different genres that evidence a clerical culture that draws the diagrams of its writing from its many disciplines.

Accretion

Early on, the overlaps of diagrams and language from the most material or graphic to the most rhetorically complex operations are explored from within both verbal and visual representation. The fifteenth-century interpretation of the main dividing lines of the maps known as T/O maps as letters forming the acronym for *Terrarum Orbis*, follows and complements the earlier idea that the T should be read as a figure of the Crucifixion, an essential symbol of sacrifice and salvation.[27] This type of map, inherited from the classical world, has its earliest extant example in Isidore of Seville's *Etymologies;* in fact the earliest surviving example is in an Iberian factitious codex binding together maps from the seventh and ninth centuries (figure 7.4). On the lower margin of the last folio of the seventh-century part of the codex there are two maps. The oldest, on the left, was drawn just a few years after Isidore's death in 636, while the one on the right imitates the lines of the left-hand map, as if rehearsing how to draw a T/O.[28] What is most interesting about this is that the graphic itself of the map is meant both to be read and to be imitated, copied to serve other purposes, other readings. The circular or spherical shape of the T/O gains symbolic force from its gestures at totality, at concentration, generating spatial oppositions within its frame that in turn produce hierarchies within the map to be exploited in a narrative actualization of these prompts. Conversely, the divisions *within* the circle—the T—suggest an opposite symbolic dynamic at work, the loss of unity or the possibility of dispersal reinforced when the earthly paradise is represented—or inserted verbally—within that circle.[29]

This basic map will incorporate other types of information that trigger new directions for narrative. For instance, the association of the three continents with Noah's three sons, Shem, Cham, and Japheth, was often layered with information brought into T/O maps from other types of medieval maps, such as Macrobian or zonal maps, which divide the world according to temperature and habitability into five zones, and with the assignation of provinces that index yet other types of information, provided elsewhere. These cartographic distributions activate social and political implications, for example in relation to the spatial distribution of labor. Honorius of Autun explicitly

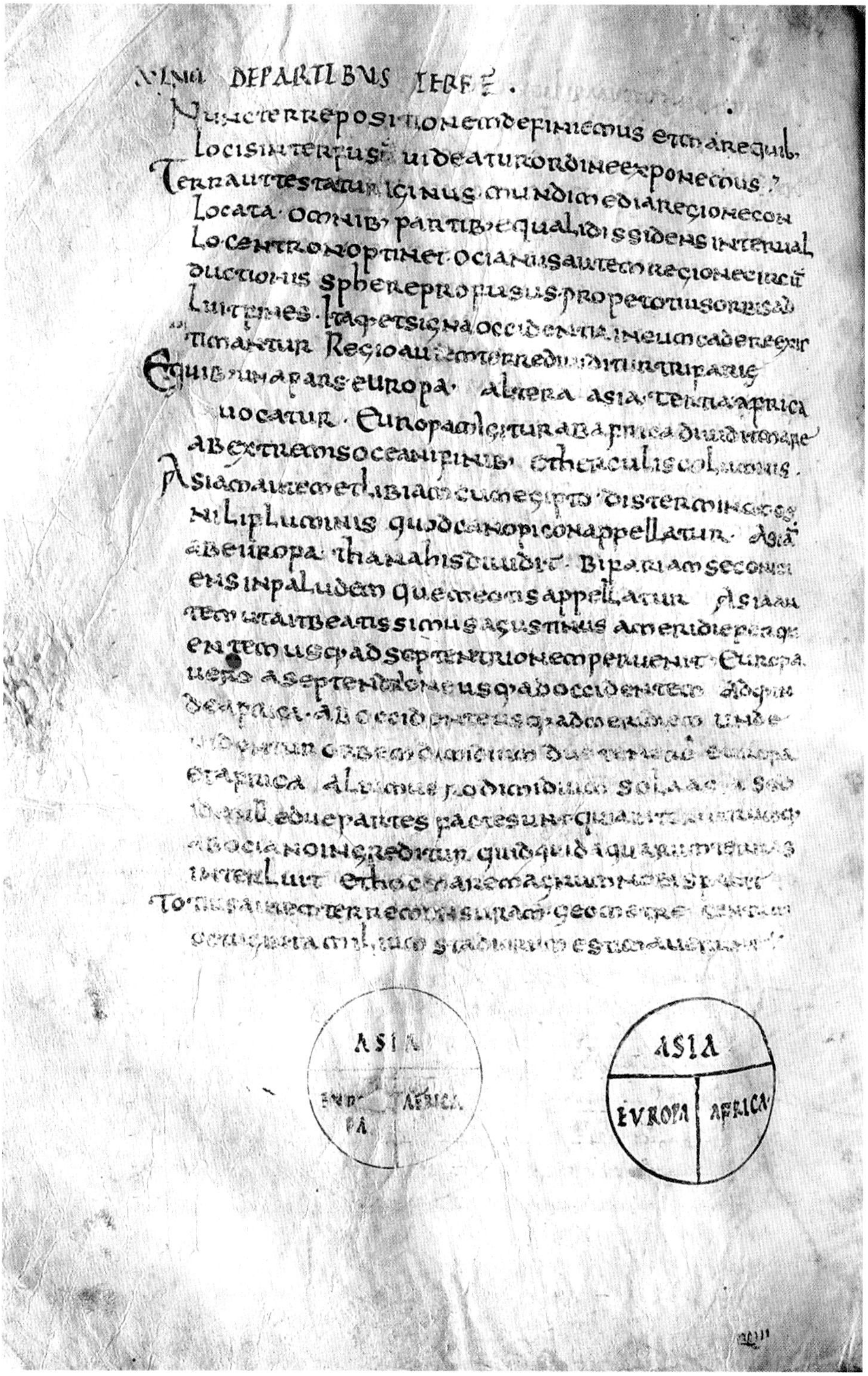

XLMa DEPARTIBUS TERRE.

Nuncterre positionem definiemus et in arequib. locis interfusi videatur ordine exponemus. Terra ut testatur Iginus mundi mediam regionem co Locata omnibus partibus equali dissidens internal lo centro optinet. Oceanus autem regione circú ductionis sphere profusus pro peto mus orbis ad Luitemes. Itaque tsigna occidentia ineum cadere ex ti mantur Regio au temernediu omni tripartie equibus una pars europa. Altera asia. tertia africa uocatur. Europa igitur ab africa diuiditur ape ab extremis oceani finibus et herculis columnis. Asia mauitem et libia cum egipto disterminates Nili fluminis quod canopicon appellatur. Asia ab europa thanahis diuidit. Bipariam secum ens in paludem que meotis appellatur. Asia tem ut ait beatissimus agustinus ameridie per a entem usque ad septentrionem peruenit. Europa uero a septentrione usque ad occidentem. Afri deapici ab occio pientes quad seuilentem unde apparent oriente dimidium due tertie europa et africa. altuus rodimidio sola ac isso lorillē due partes paetes unt qua in a ab ocia no ingreditur. quidquid aquarum eius interluit. et hoc maremagnum nobis p To tis autem terre comsi quia geometre consta oaeginta milium stadiorum esse aduerunt.

Two T/O maps

Figure 7.4

Two T/O maps with names of the continents. In *Codex Ovetensis*, fragment corresponding to Isidore, *De natura rerum*. Escorial r.II.18 24v. ©Patrimonio Nacional

makes the division of the earth among Noah's sons a division of labor in his *Imago mundi* (twelfth century), a gesture Alfonso X's *General estoria* repeats while narrating the episode of the Ark. Narration follows (and/or elicits) cartographic gestures. In a folio that binds together that earliest of Isidorian maps to a ninth-century quire, the idea of the distribution of the continents to the sons through three labels is added to that strictly geographic information of the map (figure 7.5). The names of Shem, Cham, and Japheth evidence the simultaneously narrative and cartographic production of the map, highlighting the porosity between the geographic, the historical, the spiritual, and the narrative. This first gesture of *amplificatio*, of visual and narrative digression, will multiply and expand in the centuries that follow as scribes continue to add information to the margins of the map and then to its interior, transforming the relation between diagram and text, between maps and narration, and the role of cartography itself within clerical discourses.[30]

The diagrammatic T/O presupposes a cleric who can summon from memory the information the map is there to *index*; however, the accretion of information at the margins of the map points to a narrative, expansive function of the device. As the contents accumulate and move into the maps in what we know as the encyclopedic mappaemundi, where narrative accumulation fills up the space, the map itself becomes the archive and the cleric must now forge narrative paths through and across this verbal mass. The elaborate type of map, which responded to the increasing interest in the accumulation of knowledge in simultaneous presentation, has its most famous examples in Ebstorf, Hereford (both thirteenth century), and Fra Mauro 1459 mappaemundi (figure 7.6). The Ebstorf's massive 3.5 meters in diameter represented the world as overlapping or as embodying the metaphor of man as world, except that the man is Christ as human—that is, it is God's mortal body that is figured as the earth. The head, coinciding with the East, was drawn at the top, while his hands hold the North and South and his feet, as if stepping or standing on the West, complete the figure. The world as crucified Christ thus proposed a knowledge produced by accretion and developed as *amplificatio*, where both material and spiritual meanings were elicited. From the materiality of parchment representing divine skin rematerialized on the world's surface, to the sacrifice of divinity that is the embodiment of mortality, the wor(l)d made flesh, this type of map offered to the viewer Christ and world as objects of knowledge by identification and through meditation.[31] In both extreme possibilities of medieval cartography, the diagrammatic and the encyclopedic, narrative is key to the actualization of the function of the map by the cleric, either silently and individually involved in private meditation, or as a performing cleric acting as teacher.

Maps, then, serve two distinct if often simultaneous roles related to the production of narrative. First, they harbor a treasure of materials, organizing and archiving them for later retrieval; this function is related to pedagogy and meditation. Second, they

Figure 7.5

T/O map with names of Noah's sons in margins. In *Codex Ovetensis*, fragment corresponding to Isidore, *Etymologiae*, chap. III of bk. XIV. Escorial r.II.18, fol. 25r (detail). ©Patrimonio Nacional

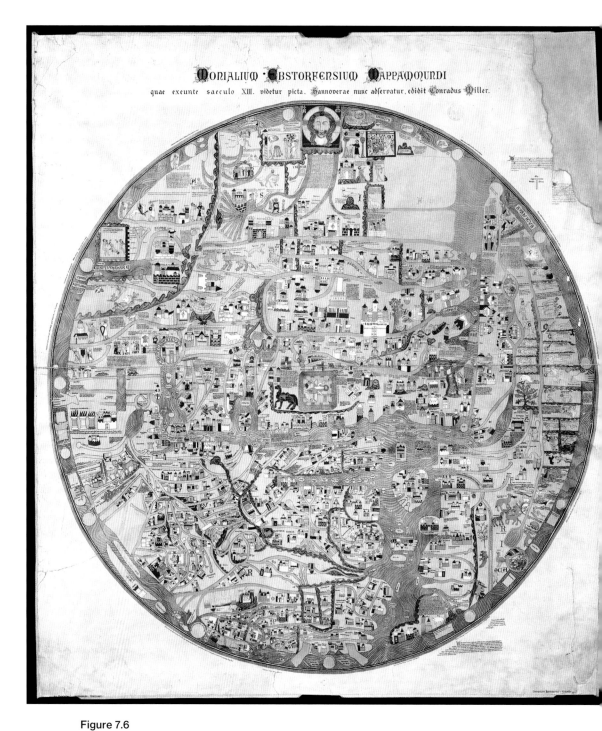

Figure 7.6

Ebstorf mappamundi. Facsimile. Bibliothèque National de France. Cartes et Plans, Ge AA 2177. With permission from the Bibliothèque Nationale de France.

produce by dint of plays on spatiality, connections, contrasts, parallels that incite or provoke narration, a function that I call compositional, inventional, and in particular contexts, fictional.[32]

History, Memory

Within diagrammatic programs, maps most often appear in scientific contexts as a stage in the progression from cosmology to geography, but at other times the relation between map and diagrams is not as straightforward. In some manuscripts, the relation between images such as cosmological spheres and eclipse drawings cannot be explained as a macro-micro relation, for the links between these is a disciplinary, thematic relation. Even less disciplinary is the relation between maps and diagrams of consanguinity and genealogical diagrams, where the oblique progression points to a symbolic dimension where both maps and diagrams play a historiographic role, visualizing human time. There, in the image of the earth that human time occupies, the border of the map, the encircling ocean, or the exterior limit simultaneously marks both the end of the time of humanity and the beginning of the time of God, the end of history and the beginning of eternal life in a metonymic link.

"Byrhtferth's Diagram" (c. 1130) is probably one of the most striking attempts to provide different types of information in a simultaneous historical format, and "one of many manifestations of the synthetic approach of medieval thinkers, who sought to combine information of different types or from disparate sources into a harmonious whole," writes Holcomb.[33] In a similar way, "through their use of stemmata or genealogical lists, the artists responsible for the Peter of Poitiers roll (ca. 1230) similarly sought to synchronize information, showing how historical events across geographic regions relate to one another." Peter of Poitiers's *Compendium Historiale in Genealogia Christi* is an exercise in *abbreviatio* of both word and image, summarizing for a pedagogical context biblical history through the narrative thread and visual strategy of the life of Christ, where map and roll exhibit the overlaps and present different strategies for the same purpose, the confirmation of Christian cosmology through geography and history.[34]

If in Peter of Poitiers the diagrammatic structures are visible, sketched out for all to see, one need not expect this of all medieval material. For we would be wrong not to intuit such cartography as a way of inventing, of composing material at any other moment: "The selection and rearrangement of *materia*," writes Douglas Kelly, "may be usefully likened to the practice of medieval cartography. Benoît de Sainte-Maure himself once saw his work as a fragmentary *mappa mundi*," and further on states that "medieval maps are thus a *figura*. Their principles of order and representation are analogous to those we have been describing for poetic invention," recalling Geoffrey of Vinsauf's evocation of the cartographer next to the architect or builder at the beginning of his *Poetria Nova*.[35] Just as Vinsauf describes a mappamundi as a basic scheme of inventional memory, he makes the image of the map equal to the image of building: "Let the mind's interior compass (*circinus interior mentis*) first circle the whole extent of the material. Let a definite order chart in advance at what point the pen will take up its course, or where it will fix its Cadiz. As a prudent workman, construct the whole fabric within the mind's citadel; let it exist in the mind before it is on the lips."[36] If Cadiz is the topographic marker for the "end" or "point of arrival" of an itinerary, it is also the rhetorical founding stone for poetic building. Whether in the shape of lists, as in the much-interpreted scholarly itineraries in cartographic key, or in genealogical enumerations, or as lists of languages or peoples, these schemes are to be read as *inventional* structures, meant not (only) to be repositories but above all to be generators, to produce in the way the verbs *to map* and *to build* make us think of.

According to Carruthers, these might make one ask, "But what can be the value of such lists? They come without any qualifying comment, a bunch of names plunked down in memorable packaging. In neither of Chaucer's poems, nor in the Inferno, do the figures named play any particular role in the stories that follow." It is perhaps a task left not for the poet, or the text, but for the reader to complete, a task of hermeneutics. Carruthers continues: "They can mean nothing *unless* the reader wishes to make something of them—perhaps, in remembering these stories and the matter of these famous works, to keep them reverberating as a potential set of comparisons and contrasts to the rest of the work. We might think of them as providing us with a foundational inventory."[37] Moving beyond these as archives, as repositories for future dissemination, we should also consider them devices that spur the reader/listener to *interpret* according to her own set of *loci*, to add, digress, to invent anew.

Device, Operation

To think of diagrams as meant to house and guard materials, but also as structuring their composition and as meant to trigger in the reader new texts, one need only look to Isidore's genealogies, or at the Beatus maps as sites for possible narration. Plotting—that backbone of narration—is what these devices set off. In the discussion of an Austrian thirteenth-century folio, Holcomb focuses on a medieval wind diagram, with its accumulation of ancient and medieval scientific knowledge in the common tool of the *rota*, in which a wind poem, attributed to Isidore of Seville, is written:

> In the manner of concrete poetry, the verses [which begin in the innermost roundel with the description of the twelve winds] continue in the spaces next to the heads; the text specific to each wind are aligned with the appropriate personification. The text in the tightly nested set of four rings surrounding the central medallion provides a succinct chart of parallel terms: the cardinal directions stacked upon the

names of the winds in both transliterated Greek and Latin. Dominating the whole scheme are the head, hands and feet of Christ that emerge from the cardinal points at the south, east, west, and north of the *rota*, a reminder of the Christian God's governing role that could not be more overt.[38]

The folio, as analyzed here, effectively brings out three different plots or discourses to work together, diagrammatic thought, poetry, and illustration, to play on oppositions and suggestions, on allusion and elaboration, on condensation and metaphorical possibilities enabled by the overlaps—spatial coincidences—that map these different potentials.

Cartography is perhaps the discipline that has attracted the most attention within diagrams because of its overt use of different discourses. Maps, as has been noted for later works, especially for late medieval and early modern works but not for the most diagrammatic of examples, the earliest, appear not only in scientific or historical works, they also gloss or illustrate works of fiction. The Ripoll collection of the Archive of the Crown of Aragon houses a copy of Gautier de Châtillon's *Alexandreis* with an extensive gloss in folio 9r triggered by the beginning of the description of Asia as Alexander disembarks on its shores. The *Alexandreis* was a textbook in medieval education, and it was principally used for the study of Latin, as many of the extant copies evidence through the extensive marginal and interlinear glosses, translations, and corrections they bear. On this copy of the *Alexandreis,* in folio 8v, the verbal gloss is restated visually by a map, a simple T/O oriented north noting the names of the continents. As Patrick Gautier Dalché has pointed out, one out of every four copies of the *Alexandreis*, a frequency that speaks to the map's role in the legend's pedagogical dissemination, features a map, whether schematic or as elaborate appendixes (figure 7.7). Maps in works considered to pertain to both history and fiction, as texts reinventing the past and actively imagining a present and future, such as the Alexander romances, suggest a more porous consideration of the role of cartography in medieval culture, one not bound to discipline or to the visual or the verbal. In this culture maps fulfill roles of abstraction and archiving; they are part of scientific knowledge and of historiography. They are most obviously relied on for their geographic function, but also, and perhaps mainly, for their rhetorical roles. Both as gloss and interpretation in relation to the text, but also because of cartography's specific actualization of *deixis*, techniques of *digressio* and *abbreviatio*, a use of allegory, and a use of figural language in general can be seen at play not only as pictorial companions to works of fiction, but as inventional elements of the text.[39] It is only, however, if we consider cartography as a general tool of the period's visual and verbal curriculum, as a constant and pervasive presence in the most basic understanding of glossing, illustration, supplementation, and invention, that one can explore in particular works the sharing of operations with the composition of fiction.

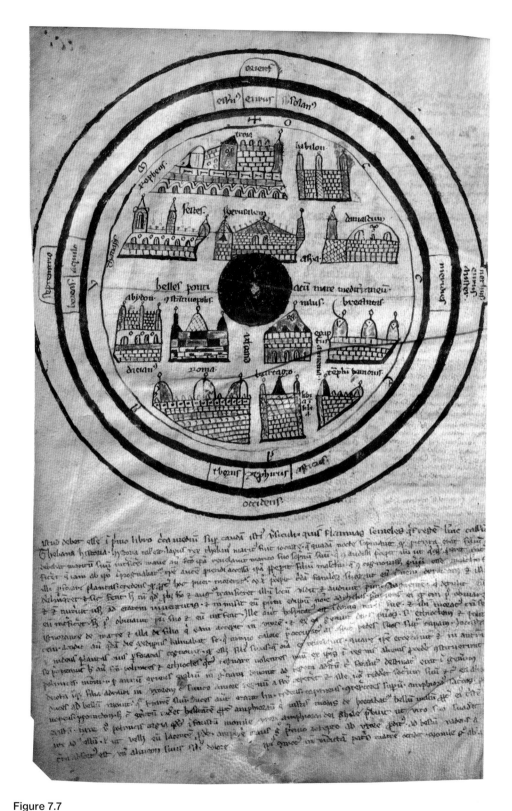

Figure 7.7

Map and text: Visual and verbal glosses. Gautier de Châtillon, *Alexandreis*, 13th c. Bibliothèque Nationale de France. Latin 8352, fol. 100v. With permission from the Bibliothèque Nationale de France.

Beyond the question of disciplines, there is also a final question of method or means that makes maps and literature coincide, one that has a particular resonance with contemporary discussions on the image and the diagrammatic. Drawing as different if not opposed to painting or writing is what describes the technique of most diagrams, and certainly those that forge a particular relation to literature. The linguistic obstacle to clearly separating drawing from writing or painting stems from the lack of a specific word for it in the Middle Ages, but this difficulty also helps us emphasize the coincidences and overlaps between these apparently distinct practices. The differences are not readily ascribable to differences in materials or tools, nor are they differences of function, and they are not attributable to a desire or lack thereof for luxury or complexity. "The distinctiveness of drawing in the Middle Ages arises from its particular visual effect," notes Holcomb, for even if some traits are general–the presence of line, for instance, or the contrast with an alternative visual aesthetic choice–these are not rules. Diagrams may often be polychrome or coexist with illumination: "Medieval drawing might best be defined in terms of its use, not of the pen, but of the parchment"–that is, drawing itself is distinct due to its spatial concerns.[40] As an approach to parchment, drawing offers a particular intimacy with the written text, with the shared operations guiding the hand that draws the line and writes the text. The coincidence between scribe and draftsman, the fact that the people who drew T/O maps, consanguinity charts, wind diagrams, or *rotae*, were not the artists of miniatures, but people who used pens to fill in spaces in and around the texts that so often they had themselves drawn, makes the argument for the commonality of diagrams as figures of thought particularly apt. Diagrammatic thought, useful for both the visual and the verbal, as existing within the same person, is what brings these operations of cartography and narrative together.[41]

Karl Whittington has started to piece together the rich discussion on the contemporary image with the medieval visual arts, articulating in compelling ways the relations between diagrams, maps, and painting, especially in the Italian *trecento*. His remarks on how the points of connection between different discourses in medieval images seem designed not to illustrate but to invite the probing of such connections is the sort of interrogation that we must bring to medieval fiction, perhaps not (only) trying to read a map behind the text, but to read the cartographic operations layered on it, the diagrammatic gestures inviting the reader to continually pursue new links between the historical, the cosmological, the divine, the human, the scientific, and the personal.[42] Maps, as perhaps the most successful of medieval diagrams, are the departure point in investigating the work of diagrammatic thought in literature.

Notes

1. Pilar Ultrilla et al., "A Paleolithic Map from 13,330 calBP: Engraved Stone Blocks from the Late Magdalenian in Abauntz Cave (Navarra, Spain)," *Journal of Human Evolution* 57, no. 2 (2009): 99–111, abstract.

2. Alexander Gerner, "Diagrammatic Thinking," in *Atlas of Transformation*, http://monumenttotransformation.org/atlas-of-transformation/index.html.

3. Jacques Rancière, *The Future of the Image* (New York: Verso, 2007), 4. But Rancière elaborates on this basic definition often, playing with similar parallels and oppositions, such as the visual and the verbal, using the term *operations*, which I will turn to as well.

4. J. Elkins and M. Naef, eds., "Introduction," *What Is an Image?* (University Park: Pennsylvania State University Press, 2011), 4.

5. Elkins and Naef, *What Is an Image?*, 5. Also see John Bender and Michael Marrinan, *The Culture of Diagram* (Stanford, CA: Stanford University Press, 2010).

6. See W. J. T. Mitchell, *What Do Pictures Want?: The Lives and Loves of Images* (Chicago: University of Chicago Press, 2005); Jean-François Lyotard, *Discours, figure* (Paris: Klincksieck, 2002); Jean-Luc Nancy, *The Ground of the Image* (New York: Fordham University Press, 2005); Marie-José Mondzain, "Can Images Kill?," *Critical Inquiry* 36, no. 1 (2009): 20–51; Rosalind Krauss, *The Optical Unconscious* (Cambridge, MA: MIT Press 2004); Georges Didi-Huberman, *Confronting Images: Questioning the Ends of a Certain History of Art* (University Park: Penn State University Press, 2004); Göttfried Boehm, *Wie Bilder Sinn erzeugen: Die Macht des Zeigens* (Berlin: Berlin University Press, 2007).

7. Sybille Krämer, "Writing, Notational Iconicity, Calculus: On Writing as a Cultural Technique," *Modern Language Notes* 118, no. 3 (2003): 518–537, here 519*ff.*.

8. See the discussion on this kinship between *pictura* and *littera*, *painture* and *parole*, in Mary Carruthers, *The Book of Memory: A Study of Memory in Medieval Culture* (London: Cambridge University Press, 2008), esp. 274–278.

9. Melanie Holcomb, *Pen and Parchment: Drawing in the Middle Ages* (New Haven, CT: Yale University Press, 2009), 116.

10. For further discussion see Michael W. Evans, "The Geometry of the Mind," *Architectural Association Quarterly* 12, no. 4 (1980): 32–55, http://www.she-philosopher.com/library.html, §§2.4, 3.1–3.2.

11. François Charette, "Illustration, Scientific," in *Medieval Science, Technology, and Medicine: An Encyclopedia*, ed. Thomas F. Glick, Stephen J. Livesey, and Faith Wallis (New York: Routledge, 2005), 266.

12. Holcomb, *Pen and Parchment*, 20. There is a vast amount of important work in this area, but see the very clear *Planetary Diagrams for Roman Astronomy in Medieval Europe, ca. 800–1500*, by Bruce Eastwood and Gerd Grasshoff (Philadelphia: American Philosophical Society, 2004). The introduction is particularly helpful, and pages 5–6 (and notes) talk about the use of ephemeral maps, drawn in dust, used for didactic purposes, complementing lectures.

13. For late medieval and early modern cartographic writing in Iberia, particularly chivalric fiction, see my *Archipelagoes: Insular Fictions from Chivalric Romance to the Novel* (Minneapolis: University of Minnesota Press, 2011), where I document shared operations between the cartographic genre of the *isolario* and the book of chivalry.

14. Holcomb, *Pen and Parchment*, 20.

15. Illumination, which included diagrams, was present in the manuscripts of a large variety of scientific disciplines, including medicine (with illustration for anatomy and physiology, but also

for techniques such as bleeding, the fabrication of medications, orthopedics, or bandages), astronomy (constellations) and astrology (zodiac signs), cosmography and cartography, natural sciences, zoology and hunting, alchemy, pneumatics and automata, and of course, the military sciences (Alain Touwaide, "Illumination," in *Medieval Science, Technology, and Medicine: An Encyclopedia*, ed. Thomas F. Glick et al. (New York: Routledge, 2005), 262). Even in medical texts, illustration was much more likely to appear in late medieval manuscripts and was generally the result not of a commission by the author of the text but as a later desire to adorn "the manuscript at significant breaks in the text with elegant pictures," sometimes with "miniaturized glimpses of the world of medicine." The frequency of images with a value in diagnosis, prognosis, or therapy is easily explained in otherwise unillustrated and unpretentious manuscripts. However, the use of diagrammatic elements in different contexts offers a more interesting challenge (see Peter Murray Jones, "Medical Illustration," in *Medieval Science, Technology, and Medicine: An Encyclopedia*, ed. Thomas F. Glick et al. (New York: Routledge, 2005), 263).

16. In placing medieval diagrams in the context of visual representation, it is not uncommon to find arguments recriminating them for a lack of realism that many assume was present in classical scientific images—a recrimination commonly repeated against medieval maps. As Touwaide reminds us, however, "The most ancient pictures were schematic ... without a synthetic perception. Realism would be a later reinterpretation of such pictures. ... In the West, observation of nature was supposedly introduced into scientific illustrations around 1300 CE. In printed books, it did not appear until 1530 and the *Herbarum vivae eicones* of Otto Brunfels (ca. 1489–1534)" (Touwaide, "Illumination," 262).

17. Evans, "The Geometry of the Mind," §7.1.

18. In Charette, "Illustration, Scientific," 266. See also Mary Carruthers and Jan M. Ziolkowski, *The Medieval Craft of Memory: An Anthology of Texts and Pictures* (Philadelphia: University of Pennsylvania Press, 2003). To rein in the discussion of the wide variety of texts she discusses, Carruthers defines diagrams as the use of pictures, both verbal and graphic, that serve "to consolidate, summarize, and fix the main subjects to be attended to" (*The Book of Memory*, 294), a subject she expands on in *The Craft of Thought* (Cambridge: Cambridge University Press, 2000) through the analysis of meditational monastic practices. Even though she amply discusses *inventio* in the complex task of recollecting as a sort of reordering of materials and elements diagrammatic thought proposes, I see diagrams as serving not only the diverse tasks of *memoria* but as invitations for innovative compositions that go beyond recollection.

19. Hugh of St. Victor, *Didascalicon* (New York: Columbia University Press, 1961), 71.

20. On *rotae*, see Kathryn Gerry's useful discussion in *Pen and Parchment: Drawing in the Middle Ages*, ed. Melanie Holcomb (New Haven, CT: Yale University Press, 2009), esp. 108–110.

21. Serafín Moralejo Alvarez, "El mapa de la diáspora apostólica en San Pedro de Rocas," *Compostellanum; revista de la Archidiócesis de Santiago de Compostela* 31, nos. 3–4 (1986): 315–340, here 315–323. Juan Sureda associates this map with the mappamundi in the Osma Beatus; J. Manuel García Iglesias goes on to relate it to the Lorvao and Oña Beatus. John Williams states there is the possibility that the mappamundi in the Real Academia de la Historia (cód. 25) was influenced by a Beatus map, "for the likelihood that Castilian scribes in the 10th century had seen one is high" (cited in John Williams, "Isidore, Orosius and the Beatus Map," *Imago Mundi* 49 (1997): 7–32, here 30).

22. Book III begins speaking of the division of the earth into three continents, and after introducing Europe in general terms, Pliny goes on to describe Hispania, first in general and then by

province. The fragment mentioning the map on the Porticus Vipsania, in chapter III, is on Baetica.

23. For more on these large-format, public mappaemundi, see Mark Rosen, *The Mapping of Power in Renaissance Italy: Painted Cartographic Cycles in Social and Intellectual Context* (Cambridge: Cambridge University Press, 2015), 43*ff.*, though he does not mention San Pedro de Rocas. On Pope Zacharias's map see Marcia Kupfer, "Medieval World Maps: Embedded Images, Interpretive Frames," *Word and Image* 10 (1994): 262–288, and also her "The Lost Mappamundi at Chalivoy-Milon," *Speculum* 66, no. 3 (1991): 540–571; on Charlemagne's tablets, see Emily Albu, "Imperial Geography and the Medieval Peutinger Map," *Imago Mundi* 57 (2005): 136–148.

24. Ernst Kitzinger, "World Map and Fortune's Wheel: A Medieval Mosaic Floor in Turin," *Proceedings of the American Philosophical Society* 117, no. 5 (1973): 344–373, here 358.

25. For the interpretation, which links the floor mosaic to the imagined floor map of Baudri in an architectural transposition of cosmography, see Barral i Altet, "Poésie et iconographie: Un pavement du XIIe siècle décrit par Baudri de Bourgueil," *Dumbarton Oaks Papers* 41 (1987): 41–54. For a general account of the relations between these floor mosaic mappaemundi and other more diagrammatic sources see Lucy E. G. Donkin, "*Usque ad Ultimum Terrae:* Mapping the End of the Earth in Two Medieval Floor Mosaics," in *Cartography in Antiquity and the Middle Ages: French Perspectives, New Methods*, ed. Richard J. A. Talbert and Richard Watson Unger, 189–218 (Leiden: Brill, 2008).

26. The imaginary status of Baudri's map, which other scholars have considered based on a real floor map, is underlined by Jean-Yves Tilliette, "La chambre de la comtesse Adèle: Savoir scientifique et technique littèraire dans le c. cvcvi de Baudri de Bourgueil," *Romania* 102 (1981): 145–171.

27. Evelyn Edson, *Mapping Time and Space* (London: British Library, 1997), 4–5.

28. This ninety-five-folio codex, known as the *Codex Ovetensis*, is copied in several hands, and it has been argued that folios 1–8 might be in Saint Eulogius of Córdoba's own hand. The codex has been augmented with materials of unknown origin that were added to an original Iberian center. It bears numerous marginal annotations, some in Arabic, adding to the intriguing collection of historical, scientific, and especially geographic works it compiles. Carlos Benjamín Pereira Mira, *El "Codex Miscellaneus Ovetensis" (Ms. Esc. R.II.18: Fuentes y bibliografía, Estado de la cuestión* (Oviedo: Universidad, Departamento de Historia, 2001), and *Éxodo librario en la biblioteca capitular de Oviedo: El Codex miscellaneus ovetensis (manuscrito escurialense R. II. 18)* (Oviedo: Trea/Ediuno, 2006) provide complete descriptions and a history of the volume.

29. See the detailed discussion of the representation of paradise in medieval cartography in Allesandro Scafi, *Mapping Paradise: A History of Heaven on Earth* (Chicago: University of Chicago Press, 2006), esp. 84–124.

30. See chapter 2 of my *The Task of the Cleric: Cartography, Translation and Economics in Thirteenth-Century Iberia* (Toronto: University of Toronto Press, 2016), detailing these rhetorical operations in the cartographic corpus of Iberia pre–thirteenth century.

31. See Evelyn Edson, *The World Map, 1300–1492: The Persistence of Tradition and Transformation* (Baltimore: Johns Hopkins University Press and Santa Fe, NM: Center for American Places, 2007), esp. chap. 1, for the context of the Ebstorf within the corpus of elaborate mappaemundi from the thirteenth century onward.

32. In rhetoric, the figure of thought serves to structure the process of thinking itself, not only that of argumentation. For Carruthers, this thinking process must include memory at its center, for which diagrams play a fundamental role: "For mnemonic purposes, diagrams, like other sorts of images in medieval books, have a combination of two functions: they serve as fixes for memory storage, and as cues to start the recollective process. The one function is pedagogical, in which the diagram serves as an informational schematic; the other is meditational and compositional" (*The Book of Memory*, 332). As intrinsic to these arts of memory, the map in the text shares a number of traits with rhetoric, most basically in that it organizes discourse (*locus, loci*, commonplaces as the most basic building block of discourse, for instance).

33. The diagram is available as an interactive, zoomable image on the British Library's website, at http://bl.uk/onlinegallery/onlineex under the name "Byrhtferth's Diagram, In A Scientific Textbook."

34. Holcomb, *Pen and Parchment*, 21–22. The (early) copy held by the Walters Art Museum is in codex form, and the genealogy is conveyed mostly through roundels (at thedigitalwalters.org, W 796), but the Cloisters' Collection copy is a roll, over a meter and a half long, covered in a complex set of diagrams that complement and supplement each other (metmuseum.org, as "Genealogy of Christ"). Many other copies are available online.

35. Douglas Kelly, *The Art of Medieval French Romance* (Madison: University of Wisconsin Press, 1992), 66.

36. Geoffroi de Vinsauf, *Poetry Nova* (Toronto: Pontifical Institute of Medieval Studies, 1967), 17.

37. Mary Carruthers, "The Poet as Master-Builder: Composition and Locational Memory in the Middle Ages," *New Literary History* 24, no. 4 (1993): 881–904, here 887.

38. Holcomb, *Pen and Parchment*, 118.

39. I elaborate on these extensively in my book *The Task of the Cleric*, with relation to a specific pre-thirteenth-century cartographic corpus and the *Libro de Alexandre*. For a rhetorical catalogue and sources for the *Libro de Alexandre* see Peter Such, "The Origins and Use of School Rhetoric," doctoral dissertation, University of Cambridge, 1978, esp. 76*ff.*. On abbreviation in the *Alexandre*, see Juan Casas Rigall, "La *abbreviatio* y sus funciones poéticas en el *Libro de Alexandre*," *Troianalexandrina* 5 (2005): 63–96.

40. Holcomb, *Pen and Parchment*, 32.

41. Suzanne Conklin Akbari is working precisely on such connections between diagrammatics and historiography in her book project *The Shape of Time*. See also Evans, "The Geometry of the Mind," note to figure 17.

42. Karl Whittington discusses a couple of items in the catalog for *Pen and Parchment* (#45 and #46), the first of which is a diagram by Opicinus de Canestris. As is frequent with Opicinus's later works, especially *Vaticanus latinus* 5435, maps are often a part of the drawings or serve as a basis for them. Whittington calls these hybrids of cartography and figurative painting "body-worlds," or other times, embodied maps, and delves into the nature of cartography as a source for operations and interactions with other disciplinary rhetorics. Whittington discusses how previous scholars consider Opicinus's drawings exceptional within medieval visual culture, but it is perhaps more interesting to see them as an extreme exaggeration, as the pushing to the end of the possibilities that diagrammatic thought made visible. I also thank Professor Whittington for sharing a copy of his talk, "Icons of Space: Grids, Maps and Pictures ca. 1300," which was of enormous help in shaping my thoughts here.

Bibliography

Albu, Emily. "Imperial Geography and the Medieval Peutinger Map." *Imago Mundi* 57 (2005): 136–148.

Altet, Barral i. "Poésie et iconographie: Un pavement du XIIe siècle décrit par Baudri de Bourgueil." *Dumbarton Oaks Papers* 41 (1987): 41–54.

Bender, John, and Michael Marrinan. *Culture of Diagram.* Stanford, CA: Stanford University Press, 2010.

Carruthers, Mary. *The Book of Memory: A Study of Memory in Medieval Culture.* London: Cambridge University Press, 2008.

Carruthers, Mary. *The Craft of Thought.* Cambridge: Cambridge University Press, 2000.

Carruthers, Mary. "The Poet as Master-Builder: Composition and Locational Memory in the Middle Ages." *New Literary History* 24, no. 4 (1993): 881–904.

Carruthers, Mary, and Jan M. Ziolkowski. *The Medieval Craft of Memory: An Anthology of Texts and Pictures.* Philadelphia: University of Pennsylvania Press, 2003.

Charrete, François. "Illustration, Scientific." In *Medieval Science, Technology, and Medicine: An Encyclopedia*, ed. Thomas F. Glick, Stephen J. Livesey, and Faith Wallis. New York: Routledge, 2005.

Donkin, Lucy E. G. "*Usque ad Ultimum Terrae:* Mapping the End of the Earth in Two Medieval Floor Mosaics." In *Cartography in Antiquity and the Middle Ages: French Perspectives, New Methods*, ed. Richard J. A. Talbert and Richard Watson Unger, 189–218. Leiden: Brill, 2008.

Eastwood, Bruce, and Gerd Grasshoff. *Planetary Diagrams for Roman Astronomy in Medieval Europe, ca. 800–1500.* Philadelphia: American Philosophical Society, 2004.

Edson, Evelyn. *Mapping Time and Space.* London: British Library, 1997.

Edson, Evelyn. *The World Map, 1300–1492: The Persistence of Tradition and Transformation.* Baltimore: Johns Hopkins University Press and Santa Fe, NM: Center for American Places, 2007.

Elkins, J., and M. Naef, eds. *What Is an Image?* University Park: Pennsylvania State University Press, 2011.

Evans, Michael W. "The Geometry of the Mind." *Architectural Association Quarterly* 12, no. 4 (1980): 32–55, http://www.she-philosopher.com/library.html.

Gerner, Alexander. "Diagrammatic Thinking." In *Atlas of Transformation*, http://monumenttotransformation.org/atlas-of-transformation/index.html.

Holcomb, Melanie. *Pen and Parchment: Drawing in the Middle Ages.* New Haven, CT: Yale University Press, 2009.

Jones, Peter Murray. "Medical Illustration." In *Medieval Science, Technology, and Medicine: An Encyclopedia*, ed. Thomas F. Glick et al. New York: Routledge, 2005.

Kelly, Douglas. *The Art of Medieval French Romance.* Madison: University of Wisconsin Press, 1992.

Kitzinger, Ernst. "World Map and Fortune's Wheel: A Medieval Mosaic Floor in Turin." *Proceedings of the American Philosophical Society* 117, no. 5 (1973): 344–373.

Krämer, Sybille. "Writing, Notational Iconicity, Calculus: On Writing as a Cultural Technique." *Modern Language Notes* 118, no. 3 (2003): 518–537.

Kupfer, Marcia. "The Lost Mappamundi at Chalivoy-Milon." *Speculum* 66, no. 3 (1991): 540–571.

Kupfer, Marcia. "Medieval World Maps: Embedded Images, Interpretive Frames." *Word and Image* 10 (1994): 262–288.

Moralejo Alvarez, Serafín. "El mapa de la diáspora apostólica en San Pedro de Rocas." *Compostellanum; Revista de la Archidiócesis de Santiago de Compostela* 31, nos. 3–4 (1986): 315–340.

Pinet, Simone. *Archipelagoes: Insular Fictions from Chivalric Romance to the Novel*. Minneapolis: University of Minnesota Press, 2011.

Pinet, Simone. *The Task of the Cleric: Cartography, Translation and Economics in Thirteenth-century Iberia*. Toronto: University of Toronto Press, 2016.

Rancière, Jacques. *The Future of the Image*. New York: Verso, 2007.

Rigall, Juan Casas. "La *abbreviatio* y sus funciones poéticas en el *Libro de Alexandre*." *Troianalexandrina* 5 (2005): 63–96.

Rosen, Mark. *The Mapping of Power in Renaissance Italy: Painted Cartographic Cycles in Social and Intellectual Context*. Cambridge: Cambridge University Press, 2015.

Scafi, Alessandro. *Mapping Paradise: A History of Heaven on Earth*. Chicago: University of Chicago Press, 2006.

St. Victor, Hugh of. *Didascalicon*. New York: Columbia University Press, 1961.

Such, Peter. "The Origins and Use of School Rhetoric." Doctoral dissertation, University of Cambridge, 1978.

Tilliette, Jean-Yves. "La chambre de la comtesse Adèle: Savoir scientifique et technique littéraire dans le c. cvcvi de Baudri de Bourgueil." *Romania* 102 (1981): 145–171.

Touwaide, Alain. "Illumination." In *Medieval Science, Technology, and Medicine: An Encyclopedia*, ed. Thomas F. Glick et al. New York: Routledge, 2005.

Ultrilla, Pilar et al. "A Paleolithic Map from 13,330 calBP: Engraved Stone Blocks from the Late Madgalenian in Abauntz Cave (Navarra, Spain)." *Journal of Human Evolution* 57, no. 2 (2009): 99–111.

Vinsauf, Geoffrey of. *Poetria Nova*. Toronto: Pontifical Institute of Medieval Studies, 1967.

Whittington, Karl. Catalogue, items 45 and 46. *Pen and Parchment: Drawing in the Middle Ages*. Ed. Melanie Holcomb. New Haven, CT: Yale University Press, 2009.

Williams, John. "Isidore, Orosius and the Beatus Map." *Imago Mundi* 49 (1997): 7–32.

8
Hybrid Maps: Cartography and Literature in Spanish Imperial Expansion, Sixteenth Century

Ricardo Padrón

Introduction

During the sixteenth century, the kingdoms of Castile and Aragon discovered that the unexpected issue of the marriage between their Ferdinand and Isabella was a commanding role in Europe and the Mediterranean.[1] Meanwhile, efforts to exploit the islands of the Atlantic had revealed previously unknown lands and launched Spain on an unplanned quest for global empire. A new world was coming into being, suddenly and relentlessly. How to comprehend it, and Spain's place within it? Spanish culture desperately needed new maps of all kinds, including new representations of territory. Luckily, it had at its disposal the resources of the European Renaissance, including the newly rediscovered *Geography* of Claudius Ptolemy, which taught its readers how to build maps based on geometric principles and using quantitative data, and in the process, gave them a new way of seeing the world. As David Woodward puts it, Ptolemy's grid allowed for the idea of a world "over which systematic dominance was possible, and provided a powerful framework for political expansion and control."[2] It was the perfect cartography for an emerging empire.

Yet Spain produced few gridded maps, whether of the Iberian Peninsula or of its overseas empire. This should come as no surprise, because Ptolemaic maps were a utopian aspiration, built from an immense archive of written materials that, from the mapmaker's perspective, suffered all sorts of limitations.[3] The texts in that archive included the reports of conquistadors, missionaries, colonial officials, ship pilots, and travelers of all kinds. These texts, moreover, were not merely sources for making maps. They were themselves maps of a very different kind, and they would do much of the work of mapping Spain's new worlds. They teach us to think, not of literature *and* cartography, but of literature *as* cartography, and as a consequence, teach us to think of Spanish cartographic culture as a space of interaction among different cartographic modes. That interaction, moreover, was actually captured by some of the period's most prominent literary works, particularly epic and lyric compositions by Alonso de Ercilla and Luis de Góngora that map the world in verse.

Ptolemy's Grid and Abstract Space

The *Geography* of Claudius Ptolemy was the basic touchstone of Renaissance cartography. At first, it appealed primarily as a reader's aid that helped one locate the places mentioned in Greco-Roman historiography.[4] Over time, however, the *Geography* began to be treated as a manual for making maps on mathematical principles. A very loose community of like-minded mapmakers and map readers emerged all over Europe. The mapmakers might be highly educated humanists or prominent printers and engravers, but sometimes they were former ship pilots of humble origins pursuing a second career, so to speak. Humanists and other intellectuals stood out among the early adopters of Ptolemaic maps, as did the rich patrons who financed the work. But the community of users grew dramatically in size and diversity as the sixteenth century transformed maps from rarities into essential elements of such varied endeavors as warfare, diplomacy, architecture, urban planning, archeology, taxation, and navigation.[5]

This community of mapmakers and map users shared ideas about what a map should be and how it should be made. According to this community, a good map faithfully represented the surface of the earth or some part of it, although it could also represent the heavens. It might engage in speculation about parts unknown, but if it did, it usually marked the frontier between the certain and the speculative. It was made in full awareness of the distortions that inevitably resulted when one depicted the spherical earth on a flat surface and used a systematic geometric projection to control for these distortions in predictable ways. It drew on a reliable set of data, either gathered specifically through surveys carried out for the purpose, or mined from sources considered trustworthy.

At the heart of the Ptolemaic map lay the coordinate grid. This was the tool that made it possible to control for the distortions produced by projecting the curved surface of the earth onto the flat surface of the map, and it was the basic framework into which one plotted locations according to their latitude and longitude. It also made manifest a particular understanding of space as an isotropic expanse substantially independent from the objects it contained. This was something relatively new. Antiquity and the Middle Ages had been familiar with abstract space, but the Renaissance fell in love with it, using it to make maps, draw pictures, design buildings, and plan cities. In the Spanish New World, the grid actually acquired material form in the orthogonal street plans of colonial cities, thereby announcing the shape of things to come, a modernity that assigned a privileged role to representations of space built on abstraction and that allowed the mapmaker, the architect, the city planner, and the like unprecedented sway over the lived space of human existence.[6] One of the central characteristics of this new spatiality, at least as far as cartography was concerned, was its close association with ocularity. The map's abstract spatiality promised to make the territory visible to the viewer, as if he could see it from a godlike height. When a map lies "before our eyes," wrote Abraham Ortelius, one of the most important map publishers of the sixteenth century, "we may behold things done, or places where they were done, as if they were present to our eyes."[7]

The Limitations of the Ptolemaic Map

Over the course of the past few decades, numerous scholars have repeatedly demonstrated that maps are never transparent representations of territory, no matter how much they seem to be, or how much their makers and users claim them to be. Maps are saturated with bias, whether cultural, ideological, or political.[8] Sixteenth-century maps were certainly not the windows on the world that their creators made them out to be, but the reasons for this were not limited to their general deconstructability. The maps of the period suffered from a series of limitations, technical and otherwise, that assured that no gridded map could ever achieve the degree of transparency to which it aspired. Some of the techniques involved in collecting data, designing a map, or producing the finished product were simple enough in principle but daunting in practice. Theoretically, it was easy to determine one's latitude, but in practice, it required environmental conditions that were not always available, and instruments a great deal more precise than those the sixteenth century had at its disposal. This is not to say that mapmaking did not improve over the course of the sixteenth century, however slowly and sometimes fitfully. In 1500, for example, European cartography lacked a projection that could be used to construct good nautical charts. The Portuguese mathematician Pedro Nunes was the first to identify the mathematical challenge in 1546, but it was not until 1569 that the Dutch mapmaker Gerhard Mercator solved it by making a usable chart, and it was not until 1599 that the English mathematician Edward Wright outlined the mathematics behind Mercator's projection, making it available for reproduction. Despite these advances, however, intractable technical limitations continued to vex mapmakers. It is truly astonishing to think the period managed to fetishize the gridded map when it had no reliable means at its disposal for measuring longitude![9]

Nevertheless the limitation that matters most to this discussion has to do with the kind of source material that mapmakers had no choice but to use. Ideally, maps were built from precise measurements of real-world spatial information carried out deliberately and systematically. Occasionally, they were actually made this way. During the 1550s, for example, Philip II commissioned the mathematician Pedro Esquivel to map his Iberian kingdoms on the basis of a survey that he was to conduct. The result was the so-called Escorial Atlas, one of the most detailed and accurate cartographic images of any European kingdom produced during the sixteenth century. But surveys of this kind were expensive, and often demanded too much from the limited store of technical expertise and institutional capacity available even to the wealthiest patrons. As a result, mapmakers had to draw their data, more often than not, from a heterogeneous archive of materials, very little of which had been prepared with the needs of the mapmaker in mind.

Some of these materials consisted of the oral reports of travelers, like the ship pilots who were sometimes interviewed by the mapmakers of Seville's *Casa de la contratación*, the royal institution charged with regulating Spain's relationship with its overseas possessions.[10] Some of the sources were cartographic in nature, perhaps nautical charts drawn by a ship's pilot, sketch maps prepared by some conquistador, or even maps drawn by non-Europeans, in the cartographic idiom of a distant culture. Most of the archive, however, consisted of written texts. In Spain, these included the logbooks of ships' pilots, the written reports of conquistadors and explorers, the responses of colonial officials to questionnaires from the crown, the letters of missionaries, or the narratives of travelers, ancient, medieval, or modern, not to mention the growing library of European historiography and geography. Texts like these were indispensible to the work of mapmaking, particularly when it involved the huge distances and enormous spaces of an overseas empire.

The vast majority of these texts were written without any thought to the needs of the mapmaker, and even without any knowledge of his existence. They included all sorts of spatial and environmental information, but often as a byproduct of their engagement with some other activity or interest, like saving souls, searching for lost cities, or guiding a ship to its destination. The information may have been gathered any number of ways, but precise measurement using appropriate instruments was rarely one of them. This meant that the data that could be garnered from such sources, from the mapmaker's point of view, was often inaccurate, imprecise, incomplete, or simply conveyed in a manner that was difficult to translate into the work of plotting locations into a coordinate grid. Travel narratives, for example, often expressed distances from one place to the next in terms of the time it took to make the journey and could be very vague about the direction of travel. How did one translate such estimates into map coordinates? Ship pilots usually recorded latitudes, compass bearings, and daily distances traveled, but were their figures accurate? And what to do with the compass bearings, when one did not understand the phenomenon of magnetic variation, the tendency of the compass needle to point away from true north? Mapmakers had to be careful readers, and sometimes good guessers. This meant that their maps were not at all scientific representations of territory built by plotting reliable quantitative data into a sophisticated geometric armature. Instead, they were visual renderings of spatial and environmental information garnered primarily from a diverse archive of writing that was almost always inadequate to the task.

Some of the most useful information could be garnered from set-piece descriptions of individual places, such as cities, islands, or even entire regions. For example, when the Venetian mapmaker Giacomo Gastaldi constructed a map of Mexico City, he seems to have used the description of Tenochtitlán found in Cortés's "Second Letter from Mexico" to modify the existing prototypes, which included the map published with the 1524 Latin edition of Cortés's text, and a 1528 derivative of that earlier image. (See figure 8.1.) Both of these maps stemmed from a lost Aztec original that probably simplified the geography of the region for symbolic purposes.[11] This is why they both depict Tenochtitlán as an island at the center of a single, roughly circular lake, in clear contradiction to Cortés's account, which speaks of two lakes, one saltwater and the other, fresh.[12] Gastaldi must have noticed the discrepancy, and favored the verbal description over the cartographic prototypes in the design of his own map. (See figure 8.2.)

Set-piece descriptions were thus quite useful for what Ptolemy called "chorography," the representation of localities in ways that captured their individual character, but they only helped with "geography," the representation of larger spaces in ways that emphasized quantitative spatial relationships, insofar as they included some sort of locational information.[13] Ideally, this meant coordinates of latitude and longitude, but these were rarely available in the mapmaker's archive. As we have seen, the best a mapmaker could hope for was a measure of latitude, and/or some indication of the distance and direction from the place in question to some known location. A particular place, therefore, could only be mapped to that of other places, as part of a web of displacements anchored, hopefully, in what was already known. To search for this sort of information in the texts of the mapmaker's archive is to take in the text as a whole, to consider its fundamental structure, and to discover how these texts were themselves maps of a sort.

The Spatiality of the Itinerary

That structure often takes the form of a route of travel, or a network of such routes. The text leads from place to place, providing some indication of the distance and direction from one to the next, with varying degrees of precision or thoroughness. At the very least, it establishes the order in which the places appear along the route. The itinerary might be fleshed out with dramatic episodes and detailed descriptions, as in the report of a conquistador, or it might exist practically on its own, bereft of anything but the locational information, as in the log of a pilot. In any case, itineraries appear all over the mapmaker's archive, at both the microlevel of individual descriptions and the macrolevel of overall textual organization.

One might think that their presence can be explained by pointing to the fact that these are written texts, and that thanks to the linearity of verbal exposition, they have no choice but to take the form of a tour, introducing one aspect of their object at a time, in succession, until they have described the whole.[14] Not all tours, however, are

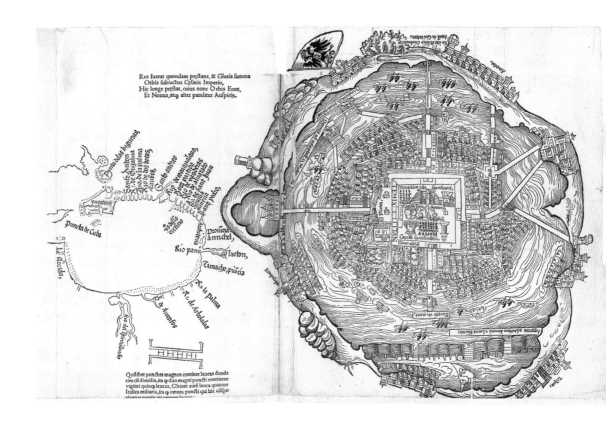

Figure 8.1

Map of Tenochtitlán and the Gulf of Mexico from *Praeclara Ferdina[n]di: Cortesii de noua maris oceani Hyspania narratio* ... Nuremberg, 1524. Courtesy of the John Carter Brown Library at Brown University.

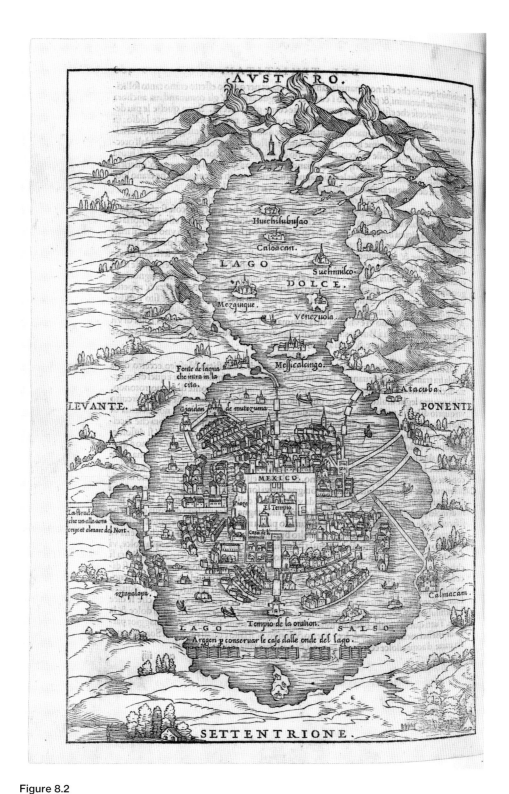

Figure 8.2

Map of Mexico City attributed to Giacomo Gastaldi from Giovanni Battista Ramusio, *Terzo volume delle nauigationi et viaggi*. Venice, 1556. Courtesy of the John Carter Brown Library at Brown University.

itineraries. Some are what psychologists of perception call "gaze tours," which lead the reader around an object in piecemeal fashion, but assume a perspective outside the object or space being described.[15] Take, for example, the following excerpt from the description of Europe in Peter Apian's 1524 *Cosmographia*: "In Europe, the first region toward the west is Spain, which the Greeks called Iberia. It is the head of the imaginary dragon-body which is Europe. Ancient writers divided Spain into three parts. ... The province adjacent to Spain is France ... which is separated from Spain by the Pyrenees on its west."[16] The passage suggests a discursive setting in which one person, the narrator, talks another person, the narratee, through a map of Europe spread out before them, moving from west to east. Both narrator and narratee remain outside, even above, the geography being described.

By contrast, the "route tour," or itinerary, brings the narratee into the space being described. It does so by traveling along a real-life route or routes that the narrator has taken, often takes, or believes can be taken. Not only is the route real, but it is made to feel that way through references to bodily experience. Here we can consider the following example from a report filed by a colonial official in 1580, describing the region of New Spain under his charge:

> The town of Tecuicuilco ... is seven great leagues from the said city [Oaxaca], where the mayor habitually resides, over very high mountains, and toward the north of the city. This town [Tecuicuilco] has another village subject to it–what is known here as an "estancia"–which is called Santa Inés Tepeque; it is one league from here. ... So is the town of Atepeque. ... It is four leagues from the said town of Tecuicuilco, on the northeast side, along rough, mountainous, twisting paths.[17]

This is a route tour. Like the gaze tour, it arranges individual places in the linear series made necessary by writing, but it follows real-life routes from one to the next, rather than simply moving from one place to the next without any regard for how a body would actually make the trip. It also brings the reader into the space by mentioning the difficulties of travel, such as the high mountains and the tortuous paths. Linearity, real-life routes, and an embodied perspective: these are the characteristics of the "route tour" or "itinerary." Of the three, only linearity is imposed by the verbal medium.

The other two characteristics may very well stem from the fact that the authors of so much of the material in the archive got to know their territory by actually traveling through it as explorers, conquerors, missionaries, merchants, or pilots.[18] Alternatively, they may respond to an overall tendency in the culture at large to treat the itinerary as the default approach to conceptualizing and representing territory. Contemporary psychological subjects tend to mix itinerary perspectives with other points of view, like the extrinsic one of the gaze map, but we must remember that these people are products of a modern, map-saturated world.[19] The authors of the texts in the mapmaker's archive were not. Until rather late in the sixteenth century, maps were rare commodities, and many of those authors would have had little or no experience with them. In fact, so new were maps to sixteenth-century culture that the word *mapa* was not even in general circulation when the century started, and by century's end, it still enjoyed some neologistic glow.

Just as the word *mapa*, and maps themselves, were new to most sixteenth-century Spaniards, so was the abstract spatiality associated with them. When Spaniards used the word *espacio* ("space"), they rarely meant a two-dimensional or three-dimensional expanse. In most instances, they would have been referring to an interval of time. For example, when a character in *La Celestina* (1499) complains about how long someone else is taking to arrive, he explains, "What a long *space* [time] the old bearded woman takes!"[20] The use of *espacio* to refer to space rather than time echoed the linearity of the temporal usage, denoting a distance rather than an expanse. In *Don Quixote*, part I (1605), a character rides to meet his companion, "who was waiting for him a good *space* [distance] from there."[21] Even among the learned such as the compiler of the period's most extensive dictionary of Spanish, *espacio* does not register as a name for expanses, but rather for intervals of time or space.[22] It is only in texts written by people with mathematical or technical training that we find *espacio* used to refer to two-dimensional expanses, and it is not until the eighteenth century that this usage becomes common to the culture as a whole. When it finally registers in the eighteenth-century *Diccionario de autoridades*, moreover, its definition makes specific reference to latitude and longitude—that is, to cartography.

For the vast majority of Spanish speakers *espacio* meant an interval of time or a distance between two points, suggesting that the itinerary was the dominant spatiality of the early modern Spanish-speaking world. As such, it could not have been more different from the abstract space of the gridded map. While the gridded map represents space as an abstraction substantially independent from, and logically prior to, the objects it contains, the itinerary builds space by connecting preexisting places along routes of travel. The grid renders space as a uniform abstraction that extends equally in all directions, undergirding everything. The itinerary creates a route or routes, but has nothing to say one way or another about the "nonspace" outside the network. The grid presents space for optical inspection, saying nothing about how knowledge of that space came into being. The itinerary memorializes the process by which the territory came to be known, physical travel. The user of the gridded map looks down on space from the disembodied perspective of Ptolemy's eye. The reader of the itinerary gets to know space by moving through it.[23]

The time has come, therefore, to recognize that the texts in the mapmaker's archive cannot be understood merely as imperfect sources for the mapmaker's data. They were the products of what Matthew Edney has called a "mode" of mapmaking, a particular conjunction of techniques, assumptions, institutions, and materials that combine to map space in recognizable, more or less consistent ways.[24] According to Edney's approach, different modes of mapping can and often do coexist within a particular place and time, and that coexistence cannot always be understood in terms of hierarchies of value or teleologies of development. In fact, the language of modes invites us to refrain from defaulting to such models, and to look for more subtle ways modes of mapping interact, displace each other, combine, repel, or simply cohabit. To say that written itineraries represent a mode of mapping space and territory, therefore, is to say that such texts are not just source texts for "real" maps, but are themselves "real maps," verbal ones that play by their own rules, with their own sense of space.

Reading Early Modern Itineraries

If this is true, and I believe it is, then we need to develop certain habits as readers, and avoid others. First, we must recognize that as twenty-first-century people, we are products of a map-saturated culture that privileges the spatiality of the grid. As a result, we can often fall into the trap of allowing a modern map, even if only a mental one, to mediate our reading of early modern cartographic literature (i.e., writing that maps places and spaces). We ask ourselves where the traveler "really is" on a modern map, and fill in the geographic details that the narrator does not provide. While it can be very useful to do such things, we must not allow ourselves to believe that we are thereby discovering the geography of the text we are actually reading.

Second, we should expect that the maps that emerge from our reading, when we refuse to simply trace the route on a modern map, will be schematic and, by the measure of modern cartography, highly inaccurate. We must learn to accept this and work with it. We would do well to remember the examples of subway maps or the sort of sketch maps we used to draw for visitors to our home, before they could find their way using Google Maps. Such maps are all about significant places and the routes that connect them. They are never drawn to scale. They get distances wrong and simplify directions. Nevertheless, they serve their purpose well, that of getting commuters to work or your guests to your barbecue. Of course, such maps are not meant to have any meaning beyond their limited purpose. The itinerary map that emerges from a verbal text, however, might very well have some meaning, or some function in the text, beyond the purpose of facilitating imaginative wayfinding on the part of the reader. It is our job as readers of cartographic literature to figure out what that meaning might be.

Third, we must acknowledge that, in the most interesting cartographic literature, the itinerary structure does not exist in isolation, and should never be considered alone, like a skeleton stripped of its flesh. Itinerary structures occur within texts that often, but not always, tell stories. To ask what the itinerary structure means is to ask what importance the diegetic space has to the narrative, and vice versa. It is to ask how movement through space informs the production of narrative meaning, and how narrative invests place and space with meanings they might not otherwise have. To map "New Spain" in and through the itinerary mode common to the reports of conquistadors and colonial officials, the letters of missionaries, and even the histories and geographies of ambitious synthesizers, was very often to construct a series of itineraries, from Veracruz to Mexico City, and from Mexico City outward to places like Oaxaca and Acapulco. Only through the mediation of a map was "New Spain" made to extend beyond the itineraries of conquest, colonial governance, and evangelization, into territory that remained less affected by the presence of Europeans.

Finally, we would do well to pay close attention to texts where the itinerary mode with its unidimensional spatiality rubs up against the Ptolemaic mode, with its two-dimensional one. This sometimes occurs in texts written by authors who, for whatever reason, had some familiarity with gridded maps but remained rooted in the world of the itinerary. Not only do their texts allow us to understand what it might mean for two modes of mapmaking to interact, but they also suggest that Spanish culture was aware, on some level, of the existence of different modes, and of the purposes to which each could be put. They allow us to glimpse how the culture not only struggled to make sense of the new world in which it found itself, but also reflected on the very tools it had available to map that world.

Competing Cartographic Modes in Ercilla and Góngora

Iberian culture of the Middle Ages, Renaissance, and Baroque seems to have had a penchant for mapping the world in verse. We can identify a series of texts written in Spain and Portugal between the thirteenth and seventeenth centuries that include what one critic has called a "mappamundi episode," a description of the world appearing as part of a verse narrative, often a heroic one.[25] The tradition has rich precedents in Greco-Roman antiquity, and reaches its culmination, as far as Iberia is concerned, with the mappamundi episodes of part II of the *Araucana* of Alonso de Ercilla y Zúñiga (1578), an epic poem about the conquest of Chile, and the *Soledad primera* of Luis de Góngora y Argote (1613), a bucolic idyll.[26] Ercilla's mappamundi provides a gaze tour of the known world meant, on the face of things, to celebrate the majesty of Philip II's far-flung empire, while Góngora's mappamundi can be understood as an itinerary tour of the world that provides a critical, even cynical answer to Ercilla's imperial map.

Alonso de Ercilla was a poet and courtier with close ties to Philip II who formed part of the military expedition sent from Peru to the south of Chile in 1557 to suppress the revolt of the Mapuche people, known as the Araucanians. According to its author,

La Araucana was born of his battlefield experiences, but it was clearly raised back in Europe, on a steady diet of overtly literary material. The imprint of the literary tradition is perhaps strongest in a series of patently fictional episodes that appear in part II of the poem's three parts. In one of them, a literary sorcerer in Araucanian costume has Ercilla, who appears as a character in his own poem, gaze into his crystal ball to see the world in its entirety.

Ercilla's vision takes the form of a poetic gaze tour of the entire world, forty-six octaves long, the *Araucana's* mappamundi episode. Most of it reads like the following stanzas, which trace a series of place names down the length of the Chilean theater of war, then follow Magellan's expedition across the Pacific to the Spice Islands:

> Vees la ciudad de Penco y el pujante
> Arauco, estado libre y poderoso;
> Cañete, la Imperial, y hacia el levante
> la Villa Rica y el volcán fogoso;
> Valdivia, Osorno, el lago y adelante
> las islas y archipiélago famoso
> y siguiendo la costa al sur derecho
> Chiloé, Coronados el estrecho
> por donde Magallanes con su gente
> al Mar del Sur salió desembocando,
> y tomando la vuelta del poniente
> al Maluco guió norduesteando.
> Vees las islas de Acaca y Zabú enfrente,
> y a Matán, do murió al fin peleando;
> Bruney, Bohol, Gilolo, Terrenate,
> Machicán, Mutir, Badán, Tidore y Mate. (27. 50–51)

(Gaze on Penco, and the mighty / Shrine of Freedom's cult, Arauco. / Glimpse the Imperial, Cañete, / Villarica's wroth volcano; / To the West, Valdivia, Osorno, / Lago; past the isles and fabled / Archipelago; on the south coast, / Chiloé, Strait of Coronados. // To the Straits, through which Magellan / Found the South Sea, steered his compass / Northward past Molucca, sailing / To Zabú and Acacan Islands, / Dying on Matán in battle, / With Bohol, Bruney, Gilolo, / Terrenate, Machian, Mate, / And Mutir, Badán, Tidore.)

The stanzas are made up primarily of place names predicated to imperative forms of the verbs *look* and *see*, occasionally interrupted by a micronarrative that serves to mediate an important transition from one part of the globe to the other, or to editorialize about some uniquely important place. The number and arrangement of the stanzas is very deliberate. Half the tour covers Europe, Asia, and Africa, and the other half covers Spain, the Americas, and the Spice Islands, to which Spain lay claim. At the center of the stanzas devoted to Spain we find the Escorial, the massive monastery-palace that Philip II constructed as a monument to his military victories, a mausoleum for his dynasty, and a tribute to the orthodox Catholicism that he hoped to defend and believed himself to embody. On the face of things, the episode appears to tell its Spanish reader that his or her world constitutes a righteous empire encompassing half the globe.

One might think that Ercilla would have drawn the material for his mappamundi episode from a contemporary map of the world, such as the "Typus orbis terrarum" of Abraham Ortelius, the famous mapmaker who had been named Royal Geographer to Philip II, but according to his modern editor, he seems to have favored textual sources, like Strabo, Pomponius Mela, Pliny, and the modern authors of Spanish Americana like Oviedo.[27] In other words, he turned to the same textual archive that nourished the production of iconographic maps, but plotted his locations into the rhythms and meter of his royal octaves, rather than the coordinates of a cartographic grid. The result is a subtle parody of that cartography. Like the Renaissance maps to which this episode implicitly alludes, Ercilla's mappamundi invites us to "look" and "see," by implicating us in the commands that the sorcerer directs toward his guest. But unlike those maps, this verse cartography cannot deliver on the promise.

The reader can imagine Ercilla gazing awestruck into the crystal ball, as the sorcerer's tour takes him around the world, place by place, but may very well have trouble imagining the vision itself, particularly if that reader is a contemporary of Ercilla's, not a modern reader accustomed to playing with Google Earth. Only the most learned sixteenth-century reader, or the one with the best library, or the best maps, could have fleshed out Ercilla's list of toponyms with the various details, geographic, historical, cultural, required to make them mean something. Most would have run into trouble at some point or another, particularly when the gaze tour reached the newly discovered reaches of the New World. At that point, we can imagine, the reader would have become aware of the fact that he or she was looking, not at the world itself, and not even at a cartographic representation of the world, but at a mere list of names. The very savviest of those readers, moreover, might even have understood what this awareness implied for a grasp of maps and globes. They would have understood that they, just like Ercilla's verse map, were nothing more than visual renderings of textual information, of names, designed in such a way as to deceive us into thinking that they somehow represented the world itself.

Something is definitely awry with the mappamundi episode and its putative celebration of Spain's global imperial destiny, particularly when we consider what happens in the rest of the poem, which depicts Spain's conduct of the war as excessively cruel and vindictive while simultaneously casting the Araucanians as classical heroes. The itinerary of conquest repeatedly becomes a frustrated act of penetration, at once territorial and sexual. But rather than dwell further on *La Araucana*, I turn to the *Soledad primera* of Luis de Góngora and its own mappamundi episode. It takes some reflection to discover how Ercilla's map of the world in verse might be understood to mock the pretensions of cartography to place Spain at the center of the world and the heart of a hemispheric empire. Góngora's map, by contrast, leaves no room for doubt about its acerbically critical attitude toward Spain's overseas expansion.

The *Soledad primera* tells the story of an unnamed man who finds himself shipwrecked on an unnamed coast, where he enjoys the hospitality of humble country folk. The simple story provides an armature in which Góngora hangs intensely wrought images of the objects and doings of everyday life. At one point, however, the text takes an epic turn. The traveler encounters an old mountaineer who speaks against the art of navigation as a manifestation of human greed and a source of human suffering. Along the way, his diatribe maps the world in verse. In fact, one of Góngora's most prominent critics has not hesitated to call the episode the poetic equivalent of the ornate world maps for which Renaissance cartography is so well known, infinitely more accomplished than what he considers to be Ercilla's sterile rhymes.[28]

In one very important sense, Góngora's verse cartography can be understood as an attempt to undo the very building blocks of Ercilla's map, and of cartography in general: place names. While Ercilla lists toponyms, and relies on the reader to recognize them and invest them with meaning, Góngora mentions very few place names, and none outside the world known to the ancient Greeks and Romans.[29] Instead, he figures places through elaborate circumlocutions built primarily out of mythological allusions. In the following passage, for example, he maps the same geography that appears in the second of the two stanzas from Ercilla quoted above:

De firmes islas no la inmóvil flota
en aquel mar del Alba te describo
cuyo número, ya que no lascivo,
por lo bello, agradable y por lo vario
la dulce confusión hacer podía
que en los blancos estanques del Eurota
la virginal desnuda montería,
haciendo escollos o de mármol pario
o de terso marfil sus miembros bellos,
que pudo bien Acteón perderse en ellos. (481–490)

(Of the immobile fleet of islands lying / at anchor in that dawn sea / I say nothing, a multitude, though not / licentious, that through its beauty and delight / and variety might arouse / the same sweet perturbation / as did the white pools of the Eurotas / the naked virginal troop / whose ravishing limbs formed reefs / as if of marble or smooth ivory / among which Acteon / could not but lose his way.) (490–501)

Ercilla names the islands of insular Southeast Asia visited by Magellan's expedition. Góngora never does. Instead, he converts them into the dazzlingly white body parts of the female entourage accompanying the goddess at her bath. Ercilla mentions that Magellan died in these islands. Góngora, instead, alludes to the figure of Actaeon, the hunter whose accidental voyeurism raised the ire of the goddess and led to his death. By investing the islands with sensuous beauty, and alluding to their discovery as an act of lascivious transgression, Góngora does not pretend to make them visible, as a map would, or to nudge his reader into reflecting on the problems of visibility, as Ercilla's map does. Instead, he makes the world intelligible, in terms at once geographic, moral, political, and historical, while simultaneously playing with the very notion of visibility and its relationship to desire and truth.

Throughout his mappamundi, Góngora rails against the Apollonian perspective of cartography, with its pretension to see it all and know it all. One of the ways he does this is by constructing his poetic cartography, not as a gaze map, but as a route map. Rather than adopt a commanding perspective external to the geography he traces, like the one provided by the crystal ball in *La Araucana*, Góngora's text follows in the wake of Columbus, da Gama, Magellan, and other figures from the major expeditions of the so-called Age of Discovery. He weaves their many historical journeys into a single allegorical itinerary, the round-the-world voyage of Greed personified. Greed never flies into the heavens to see the world from above, not even when his ship, in Góngora's complex figures, becomes Apollo's chariot. It remains earthbound, or rather seabound, cutting through crystalline waves rather than soaring into the celestial spheres.

Like the islands in the passage above, every place along the way becomes one version or another of Diana's nymphs, a feminine figure who attracts the lascivious gaze of a decidedly masculine explorer bent on possession even at the cost of transgression, violation, and sacrifice. While Ercilla describes the world, Góngora sails through it, attaching geographic knowledge to the insatiable desire and violent striving of empire. It is an itinerary that can lead only to death, the death of the mountaineer's son on the far side of a world that has come to be known, in various senses of the word, at tremendous expense. Góngora's itinerary thus denounces empire, and simultaneously mocks the pretense of cartography to show us the world.

These are only two examples from a vast cartographic literature that emerged out of Spain's internal transformation and external expansion during the early modern period, but they are particularly interesting examples. They are not just modes of mapping, like a written itinerary or gridded map, but reflections on the nature, limitations, and purposes of mapping itself, at least in the context of Spain's imperial expansion. On the one hand, we cannot understand them without first getting a handle on the different modes of mapping available in early modern Spain, and the very different spatialities they favored, but on the other, we cannot reduce either to a simple manifestation of the two modes and spatialities I have charted. Their very existence points to a key feature of the Spanish effort to map the new world in which Spain found itself as a result of both its internal transformation and its external expansion. Some Spanish writers, at least, were aware of the profound difficulty of the challenge, and of the inadequacy of the tools at hand. They were familiar with maps of all kinds, but they were aware of the limitations their maps presented, and of the history of violence with which some of them were intertwined.

Notes

1. This chapter summarizes arguments made previously in Ricardo Padrón, *The Spacious Word: Cartography, Literature, and Empire in Early Modern Spain* (Chicago: University of Chicago Press, 2004), and Ricardo Padrón, "Against Apollo: Góngora's *Soledad Primera* and the Mapping of Empire," *MLQ* 68, no. 1 (2007): 363–393. For a different perspective, see Simone Pinet, "Literature and Cartography in Early Modern Spain: Etymologies and Conjectures," in *The History of Cartography*, ed. J. Brian Harley et al., vol. 3, pt. 2, 469–476 (Chicago: University of Chicago Press, 2007).

2. David Woodward, "Maps and the Rationalization of Geographic Space," *Circa 1492: Art in the Age of Exploration*, ed. Jay A. Levenson, 83–88 (New Haven, CT: Yale University Press, 1991).

3. I use the term *archive* primarily in its Foucauldian sense, as a name for all the material traces left behind by a particular period and culture, although it should be noted that there were also literal, institutional archives involved, like that of Seville's *Casa de la contratación*.

4. Patrick Gautier Dalché, *La géographie de Ptolémée en Occident (IVe–XVIe siècle)* (Turnhout, Belgium: Brepols, 2009).

5. David Buisseret, *The Mapmaker's Quest: Depicting New Worlds in Renaissance Europe* (Oxford: Oxford University Press, 2003).

6. On Spanish New World cities, see Angel Rama, *La ciudad letrada* (Hanover, NH: Ediciones del Norte, 1984), 1–39. On abstract space and modernity, see Henri Lefebvre, *The Production of Space*, trans. Donald Nicholson-Smith (Oxford: Blackwell, 1991), 229–291.

7. As quoted in John R. Short, *Making Space: Revisioning the World, 1475–1600* (Syracuse, NY: Syracuse University Press, 2004), 74. For more on Ptolemaic space, see Jean-Marc Besse, *Les grandeurs de la Terre: Aspects du savoir géographique à la Renaissance* (Lyon: Ens, 2003), 112–130.

8. J. B. Harley and Paul Laxton, *The New Nature of Maps: Essays in the History of Cartography* (Baltimore: Johns Hopkins University Press, 2001).

9. See Dava Sobel, *Longitude: The True Story of a Lone Genius Who Solved the Greatest Scientific Problem of His Time* (New York: Walker, 1995).

10. For more on maps of the *Casa de la Contratación*, see Alison Sandman, "Spanish Nautical Cartography in the Renaissance," in *The History of Cartography*, vol. 2, pt. 2, 1095–1449. For more on Spanish cartography, see the essays by David Buisseret in the same volume.

11. Barbara Mundy, "Mapping the Aztec Capital: The 1524 Nuremberg Map of Tenochtitlan, Its Sources and Meanings," *Imago Mundi* 50 (1998): 11–33.

12. Hernán Cortés, *Letters from Mexico*, trans. A. Pagden (New Haven, CT: Yale Nota Bene, 1971), 102.

13. For the distinction, see Ptolemy, *Ptolemy's Geography: An Annotated Translation of the Theoretical Chapters*, ed. and trans. J. Lennart Berggren and Alexander Jones (Princeton, NJ: Princeton University Press, 2001), 57–59.

14. See Willem J. M. Levelt, "Cognitive Styles in the Use of Spatial Direction Terms," in *Speech, Place, and Action: Studies in Deixis and Related Topics*, ed. Robert J. Jarvella and Wolfgang Klein, 251–268 (New York: Wiley, 1982).

15. Holly A. Taylor and Barbara Tversky, "Perspective in Spatial Descriptions," *Journal of Memory and Language* 35, no. 3 (June 1996): 371–391, here 376.

16. Peter Apian and Frisius Gemma, *Libro dela cosmographia* (Antwerp: Gregorio Bontio, 1548), 32r. Google Books. My translation.

17. Rene Acuña, *Relaciones geográficas del siglo XVI* (Mexico City: Instituto de Investigaciones Antropológicas, Universidad Nacional Autónoma de Mexico, 1982), vol. 3, 87–88.

18. This is how Mundy and Scott explain the presence of itineraries in colonial maps and geographic descriptions. See Barbara Mundy, *The Mapping of New Spain: Indigenous Cartography and the Maps of the Relaciones Geográficas* (Chicago: University of Chicago Press, 1996), 35–37; Heidi Scott, *Contested Territory: Mapping Peru in the Sixteenth and Seventeenth Centuries* (Notre Dame, IN: University of Notre Dame Press, 2009), 76–77.

19. On mixed perspectives, see Taylor and Tversky, "Perspective in Spatial Descriptions."

20. Fernando de Rojas, *La Celestina*, ed. Dorothy Severin (Madrid: Cátedra, 1989), 138.

21. Miguel de Cervantes Saavedra, *Don Quijote de La Mancha* (Barcelona: Instituto Cervantes, Crítica, 1998), vol. 1, 8.

22. The history of Spanish dictionary definitions can be consulted using the *Nuevo tesoro lexicográfico de la lengua española*, available at http://buscon.rae.es/ntlle/SrvltGUILoginNtlle.

23. For more on this contrast, see Michel de Certeau, *The Practice of Everyday Life*, trans. Steven Rendall (Berkeley: University of California Press, 1984), 120.

24. On cartographic modes, see Matthew H. Edney, "Cartography without 'Progress': Reinterpreting the Nature and Historical Development of Mapmaking," *Cartographica* 30, nos. 2–3 (1993): 54–68.

25. James R. Nicolopulos, *The Poetics of Empire in the Indies: Prophecy and Imitation in La Araucana and Os Lusíadas* (University Park: Pennsylvania State University Press, 2000), 221–269.

26. Alonso de Ercilla, *La Araucana*, ed. Isaías Lerner (Madrid: Cátedra, 1993), and *The Araucaniad: A Version in English Poetry of Alonso de Ercilla y Zúñiga's La Araucana*, trans. C. M. Lancaster and P. T. Lancaster (Nashville: Vanderbilt University Press, 1945); Luis de Góngora y Argote, *Soledades*, ed. Robert Jammes (Madrid: Castalia, 1994), and *Selected Poems of Luis de Góngora*, ed. and trans. John Dent-Young (Chicago: University of Chicago Press, 2007).

27. Ercilla, *La Araucana*, 736n7.

28. Robert Jammes, "Historia y creación poética: Góngora y el descubrimiento de América," in *Hommage à Claude Dumas: Histoire et Création*, 53–66 (Lille: Presses universitaires de Lille, 1990), here 61.

29. Ibid., 56–57.

Bibliography

Acuña, Rene. *Relaciones geográficas del siglo XVI.* Mexico City: Instituto de Investigaciones Antropológicas, Universidad Nacional Autónoma de Mexico, 1982.

Apian, Peter, and Frisius Gemma. *Libro dela cosmographia.* Antwerp: Gregorio Bontio, 1548.

Besse, Jean-Marc. *Les grandeurs de la Terre: Aspects du savoir géographique à la Renaissance.* Lyon: Ens, 2003.

Buisseret, David. *The Mapmaker's Quest: Depicting New Worlds in Renaissance Europe.* Oxford: Oxford University Press, 2003.

de Certeau, Michel. *The Practice of Everyday Life.* Trans. Steven Rendall. Berkeley: University of California Press, 1984.

Cervantes Saavedra, Miguel de. *Don Quijote de La Mancha.* Barcelona: Instituto Cervantes/Critica, 1998.

Cortés, Hernán. *Letters from Mexico.* Trans. A. Pagden. New Haven, CT: Yale Nota Bene, 1971.

Edney, Matthew H. "Cartography without 'Progress': Reinterpreting the Nature and Historical Development of Mapmaking." *Cartographica* 30, nos. 2–3 (1993): 54–68.

de Ercilla, Alonso. *La Araucana.* Ed. Isaías Lerner. Madrid: Cátedra, 1993.

de Ercilla, Alonso. *The Araucaniad: A Version in English Poetry of Alonso de Ercilla y Zúñiga's La Araucana.* Trans. C. M. Lancaster and P. T. Lancastar. Nashville: Vanderbilt University Press, 1945.

Gautier Dalché, Patrick. *La géographie de Ptolémée en Occident (IVe–XVIe siècle).* Turnhout, Belgium: Brepols, 2009.

de Góngora y Argote, Luis. *Selected Poems of Luis de Góngora.* Ed. and trans. John Dent-Young. Chicago: University of Chicago Press, 2007.

de Góngora y Argote, Luis. *Soledades.* Ed. Robert Jammes. Madrid: Castalia, 1994.

Harley, J. B., and Paul Laxton. *The New Nature of Maps: Essays in the History of Cartography.* Baltimore: Johns Hopkins University Press, 2001.

Jammes, Robert. "Historia y creación poética: Góngora y el descubrimiento de América." In *Hommage à Claude Dumas: Histoire et Création*, 53–66. Lille: Presses universitaires de Lille, 1990.

Lefebvre, Henri. *The Production of Space.* Trans. Donald Nicholson-Smith. Oxford: Blackwell, 1991.

Levelt, Willem J. M. "Cognitive Styles in the Use of Spatial Direction Terms." In *Speech, Place, and Action: Studies in Deixis and Related Topics*, ed. Robert J. Jarvella and Wolfgang Klein, 251–268. New York: Wiley, 1982.

Mundy, Barbara. "Mapping the Aztec Capital: The 1524 Nuremberg Map of Tenochtitlan, Its Sources and Meanings." *Imago Mundi* 50 (1998): 11–33.

Mundy, Barbara. *The Mapping of New Spain: Indigenous Cartography and the Maps of the Relaciones Geográficas*. Chicago: University of Chicago Press, 1996.

Nicolopulos, James R. *The Poetics of Empire in the Indies: Prophecy and Imitation in La Araucana and Os Lusíadas*. University Park: Pennsylvania State University Press, 2000.

Padrón, Ricardo. "Against Apollo: Góngora's *Soledad Primera* and the Mapping of Empire." *MLQ* 68, no. 1 (2007): 363–393.

Padrón, Ricardo. *The Spacious Word: Cartography, Literature, and Empire in Early Modern Spain*. Chicago: University of Chicago Press, 2004.

Pinet, Simone. "Literature and Cartography in Early Modern Spain: Etymologies and Conjectures." In *The History of Cartography*, ed. J. Brian Harley et al., vol. 3, pt. 2, 469–476. Chicago: University of Chicago Press, 2007.

Ptolemy, J. *Ptolemy's Geography: An Annotated Translation of the Theoretical Chapters*. Ed. and trans. Lennart Berggren and Alexander Jones. Princeton, NJ: Princeton University Press, 2001.

Rama, Angel. *La ciudad letrada*. Hanover, NH: Ediciones del Norte, 1984.

de Rojas, Fernando. *La Celestina*. Ed. Dorothy Severin. Madrid: Cátedra, 1989.

Sandman, Alison. "Spanish Nautical Cartography in the Renaissance." In *The History of Cartography*, ed. J. Brian Harley et al., vol. 3, pt. 2, 1095–1449. Chicago: University of Chicago Press, 2007.

Scott, Heidi. *Contested Territory: Mapping Peru in the Sixteenth and Seventeenth Centuries*. Notre Dame, IN: University of Notre Dame Press, 2009.

Short, John R. *Making Space: Revisioning the World, 1475–1600*. Syracuse, NY: Syracuse University Press, 2004.

Sobel, Dava. *Longitude: The True Story of a Lone Genius Who Solved the Greatest Scientific Problem of His Time*. New York: Walker, 1995.

Taylor, Holly A., and Barbara Tversky. "Perspective in Spatial Descriptions." *Journal of Memory and Language* 35, no. 3 (June 1996): 371–391.

Woodward, David. "Maps and the Rationalization of Geographic Space." In *Circa 1492: Art in the Age of Exploration*, ed. Jay A. Levenson, 83–88. New Haven, CT: Yale University Press, 1991.

9
Bend of the Baroque: Toward a Literary Hydrography in France

Tom Conley

Introduction

The Baroque has a penchant for rivers. In visual culture that pullulates throughout the sixteenth century, especially in its later years when the technologies of woodcut and copperplate illustration yield an unforeseen production of images, maps make their way into books and get assembled in atlases. They are painted, placed, even plastered on walls of homes and offices; they circulate in the paper commerce of urban centers; and, as I argue below, they are everywhere in the creative worlds of poets and writers. Much as roads are traced in exaggerated lines in today's road atlases or across the liquid crystal of computer screens or iPhones, in early modern cartography rivers take the form of arteries and veins of commerce. Herein they inspire literary fantasy. To the delight of amateurs and specialists alike, rivers cause the maps on which they are drawn to give the impression of a flow and even pulsating body. Their hydrography cannot be dissociated from what we believe concurrent writers and artists were making of them when printed maps were proliferating, not coincidentally, at a time synchronous with the advent of what historians call a Baroque sensibility.[1] Without rehearsing time-honored principles of its aesthetics, the aim of this chapter is to see how, why, and with what effects and implications the French Baroque shares uncommon synergy with the mapping of rivers.[2]

    The remarks that follow will pass through six successive literary iterations, each implicitly or explicitly tied to maps of hydrographic facture, in a progression designed to discern the emergence of a Baroque sensibility where map and writing are conjoined. First, as a point of departure, a topographic map that Oronce Fine draws and presents to Henry II in manuscript in 1549 (before appearing in print in 1551) makes clear the importance of rivers in a manual of equally theoretical and practical inflection. Second, a fluvial poem by Pierre de Ronsard, assumed to be of a "mannered" signature, sets classical forms in a topography celebrating the wealth and beauty of the Touraine in 1555. By contrast, third, the drawing of a river in a diagram and architectural view in the *Amadis de Gaule,* a commanding work in the middle third of the century, indicates how a fluvial style brings together literature, hydrography, and architecture, and not least in Jacques Androuet du Cerceau's two volumes of drawings of the great châteaus and edifices of France (1576), whose implicit hydrography will be juxtaposed, fourth, to descriptions of rivers in Agrippa d'Aubigné's *Tragiques* (begun in 1577 and published in 1616), an epic overview of the Wars of Religion. Fifth, in a similar context, in Montaigne's *Essais* (1580–1595), which he describes as a work of flux and flow, finds reflection in his description of his neighboring River Dordogne. Its cartographic latency is brought forward, finally, in contrast to Nicolas Sanson's map of the rivers of France (1641), a cartographic object rife with literary innuendo, that stands in contrast to the most memorable of all pieces of hydrographic literature, Madeleine de Scudéry's *Clélie* (1654–1660), in which the *Carte de Tendre*, an allegorical map decisive for the Baroque literary canon, first appears. The stake is thus to follow a course that, like a river, meanders forward and back again between words and images, and between the arts of allegory and of new and emerging science.

Oronce Fine: River Lines in *L'esphere du monde*

In 1549 cartographer Oronce Fine (1494–1555) presents to Henri II, recently crowned king of France, an elegantly drawn manuscript of *L'esphere du monde,* a summation and translation of his earlier *Cosmographie*, an authoritative cosmography of the decade before. The intention might have been to inform the monarch of an accomplished cosmographer's talents for which, underpaid professor of mathematics that he was, he felt had gone without notice. Cartographer, astrologer, artist, man of science and letters nearing the end of a distinguished career, Oronce needed symbolic and– urgently–monetary remuneration. Having subsisted since 1531 as Royal Professor of mathematics at the Collège de France, to make ends meet Oronce had been translating into French scientific works of his own signature. Based on his own *De mundi sphæræ* (1542), *L'esphere du monde* appeared first in an illustrated manuscript on vellum (1549) and, two years later, replete with woodcut images, in a handsome printed edition by Michel de Vascosan. Both the manuscript and the book begin from an eight-line *captatio* begging forgiveness on the part of gentle readers for whatever errors they may find in a work aimed for common and collective edification. The presentation copy ends, sixty-nine folios *infra,* with an *explicit* in *cul-de-lampe,* verifying that the work above has been "newly composed in French, written, depicted and portrayed by Oronce Fine, native of Dauphiné, reader in mathematics to the King." He underscores, first, the wounded virtue of the indigent scholar that he is.[3] There follows praise of the quadrivium and of their "subalterns, geography and perspective," but in addressing his king, he underscores the importance of rivers in his topographic projections:

> I have edited and translated into French one of the most delectable parts that there is between the said mathematics, that is, the universal description of the entire world with the most notable things that originate in our sphere for reason of the first and regular movement of the heavens that is called cosmography, and the principles and rudiments of geography and hydrography that deal with the marine world.[4]

The letter draws attention to the way that water is vital to the dynamic equilibrium and commerce of the world, and that hence, by implication (or by way of Aristotle), the world, or "terraqueous globe," and its constituent parts move in concert with forces of attraction and repulsion, its ever-shifting telluric and watery elements reacting to the effects of heat and cold and humidity and aridity. In the first book a diagram of the four elements and their relations establishes a field of tension in which degrees of adhesion and repulsion are drawn in what would be the quadrature of a circle (figure 9.1).

Further explanation follows at the end of the fifth book, on geography and hydrography in the context of the "composition of geographic maps, of provinces and particular regions" (chapter 6) before attention is drawn to the representation of the globe in two dimensions and, finally, "the distinction of the winds, according to hydrographers, and of the true composition of what are called marine maps" (chapters 7 and 8). At this point, in the mode of a "scientific" literature, a didactic discourse melds with visual matter that both illustrates the conceptual material and fosters fantasy, especially where, in his words on topography, Oronce describes the flow and passage of water. Which, early in the last book, he gets at when situating the city of Paris in the eighth climate (or line of latitude) in what is inferred to be its spatial relation with the elements that distinguish topography from geography: "And it is worth noting the said climates are shown in the name of the most notable cities, or great rivers, or mountains, which are in them: thus its pertinence for the pleasure and fantasy of each and every geographer."[5] The perimeters of a regional map, Oronce later describes, are charted with respect to an originary point from which lines of latitude (from what would be an *x*-axis) and longitude (the corresponding *y*-axis) are measured and drawn to form a grid. He supplies a magnificent topographic, gridded map of southern France, the Alps of Provence, the Rhône Valley, and southwestern Switzerland and northern Italy, projected as a trapezoid to simulate the curvature of the globe, whose point of origin is adjacent to his birthplace near Briançon (figure 9.2).[6] Noting that the toponyms are "expressed by their own common names" (f. 54r°), Oronce suggests that in question is a map less of *Gallia* than of territories of France.

As he had done in his great map of *Gallia* of 1535, the regions included in this topography are defined by what appears to be a chimerical swath of waterways, all broadly drawn and colored in green ink. Cutting its way across the Valais and joining a squat Lake Geneva before bending toward Lyon where it meets the Saone, the Rhône flows directly south. Meeting the Durance below Avignon, it finally bifurcates and empties into the Mediterranean. Fed by veins of tributaries, the Po flows by Parma and turns north to Milan. Its source extends toward Briançon, located at the vanishing point of the map, not far from the author's birthplace. Separated by what seems to be a short portage in the Alps, the proximity of the French and Italian fluvial networks suggests that the nations are in close and active exchange, the waterways seemingly drawn to underscore how fluvial commerce feeds the nation. The network of

❡ De lesphere.

ciel: C'est assavoir les quatre simples elements, qui sont le feu, lair, leau, et la terre, conuenans incessament a la generation et corruption de toutes choses: Auec la diuerse et innumerable espece de tous les corps tant perfaictz que imperfaictz, engendres materiellemēt par la naturelle conuiction, et vertu des dictz quatre elements. ❡ Lesquelz elements, ne peunēt estre plus de quatre. C'est assavoir autant, et non plus ne moins, qu'il y ha de premieres qualitez predominantes en icenlx, qui sont chaleur, humidite, froidure, et secheresse: et qu'il y ha de conuenances ou combinations d'icelles qualitez, qui peuuēt estre ensemble en ung corps elementaire. C'est assavoir chauld et sec, qui sont au feu, chauld et humide, qui sont en lair, froid et humide, qui sont en leau, et froid et sec qui sont en la terre. Dont la chaleur domine au feu, l'humidite en lair, la froidure en leau, et la secheresse ou siccité en la terre. Car la chaleur et froidure, qui sont qualitez actiues: semblablemēt l'humidite et secheresse, qui sont qualitez passiues: sont totalemēt contraires, et ne peunēt estre ensemble en ung mesmes subiect elementaire. Qui est la cause, que le feu et leau, pareillement lair et la terre: sont totalemēt contraires. Cōme demõstre ceste fig.

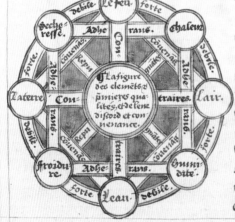

Figure 9.1

Oronce Fine, diagram of the relation of the four elements of the world, *L'esphere du monde* (Paris, 1549), f. 7 r°. Houghton Library, Harvard University.

¶ De lesphere.   F 65.

dirente declairée au iiij.e chapitre precedent. Lexemple
des choses dessusdites, peult estre prins par la figure qui se
suyt, contenant une partie de la gaulle Narbonnoyse, de 8/
degrez de longitude, et 5/degrez de latitude. En laquelle fi=
gure, le parallele septentrional est a b, le meridional c d:
le meridien occidental a c, et le oriental b d. Le moyen entre
les susditz meridiens, e f: on quel ung degré divisé a part
en 60/minutes, est g h. Desquelles 60/minutes, chascun
degré du parallele a b/en contient quasi 41; et du parallele
c d/ 44, avec 35/secondes. Les lieux descriptz en icelle carte,
par forme dexemple, sont exprimez par leurs propres noms.

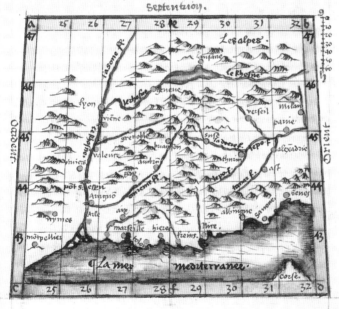

Figure 9.2

Oronce Fine, topography and hydrography
of southern France and northwestern Italy,
in *L'esphere du monde* (Paris, 1549), f. 54 r°.
Houghton Library, Harvard University.

inklines indicates a force of hydrographic fantasy that possibly gives rise to a sense of a national geography.

Mannered Waters

Oronce's topographic projection belongs to a manual of science written in pellucid vernacular. By contrast, Pierre de Ronsard's "Voyage de Tours ou amoureux" (1560), a poem that might be called a fragment of lyrical topography, tells of a trip on the part of Thoinot (Jean-Antoine de Baïf) and his friend Perrot (Pierre de Ronsard), in longing pursuit of their beloved Francine and Marion. Narrated tongue in cheek, the lyric takes an ironic distance from collections of *Amours* that both authors dedicated to the women who figure in this longer piece of verse. Like butterflies and little bees (*avettes*), fluttering and buzzing about, pollinating flowers in the gardens of the Touraine, fragile creatures that they are, the two men worry about the future of their race. Perrot loves Marion (Marie de Bourgueil, Ronsard's object of desire in *La Continuation des Amours* of 1555) and Thoinot, Francine (a fictive female, avatar of the same poet's Méline of 1552). A point of reference is the Clain, a river near Châtelleraut that empties into the Vienne where, at the sight of Francine, head over heels, Thoinot discovers her name engraved in his heart. The voyage begins from Coustures, a village on the River Loire, located near Ronsard's La Possonière, before passing south, by the author's cherished Forest of Gastine, then along the same axis, down to nearby Marray. En route they pass through Beaumont-la-Ronce and stop by Langennerie before sighting, to the west of the city of Tours, the tower of the priory of Saint-Cosme. Having consulted a soothsayer in "Crotelles" (Croutelle), to the south and not far from the River Clain, Thoinet learns that he is under Love's jurisdiction and in the throes of Eros:

> Why, tired of dance, recumbent amidst flowers,
> Haven't I leaned on your lap or your head,
> Or set my eyes on yours, or my mouth
> Upon your two breasts born of snow and ivory?
> Do I seem too old for you? A tender beard
> Only now begins to spread over my cheek,
> And your mouth, more beauteous than coral
> Were it to kiss me, would be my bounty.[7]

Like Thoinet, his desires are unrequited. From a high rock Perrot decides to throw himself into the Loire—less to drown and be done with life and love than to cool his desire and to cleanse himself of his feigned sorrow. From a sandy shore he watches Marion and her mother sail downward, into the landscape he would wish to see transformed, its gravelly riverbed turned into sheets of rubies and pearls, its muddy shores suddenly dappled with flowers. He wishes he could become the river itself:

> They say that in times past some people changed
> Their form into a river, and they even swam
> In the current that drop by drop sallied from their blood
> When their body, distilled, transformed into water.
> Why can't I change my human resemblance
> Into the form of water that carries this craft?
> Murmuring, I would go under the vessel's hull,
> I'd go all around, and my loving water
> Would kiss her hand, and then her gracious mouth,
> Following her up to the white Chapel,
> Then leaving my canal to do what I would,
> Following her course I'd go as far as Bourgueil,
> And there, below a pine, reposing on the grass,
> I'd wish to be embodied in my original figure.[8]

The pine tree under which he recovers human form stands as a geographic marker on the hillside of the north shore of the Loire in which he has flowed downstream. Ovidean metamorphosis takes over when allusion is made to Glaceus and Hippomene's strategem to capture swift Atalante by setting golden apples along the path of a foot-race, but in the end he praises how, *as water*, he would caress her and finally, when the voyage ends, he will ultimately engrave Marion's name on a tree that his aqueous form will have nourished. In dreaming of their concourse he would marry her Angoumois origins with those of his own Vendée. The Pont Guyet (near Saint-Nicolas-de-Bourgueil), the final toponym noted in the poem's itinerary, becomes the happy locus where he imagines himself consummating their love, on a grassy slope where, like the end of the poem just below, the couple will expire in pleasure.

In his astute and erudite commentary appended to the poem, Rémy Belleau, fellow traveler of the Pléiade, notes how the voyage owes much to Theocritus, "an imitation of the Syracusian women in the fashion that little love-angels flew around the corpse of Adonis, much as nightingales fly from branch to branch." Belleau immediately dilates on the place-names for the purpose of creating a poetic topography: *Coustures* is Ronsard's birthplace, *Gastine* the name of a forest growing out of a wasteland, while *Mauré, Beaumont-la-Ronce*, and *Langennerie* specify hamlets, and so on, until the poet reaches *Pont Guyet*, Marie de Bourgueil's birthplace, when the final toponym is set in counterpoint to the site of departure. In the gloss Belleau adds that the *tombeau de Turnes* refers to "Turnus, founder of Tours," who is "buried under the city's castle that is washed with the waters of the Loire that we still see today near the bridge by the battlements of the castle."[9]

An amorous fantasy turns into an itinerary or, as Ronsard might have wagered, into a pilgrimage that his future admirers would wish to follow. Metamorphosis of

Ronsard's body into a watery current qualifies the poem as hydrography. A tour-de-force mythologizing the poet's *terroir*, "Le Voyage de Tours" constructs the illusion of the Touraine as the "garden of France": far from wilderness or urban centers, the rivers set the landscape into motion, in meander and flow to and from classical and contemporary time. It cannot be denied that early in the poem, the passing mention of butterflies and bees refers to Grecian pastoral, but also to a map of a region at a time when Ronsard shifts from an amorous to a political mode, in his elegies of 1559 and, three years later, in his *Discours* that will draw the Wars of Religion into his verse. The "Voyage de Tours" may be Ronsard's last piece of a "mannered" mapping of a world, close to home, that becomes a site and situation that Baroque poets will invert.

From *Amadis* to Androuet

Not far from Ronsard's stomping grounds, the castle of Chambord, built to serve the king's taste for cynegetics, is set in a marshland and surrounding forest, amid a veinous network of rivers, the currents mirroring the drive and force of venery itself—the lust it inspires, as the Pléiade had made clear over and again, serving the ends of poetry. Set in the context of a river, the plan of the château appears in an illustrated edition of *L' Amadis de Gaule*, the *roman fleuve* that in 1525–1526 Francis I brought to France after having it read to him while incarcerated for a year in Madrid. In French translation it became one of the most popular and best read of all serial novels in midcentury France and beyond. In the fourth book, Nicolas Herberay des Essarts's fluvial prose finds a visual correlative in a topographic plan of the château and its environs (figure 9.3).

Two woodcuts in double folio punctuate the narrative. In the immediately preceding text, the hero, Amadis, is said to have heard his desired Grasinde graciously praise Dame Oriane. She blushes, no sooner responding that she, "as you see me, a simple disinherited Gentlewoman," holds Grasinde, her elegant friend, in infinite admiration. Ambling along a garden path, Grasinde and Oriane continue their exchange until they reach the Palace of Apolidon, from which, adds the narrator, Princess Grasinde was a descendent, "and because it was one of the most sumptuous buildings in the world, it seemed to me worthy to set it down in writing."[10]

Turning the page, the reader happens not upon a description but, on the verso folio, a plan, the "map of the Closed Island," whose measurements and columnar designs are indicated in geometric drawings in the outer margins. The ground plan of a château and a garden are set amid landscapes viewed in eight squares, four of thick or managed forests and two of pastures. Included are rabbits (upper left), a hound and a roebuck (upper right), and two unicorns (middle right). A river surges through the idyllic space. At the upper-central juncture of the four enclosed landscapes in the upper areas the waters flow into a lake in which a mermaid swims. The river trifurcates, rushing downward in channels that open and close on both sides of the plan of the château and garden before exiting the field of the image. In conjunction with the inner borders ten bridges span the river at different points over the borders of the squares and beyond the frame of the woodcut. A slender male, nude and holding a staff, stands at the center of the plan of a labyrinth in the square above and to the right of the garden, in whose center, a minuscule goddess—no doubt a Venus—emerges from a fountain.

On the folio to the right is a scenographic view of courtyard and elevation of what a trained eye identifies as the Castle of Chambord. The narrative stops, the images intervene, and in the following chapter a detailed "Description de l'ignographie & plant du palays qu' Apolidon avoit fait construyre en l'Isle Ferme" ("Description and Ichnography of the Palace That Apolidon Had Built on the Closed Island") further interrupts the dialogue. In reality, the canal from the river Cosson fills the moat around the château, but in the woodcut the river becomes a churning mass that mobilizes the map and the architectural view, as if it were a complement or correlative to an endless narrative passage that spawns the imagination of metamorphosis and transformation. The churning waters in the image on the left lend a sense of flow to the text of a novel that continually unwinds, unfolds, bifurcates, and meanders.

Not immediately associated with cartography or literature as such, but a document whose copperplates and descriptive prose provide a strong sense of the interrelation of the one and the other, Jacques Androuet du Cerceau's *Plus excellents bastiments de la France* (1576) integrates classical Italian and French architectural idioms in its survey of the great châteaus. It pays special heed to the rivers and aquifers that feed them. Visual or textual mention is made of their fluvial design and of where they figure in the nation's hydrographic network. In the sum of ichnographic, orthographic, and bird's-eye views the châteaus are generally shown adjacent to rivers that serve their residents and irrigate their moats. A series of copperplate drawings meant to be read contemplatively, *Les plus excellens bastiments* can be appreciated as a blueprint or "map" for what might become a hydrographic literature. In some of the depictions Androuet inserts minuscule staffage, elegantly dressed men and women ambling about arm in arm on terraces and under arcades, carriages delivering dignitaries to the entryways, couples in loggia whispering to each other, even strange figures leaning over balustrades—as if, more than enlivening the décor, they attested to the beauty of a courtly society in milieus worthy of their presence. Yet in the dedicatory letter to Catherine de Medici, the author writes: "I felt I could only bring forward this first volume of the exquisite buildings of our kingdom, hoping that our poor French subjects (in whose eyes and ears are only desolation, ruin and rapine that the recent wars have brought) can now draw a breath to take pleasure and happiness contemplating some of the beautiful and excellent edifices that for now France is enriched."[11] The situation

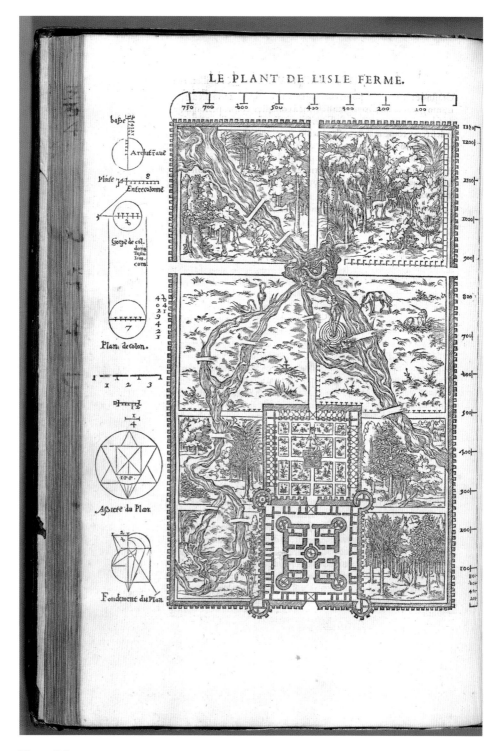

Figure 9.3

"Portrait de l'isle ferme" (ichnography and scenography) in Herberay des Essarts, *Amadis de Gaule*. Houghton Library, Harvard University.

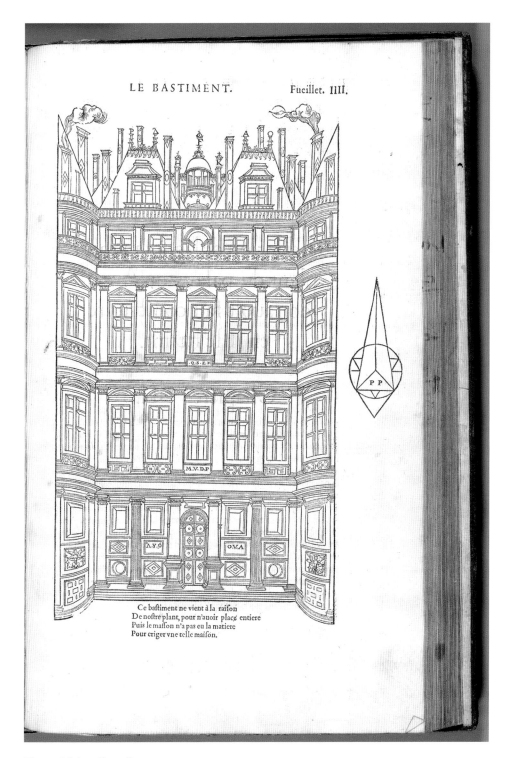

Figure 9.3 (continued)

in which the images are shown is far from what they portray. War, avows the author, gives the lie to the uncanny or, as if in a draftsman's prayer, a wished-for calm in his drawings of the virtual monuments and estates in the *Plus excellents bastiments.*

In his appreciation of Androuet's assemblage Yves Pauwels notes that the architect, also a draftsman and an on-site surveyor, works in the same way as Ronsard, favoring a virtual or likely—*vraisemblable*—representation of his object over one that would be exacting and truthful. As the title suggests, the buildings and their settings, drawn to the highest degree of excellence, beyond real measure, are comparable in fact to Ronsard's epic *Franciade,* the great poem of the same moment (published in 1572, but sketched out and rehearsed in fragments he read aloud at the court from the early 1560s, shortly after "Le Voyage de Tours").[12] Androuet cannot be understood, notes Pauwels, without considering that he is a graphic poet and not a builder. The latter, contrary to verbal and visual artists in the mode of Philibert de L'Orme, never knows "at once how to mask, wash, shade, color" and to compass their creations.[13] Like the *Plus excellents bastiments*, Ronsard's poem is written for Catherine de Medici and the young Charles IX. Aiming high and low, the epic is set in the thick of the Wars of Religion, and as such the latter become an absent presence, a virtual veil cast over the story of Francus, much as had Aeneas in his voyage to Rome, a heroic personage and refugee from the fall of Troy, who travels to Gaul to found the French nation. In *La Franciade* mythography meshes with cartography, while in Androuet du Cerceau the sublime views—ichnographic, scenographic, and bird's-eye—stress the calm and enduring excellence that on occasion fractures where the descriptive prose mentions conflict or depredations that impede the completion of the buildings as they are shown. The châteaus are inside and outside of time, within history, while the excellence of their often fluvial settings remains virtual.

Described in the context of rivers, the buildings and their surroundings become sites where cartography, history, and epic poetry converge. Chenonceau, Androuet du Cerceau notes, "is situated in the Touraine, set upon a bridge on the River Cherff, and even on one of its ends it is only a mass without a courtyard, enclosed by diverse separations of pavilions. Finding the site much to her liking, the Queen Mother bought and has since enlarged it with several buildings, intending to have it developed according to the design that I have drawn for you with a map [*plan*]." Two ground plans are set before two complementary profiles of the château, on opposite sides, from complementary points of view, one upstream and the other down (figure 9.4).

The fortress-like aspect is given in the four towers, aerated with windows fashioned with cruciform mullions that replace old-fashioned *meurtrières* or arrow loops. A virtual plan of *addimenta*, "augmentations du bastiment deliberees par la Royne mere du Roy" ("extensions to the building the Queen Mother had countenanced"), includes a trapezoidal plan of the arcades, four garden plots, and adjoining pastures that offer an idealized plan attached to the structure built over the river. Yet at the time of its publication the scenography carries uncanny undercurrents. From upstream the river is of light texture, flowing serenely, while from the other side the tight hatchings describing the current are dense and dark. The waters flowing under the bridge by the pavilion recall the pastoral geography of the *Amadis*. Neither roiled nor troubled, their calm underscores an unlikely tranquility. They seem to shroud the presence of the wars that the architect Androuet bemoans in the dedicatory preface to the volume of 1576.

D'Aubigné's Hydrography

The currents in Androuet's image remind viewers, be they then or now, of fluvial matter that in his *Tragiques* Agrippa d'Aubigné (1552–1630) crafts in a manner willfully opposed to Ronsard or the restorative ambiance of *Les bastiments*. Reputedly begun in 1578, in the thick of the Wars of Religion, and published in Geneva in 1616, eighteen years after their conclusion was said to have come with the Edict of Nantes, the epic is saturated with blood, bodily carnage, and sullied waters. Strongly marked by Protestant cartography and D'Aubigné's ongoing chronicle of the Wars of Religion, *Histoire universelle* (in eleven volumes, 1616–1630), the epic poem overlooks France at war from God's point of view. In its seven books, each ranging in length from 1100 to 1500 Alexandrine lines, *Les Tragiques* is a work of monumental strife, ending only with the author's proclaimed advent, at the end of time, of the Protestant church. A reading in its sequence has effects comparable to the experience of a passion play. Composed of stations set in an order, at once rigorous and helter-skelter, the epic moves from the misery of the wars to a vicious satire of the Valois court, then to a chronicle of the major massacres and battles, and finally to the "accomplishment of divine will at first in, then outside of history."[14] Like the six other books, *Les Fers,* the fifth, is composed of "tableaux" whose earliest flashes are taken from the Conjuration of Amboise (March 16, 1560), an event precipitating the wars that officially begin in 1562, before praise is made of God, the omnipotent witness-cartographer who has been obliged to let the murderous conflict take place in view of a greater plan in which catastrophe gives way to apocalypse and final judgment. Implicit appeal is made to a cartographic model: the tableaux are "image maps" that configure a mobile composition that might be called a horrific "Tour de France."

Three points of reference are key: first, at the outset of the book of 1,565 Alexandrines in rhymes paired in alternating units of masculine and feminine endings, reference is made to the royal court's Tour of the Kingdom in 1564–1566, led by regent Catherine de Medici and her son, Charles IX, conceived and executed to shore up national unity. Departing from Paris and going east and south, then west, and finally

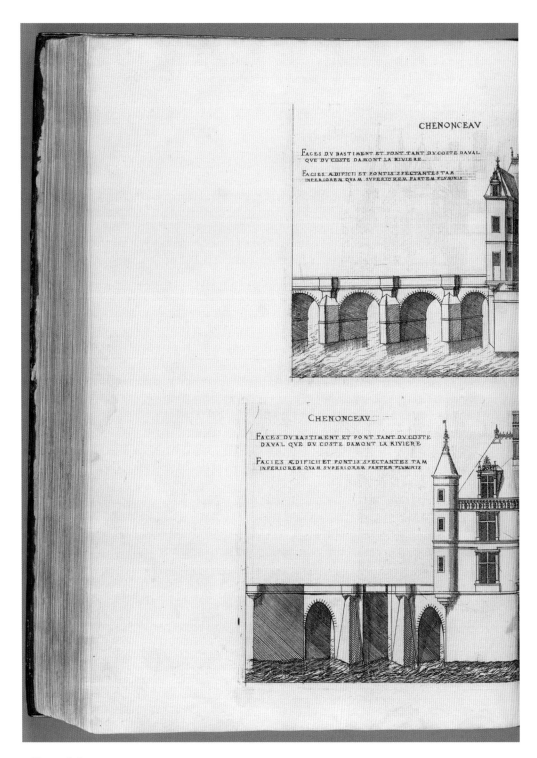

Figure 9.4

Two views of Chenonceau in Jacques Androuet du Cerceau, *Les Plus excellents bastiments de la France*. Houghton Library, Harvard University.

Figure 9.4 (continued)

north again, the Valois court charted an itinerary through the nation's geographic divisions as defined by its major river valleys. A second point of departure is the Louvre and the adjacent Palace of the Tuileries (lines 192–212). The château is a nightmarish place bordering the Seine that, like the repressed, returns for reason that the church of Saint-Germain-l'Auxerrois, close to the Louvre, in the night of August 19, 1572, first tolled the bell announcing Saint Bartholomew's Massacre. Third, correlatively, the rivers of France become the bleeding veins of the national body. In its mosaic composition in which the "tableaux" make appeal to *enargeia,* visual evidence of history, an informed sense of hydrography saturates verse recalling, on the one hand, allegorical maps (such as Jean-Baptiste Trento's *Mappe-monde nouvelle papistique* of 1565) and, on the other, to those of Protestant inflection that Gabriel Tavernier engraved for Maurice Bouguereau's *Théâtre françoys* (1594). Even if they do not stand "behind" the verse, these documents serve as guides for what D'Aubigné puts before his readers' eyes.

Les Fers begins when Catherine and Charles return to the throne after having shown to the nation at large their power of authority. The Devil and God intervene in heated debate over who will win the war being waged in the world below. God looks upon the Louvre, which at the moment, in *Les plus excellents bastiments,* Androuet du Cerceau remarked in the prefatory matter, was being readied for renovation prior to beginning his book with a scenography of the pavilion.[15] D'Aubigné returns to the site near the end of the poem, describing how, following the ringing of the tocsin,

> Dawn frightens the eye when the madman discovers
> The crows blackening the pavilion of the Louvre.[16]

Tableaux of this kind that comprise much of *Les Fers* follow the logic of hydrographic atlases, like Bouguereau's, that include maps of France stressing the wealth of its waterways, whether Guillaume Postel's map in the opening pages (figure 9.5) or, later, François de la Guillotière's great "Charte de la France" that was begun during the wars and finally published in 1616.

In the fifth book of *Les Tragiques,* however, rivers are reddened with the blood of decapitated martyrs.

> All that the Loire, Seine and Garonne nourish
> Was in turn depicted in each of the rivers' flow,[17]

he writes, adding that the Rhône (or, given its etymon *rhodanus,* the "Red" River) is not exempt from the spectacle that began at Amboise, where the Catholics captured the Huguenot conjurors, hanged them, and left them dangling from the balustrades of the courtyard, one of whom was put on display on the bridge over the River Loire before being chopped into pieces and deposited at the entry to the city. Agrippa's father, he

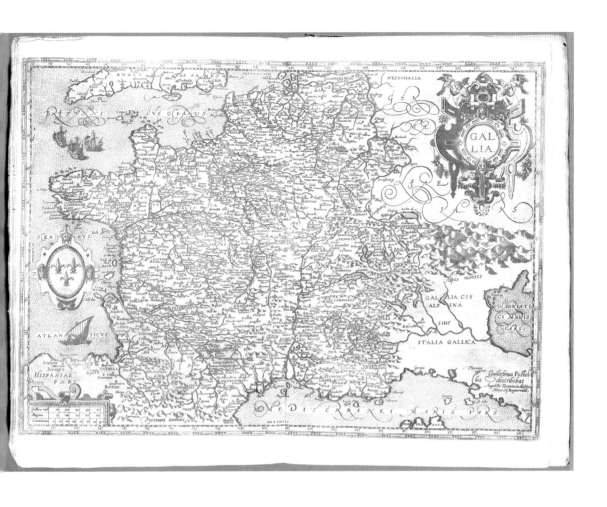

Figure 9.5

Guillaume Postel, map of France in Maurice Bouguereau, *Le Théâtre françoys* (1594). Houghton Library, Harvard University.

reported elsewhere, led him, then a child of eight in 1560, to behold the spectacle that immediately spurred him to take revenge. The poem soon charts a hydrography of terror so violent that the rivers of France become roiled with blood and choked with cadaverous flotsam:

> We see the Loire, never known so wild, wash
> The feet of a city that had just left
> Sixteen hundred victims stabbed, dozens hanged:
> The Orleans palace saw its rooms stuffed full,
> Where piles of bodies were an island, a road, a mountain,
> Causing the river to reverse its course.
> Up and down, and both the hands and the cities
> That had not been awash in the civil wars
> Now stirred and roiled the Loire with a new tint,
> Each one in its order becoming a tableau.[18]

He goes on to map the violence in accord with riverine cities whose enumerations are far less historical or chronologically sequential than synchronically geographic. Such is the descent along the Rhône from Lyon, where "tous les lions refuserent l'office" (line 1073) ("all the lions refused their duties") and southward: "Voilà Tournon, Viviers et Vienne et Valence/Avec terreur de Lyon l'insolence" (lines 1081–1082) ("There you see Tournon, Viviers, and Valence, with the insolence of Lyon's terror"), followed by the Seine, before a cavalcade of city names upsets the cartographic template, exactly when the poet implores the reader to see the words and the ear to take inspiration in hearing the breath that carries the crazed bard's description. Here and elsewhere d'Aubigné implicitly calls into question a tradition of verse that Ronsard and Du Bellay had drawn from classical sources.[19]

Yet without contemporary maps and their gazetteers the evocation could not be so complete, so compelling, or as if in its inspiration it were scanning a map, in a literal sense, so moving. Guillaume Postel's *Gallia* (figure 9.5), which held authority for over two or three decades, that Bouguereau sets in the first place among his three maps of France (by Petrus Plancius and Jean Jolivet) in his Protestant atlas designed for Henri IV, would be a correlative. Its exaggerated network of ever-ramifying rivers whose shorelines and borders are blackened with profusions of toponyms allows the eye to construct a montage of memory images—"tableaux"—built from events affiliated with places. The "spatial story" that d'Aubigné relates through appeal to the rivers of France turns a map of Postel's facture into a spatial arrangement of history in which events, their places, and different moments in the Wars of Religion are seen at once, on the same plane, and, if a figure describing the drip paintings of Jackson Pollack is fitting, *all over*.

The Dordogne and Garonne: From Montaigne to Sanson

Reaching back to the traumatic memories of what he had witnessed at Amboise where the Loire was clotted with the blood of the Protestants whom their Catholic enemies had butchered, d'Aubigné almost literally remaps the event in a hydrography analogous to the channels that, were it restored to health, would nourish the national body. Writing his *Essais* (1580–1595) at roughly the same moment, overwhelmed by the devastation the successive wars were wreaking on his countryside, in the midst of apparent confusion and trauma, Montaigne avows that in his memory he cannot make heads or tails of their succession. At the end of a tale he tells of his fall from a horse when riding outside of his château during the "third or second" of the conflicts (avowing, "I can't remember which"), he plots a virtual itinerary that leads him everywhere and nowhere. Montaigne returns to his château to find solace and pleasure in the introspective labor of self-study. Inking a long annotation in the margins of his personal copy of the 1588 edition, he adds, "It's a thorny enterprise, and more than it seems, to follow an allure so vagabond as that of our mind; to penetrate opaque depths of its inner folds; to choose and to arrest the air of so many of its slightest agitations."[20] Here and at other junctures the essay becomes a journey whose wavering itinerary he draws as he writes, the axis or vanishing point being the self, the *moi*, that continually relocates and revises itself in the movement of its diction. The analogy of the writing to the course of a river is noted in an annotation added to "De l'art de conferer" ("Of the Art of Conferral") where he asserts that the life he represents is so fluvial in nature that in the end, "in sum, we must live among the living, and without our worrying or without our alteration, let the river flow under the bridge."[21] Yet, otherwise, in "Des cannibales" ("Of Cannibals") (I, xxxi), following a brief description of his nearby Dordogne, a river vital for commerce between the inland regions as far east and north as Auvergne, Montaigne considers the disastrous aftermath of the Columbian encounters and colonization of the New World. Hardly overlooking the internal wars plaguing France, the character of the river allows him to call into question the worldviews of classical geographers and, in turn, to consider the distinction Ptolemaic geographers make between cosmography and topography. In the 1588 edition of the *Essais*, Montaigne inserts a description of the nearby waterway between a critique of Plato's myth of Atlantis and Aristotle's intuitions about the shape of the world:

> When I consider the impression that my Dordogne River has made in my time downstream on the right bank, and that in twenty years it has gained so much ground and pulled away the foundations of several buildings I surely see how this is an extraordinary agitation: for, were it always to go this way or thus too in the future the figure of the world would be turned topsy-turvy. But rivers undergo changes: sometimes they extend on one side, and then on another; at times they are contained. I'm not speaking of sudden floods whose causes we manage. In Médoc,

on the seaside, my brother, the Lord of Arsac, sees his land buried under the sands that sea vomits before it; the rooftops of a few buildings still appear; his estate and holdings have been turned into meager pasturage. The locals are saying that for some time the sea is pushing so strongly toward them that they've lost four leagues of land. These sands are its forerunners: and we see great dunes of moving sand that are marching a half-league ahead and winning over the land.[22]

The description of the Dordogne's ever-changing character would appear to mirror that of the self, the very matter of the essays. Were the floods and inundations removed from the context of local experience, the river would be a sign of a world thrown topsy-turvy, of a hydrography of war in which the rising sea vomits heaps of moving sand that conquers the shores and flatlands above the juncture of the Dordogne and Garonne. The churning river implies the presence of cosmic and even millenarian events. Implied nonetheless is that the locals of the area are aware of a phenomenon that sometimes appears on contemporary maps. Without naming it as such, Montaigne evokes the phenomenon of the "Mascaret," a tidal flow that periodically floods the area around Libourne and upstream. Spinning off *mascarer*, what in 1611 he registers as "to blot, soyle, blurre, sullie, disfigure," Randle Cotgrave notes, "Mascaret d'eaux: A huge, and sudden ravage, or inundation of waters," then citing, in French: "Mascaret is the name for a great mountain of water that surges up in the River Dordogne, toward the plains of Libourne; in the summer, in the more peaceful times, all of a sudden it rises and makes a path as long as the water, and sometimes less so, capsizing boats."[23] Known to sailors and managers of river traffic, a sign of the extent of the Mascaret appears on the shaded areas of the Gironde, Garonne, and Dordogne (reaching as far as Libourne) on François de la Guillotière's depiction of the Bordelais (figure 9.6).[24]

The paradox in Montaigne's description of the flooding is that what is highly local, belonging to the topography of his homeland, is treated cosmographically and, further, that only sentences later, rejecting the likes of the same scientists (implied is André Thevet, then the cosmographer to three kings of France), the essayist asserts, "we need topographers to provide us with detailed descriptions of the places where they have been."[25] The topographer counts among those who might have "a specific knowledge or experience of the nature of a river or of a fountain, moreover who knows only what everyone knows."[26] Local knowledge of the Dordogne prompts the essay—almost literally—to ramify and become a founding study in early modern ethnography. Anyone who looks at the maps of the moment quickly remarks that they too hover between analogy and the fruits of firsthand observation. Guillotière's rendering of the mouth of the Garonne implies that it speaks to the great fish in the Bay of Biscaye—perhaps a giant shad—that prepares to swim upstream. Although the map is not drawn in the fashion of an allegory, the fact that the mouth or *embouchure* is as it is suggests that it "speaks," thus that its two shores that give onto the bay would be akin to two lips and the Gironde itself not without comparison to a throat or an esophagus.

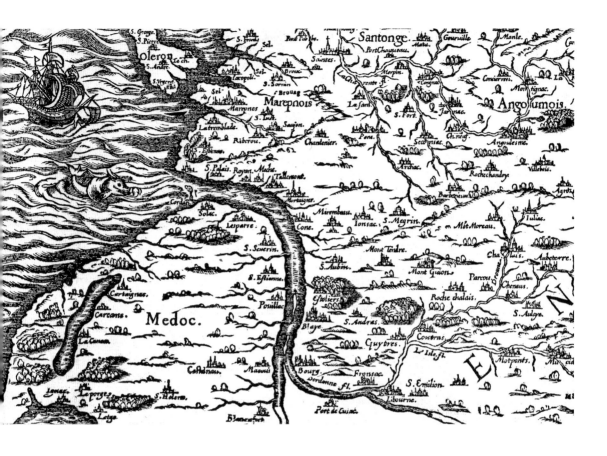

Figure 9.6

François de la Guillotière, "Charte de la France,"
detail of the Bordelais.

In a later and, except for its title, a putatively scientific map, Nicolas Sanson's *Carte des rivières de France curieusement recherchée* (1631 and 1641) (figure 9.7), the mouth of the Garonne is of the same contour. Meeting the Dordogne (where it loses its name) at the estuary of the Gironde, it flows northwest and then bends westward to empty into the sea, where it would seem to utter or ventriloquize its name (in lowercase and smaller type size) at its very mouth, before yelling loudly, to the left of the toponym, "Mer de Gascogne" (in uppercase) (figure 9.8).

Further inspection reveals that river names, the only toponyms on the map, set parallel to the lines tracing the course of the waterways, when seen from afar, seem to fashion a sense of relief and an orography. Printed as a complement to Louis Coulon's *Les Rivières de France*, the map becomes the setting for both literature and legend. A piece of literary cartography unto itself, Coulon's two-volume compendium can be countenanced as a fluvial encyclopedia by which the rivers become the organizing principle for names, history, and phenomena discussed and enumerated along the itineraries the waterways are plotting from their origin to their end. The result is a fluvial order whose twists and meanders might be qualified as Baroque. The chapter on the Dordogne begins with the name, where the rivulets *Dor* and *Donne* meet, then become the *Dordonne* that flows from the hills of Auvergne into the Périgord, along whose course Coulon appends tales of times past and present to descriptions of the place names that mark the itinerary. Sanson's map serves as a visual index for the ample prose that "speaks" as the reader's eye follows the course of the rivers.[27] Likewise and tellingly, in his treatment of the River Lignon in the Forez, the setting for Honoré d'Urfé's *L'Astrée*, the *roman fleuve* said to have captured the literary imagination for much of the seventeenth century, Coulon and his name become synonymous with the setting in which the distraught hero throws himself into the river. Attesting to the celebrity of the fiction, he writes:

> Under the Marquis d'Urfé's plume the *Lignon* flows more softly than at the feet of the *Noire, Estable* and *Saint Didier*, and is believed to be more glorious for having been chosen as the confidant of the loves of Astraea and Celadon than for watering the delightful gardens of the *Bastie* where, joining the Loire, it crowns it with flowers [*fleurs*] in view of the *Donzi*, of the *Neironde*, three other little rivers that borrow their names from the places through which they flow, next to Feurs, and empty into the Loire. The *Lignon* draws its source from the spring on Mount Loule where there are three robust fountains, whence it descends to Sauvin, Saint Georges, Couvant, and to the Crevé Bridge, where it greets the River S. Turn and acquires the name Lignon, and once past, with some impetuosity, above Bovin at Boteresse, at Bonlieu, Monverdun la Bastie, and after the flatlands of the country, it loses its name near Feurs, when it enters the Loire. Its length is of seven leagues, and it bears boats near

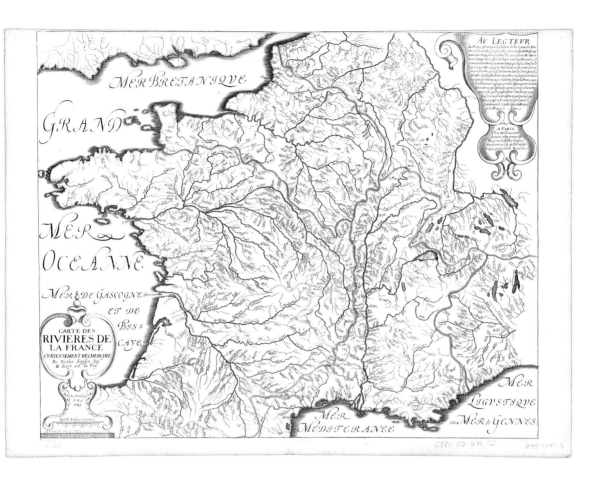

Figure 9.7

Nicolas Sanson, "Carte des rivières de France curieusement recherchée" (1641). Houghton Library, Harvard University.

Figure 9.8

Nicolas Sanson, "Cartes des rivières de France," detail. Houghton Library, Harvard University.

the Loire with a quantity of fish: trout, salmon, catfish, lampreys and others taken in its clear and lively waters celebrated by one of the most illustrious authors of our time.[28]

The relation between Coulon's text and Sanson's map indicates that a force of imagination motivates factual depiction of the rivers of France. It is especially felt on the map where the cartouches in the lower-left and upper-right corners fill out, respectively, the area of the Bay of Biscaye and Germany to the east of the Rhine. At the top, the cartouche that explains the content to the reader features strapwork designed to follow the contour of the river itself, bending and turning as if to mediate the difference between right angle of the corner of the frame and the course of the waterway. The cartouche at the bottom is set on reverse curves of a base whose outline melds the border of the nation with the decorative matter. Seen from afar, viewed in concert with the English Channel, the Mediterranean Sea, and the "Mer Oceanne," the cartouches offer a full, integral, and isolated image of the nation. These "curiously researched" rivers bear witness to its fluvial and commercial wealth.

Conclusion

It is worthwhile to consider Sanson's map and its correlative text in conjunction with the great "Carte de Tendre," an allegorical hydrography that graces *Clélie*, Madeleine de Scudéry's novel of eleven volumes (1654–1660), a beacon for the culture of the *salon* in the middle years of the seventeenth century (figure 9.9).

Showing how, within the codes of civility and *bienséance*, its user can overcome obstacles en route to obtaining new and enduring friendship, the map would seem to be the inverse of topographies that under Henri IV, Louis XII, and soon after, royal engineers (*ingénieurs du roi*) had designed to secure the borders of the French nation. Instead of indicating how best to wage war, in correlation with the author's suave and sinuous prose, the map would be a place where the user would find a new and sensuous geography of feminine virtue.[29] And, as many viewing readers have argued, the map could be of such strong *feminine* valence that the waterways imply the presence of a uterus and, stressed by the hatchings on either side that appear to embed the river into the flesh of the land, a fallopian tube. If so, the map attests to a force of generation that complements the watery wealth of Sanson's map of 1641. At the same time, the sea to the left is reminiscent of Sanson's Bay of Biscayne, while the peninsula above resembles the Armorican peninsula. The "Mer dangereuse" can be likened to the English Channel, and the stagnant "Lac d'Indifférence" to Lake Geneva. The town of Nouvelle Amitié seems close to Marseille. The outcropping on the lower right from which four spectators look onto the map recalls the upper edge of the

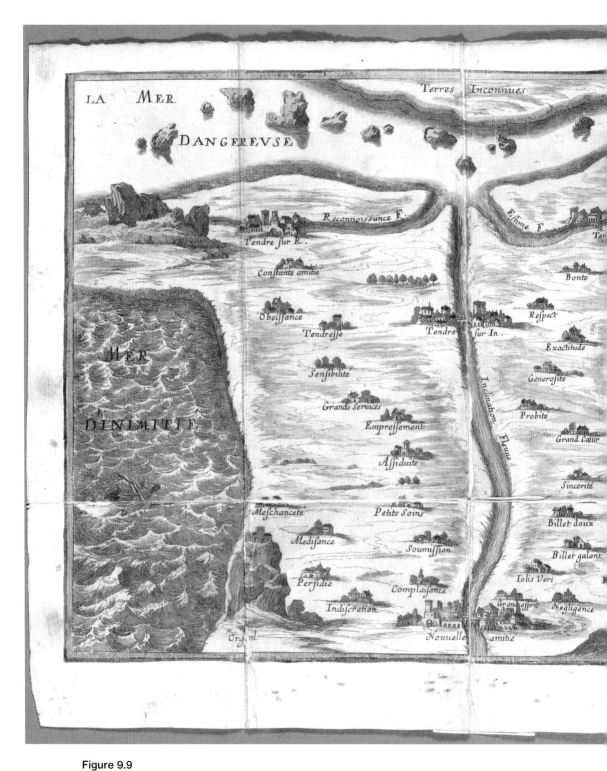

Figure 9.9

"La Carte de Tendre," in Madeleine de Scudéry, *Clélie* (1660). Houghton Library, Harvard University.

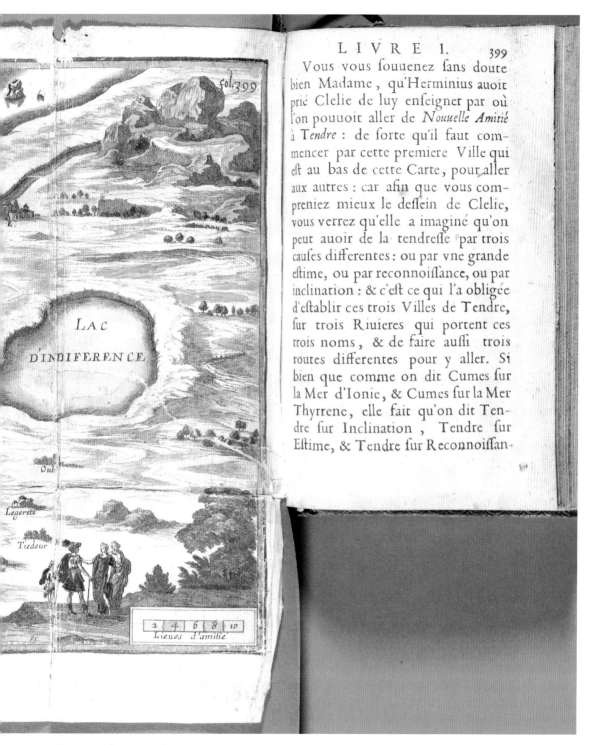

Figure 9.9 (continued)

LIVRE I.

Vous vous souuenez sans doute bien Madame, qu'Herminius auoit prié Clelie de luy enseigner par où l'on pouuoit aller de *Nouuelle Amitié* à *Tendre* : de sorte qu'il faut commencer par cette premiere Ville qui est au bas de cette Carte, pour aller aux autres : car afin que vous compreniez mieux le dessein de Clelie, vous verrez qu'elle a imaginé qu'on peut auoir de la tendresse par trois causes differentes : ou par vne grande estime, ou par reconnoissance, ou par inclination : & c'est ce qui l'a obligée d'establir ces trois Villes de Tendre, sur trois Riuieres qui portent ces trois noms, & de faire aussi trois routes differentes pour y aller. Si bien que comme on dit Cumes sur la Mer d'Ionie, & Cumes sur la Mer Thyrrene, elle fait qu'on dit Tendre sur Inclination, Tendre sur Estime, & Tendre sur Reconnoissan-

Mediterranean Sea. Sanson's "curiously researched" fluvial map, "La Carte de Tendre" possibly inspired a map of a feminine France.

An allegory, to be sure, its fluvial virtue is drawn from earlier maps of uncommon literary latency. The projection carries memories of a century of hydrographic writing, a mix of map and literature that tends to be associated with things Baroque: fluvial, serpentine, bending, ever-transforming and transformative, often perverse, yet of scientific mettle. A latent development can be imagined in considering, first, what cartographer Oronce Fine does in depiction and description of the rivers, and then in Ronsard's mannered treatment of fluvial desire in verse whose settings can be contrasted to the relation of map and text in the *Amadis de Gaule*, and further, in Androuet du Cerceau's visual depictions that, although neither immediately of cartographic or literary facture, become the setting on which, later, in his *Tragiques* Agrippa d'Aubigné draws his tableaux of hydrographic horror. We see that in the *Essais* Montaigne mixes observation, myth, and history when he appeals to his nearby Dordogne to consider change and upheaval following the Columbian discoveries, and that finally, in one of the great hydrographic maps of France, royal geographer Nicolas Sanson is firmly tied to a literary tradition of fluvial description, possibly a description that inspires the shape and form of "La Carte de Tendre," the most famous of all allegorical maps of the Baroque era, a moment when, rife with rivers, literature and cartography draw force and inspiration from each other.

Notes

1. Robert Karrow notes that, by virtue of print culture, well over a million maps circulated in Europe in 1600 ("Centers of Map Publishing in Europe, 1472–1600," *The History of Cartography*, Vol. 3: *The European Renaissance*, ed. David Woodward (Chicago: University of Chicago Press, 2007)).

2. Works concerning Baroque aesthetics informing this chapter include Heinrich Wölfflin, *Renaissance und Barock: Eine Untersuchung über Wesen und Entstehung des Barockstils in Italien* (Munich: Bruckmann, 1908); Jean Rousset, *La Littérature de l'âge baroque en France: Circé et le paon* (Paris: José Corti, 1953) and his *Anthologie de la poésie baroque en France,* 2 vols. (Paris: Armand Colin, 1961); Arnold Hauser, *Mannerism: The Crisis of the Renaissance and the Origin of Modern Art* (Cambridge: Belknap Press of Harvard University Press, [1965] 1986); Marcel Raymond, *Baroque et Renaissance poétique: Préalable à l'examen du baroque littéraire français* (Paris: José Corti, 1955); Gisèle Mathieu-Castellani, *Mythes de l'eros baroque* (Paris: PUF, 1981), and her anthology, *La Poésie amoureuse de l'âge baroque: Vingt poètes maniéristes et baroques* (Paris: Livre de Poche, 1990); Gilles Deleuze, *Le Pli: Leibniz et le baroque* (Paris: Éditions de Minuit, 1988); Helen Hills, ed., *Rethinking the Baroque* (Burlington, VT: Ashgate, 2011).

3. He writes that mathematics are studied for the love of themselves, and that "me voyant privé long temps de subside paternel, et assés mal traictez des biens de fortune, je me suis totalement soubmiz alestude mathematique, suyvant ma naturelle inclination. En faveur duquel, iay despendu ce peu de patrimoine que Dieu m'avoyt donné," as well as that after thirty years of

teaching, "Dont ie attends encore la recompense, laquelle ne puis esperer (après Dieu) que de vostre faveur, et liberale clemence" ("For a long time seeing myself deprived of paternal subsidy, and having been ill-treated in the blessings of fortune, I have dedicated myself entirely to the study of mathematics, following my natural inclination and for which I have spent the remaining patrimony that God had bequeathed to me," and that, after thirty years of teaching, "I still await remuneration, for which, after God, I can only hope through your favor and generosity") [f. 3r°]. In the printed edition (see note 5 below), in a poem addressed to the Duchess of Valentinois, he begs her to remind the king that he has not been paid for his labors: "Tu luy pourras faire la remonstrance / Comment je n'ay de travailler cessé / Depuis trent'ans en mon art & science, / Dont nay esté encore recompense" ("You may wish to remind her that I have given myself endlessly to my art and science for which reward is still yet to come") (Oronce Fine, *L'esphere du monde: Proprement dicte Cosmographie* (manuscript with dedication to Henri II of France; Paris, 1549) (Houghton Library, Harvard University, MS. Typ 57, f. 3r°)).

4. In the printed edition (see note 5), "Je vous ay redigé, & mis en Françoys, une des plus belles & delectables parties qui soit entre lesdittes mathematiques. C'est à' sçavoir, la description universelle de tout le monde, avec les choses plus notables qui proviennent ça bas, à cause du premier & regulier movement de tout le ciel, que lon appelle cosmographie, & les principes & rudiments de la geographie & hydrographie concernant le fait de la marine" (f. aa iii v°).

5. "Et convient noter, que lon nomme lesdits climats, du nom des plus notables villes, ou fleuves, ou montaignes, qui sont en iceux: ainsi comme il advient à propos, *au plaisir & fantasie d'un chacun geographe*" (Oronce Fine, *L'esphere du monde, proprement ditte, cosmographie, composee nouvellement en françois & divisée en cinq livres: comprenans la premiere partie de l'astronomie, & les principes universels de la geographie & hydrographie: avec une épistre, touchant la dignité, perfection & utilité des sciences mathématiques* (Paris: Michel de Vascosan, 1551), f. 47v°, emphasis added).

6. Such would be the autobiographical "secret" of the map, its point of origin and growth, indeed the cartographer's signature. I develop the point in *The Self-Made Map: Cartographic Writing in Early Modern France* (Minneapolis: University of Minnesota Press, [1996] 2011), 131.

7. Pourquoy lasse du bal entre ces fleurs couchée
   N'ay-je sur ton giron ou la teste panchée
   Ou mes yeux sur les tiens, ou ma bouche dessus
   Tes deux tetins de neige et d'yvoire conceus?
   Te semblay-je trop vieil? encor la barbe tendre
   Ne fait que commencer sur ma joue à s'estendre,
   Et ta bouche qui passe en beauté le coural,
   S'elle veut me baiser, ne se fera point mal.

  (In *Les Œuvres de Pierre de Ronsard* (Paris: Chez Nicolas Buon, 1623), 161, col. 1, http://www.bvh.univ-tours.fr. See also Pierre de Ronsard, *Œuvres complètes,* vol. 1, ed. Jean Céard, Daniel Ménager, and Michel Simonin (Paris: Éditions Gallimard/Pléiade, 1994), 206–207 (lines 117–124).)

8. On dit au temps passé que quelques uns changerent
   En riviere leur forme, & eux-mesmes nagerent
   Au flot qui de leur sang goutte à goutte sailloit,
   Quand leur corps transformé en eau se distilloit.

> Que ne puis-je muer ma resemblance humaine,
> En la forme de l'eau qui cette barque emmeine?
> J'irois en murmurant soubs le fond du vaisseau,
> J'irois tout alentour, & mon amoureuse eau
> Baiseroit or' sa main, ore sa bouche franche,
> La suyvant jusqu'au port de la Chappelle blanche.
> Puis laissant mon canal pour iouyr de mon vueil,
> Par le trac de ses pas j'irois jusqu'à Bourgueil,
> Et là dessous un Pin, couché sur la verdure,
> Je voudrois revestir ma premiere figure.
>
> (Ronsard, *Les Œuvres*, 162–163, cols 2, 1 (Gallimard/Pléiade, vol. 1, 209, lines 217–230))

9. Ronsard, *Les Œuvres*, 165, cols. 2, 1.

10. Nicolas d'Herberay des Essarts, trans., *Le Quatriesme livre d'Amadis de Gaule, auquel on peult voir quelle issue est la guerre entreprises par le roy Lisvart contre Amadis* (Paris: Jean Longis, 1550), f. 3r° (Houghton Library, Harvard University, Typ 515.46.138F).

11. "J'ay pensé ne pouvoir mieulx à propos mettre en lumiere ce premier Livre des Bastimens exquis de ce Royaume: esperans que nos pauvres François (és yeux & entendemens desquels ne se presente maintenant autre chose que desolations, ruines & saccagements, que nous ont apporté les guerres passees) prendront, peult estre, en respirant, quelque plaisir et contentement, à contempler icy une partie des plus beaux & excellens edifices, dont la France est encore pour le jourd'huy enrichie." *Le Premier volume des plus excellents Bastiments de France. Auquel sont designez les plans de quinze bastiments, & de leur contenu: ensemble les elevations & singularitez d'un chascun*. Par Jacques Androuet du Cerceau, architecte. A Paris: Pour ledit Iacques Androuet du Cerceau. 1576, f. A. ii.r°. See also Guillaume Fondenell, "Catherine de Médicis et les jardins," in *Jardins de châteaux à la Renaissance*, ed. Élisabeth Latrémolière and Pierre Gilles Giraut (Paris: Editions Gourcuff Gradenigo, 2014), 117–124.

12. Phillip Usher outlines the history of the composition and piecemeal presentation of the poem in the introduction to his annotated translation, *The Franciad* (New York: AMS Press, [1572] 2010), xx–xxi.

13. Preface (dated 2006) to the entry for Androuet du Cerceau's *Les Plus excellents bastiments de la France*, in "De architectura," http://architectura.cesr.univ-tours.fr. I have consulted the edition in two volumes, dated 1576 and 1579 respectively, at the Dumbarton Oaks Research Collection (DO Rare RBR 0–2–5 AND).

14. Frank Lestringant, preface to his edition of *Les Tragiques* (Paris: Éditions Gallimard/Poésie, [1995] 2003), 39.

15. Androuet du Cerceau obliquely indicates the terrible condition of things in view of the death of the king and the incompleteness of a great edifice elegantly set by the River Seine. Under the orders of Henry II, the building was to be redesigned but in the 1570s was far from finished, leaving the artist only to imagine what remains for him to draw: "Duquel toutefois je n'ay fait aucun plan icy, pour l'esperance que j'ay, qu'avec le temps l'œuvre nouveau se parachevera, me contentant d'avoir representé celuy des susdits premier & second estages neufs" ("For which I have not here included a view in the hope that the new edifice will be completed, having only represented that of the aforesaid new first and second floors") (Androuet, *Les plus excellens bastiments*, f. A.iii. r°).

16. Le jour effraie l'œil quand l'insensé découvre
    Les corbeaux noircissans le pavillon du Louvre.

   (D'Aubigné, *Les Tragiques*, [1995] 2003, 257, lines 1017–1018. In his critical edition of *Les Tragiques*, Jean-Raymond Fanlo notes how "the crow perched on a house was considered to be a presage of death to befall the inhabitant" (Paris: Honoré Champion, 1995), vol. 1, 518n.)

17. Tout ce que Loire, Seine et que Garonne abreuve
    Estoit par rang despeint comme va chaque fleuve. (lines 665–666)

18. On void Loire, inconnu tant farouche, laver
    Les pieds d'une cité qui venoit d'achever
    Seize cent poignardés, attachez à douzaines:
    Le palais d'Orleans en vid les salles pleines,
    Dont l'amas fit une isle, une chaussee, un mont,
    Lequel fit refouller le fleuve contremont,
    Et dessus et dessoubs; et les mains, et les villes,
    Qui n'avoyent pas trempé dans les guerres civiles,
    Troublent à cette fois Loire d'un teint nouveau,
    Chacun ayant gaigné dans ce rang un tableau. (lines 1063–1072)

19. With exceptions: Kathleen Long studies metaphors from Ovid's *Metamorphoses* in the descriptions of the rivers in *Les Fers*, further confirming the persistence of his relation with Ronsard, noted above, who became water in "Le Voyage de Tours." See "Les Rivières, sites de massacre et mémoire dans les *Tragiques*," in *Illustrations inconscientes: Écritures de la Renaissance*, ed. Bernd Renner and Phillip Usher, 439–454 (Paris: Éditions Classiques Garnier, 2014), 444–445.

20. "C'est une espineuse entreprinse, et plus qu'il ne semble, de suyvre une alleure si vagabonde que celle de nostre esprit; de penetrer les profondeurs opaques de ses replis internes; de choisir et arrester tant tant de menus airs de ses agitations" (Michel de Montaigne, *Les Essais*, ed. Pierre Villey, updated by Verdun-L. Saulnier (Paris: PUF/Quadrige, 1988), 378); the same edition is reproduced in the Montaigne Project, https://www.lib.uchicago.edu/efts/ARTFL/projects/montaigne.

21. "Somme, il faut vivre entre les vivants, et laisser courre la riviere sous le pont sans nostre soing, ou, à tout le moins, sans nostre alteration" (Montaigne, *Les Essais*, 929).

22. "Quand je considere l'impression que ma riviere de Dordoigne faict de mon temps vers la rive droicte de sa descente, et qu'en vingt ans elle a tant gaigné, et desrobé le fondement à plusieurs bastimens, je vois bien que c'est une agitation extraordinaire: car, si elle fut tousjours allée ce train, ou deut aller à l'advenir, la figure du monde seroit renversée. Mais il leur prend des changements: tantost elles s'espendent d'un costé, tantost d'un autre; tantost elles se contiennent. Je ne parle pas des soudaines inondations de quoy nous manions les causes. En Medoc, le long de la mer, mon frere, Sieur d'Arsac, voit une siene terre ensevelie soubs les sables que la mer vomit devant elle; le feste d'aucuns bastimens paroist encore; ses rentes et domaines se sont eschangez en pasquages bien maigres. Les habitans disent que, depuis quelque temps, la mer se pousse si fort vers eux qu'ils ont perdu quatre lieuës de terre. Ces sables sont ses fourriers: et voyons des grandes montjoies d'arène mouvante qui marchent d'une demi lieue devant elle, et gaignent païs" (Montaigne, *Les Essais*, 204).

23. "L'on appelle mascaret une grande montaigne d'eau qui se fait en la Riviere de Dordonne, vers les contrées de Libourne; au temps d'esté, es saisons les plus paisibles, & tout en un moment elle se forme, & fait une course quelquesfois bien longue de l'eau, & quelques fois plus courte, renversant les bateaux" (Randle Cotgrave, *A Dictionarie of the French and English Tongues* (London: Adam Inslip, 1611)).

24. The Mascaret figures prominently, in Louis Coulon, *Les Rivieres de France, ou description geographique & historique du cours & débordement des Fleuves, Rivieres, Fontaines, Lacs & Estangs qui arrousent les Provinces du Royaume de France. Avec un Denombrement des Villes, Ponts, Passages, Batailles qui ont esté données sur leurs rivages, & autres curiositez remarquables dans chaque Province* (Paris: François Clousier, 1644), vol. 2, 529. Père François de Dainville notes that Coulon's text "constitues a veritable commentary" to Sanson's map of 1641, in *Le Langage des geographes* (Paris: Picard, [1964] 2002), 133.

25. "Il nous faudroit des topographes qui nous fissent narration particuliere des endroits où ils ont esté" (Montaigne, *Les Essais*, 205).

26. "Quelque particuliere science ou experience de la nature d'une riviere ou d'une fontaine, qui ne sçait au reste que ce que chacun sçait" (ibid., 205).

27. In "At the Mouth of the Gironde: Nicolas Sanson after Montaigne," *English Language Notes* 52, no. 1 (Summer 2014): 31–43, I have sought to develop the point in greater detail, where emphasis is placed on the force of analogy in protoscientific cartography and literature.

28. Coulon, *Rivières*, vol. 2, 252 (my translation). In the pages on the Dordogne, Coulon writes of the dangers facing pilots who confront the force of the Mascaret.

29. In *Mapping Discord: Allegorical Cartography in Early Modern French Writing* (Newark: University of Delaware Press, 2004), Jeffrey Peters shows how, in the construction of its allegory of friendship, the map "borrows all of its basic visual features from a language of contemporaneous cartographic practices" (p. 84).

Bibliography

Androuet du Cerceau, Jacques. *Le Premier volume des plus excellents Bastiments de France. Auquel sont designez les plans de quinze bastiments, & de leur contenu: Ensemble les elevations & singularitez d'un chascun.* Par Jacques Androuet du Cerceau, architecte. A Paris: Pour ledit Iacques Androuet du Cerceau. 1576–1579. Dumbarton Oaks Research Collection DO Rare RBR 0–2–5 AND.

d'Aubigné, Agrippa. *Les Tragiques*. Ed. Frank Lestringant. Paris: Éditions Gallimard/Poésie, [1995] 2003.

d'Aubigné, Agrippa. *Les Tragiques*. Ed. Jan-Raymond Fanlo. Paris: Honoré Champion, 1995.

Conley, Tom. "At the Mouth of the Gironde: Nicolas Sanson after Montaigne." *English Language Notes* 52, no. 1 (Summer 2014): 31–43.

Conley, Tom. *The Self-Made Map: Cartographic Writing in Early Modern France*. Minneapolis: University of Minnesota Press, [1996] 2011.

Cotgrave, Randle. *A Dictionarie of the French and English Tongues*. London: Adam Inslip, 1611.

Coulon, Louis. *Les Rivieres de France, ou description geographique & historique du cours & débordement des Fleuves, Rivieres, Fontaines, Lacs & Estangs qui arrousent les Provinces du Royaume de France. Avec un Denombrement des Villes, Ponts, Passages, Batailles qui ont esté données sur leurs rivages, & autres curiositez remarquables dans chaque Province.* Vol. 2. Paris: François Clousier, 1644.

Dainville, Père François de. *Le Langage des geographes.* Paris: Picard, [1964] 2002.

Deleuze, Gilles. *Le Pli: Leibniz et le baroque.* Paris: Éditions de Minuit, 1988.

Fine, Oronce. *L'esphere du monde: Proprement dicte Cosmographie.* Manuscript with dedication to Henri II of France; Paris, 1549. Houghton Library, Harvard University, MS. Typ 57.

Fine, Oronce. *Le sphere du monde, proprement ditte, Cosmographie: Composée nouvellement en françois & divisée en cinq livres: comprenans la premiere partie de l'astronomie, & les principes universels de la geographie & hydrographie: avec une épistre, touchant la dignité, perfection & utilité des sciences mathématiques.* Paris: Michel de Vascosan, 1551. Houghton Library, Harvard University, *FC5.F494.Eh551s.

Fondenell, Guillaume. "Catherine de Médicis et les jardins." In *Jardins de châteaux à la Renaissance*, ed. Élisabeth Latrémolière with the collaboration of Pierre Gilles Giraut, 117–124. Paris: Editions Gourcuff Gradenigo, 2014.

Hauser, Arnold. *Mannerism: The Crisis of the Renaissance and the Origin of Modern Art.* Cambridge, MA: Belknap Press of Harvard University Press, [1965] 1986.

Herberay, Nicolas de Essarts, trans. *Le Quatriesme livre d'Amadis de Gaule, auquel on peult voir quelle issue est la guerre entreprise par le roy Lisvart contre Amadis.* Paris: Jean Longis, 1550. Houghton Library, Harvard University, Typ 515.46.138F.

Hills, Helen, ed. *Rethinking the Baroque.* Burlington, VT: Ashgate, 2011.

Karrow, Robert. "Centers of Map Publishing in Europe, 1472–1600." In *The History of Cartography*, Vol. 3: *The European Renaissance*, ed. David Woodward, 611–621. Chicago: University of Chicago Press, 2007.

Long, Kathleen. "Les Rivières, sites de massacre et mémoire dans les *Tragiques*." In *Illustrations inconscientes: Écritures de la Renaissance*, ed. Bernd Renner and Phillip Usher, 439–454. Paris: Éditions Classiques Garnier, 2014.

Mathieu-Castellani, Gisèle. *La Poésie amoureuse de l'âge baroque: Vingt poètes maniéristes et baroques.* Paris: Livre de Poche, 1990.

Mathieu-Castellani, Gisèle. *Mythes de l'eros baroque.* Paris: PUF, 1981.

Montaigne, Michel de. *Les Essais.* Ed. Pierre Villey, updated by Verdun-L. Saulnier. Paris: PUF/Quadrige, 1988.

Peters, Jeffrey. *Mapping Discord: Allegorical Cartography in Early Modern French Writing.* Newark: University of Delaware Press, 2004.

Raymond, Marcel. *Baroque et Renaissance poétique: Préalable à l'examen du baroque littéraire français.* Paris: José Corti, 1955.

Ronsard, Pierre de. *The Franciad.* Ed. and trans. Phillip John Usher. New York: AMS Press, 2010.

Ronsard, Pierre de. *Les Œuvres de Pierre de Ronsard.* Paris: Chez Nicolas Buon, 1623.

Ronsard de, Pierre. *Œuvres complètes*. Vol. 1. Ed. Jean Céard, Daniel Ménager, and Michel Simonin. Paris: Éditions Gallimard/Pléiade, 1994.

Rousset, Jean. *Anthologie de la poésie baroque en France*. 2 vols. Paris: Armand Colin, 1961.

Rousset, Jean. *La Littérature de l'âge baroque en France: Circé et le paon*. Paris: José Corti, 1953.

Wölfflin, Heinrich. *Renaissance und Barock: Eine Untersuchung über Wesen und Entstehung des Barockstils in Italien*. Munich: Bruckmann, 1908.

10
Goethe and the Cartographic Representation of Nature around 1800

John K. Noyes

Only One Nature Everywhere

Elective Affinities (1809), the novel Goethe once referred to as his best,[1] tells the story of a marriage falling apart. Edward and Charlotte, happily married, are forced to rethink the foundation of their relationship with the arrival of Charlotte's niece Ottilie and Edward's old friend, the Captain. As Edward and Charlotte witness how the stirrings of desire threaten their bond, Goethe poses the question whether human nature can be understood as governed by intrinsic rules that determine relationships—a test of Leibniz's ideas on self-development and predetermined order. Is the bond of marriage like the bonds of chemical compounds, driven by their own predetermined relationships and thus in a way ruled by fate (or at least by mathematical principles), or are the forces of reason and social convention stronger than nature? In posing this question, Goethe will complicate the problem in two ways: first, by his underlying conviction (voiced in his own announcement of the novel's publication) that "there is indeed only *one* Nature everywhere, and that the traces of obscure, passionate necessity run irrepressibly through the bright freedom of reason";[2] and second, by showing how the process of obscure passion bleeding into bright reason is carried through into the realm of representation. It is here that the topic of cartography is raised.

In book one, the Captain is occupied, among other things, with drafting a "topographic map on which the estate and its surroundings had been drawn in pen and wash—with graphic accuracy in a relatively large scale, its precision thoroughly checked by the Captain's trigonometric measurements."[3] The fate of this map is marked by a decisive moment. One evening, after Edward, Charlotte, Ottilie, and the Captain have been out walking, they examine the map together. They "traced the path they had taken on the map and noted ways for improving it at certain spots. They talked over all their earlier suggestions and compared them with their newest plans, agreeing once again on the location of the new building opposite the manor and thus completing the circular arrangement of paths."[4] As they examine the map and consider a location for the new building, Ottilie remains silent, but when Edward asks her opinion, she suggests an alternative location. Edward, driven by his enthusiasm over Ottilie's inspiration, defaces the map with a broad coarse pencil mark. "'She's right!' exclaimed Edward. "Why didn't it occur to us? That's what you mean, isn't it, Ottilie?" He took a pencil and drew in rough, bold strokes a long rectangle on the rise. The Captain was taken aback; he didn't like to have his careful, cleanly drawn map disfigured in this way."[5] What exactly has happened?

The mapping of Edward's estate and the plan for the new building continue the general project, explained at the beginning of the novel, of domesticating nature and rendering it productive. As the novel opens, we meet Edward "spending the finest hours of an April afternoon grafting freshly cut shoots onto young rootstocks."[6] Charlotte is building a hut with a view over the valley. In this way, nature becomes a prospect for pleasure while at the same time providing the financial means for sustaining a life of leisure. This project involves removing all possible dangers from the landscape. Contemplating possible risks, "they provided themselves with all sorts of necessary life-saving equipment, the more so, in that with so many lakes, rivers, and dams nearby accidents of this kind occasionally did happen."[7] In the course of the novel, this other nature, untamed by human labor, will claim the life of Edward and Charlotte's child.

The Captain's map is part of the project of holding this other nature at bay. In the moment of Edward's impulsive act, the map appears as an aesthetic object and an object of struggle. The science of cartography as practiced by the Captain has shown itself to be in league with an aesthetic practice that renders nature subordinate to the laws of mathematics. The mathematical taming of nature emerges at odds with the untamed nature it would seek to domesticate, embodied in Ottilie. And it is at odds with the workings of desire, embodied in Edward, and itself equated with untamed nature. Edward's impulsive act reveals how cartography straddles the boundary between the mathematical organization of nature and that other nature that is always escaping control. But something else is at stake too. Defaced by Edward, the Captain's map marks even more clearly the difference between the two modes of rationality that dominate the novel: the measuring of nature to give it value, and the distanced regard that converts value to pleasure; nature for business and nature for leisure; instrumental rationality and aesthetic rationality. The Captain intends his map as a machine for bonding instrumental reason with ownership, and thus with surplus value. But Edward's refusal to respect the map's careful drafting shows that, as an aesthetic object, it embodies the discrepancy between instrumental reason and aesthetic objects, between ownership and its representations. In the context of this novel, it also embodies the tensions between the objects of knowledge and their metaphorical formulations.

The defaced map in *Elective Affinities* makes visible a constellation of ideas important for understanding literature and cartography in the late eighteenth and early nineteenth centuries. This is to be understood not only in the simple sense that Goethe found it useful to place cartographic representation at the center of the decisive conflict that structures his book. It also needs to be explored in the deeper sense in which cartographic representation was highlighting some of the same representational problems that were experienced in literature, and that Goethe repeatedly thematized in his novels. This chapter sets out the terms of such an exploration. I will show that the map exists as an object with both mathematical and aesthetic dimensions, and that these work against one another. As such, the map combines what the characters in the novel seek to separate: the world of objective principles and the world of desire. This has far-reaching consequences for an aesthetic form that made such strong claims on objectivity and that seemed to cement its claims to territorial possession. Unraveling these claims, Goethe shows how objectivity in cartography is always threatened by the workings of desire.

Goethe's Maps: Instrumental and Aesthetic Objects

First, consider for a moment how important maps were for Goethe. They rarely find their way into his fictional writings, but when they do, they tend to be plunged into the center of the representational problems that drive his fictional narratives. In the *Conversations of German Refugees* (1796), a map shows Fritz that the fire lighting the night sky is his aunt's house burning, prompting him to an untimely meditation on inexplicable natural phenomena, such as the "sympathy among wood cut from the same tree."[8] In *Wilhelm Meister's Apprenticeship*, maps promise to find Wilhelm's mysterious savior, and they help Mignon orient her desire for her lost birthplace.

Goethe was a map collector. His collection at the *Nationalmuseum* in Weimar contains an impressive array of maps: various maps of the world showing the latest voyages of discovery and the newly charted lands, the Empire of Germany and the Kingdom of Prussia, detailed maps of other regions of Germany, a route map from Leipzig to Vienna via Prague, another one from Leipzig to St. Petersburg, maps of canals in Amsterdam and the great canal linking Lake Erie to the Hudson River, a political map of France and her colonies, a map of the wars of independence in America, and so on. He also had a sustained interest in geological maps, such as the one shown in figure 10.1, as evidenced in his notebooks: "As a result of [Carlos de] Gimbernat's investigations, the areas around Baden gave rise to increasing interest, and his geological map of the region ... more than met the requirements of my studies at the time" (*Tag- und Jahreshefte*, "Daily and Annual Notebooks," 1817).[9]

He augmented the geological map drawn in 1782 by the mining inspector Johann Friedrich Wilhelm Charpentier, the "Mineralogische Geographie der Chursächsischen Lande" (figure 10.2). Writing to Merck in November 1782, he expressed his desire to advance the mineralogical mapping of all Europe.[10]

In 1822 he wrote a very positive review of Christian Keferstein's 1821 geological map of Germany, published as part of Keferstein's project *Deutschland, geognostisch-geologisch dargestellt*.[11] In his *Tag- und Jahreshefte* of the same year Goethe notes that "the continuation of Keferstein's 'Geognostisches Deutschland' was likewise very

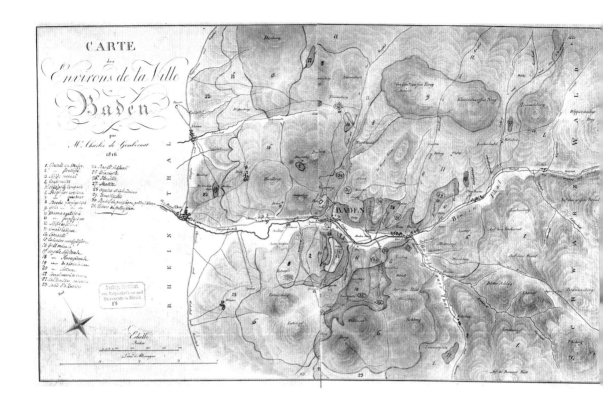

Figure 10.1

C de Gimbernat, *Mapa geològica de los alrededores de Baden* (1816)

Figure 10.2

Johann F. Charpentier, *Mineralogische Geographie der Chursächsischen Lande*, Detail (Leipzig: Crusius, 1778). Kartensammlung der Staatsbibliothek Berlin.

helpful and would have been more so with more careful coloration. In cases such as this, it will probably be necessary to repeat often that, where color is intended to differentiate, it should itself be differentiated."[12]

Goethe was also well acquainted with General Friedrich von Müffling, who had assisted in the trigonometric survey of Westfalia in 1803, resulting in Major General Carl von Le Coq's *Große Karte von Westphalen*, completed in 1813 (figure 10.3). In a letter to Duke Carl August of Weimar, Goethe refers to Müffling as "this hard-working young man, whose work is so exact and reliable."[13]

Friedrich Kittler claims that Müffling was the model for the Captain in Goethe's novel.[14] If this is so, we can see what kind of map Goethe might have had in mind when he let Edward step forward with his coarse pencil. It seems that maps in general have, in Goethe's mind, an obligation to present knowledge in an aesthetically pleasing manner. This includes aligning their representational strategies carefully with the knowledge they intend to convey—ensuring that the colors are strong enough to carry the distinctions the map intends, but that the marks are exact and reliable, like the disposition of the person who drafts them. But in *Elective Affinities*, the Captain's map is asked to do more. It reveals the separation of the instrumental and the aesthetic. In the lives of those who use this map, this separation is understood as the distinction between work, which the novel calls *Geschäft* (business), *Ernst* (seriousness), *Strenge* (discipline); and its opposite, leisure, or *Willkür*. *Willkür* is a key word in the novel, since it combines the will with the idea of choice. *Kür* is an older way of saying *Wahl*, which provides part of the title of Goethe's novel. Immediately on completion of the map, Edward suggests to his friend that they now "move on to our next task, the description of the property. For this, we must make a sufficient start on the work to be able to draft leases and suchlike later on. Only let us establish one principle: separate all business matters from life. Business demands earnest discipline [*Ernst und Strenge*], life demands spontaneity [*Willkür*]; business demands clear logic, but life often finds inconsistency necessary—indeed it can be desirable and entertaining. If you are sure in the one, then you can be all the freer in the other, instead of finding freedom undermining certainty and cancelling it out."[15] The map is built on one of the novel's central conflicts, that between self-curtailment in the name of social relations (*Ernst und Strenge*) and self-realization in the name of truth (*Willkür*, which here is shorthand for the Leibnizian idea of *entelechy*).

Charlotte is the custodian of social relations and earnest discipline. Edward is committed to exploring the limits of *Willkür*. As a result, the world becomes a testing ground for his personal desires, and their relationship becomes a testing ground for the ability of discipline and renunciation to contain the forces of nature. When Edward and the Captain are making plans for building the summerhouse, she reminds them of the expense. Edward admonishes her: "'You don't seem to have too much trust in us,' said Edward. 'Not when it comes to your personal desires [*in willkürlichen Dingen*],'

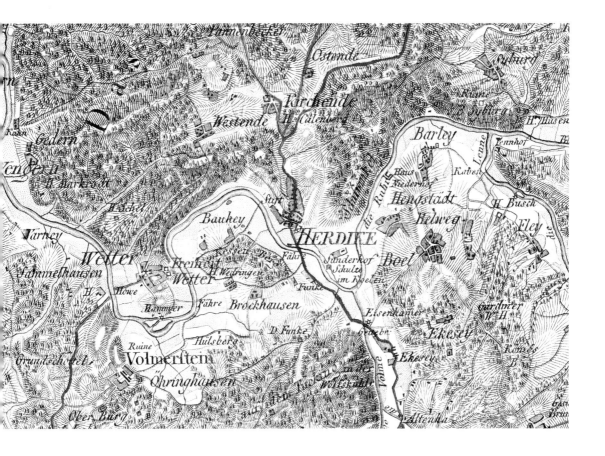

Figure 10.3

Carl von Le Coq, *Topographische Karte von Westphalen* (1805). Detail. Kartensammlung der Staatsbibliothek Berlin.

responded Charlotte. 'We women can control our personal desires [*Willkür*] better than you.'"[16] Seen through Charlotte's reasoning eyes, Edward's pursuit of self-fulfillment looks like an uncontrolled surrender to desire. *Willkür* becomes another word for caprice. In Goethe's novel, the distinction between desire and earnest discipline reveals itself as tenuous, and the world of representation is where it begins to collapse. The map partakes of this process. It marks the moment when Charlotte realizes she might lose the Captain and sees her inner feelings, where she realizes the personal cost of trying to separate *Ernst und Strenge* from *Willkür*. The count, visiting the estate, admires the Captain's cartographic skills and mentions the prospect of finding him employment elsewhere. "Charlotte felt inwardly torn. Taken aback by his suggestion, as also by her own reaction, she could not utter a word. Fortunately, the count went on discussing his plans for the Captain, and their advantages were only too apparent to Charlotte. At this point the Captain came back and unrolled his map for the count. With what new eyes she saw the friend she was about to lose!"[17] What Charlotte sees is that representation, no matter how carefully drawn and mathematically organized, is bound in a context of desire.

Early on, the narrator makes explicit this contextualization of representation. The Captain's topographic map is the initial expression of Edward's "long-cherished desire to become more familiar with his estate and make better use of it."[18] When Edward expresses this desire to his newly arrived friend, the Captain immediately understands what is to be done: "The first thing we should do ... is to survey the whole properly with a compass. This is an easy and enjoyable task, and even though it does not give perfectly accurate results, it is still useful and makes a satisfying start; in addition, it can be done without much assistance and you can be sure of completing it."[19] The Captain possesses the necessary instruments and know-how, and he instructs "a number of hunters and peasants" in his craft. He puts the measurements to paper, drawing, crosshatching, and tinting with watercolors. When Edward sees his estate take shape on paper, it is "like a new creation. He felt that he was seeing it now for the first time; and for the first time it really seemed to belong to him."[20]

Drafting Plans: Representation and Possession

In the history of cartography, this desire for possession cemented in the map's representational strategies has a strong political dimension. This has received a great deal of attention in the area of European expansionism. Here it has become commonplace to associate the map's representation of ownership with the suppression of the map's genesis and representational strategies. This interpretation reaches back at least to Marlow's desire in the *Heart of Darkness* to (in the words of Edward Said) "fill in the great blank spaces on the map."[21] Paul Carter in *The Road to Botany Bay* speaks of how the explorer's map redefines the significance of blankness: when Cook charted new land with blank areas representing those regions he had not yet surveyed, "the blank spaces of the map were active, locating future histories."[22] This recording of blank space on the Euclidean field of the map was, as he argues, conducive to the allotment of land in the settler colonization of Australia.[23]

Here, mathematics aligns cartography with an entire set of representational practices that convert experience into a discourse of truth about colonial ownership. Mary Louise Pratt argues that the explorer's discovery is only made real "after the traveler (or other survivor) returns home, and brings it into being through texts: a name on a map, a report to the Royal Geographical Society, the Foreign Office, the London Mission Society, a diary, a lecture, a travel book."[24] Michel de Certeau notes the historical process by which, "between the fifteenth and the seventeenth centuries, the map became more autonomous." At first, cartographers include traces of the process giving rise to the map, such as "the sailing ship painted on the sea [which] indicates the maritime expedition that made it possible to represent the coastline." However, as the map develops, it "gradually wins out over these figures; it colonizes space; it eliminates little by little the pictural figurations of the practices that produce it." What this means is that the map can only give a totalized representation of geographic knowledge because it "pushes away into its prehistory or into its posterity, as if into the wings, the operations of which it is the result or necessary condition."[25] In this reading, the map is not only a plan for defining and refunctionalizing territory, it is also a representational practice built on hiding desire. The refunctionalization Certeau speaks of is achieved at the cost of representational amnesia: the presentation of the map is dissociated formally from the process of its production. In his example this refunctionalization is historically understood: it is the history of cartographic representation alongside the history of European expansionism. In Goethe's story, the map appears as an object in which this refunctionalization has already taken place. The Captain's geometric techniques disappear beneath the aesthetically pleasing surface. However, when the representational amnesia of the map is revealed by Edward's coarseness, what is revealed is not, in the first instance, a process of historical but of subjective amnesia. The map's claim to truth rests on pushing away into its prehistory or posterity (to borrow Certeau's words) the desires that brought it into being, whereas Edward's act reveals that the clean, carefully constructed representational surface of the Captain's map is an expression of a carefully managed regime of desire. In *Elective Affinities*, desire is the engine of cartography, and it is cartography's repressed truth.

The entanglement of representation and desire is reflected in Goethe's language. He uses the word *Plan* interchangeably with *Karte*, but he also uses *Plan* in the sense of an intention, even a desire, or one might say *Willkür*. Plans are not only about objective representations of territory, they are about ownership, of land by people, of people by other people. Ottilie's visit does not fit into Charlotte and Edward's plans;[26] Edward's and the Captain's plan for changing Charlotte's layout of the estate is too ambitious and does not take account of the costs;[27] Luciane's plan is "to charm men of importance, rank, reputation, fame or some other significance, and to put prudence and good sense to shame by winning over even the most cautious to her strange, wild ways."[28] The semantic fuzziness surrounding plan, map, intention, and desire is most marked in the moment before Edward defaces the Captain's map. As the three friends discuss the possibilities of building a moss hut and creating as pleasant a prospect as possible for binding it to the palace, Charlotte, prosaic as always, reminds the men of the cost. Edward makes a practical suggestion (felling some of the trees and selling the lumber), the Captain makes a plan, and Edward counters with a more practical proposal. All are satisfied that they have acted rationally, and their satisfaction allows them to imagine the results before they have been realized: "This sensible, moderate arrangement was certain of everybody's approval, and soon the whole company saw in their mind's eye the new paths meandering through the landscape and envisaged themselves discovering the most delightful resting-places and vantage points."[29] There is an uncontested faith here in the capacity of reason to exploit nature for the production of wealth and, parallel to this, to carve out a place where nature's prospect expresses leisure and pleasure: the same leisure and pleasure enabled by the exploitation of nature.

As a way to realize their plans and bring imagined forms to the landscape, they consult the Captain's map. When Edward asks Ottilie's opinion, he turns the map (Goethe uses the word *Plan*) away from Charlotte and toward Ottilie. The Captain's and Edward's plan is about to become Ottilie's desire; Ottilie's desire will infect Edward, then in turn the entire company; and its medium will be the Captain's map. Ottilie's desire emerges as an unsettling force in the reasoned control of nature. Goethe leaves no doubt about this:

> Putting her finger on the highest part of the rise, Ottilie said: "I would build the summer house here. You wouldn't see the manor from there, of course, since it would be hidden by the clump of trees; instead, you would be in a new and different world, with the village and the houses hidden from sight. The view of the lakes, towards the mill, the hills, mountains and countryside, is extraordinarily beautiful; I noticed it as we went past."[30]

This other face of nature, the byproduct of reasoned nature, will become in the course of the novel the place of death. But for now, Ottilie is asking the group to change their understanding of what it means to take rest, to take leisure as the opposite of production. This has an impact not only on the way they live their lives, but also on representation. After Edward has seized on Ottilie's desire and made it his own, the map will never be the same again. It has lost its transparency, clouded with desire. To use terms introduced by Sybille Krämer, the idea of a transparent map has been unsettled by the performance of its opacity: its inextricable entwinement with representational and social conventions.[31] With a single stroke of the pencil, Edward has revealed the double secret that the map needs to keep hidden if it is to function as a tool of possession: representation of territory is an expression of desire, and represented territory is not only being measured, it is being planned, created. It could even be argued that, in the course of the novel, the interchangeability of the plan as a map of landscape is increasingly sidelined by plans as intentions or desires.

Seen like this, the map becomes another one of the scientific discourses in which representations of nature are clouded by the desires of those for whom there is something at stake. Cartography, like chemistry, shows the problem of metaphor as a bearer of desire. Scientific discourse struggles with its own representational conventions to the extent that these conventions cannot free themselves from the desires in which they are embroiled. The problem is that it is never clear how to distinguish the forces that shape nature into an objective landscape from the forces that shape a person's life through the pressure of their personal desires. Scientific representations tend to rely on separating these two natural forces. But Goethe is interested in where this separation fails. This is clear in the unfolding of the novel's central chemical metaphor in the same chapter as Edward's defacing of the Captain's map. As chemistry promises a model for understanding relationships, the individual actors' desires render understanding impossible.

The failed separation of desire and truth, of obscure passions and bright reason, appeared in contemporary cartography in the form of interference between the diagram of territory and the plan for its social and political organization. The Captain used his compass and trigonometric expertise to produce his plan at a decisive moment in the history of cartography. Nicholas Boyle argues that Goethe's novel is set in the years 1806–1807—that is, just five years after Napoléon had begun to put into effect his ambitious plan of mapping all Europe by extending the triangulation of France into the Rhineland under the direction of Jean Joseph Tranchot.[32] Although it had always been linked to the conquest of territory, in Napoléon's hands cartography not only enabled conquest by providing reliable knowledge of the ground on which battles were to be fought and laying out the best routes from one place to another. It was also beginning to provide ideological ground for conquest by representing a field in which claims to national rights to territory were shown to be true. Mapping possession onto territory meant the question of the legitimate ownership of land, and this meant presenting some kind of argument for the natural place of the owner. Goethe remained committed to this idea (which he had received through Herder and Buffon). As late as 1829 he speaks to Eckermann of how "the oldest tribes took possession of the land that pleased them most, and where the region was already in harmony with the innate character of the people."[33]

Goethe places cartographic expertise in the hands of an officer who has relocated to civilian life, and his tools are immersed in the current controversies surrounding the representation of territory and its possession. The Captain may not have understood his cartographic work as an explicit confirmation of ownership, nor is there any indication that his craft is enabled by military technology, but Edward instinctively sees that what he has before him on the table is a visualization of his territorial possession. And what unfolds in the course of the novel is very closely tied to the question raised by the Napoleonic cartographers: if nature can be represented on the page of a map (or in the pages of a book), what is it about this representation that legitimates ownership? Goethe extends this alliance between representation and ownership into the realm of human life, where desire similarly corrodes the legitimacy of ownership. What does Edward see in the world of representations that causes him to believe that Ottilie could become his possession, and that this acquisition will correspond to a natural process? He states: "Many a comforting omen, many a positive sign had confirmed my belief, my delusion that Ottilie could be mine."[34] For Edward, possession is a manipulation of symbols, and, because symbols are caught between truth and desire, his interpretations are doomed to failure.

Goethe makes it clear in this novel that the question of ownership thus posed is a scientific question. As such, it can be positioned in a number of discourses. One such discourse tied the visibility of territory to conquest and ownership—witness the scenes in the final act of *Faust* leading up to the forced removal of Philemon and Baucis and the theft of their property:

> I want those lindens part of my estate;
> The few trees spoil, because I do not own them,
> Everything that I possess on earth.
> Among their branches I would like to build
> A platform with a panoramic view
> Of all that I have now accomplished[35]–

The Captain's topographic chart can be seen to recast Faust's relationship of representation to ownership in a sublimated form. Goethe held a lasting, if low-level interest in the relationship of ownership, military activity, and cartographic representation. In the *Tag- und Jahreshefte* of 1815, he recounts how, on a visit to Bieberich, he was received by Archduke Charles of Austria, the brilliant military strategist who had fought in the Napoleonic Wars. Charles described his campaigns to Goethe and showed him his "extremely accurate and cleanly drawn maps." Goethe responds by explaining to the archduke how "a good military map is the most useful for geognostic purposes. Neither the soldier nor the geognost asks who owns the river, land or mountain; the former asks in what way it can be rendered useful for his operations, while the latter asks how it can serve to supplement and confirm his observations."[36] Two years later he notes how he has gained a "not insubstantial insight into geology and geography" from the "map of European mountains by Sorriot [figure 10.4].

Thus for example the land and prospects of Spain, so vexatious for the field commander and so favourable for guerrillas, became at once clear to me. I drew the main watersheds on my map of Spain, and in this way all the routes of travel as well as any campaign or any regular and irregular initiative of this kind became clear and comprehensible to me. Anyone who uses this colossal map as the basis of their geognostic, geological, geographical and topographical studies, will find themselves greatly advantaged" (*Tag- und Jahreshefte*).[37]

Similarly, in recasting representations of ownership as representations of scientific knowledge, the Captain's map reveals a link between ownership as Edward conceives it and the cartographic representation of natural habitats. If it is possible to argue that there is a force of nature that acts as fate, drawing individuals together so they might possess one another or be possessed, then it must be possible to represent all of life as held under the sway of similar forces of nature. This is the purport of the *Elective Affinities* metaphor. Here, the question at stake is whether it is possible to conceive of a life force in nature that legitimates a love relationship as a mutual possession of two people. Is it possible to represent this life force in metaphorical terms without losing its scientific legitimacy, and doesn't the metaphor itself disqualify its own claims to truth? If we take the long view, this is part of the same process whereby the empirical evidence derived from organic life was slowly unseating, or at least unsettling, philosophical and scientific claims to be able to determine regularities in the structures of experience. This is very clearly set out by Jennifer Mensch, who traces this tendency from the various challenges to taxonomy, from Buffon through to Kant, where it played a major role, she claims, in the development of his critical philosophy.[38] What does this mean for cartography?

Representing the Order of Nature: Science at Odds with Aesthetics

In the realm of cartography, the idea of a natural force affecting the organization of life forms had been gaining ground for several decades when Goethe published *Elective Affinities*. Cartographic knowledge expressed this force by determining a proper place for related categories of organisms and displaying them in their proper places. Its most important forms were botanical and zoological, but also ethnographic cartography. In demarcating proper places, however, it was entering the natural history debates, where, in Mensch's words, "the previously parallel investigations into system and process converged in Buffon's natural history [which Goethe had read in 1780, JKN] to produce both a new view of organic life and the basis for redefining taxonomy."[39] Cartographic representation would dramatize the necessity of freezing processes and obscuring genealogies in order to establish proper places.

Around the same time he was writing *Elective Affinities*, Goethe was also expressing interest in problems of taxonomy and cartographic representation. In 1812 and

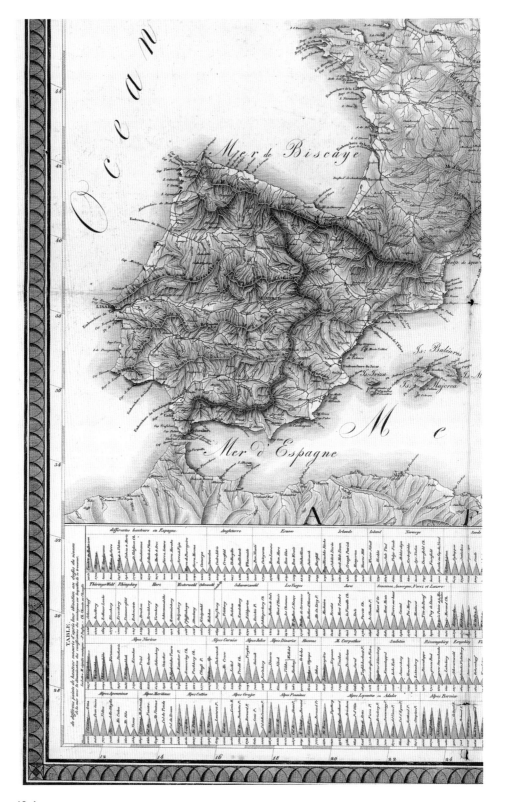

Figure 10.4

Andreas Sorriot de L'host, *Carte générale Orographique et Hydrographique de l'Europe, qui montre les principales ramifications des montagnes, fleuves, et chemins* (1816). Kartensammlung der Staatsbibliothek Berlin.

1813 he corresponded with Wilhelm von Humboldt, not only about the latter's general investigations on the role of language in the history of humanity, but more specifically about the possibility of collaborating on a cartographic representation of linguistic diversity in the world. In the *Tag- und Jahreshefte* of 1813, he notes how, together with Wilhelm von Humboldt, he "revised, refined and illuminated" what he termed "geographical maps for the sensory representation of languages distributed throughout the world."[40] On August 31, 1812, Goethe had written to Humboldt expressing his regret that during their recent meeting in Carlsbad he had not had the opportunity to make a complete record of Humboldt's "beautiful representation of how languages are distributed throughout the world." He then continues: "If you wished to do me a special favour, you could send me the overview in writing and I would illuminate a corresponding hemisphere map as an addendum to Lesage's atlas" (figure 10.5).[41]

In this connection, Goethe mentions an idea of compiling a "succinct and tabular traveling library" whereby Humboldt's "language map would serve in many cases as a memory booster and a guide in many of my readings."[42] Humboldt tells Goethe on September 7 that "the project you want is connected with a certain difficulty. To construct it with only minimum exactitude, skipping the details, is extremely easy and could almost be done standing on your head, but it is also not particularly rewarding. With exactitude one encounters some issues which are difficult to resolve. But I will gladly draft a table to keep the mid-ground between too timid and too general a determination, and I will send you world region after world region, beginning with Europe, the easiest."[43] He then elaborates on the central representational difficulty presented by the need for both detail and generalization. In this elaboration it becomes clear that the problem goes deeper. The diverse languages themselves "must be seen as part of the history of the human race and as the most important agent in the economy of intellectual natures, and for this reason the central factors of all investigations into national languages and the division of the human race into tribes and nations should be a part of these investigations."[44]

Humboldt seems to be pleading for a methodological reflexivity that sees the decision to divide humanity according to linguistic categories as a factor in deciding what these categories should be. The problem is that the differentiation of the nations and peoples has itself emerged from a history of linguistic differentiation. Why should this be a methodological problem? Isn't it simply what all cartographers of the world languages would have to take as their starting point—that today's linguistic differentiation has emerged through the history of languages and the cognitive horizons they embody? Humboldt goes on to explain that the cartographer's (or the classifier's) real problem is that the broad categorization of languages relies on an imposition of regularities that may have little to do with language itself:

Figure 10.5

Lesage [Emmanuel conte de Las Cases], *Atlas historique généalogique chronologique* (1808), p. 29: Mappe monde historique, detail. Kartensammlung der Staatsbibliothek Berlin.

In the study of actual languages, there is also little point in all the knowledge of the way languages have determined the mind and the qualities of the nations unless one can at the same time recognize the constitution of the components upon which this determination rests. But here is where the difficulties begin. Since one's impression is always a total impression, being the one point where an endless number of points converge, that which is intrinsic to each individual element is almost unnoticeable. It is here in particular that the *raisonnements a priori* have little or no effect, since there is more to be achieved through the comparison of many languages and their mutual interrelations.[45]

Humboldt recognizes that the cartographer is forced by the formal requirements of the map to impose his own a priori generalizations on the map itself. He could just as well be speaking of Edward's desire and the way it imposes his own categories of objectivity on the world.

On November 15, Humboldt sent Goethe his essay "Anleitung zur Entwerfung einer allgemeinen Sprachkarte" ("Instructions for Drafting a General Linguistic Map"). His essay shows "which region each linguistic group occupies and which languages are to be found in each country (according to the usual categorization). Using both of these, it will be easy to construct a map." However, as the earlier letter makes clear, this ease of construction is bought at the price of glossing over the epistemological problems he outlines there. Goethe responds with enthusiasm, disregarding any such difficulties. He promises Humboldt that, as soon as he has time in March, he will turn his attention to the map. He has already obtained from Justus Bertuch some sheets of Europe, which he intends to mount on a board and "to note the main languages and, as far as it is possible, the dialects, with glued-on labels."[46] This "as far as it is possible" indicates that Goethe realizes he is beginning to set foot on that dangerous epistemological ground where the guiding categorization of the map begins to become lost in the empirically noted differentiation. Dialects are the first sign that a national language might not be the best way to describe the way individuals relate their language to territory.

Goethe's map is lost. What was this map? In the first decades of the nineteenth century, it appeared that there were some useful models for dividing and categorizing the world's languages. Humboldt relied extensively on J. C. Adelung's multivolume work, *Mithridates*, begun in 1807, for his "Instructions for Drafting a General Linguistic Map." Adelung's ambitious project attempted a classification of the world's languages on the basis of the Lord's Prayer. Almost ten years later, Goethe mentions Humboldt's attempts to further develop the *Sprachcharte*, stating that he (Goethe) "has never ceased continuing to think about, to collect, and to work on the topic of the world and humanity [*über Welt und Menschen fortzudenken, zu sammeln, zu arbeiten*]."[47] Humboldt replies that his interest in language is aimed at the inner connection to thought [*innere Zusammenhang mit dem Gedanken*], the dependence or independence of this and all cognitive development and expression [*aller geistigen Bildung*] on language, which has only received its organism in the smallest part from those who now speak it, and who have experienced their fate in the same way as any other historically formed being. For it cannot be denied that not only the grammatical rules, upon which free and variable usage so powerfully depends, but also the gender of nouns, which determines the concept (itself vague) in a specific manner, transmit sensation. From the very beginning of speech, they form a series in themselves, which even we are able to recognize to a certain extent. This very problem is the most difficult, and so I too have to realize that I almost always circumvent it and, to avoid pointless labour, often stop work on projects that can at best be called preparatory.[48]

It seems to me that what Humboldt is trying to express in this extremely convoluted paragraph is the tabulatory difficulty that Goethe himself is all too aware of: once you begin to write the names of diverse languages on a map of the world, you have to face the fact that languages are historically transmitted sets of grammatical rules whose regularities determine and are determined by the physical sensations (the organic nature) of those who speak them, of individuals. To record the classes of regularities themselves is to obliterate the individual experiences that work against these rules, but also the dynamic development of language itself, its historical dimension. The act of categorization finds itself at odds with the representational conventions of the map—up to the point where the labels for different languages begin to deface the aesthetic object in the same way Eduard defaced the Captain's map. Goethe's desire for scientific order is like Edward's desire for the proper place of leisure.

There was a general understanding among scientists like Goethe, Herder, and Humboldt that the problem with language was a general problem in the classificatory sciences and that it extended to other fields of specialist cartography and to representation itself. In book six of his *Ideas for a Philosophy of History* (1785), Herder claims that the best way to represent the diversity of humanity was by collecting "faithful paintings of the differences in our species," and laying the foundation for a "speaking doctrine of nature and physiognomy of humanity." This would be the most philosophical application for art and its product would not be idealized fantasies of other nations, but a scientific map of nature's diversity and the unity of humankind. What Herder has in mind is "an anthropological map of the earth, such as Zimmermann's attempt at a zoological map, a map that would aim to depict nothing but the diversity of humanity, but doing so in all its appearances and situations. This would crown the philanthropic project."[49] Herder is referring to the *Tabula Mundi Geographico Zoologica*, published around 1760 in Augsburg by the geographer and professor of mathematics in Braunschweig, Eberhard Zimmermann (Figure 10.6).

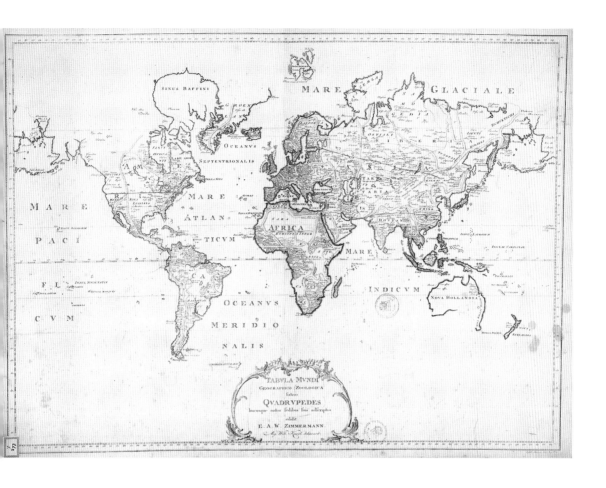

Figure 10.6

Eberhard Zimmermann, *Tabula Mundi Geographico Zoologica* (1783). Thomas Fisher Rare Book Library, University of Toronto.

Zimmermann's map was an ambitious attempt to reflect the latest developments in the categorization of species exemplified by Linnaeus. On each continent he inscribes the Latin names of the species found in a particular region, as shown in figure 10.7, at times crowding the face of the map, and in a way illustrating the problem Humboldt later described.

Herder apparently didn't realize that the French geographer Robert de Vaugondy had published an anthropological map of the world in 1778 ("Mappe-Mond suivant la projection des cartes reduites," brought out by Delamarche in Paris), shown in figure 10.8.

The key to the legend shows Vaugondy dividing humanity into four religions: Jews, Christians, Mohammedans, and Pagans. He also identifies four skin colors: white, brown, yellow, and olive. Furthermore, he presents a fourfold division of physiognomies: Europeans, flat faces with oval eyes, bear faces, and Africans. Herder, following Leibniz's concept of infinitesimal differentiations in nature, had no time for this kind of classification. He also had no hesitation in bringing it into direct association with contemporary movements of colonialism and imperialism.

It has, as noted above, become common to understand the emptiness of the Ptolemaic map in terms of the history of colonialism. This rests on the observation that the map of Ptolemy introduced the idea of geometric representation. This created what Padrón calls a positive emptiness—that is, a sense of ubiquitous symbolic control of territory that disqualifies indigenous knowledge and opens the way for territorial conquest.[50] Goethe added an important nuance to this idea when he spoke of the blank spots on maps as opening the way for scientific representation, not only as a discourse of truth, but as one of mythologization. In this way, Goethe understood that the representation of scientific innovation in prose is related to that in cartography. Under the heading *Lücke*, he begins the third part of the *Farbenlehre* by noting how, on maps of Africa, geographers used to draw "an elephant, lion or a monster of the desert wherever mountains, rivers or cities were missing, without being chastised for it. In the same way, no one will hold it against us if we insert into the large gap where gratifying, living, advancing science abandons us, a few observations which will serve the purpose of future reference."[51] Goethe continues in this section, making oblique reference to the act of mythologization that determines how this kind of gap-stopping formulation is transmitted to future scientists, and how individual researchers are put in the position of having to resist the received notions of nature and tradition.[52]

In this way, Goethe envisages a scientific activity that, like Edward's willful act of desire, unsettles claims to objectivity in scientific discourse. The scientist faces the fact that Edward cannot confront: that representations are neither true nor untrue, but embodiments of the desires of those who use them. Goethe's exploration of cartography in *Elective Affinities* thereby demonstrates how, at a moment in history when cartographic representation was having an ever-greater claim on territorial possession,

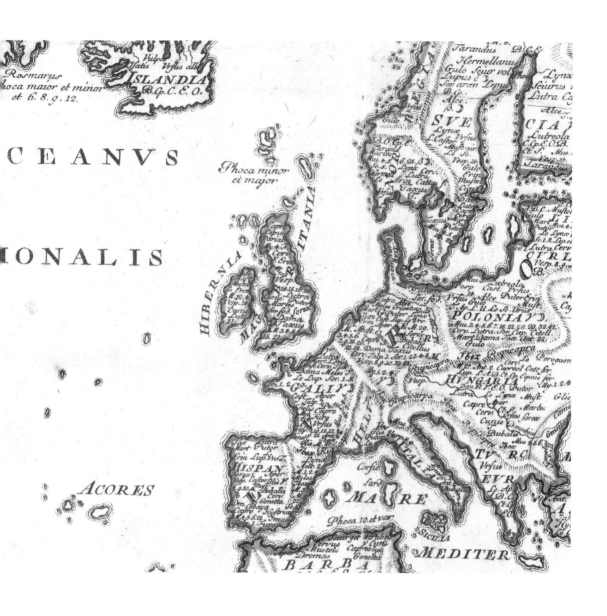

Figure 10.7

Eberhard Zimmermann, *Tabula Mundi Geographico Zoologica* (1783). Detail.

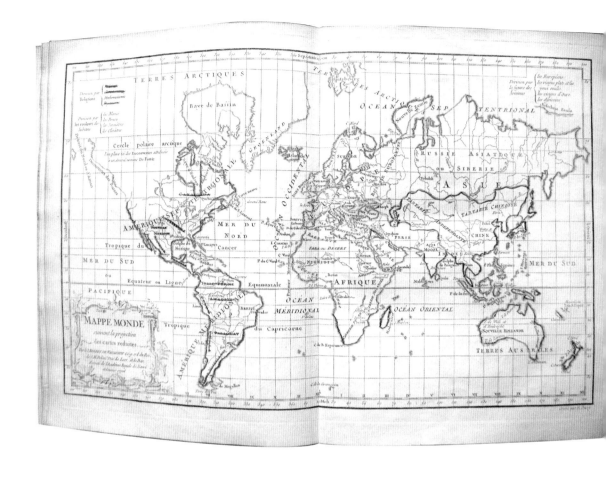

Figure 10.8

Robert de Vaugondy, *Mappe-Mond suivant la projection des cartes reduites* (Paris: Delamarche, 1778). Thomas Fisher Rare Book Library, University of Toronto.

the mathematical purity imputed to this form of representation was always at odds with the form itself. Embedded deeply in the aesthetic form of the map was an economy of desire that unsettled all claims to objectivity and truth.

Notes

All translations from German sources are my own.

1. Johann Wolfgang von Goethe, *Goethes Gespräche*, ed. Woldemar Freiherr von Biedermann (Leipzig, 1896), vol. 2, 292.

2. Johann Wolfgang von Goethe, *Werke*, ed. Erich Trunz et al. (Hamburg: C. Wegner, 1982), vol. 6, 661.

3. Johann Wolfgang von Goethe, *Elective Affinities*, *Goethe's Collected Works*, ed. Victor Lange, Eric A. Blackall, and Cyrus Hamlin (Boston: Suhrkamp, 1982), vol. 11, 109–110.

4. Goethe, *Elective Affinities*, 129.

5. Goethe, *Elective Affinities*, 129.

6. Goethe, *Elective Affinities*, 93.

7. Goethe, *Elective Affinities*, 111.

8. Johann Wolfgang von Goethe, *Poetische Werke: Kunsttheoretische Schriften und Übersetzungen* (Berlin and Weimar: Aufbau, 1976), vol. 12, 319. Cited as *Berliner Ausgabe*.

9. Goethe, *Berliner Ausgabe*, vol. 16, 269.

10. Johann Wolfgang von Goethe, *Goethes Werke* (Munich: Deutscher Taschenbuch Verlag, 1987), vol. 4, sec. 6, 82. Cited as *Weimarer Ausgabe*.

11. Johann Wolfgang von Goethe, *Sämtliche Werke, Briefe, Tagebücher und Gespräche*, ed. Dieter Borchmeyer, et al. (Frankfurt: Dt. Klassiker-Verlag, 1985), vol. 1, sec. 25, 585–587. Also see Christian Keferstein, *Teutschland, geognostisch-geologisch dargestellt und mit Charten und Durchschnittszeichnungen erläutert* (Weimar: Landes-Industrie-Comptoirs, 1821–1831).

12. Goethe, *Berliner Ausgabe*, vol. 16, 333.

13. Goethe, *Weimarer Ausgabe*, vol. 4, sec. 41, 256.

14. Friedrich A. Kittler, "Ottilie Hauptmann," *Dichter–Mutter–Kind* (Munich: Fink, 1991), 139.

15. Goethe, *Elective Affinities*, 110.

16. Goethe, *Elective Affinities*, 125.

17. Goethe, *Elective Affinities*, 143.

18. Goethe, *Elective Affinities*, 106.

19. Goethe, *Elective Affinities*, 106.

20. Goethe, *Elective Affinities*, 106.

21. Edward Said, *Culture and Imperialism* (New York: Vintage, 1993), 166.

22. Paul Carter, *The Road to Botany Bay: An Exploration of Landscape and History* (Chicago: University of Chicago Press, 1989), 24.

23. See Carter, *The Road to Botany Bay*, 204.

24. Mary Louise Pratt, *Imperial Eyes: Travel Writing and Transculturation* (London: Routledge, 1992), 200.

25. Michel de Certeau, *The Practice of Everyday Life*, trans. Steven Rendall (Berkeley: University of California Press, 1984), 121.

26. Goethe, *Elective Affinities*, 99.

27. Goethe, *Elective Affinities*, 263.

28. Goethe, *Elective Affinities*, 187.

29. Goethe, *Elective Affinities*, 129.

30. Goethe, *Elective Affinities*, 129.

31. Sybille Krämer, *Medium, Bote, Übertragung: Kleine Metaphysik der Medialität* (Frankfurt: Suhrkamp, 2008), 300–306.

32. Nicholas Boyle, "The Composition of *Die Wahlverwandtschaften*," *Publications of the English Goethe Society* 84, no. 2 (2015), 93–137.

33. Goethe, *Goethes Gespräche*, vol. 7, 42.

34. Goethe, *Goethe's Collected Works*, 233.

35. Goethe, *Faust*, in *Goethe's Collected Works*, vol. 2, ed. and trans. Stuart Atkins (Boston: Suhrkamp, 1984), 284.

36. Goethe, *Berliner Ausgabe*, vol. 16, 254.

37. Goethe, *Berliner Ausgabe*, vol. 16, 269.

38. Jennifer Mensch, *Kant's Organicism: Epigenesis and the Development of Critical Philosophy* (Chicago: University of Chicago Press, 2013).

39. Mensch, *Kant's Organicism*, 5.

40. Goethe, *Berliner Ausgabe*, vol. 16, 243.

41. Goethe, *Weimarer Ausgabe*, vol. 4, sec. 23, 84.

42. Goethe, *Weimarer Ausgabe*, vol. 4, sec. 23, 84.

43. Johann Wolfgang von Goethe, *Goethe's Briefwechsel mit den Gebrüdern von Humboldt* (Leipzig: Brockhaus, 1876), 243.

44. Goethe, *Goethe's Briefwechsel*, 244.

45. Goethe, *Goethe's Briefwechsel*, 244.

46. Goethe, *Weimarer Ausgabe*, vol. 4, sec. 23, 278.

47. Goethe, *Weimarer Ausgabe*, vol. 4, sec. 23, 289.

48. Goethe, *Goethe's Briefwechsel*, 265.

49. Johann Gottfried Herder, *Sämtliche Werke*, ed. Bernhard Suphan (Hildesheim: G. Olms, 1868), vol. 13, 251.

50. Ricardo Padrón, "Mapping Plus Ultra: Cartography, Space, and Hispanic Modernity," *Empires of Vision*, ed. Martin Jay and Sumathi Ramaswamy (Durham, NC: Duke University Press, 2014), 214–215.

51. Johann Wolfgang von Goethe, *Gedenkausgabe der Werke, Briefe und Gespräche*, ed. Ernst Beutler (Zurich: Artemis, 1960), vol. 16, 339.

52. Goethe, *Gedenkausgabe*, 343.

Bibliography

Boyle, Nicholas. "The Composition of *Die Wahlverwandtschaften*." *Publications of the English Goethe Society* 84, no. 2 (2015): 93–137.

Carter, Paul. *The Road to Botany Bay: An Exploration of Landscape and History*. Chicago: University of Chicago Press, 1989.

De Certeau, Michel. *The Practice of Everyday Life*. Trans. Steven Rendall. Berkeley: University of California Press, 1984.

Goethe, Johann Wolfgang von. *Gedenkausgabe der Werke, Briefe und Gespräche*. Ed. Ernst Beutler. 24 vols. Zurich: Artemis, 1948–1960. Cited as *Gedenkausgabe*.

Goethe, Johann Wolfgang von. *Goethe's Briefwechsel mit den Gebrüdern von Humboldt*. Leipzig: Brockhaus, 1876.

Goethe, Johann Wolfgang von. *Goethe's Collected Works*. Ed. Victor Lange, Eric A. Blackall, and Cyrus Hamlin. 12 vols. Boston: Suhrkamp, 1982–. Cited as *Collected Works*.

Goethe, Johann Wolfgang von. *Goethes Gespräche*. Vols. 1–10. Ed. Woldemar Freiherr von Biedermann. Leipzig, 1889–1896.

Goethe, Johann Wolfgang von. *Goethes Werke*. 143 vols. Munich: Deutscher Taschenbuch Verlag, 1987. *Weimarer oder Sophienausgabe*. Cited as *Weimarer Ausgabe*.

Goethe, Johann Wolfgang von. *Poetische Werke: Kunsttheoretische Schriften und Übersetzungen*. 22 vols. Berlin and Weimar: Aufbau-Verlag, 1960–1978. Cited as *Berliner Ausgabe*.

Goethe, Johann Wolfgang von. *Sämtliche Werke, Briefe, Tagebücher und Gespräche*, ed. Dieter Borchmeyer et al. 40 vols. Frankfurt: Dt. Klassiker-Verlag, 1985.

Goethe, Johann Wolfgang von. *Werke*. Ed. Erich Trunz et al. Hamburg: C. Wegner, 1961–1964.

Güttler, Nils. *Das Kosmoskop: Karten und ihre Benutzer in der Pflanzengeographie des 19. Jahrhunderts*. Göttingen: Wallstein, 2014.

Herder, Johann Gottfried. *Sämtliche Werke*. Ed. Bernhard Suphan. 33 vols. Hildesheim: G. Olms, 1867–1868.

Keferstein, Christian. *Teutschland, geognostisch-geologisch dargestellt und mit Charten und Durchschnittszeichnungen erläutert*. Weimar: Landes-Industrie-Comptoirs, 1821–1831.

Kittler, Friedrich A. "Ottilie Hauptmann." In *Dichter–Mutter–Kind*. Munich: Fink, 1991.

Krämer, Sybille. *Medium, Bote, Übertragung: Kleine Metaphysik der Medialität*. Frankfurt: Suhrkamp, 2008.

Mensch, Jennifer. *Kant's Organicism: Epigenesis and the Development of Critical Philosophy*. Chicago: University of Chicago Press, 2013.

Padrón, Ricardo. "Mapping Plus Ultra: Cartography, Space, and Hispanic Modernity." In *Empires of Vision*, ed. Martin Jay and Sumathi Ramaswamy. Durham, NC: Duke University Press, 2014.

Pratt, Mary Louise. *Imperial Eyes: Travel Writing and Transculturation*. London: Routledge, 1992.

Said, Edward. *Culture and Imperialism*. New York: Vintage, 1993.

Weidmann, Marc, and Luis Sole Sabaris. "Noticia de Carlos Gimbernat y de sus mapas geológicos de Europa Central, Alpes, Francia e Italia, a principios del siglo XIX." *Acta Geològica Hispànica* 18, no. 2 (1983): 75–86.

11
Conceptualizing the Novel Map: Nineteenth-Century French Literary Cartography

Patrick M. Bray

Introduction

With the tumult of the French Revolution and the beginning of heavy industrialization, the ways French national space was experienced and represented in literature profoundly changed in the nineteenth century. At the threshold of the century, Madame de Staël's emblematic essay *De la littérature* set out to describe how the climate and politics of nations shaped the character of their literatures, and how, in turn, literature could play a role in the future progress of humanity after the French Revolution.[1] Combining Montesquieu's emphasis on climate in *L'Esprit des lois* and Rousseau's notion of perfectibility, Staël captured the tensions and contradictions that would define French literature in the nineteenth century and beyond, between on the one hand spatial and national configurations of identity and on the other the emancipatory promise of aesthetic production. Nineteenth-century French literature, in particular the novel, portrayed representations of the spatial and political upheavals of the era at the same time it tried to find alternative spaces of resistance. Through close readings of novels by Madame de Staël, Jules Verne, and Honoré de Balzac, I will sketch out what I call "the novel map," a concept that describes these contradictory directions taken by French nineteenth-century novels as they projected a holistic image of the self inscribed in the synthetic spaces of a fictional text.

The Novel Map as Concept

Staël's essay, published in 1800, coincided with two new developments that make nineteenth-century French literature a particularly rich terrain for studying the novel map: the idea of literature as the art of writing and the invention of a French national space. The revolution rationalized and organized France, no longer as an assemblage of historical territories each with its own culture, but now as a unified whole producing citizens who were supposed to identify more with the abstract concepts of the Republic than with the local *terroir* of their ancestors. The Cartes de Cassini (figure 11.1) and later the États-Majors maps were completed at the turn of the nineteenth century, achieving a remarkably complete cartographic representation of the whole of France at a scale of 1:86,400 (twice as detailed as contemporary Michelin regional maps at 1:150,000).

The First French Republic used these maps as a starting point to redraw the internal borders of the country by geographic and natural boundaries, erasing the historical regions and replacing them with the rational, but also political, *départements*.[2] Traditional topographic knowledge of the regions as well as local toponyms were classified and subsumed under a new national ordering of space. In the process of renaming old places, traditional knowledge was lost and the individual was alienated from his or her affective and historical relationship with place, a problem only compounded by the development of industrial capitalism. Haussmanization was only the most visible example of the transformation of city space.[3] Space could now be commoditized, facilitated by the standardization of time zones at the end of the century.[4]

Just as individuals were now in theory liberated (or deterritorialized) from their historical bonds with a land that defined them metonymically, so literature became a way of defining writing as the free play of meaning, unbound by the strictures of representation. For Jacques Rancière, "literature" as a concept emerged after 1800 in the wake of Staël's *De la littérature* and developed into what he calls "the historical mode of the visibility of the art of writing."[5] According to Rancière, representation no longer determined how we evaluate literary works; instead representation became subservient to the (visibility of the) art of writing. The accuracy or perhaps verisimilitude of characters, events, and spaces mattered less than the ability of the novel to render these representations literary, which is to say in the service of style. For the first time, a novel could have anything at all as a subject—the great example being Victor Hugo's *Notre-Dame de Paris*, which features at its center a building, as opposed to the travails of a hero. As Hugo famously wrote in his novel, "ceci tuera cela" ("this will kill that"), the printing press will kill the cathedral, books will take over from architecture, the democracy of the written word will replace ancient hierarchies. Since the cathedral in Hugo's novel was a cathedral of words, conceiving of the novel as the visibility of the art of writing meant that its materiality, its visibility, was embodied in its typeset. Caught between the visibility of the novel's writing and the readability of the text, the reader of Hugo's novel is left, like the character Claude Frollo, to stare in amazed wonder puzzling over the hieroglyphs of Hugo's textual cathedral.

Literature's need to indicate the visibility of its own status as an art means that its materiality as words on a page is forever at odds with its indexicality, its ability to reference a world outside the text. Similarly, Christian Jacob has argued that maps present the contradiction of materiality and representation: "[A map] is a problematic

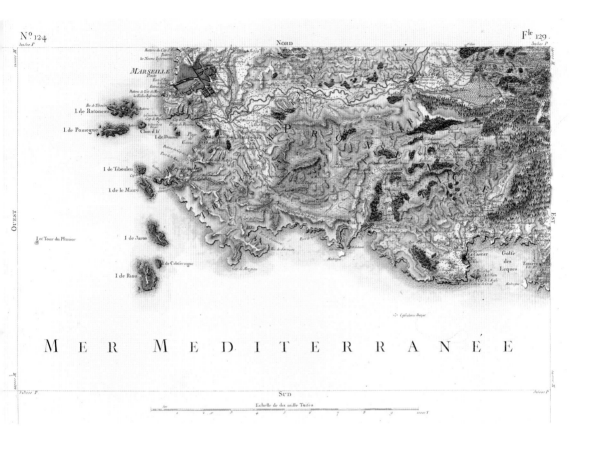

Figure 11.1
Carte générale de la France, dite Carte de Cassini, no. 124, feuille 129. Courtesy of Library of Congress, Geography and Map Division, G5830 s86 .C3 Vault.

mixture, where the transparency of a referential illusion coexists with the opacity of a medium that materializes this image."[6] Literary works function cartographically in that they point us to a distant "there" by way of the illusion of a "here." But while maps tend to render their medium as opaque as possible to privilege the referential illusion of their spatial image, literature as an art must continually remind the reader of the very presence of this illusion.

When a novel contains conventional maps, a complex play of references brings the tensions between visibility and readability to the fore. A map in a realist novel, for instance, presents not only a visual representation of a real place, such as Paris, but also the fictional place within the text. The "Paris" of the map thus exists in three iterations, as map, as text, and as real place. Furthermore, the map's placement on the page anticipates, interrupts, or prepares the textual description of the place and therefore the map participates in the novel's narrative structure, even as it distracts the reader from the linear flow of words. The act of "reading" a map invites the eye to wander and the hand to turn pages, reinforcing the materiality of the literary work and the freedom of the reader to interpret. Yet a map privileges what Michel de Certeau called "the law of the 'proper'" in relation to places, since for Certeau, a place is "an instantaneous configuration of positions. It implies an indication of stability."[7] While a map provides an image of stability for a novel's places and also the characters that inhabit them, it works against the narrative space through which characters move and in which places change over time. The confrontation between the space of a narrative and the place of the map structures to what degree the novel can represent the real and how it imagines new configurations of time and space.

The novel map is a concept that I use to describe how novels invent new, literary ways of mapping that alternate between narrative and cartography, readability and visuality, space and place, singular events and becoming in time.[8] While most nineteenth-century French novels did not contain conventional maps, I would argue that they all present in one form or another novel maps that function like conventional maps while also portraying an impossible harmony between an individual rooted in a specific place and moment, and the freedom of becoming other across space and time. The novel map can be thought of as a cartographic projection of subjectivity, necessarily imaginary or fictional, but no less powerful for the promise of self-realization it offers to writer and reader alike.

In the pages that follow, I describe three general aspects of how novel maps chart subjectivity and how they function in a range of French works from different literary movements throughout the nineteenth century: maps of national identity (Madame de Staël), maps of technological progress (Jules Verne), and maps of literature itself (Honoré de Balzac). Given the polysemic nature of the novel map, it can present all of these aspects in a single novel, as the subject is represented within the nexus of national, technological, and aesthetic discourses. Moreover, novel maps appear in all the major literary movements in France in the nineteenth century, from romanticism to realism, naturalism to symbolism. Studying how novels structure narrative and visual space offers a new way to understand literary history, not as a linear progression, but as a succession of different responses to the same original question of novel mapping.

The Mapped Self

The autobiographical mode emerged nearly at the same time as the concept of literature in the late eighteenth century and thereby invented a way of knowing the self by inscribing it in a literary text. But this new, textual self presented to the reader as a true copy of the author hints at the transformations the author went through during the process of writing this very text and can only speculate what the totality of the author's life will resemble after death (the fantasy epitomized by Chateaubriand's *Mémoires d'Outre-tombe*).[9] While maps in autobiographical novels anchor authorial presence (or rather the character identified with the author) on a page, this mapped self is limited to whatever specific event the map represents. For instance, in the manuscript for Stendhal's autobiographical novel the *Vie de Henry Brulard* from the 1830s, dozens of hand-drawn sketches and maps show the young Henry as an X or an H occupying various places in the author's memory. Yet Stendhal claims in the text not to be able to have a clear, synoptic vision of himself—a "view" that only the fluidity of the text, of the novel itself as map, can project. Throughout the century, writers exploited the oppositions between the mapped self and the textual self to discover new aspects of subjectivity.

Seven years after *De la littérature*, Staël published her novel *Corinne ou l'Italie*, a work whose form challenged and redefined what literature could do well beyond the scope of her earlier book by breaking away from a rigidly deterministic view of the relationship between self and geography. The novel tells the tragic story of Corinne, an Italian "improvisatrice" who falls in love with the Englishman Oswald, Lord Nelvil, only to lose him and her own life when he prefers her younger English half-sister. Along the way, we encounter perfidious Frenchmen reeling from the destruction of the French Revolution, feminized Italian men, strict English ladies, and the beauty of Italian art and Italian ruins. This truly transnational book also transcends genre distinctions, combining elements of autobiography, art history treatise, political philosophy, and travel guide.

Given the apparent genre confusion and the prominence given to Italy in the title, it is no wonder that the Bibliothèque Nationale de France first cataloged the novel as a travel guide to Italy.[10] In many ways the novel describes what happens when a Northern

European (the character Oswald or the reader of the French novel) is immersed in the strangeness of Italy. While we readers undoubtedly learn much about the art, history, and geography of Italy, we are also led to experience the uncomfortable feeling of losing our bearings in the ruins of Rome or on the slopes of Vesuvius while reading about the hypnotic powers of Corinne's poetry. This progressive loss of self as well as the lack of clear genre distinctions also reflect the fragmentation of national identity practiced by Staël's novel. Just as Marcel Proust divided up his own traits and attributed them to all the characters in his novel, Staël lends important facets of her own life story to the characters in her novel: Corinne has Staël's talent for improvisation and conversation, Oswald at first disdains Italian culture and mourns the loss of his overbearing father as had Staël, and the disagreeable French comte d'Erfeuil shares Staël's mother tongue and taste for courtly intrigue.

As the novel progresses, the characters begin to reveal secrets about their past that put into question notions of national character. Lord Nelvil is always described as "English," yet hails from Scotland and was embroiled in a love affair in Paris during the revolution. Corinne, who is presented in the title of the novel as embodying Italy, had an Italian mother and an English/Scottish aristocratic father like Oswald. Raised in Italy, she was sent to Scotland at the death of her mother, where she revolted against the strictures of polite aristocratic society and thereby convinced Oswald's father (before she ever met Oswald) of her unsuitability for marriage. Corinne's half-sister Lucille, however, is not at all Italian and we learn eventually that Oswald's father sought to arrange their marriage before his death. The convoluted plot not only heightens the sentimental drama, it makes it impossible to assign a stable national character to any person (or any work of art).

The most inspiring moments of the novel occur during Corinne's public performances where the textual descriptions of her celebrated improvisations manage the feat of embodying the specificity of a place, be it Rome, Naples, or Florence, and at the same time opening up the space of the novel to allow for a greater freedom of identity. Corinne's performances serve as novel maps since they show how literature creates a new space that respects the uniqueness of place and of the past, but makes a liberated future possible. After the start of her romance with Oswald, Corinne decides to stage a production of her own Italian translation of Shakespeare's *Romeo and Juliet* in Oswald's honor. An English play set in Italy by the foremost English playwright who nevertheless understood the Italian soul, Staël claims that Corinne's Italian translation thus returned the play to its native language.[11] Yet the narrator only cites the original English and translates the lines into French, leaving the reader to imagine what Corinne's Italian Shakespeare would sound like. Throughout the description of the play, Corinne becomes indistinguishable from Juliet, Verona from Rome, the Renaissance from the present, and Oswald from Romeo, even though he is only a jealous spectator. At one point the emotion of seeing Juliet-Corinne awaken to a dead Romeo leads

Oswald to faint; when he comes to, he can no longer distinguish truth from fiction, and repeats Romeo's words to Juliet. Corinne's performance succeeds in transforming Oswald from a duty-bound Englishman into a passionate Italian lover in the space of a few pages.

At the end of the novel, as Corinne dies an agonizingly slow death from a broken heart, Oswald and his wife, Corinne's half-sister, visit her in Florence with their not-so-subtly named daughter Juliette. Corinne tells her sister that she wishes to live on in the memory of Oswald through her and her daughter, passing on her love of art and of Italy. For her final performance in front of Oswald, she has a prepared text read by a young girl dressed in white and wearing flowers, reminiscent of Shakespeare's Juliet. The contrast of the dying Corinne's "somber" words read by the incarnation of youth produces a "serene" effect that elevates the performance beyond Corinne's personal agony.

In Staël's novel, the uniqueness of Corinne and of Italy, Corinne or Italy, shines through on every page as the text describes in detail the extraordinary life of the heroine and the country she personifies. Yet the most literary passages in the novel, where the power of literature itself comes to the fore, approach a timeless universality beyond the narrow confines of the self. Throughout the nineteenth and twentieth centuries, writers such as Stendhal, Nerval, George Eliot, George Sand, Barrès, Proust, and Gide followed in the wake of Staël and redrew novel maps of the self, often in Italy, looking to literature to understand and transcend their own subjectivity.

Mapping Progress

The accelerating pace of political, technological, and spatial change after the French Revolution meant that rooting the self in a stable place, mapping the self in a place of origin, became increasingly difficult. As Charles Baudelaire lamented in his 1860 poem dedicated to Hugo, *Le Cygne*, "The form of a city changes faster, alas, than the heart of a mortal." For Baudelaire as for so many of his contemporaries, any understanding of a place can only be found in memory and art, since the most lasting material trace of the constantly changing city may very well be the virtual space of a poem. David F. Bell has argued that nineteenth-century French realist novelists sought to capture the dangers and possibilities of these changes in their texts, using speed as a structuring principle of the narrative.[12] Novelists like Balzac and poets like Baudelaire would take on the ephemeral nature of modern life as their subject and their style, harnessing technology in the service of art. Both literature and technological progress transgress traditional borders to create new ways of moving through space, yet novel maps push against the spatialization of time brought about by industrial capitalism.

Jules Verne may be synonymous with the naive optimism of the nineteenth century's faith in technological progress, but his novels exhibit a profound ambiguity to scientific triumphalism and the constraints of bourgeois subjectivity. The dozens of novels that make up his *Voyages extraordinaires* from the 1850s to the end of the century followed both his own literary ambitions and also his editor Hetzel's pedagogical imperative to instruct the reading public about science and geography (Hetzel published the series in his "Bibliothèque d'éducation et de récréation"). Verne's popular novels (often significantly abridged and simplified to appeal to ever-broader audiences) combine scientific realism with the necessary creativity of the written word. As Émile Zola wrote in *Le Docteur Pascal*, the final volume in his series *Les Rougon-Macquart*, "There is [in emerging scientific fields] a margin which belongs to them [poets], between conquered, definitive truth and the unknown, from which will be torn the truth of tomorrow."[13] Verne's uncanny ability to foresee the technological marvels of the twentieth century, "the truth of tomorrow," no doubt stems in part from his probing the depths of the scientific mind, the force of desire at the origin of invention. Accurate maps of the earth, the oceans, and the moon, beautifully illustrate Verne's novels, significantly adding to their pedagogical effectiveness. But his prose also maps the invisible, the unknowable, and the unconscious, reserving the power of creativity to the chance encounter of words on the page.

Roland Barthes's short essay, "'Nautilus' et 'Bateau ivre,'" brought Verne back to critical attention by contrasting Verne's *Vingt mille lieues sous les mers* with the Arthur Rimbaud poem it inspired, "Le Bateau ivre." Barthes observes that Verne's universe is closed, whereas Rimbaud's poem is open to exploration.[14] For Barthes, there is something inherently childlike (and bourgeois) in Verne's recreating the world in a comfortable cocoon such as Nemo's *Nautilus*, a safe place from which to contemplate and appropriate nature. The ultimate sign of Verne's bourgeois will to possess and dominate the world, according to Barthes, is his daring novelistic invention: "Gage space by time, ceaselessly conjoin these two categories, risk both of them on the same throw of the dice or on the same always successful sudden impulse." Moreover, the occasional "desperado" in Verne's fiction highlights the serenity of the bourgeois hero. I would argue that while Barthes's analyses about space and time are valuable, he reads Verne's characters too literally, missing the irony that puts what Barthes elsewhere calls the bourgeois *doxa* into question.

The titles of many of Verne's novels certainly suggest that Barthes was right about the writer's conflation of space and time: *Cinq semaines en ballon*; *Vingt mille lieues sous les mers*; *De la terre à la lune, trajet direct en 97 heures 20 minutes*; *Le Tour du monde en quatre-vingts jours*. These serial novels announce their narrative space in their very titles, since readers could follow how many installments were left according to the diegetic chronology. Within this closed time frame (for example, the journey around the world cannot take eighty-one days), the narrative attempts to fill the pages with as much action as possible, reminiscent of classical French comedies such as Beaumarchais's *La Folle journée ou le Mariage de Figaro*, where the constraints of theater required that all narrative action occur in a single twenty-four-hour period. In the original Hetzel edition (initially serialized in Hetzel's magazine *Magasin d'Éducation et de Récréation* and then with the same illustrations in book form) of *Vingt mille lieues sous les mers*, the two maps of the *Nautilus*'s journey (out of 111 total illustrations) divide up the oceans by hemisphere and the narrative into two parts. Each map is placed about a third of the way into each half of the journey, meaning that the reader knows in advance where the submarine will travel. But while the maps show the world as it was conventionally understood in the mid-nineteenth century, none of the fictional mysteries of the deep that constitute the adventures of the submarine can be represented on the map. Tellingly, the second map, showing the Western Hemisphere and the end of the voyage, is placed right before the chapter titled "Un continent disparu" about the mythical civilization of Atlantis, a continent missing on the map (figure 11.2). And while the map portrays the submarine's itinerary, this is yet another spatialized representation of time, and one that does not include the mysteries recounted in the narrative.

In the submarine, the experience of time and space varies according to the subjects perceiving them: Captain Nemo looks to the infinite even as he supplies aid to revolutionary struggles; Arronax (the scientist and narrator) is torn between his desire to explore forever the vastness of the sea and his loyalty to the other prisoners; the harpooner Ned Land revolts against his captivity and finds every hour on board a torment. The inability of the *Nautilus* to produce its own oxygen punctuates the narrative with periodic returns to the surface and occasional panic. The ideal reader, too, is drawn into the fast pace of the serial narrative, but knows from the title and the maps that it must come to an end.

While these representations of speed (a configuration of time and space) structure the narrative, other moments in the text emphasize the irreducibility of time to space, of identity to place, of historical event to understanding. Nemo seems to revel in perplexing his captive guests by connecting heterogeneous places, as when he finds a shortcut under the Sinai Peninsula or when he takes them to an underwater volcanic crater or when he takes them on an underwater hunt. His discoveries of the lost Atlantis or the wreckage of the French Revolutionary ship *Le Vengeur* suggest that, at least in the otherworldliness of the deep ocean, time is in suspension and resists any attempt at appropriation or understanding. The novel is in the form of a memoir by Arronax that functions as a corrective to his earlier scholarly account of the oceans, acknowledging in a more poetic style the limits of scientific knowledge. Nemo, too, has written a manuscript, "in several languages," revealing his name, his story, and all the

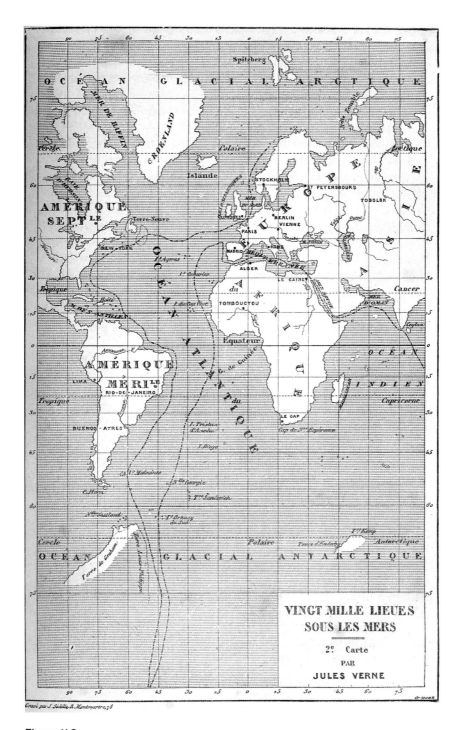

Figure 11.2

Vingt mille lieues sous les mers, Jules Verne, 2e carte. Wikimedia Commons (public domain)

secrets of the oceans, which he will one day place in a sealed, floating container to drift in the open seas where it may or may not ever be found.[15] This manuscript, written by an author whose name means "Nobody," works as a sort of shadow text to the one we are reading, promising fantastic tales and a virtual map of a world that lies just beyond our imagination.

In the much more satirical novel *De la terre à la lune*, Verne's irony reaches escape velocity as he reveals the inextricable links between military technology, political will, and the quest for ever-greater speed. The novel recounts how at the end of the American Civil War, the weapons enthusiasts of the Gun Club attempt to shoot a projectile at the moon, since they fear there will no longer be any wars in the near future. The first half of the novel is split between the details of this feat of engineering (the velocity that must be attained, the size of the cannon, the amount of gunpowder) and the poetic history of our relationship with the moon (superstitions, other fantastic journeys to the moon, lunatics). Though amazingly prescient about the realities of space travel (at least in regard to the Apollo missions), the most remarkable aspect of the novel occurs in the second half, with the arrival of the French adventurer Michel Ardan, based on Verne's friend the photographer, inventor, and death-defying balloonist Nadar. In an inspired speech to the Gun Club, Ardan demonstrates a keen understanding of the trajectory of modern technological progress that prefigures the contemporary thinker Paul Virilio, declaring that in the near future, "Distance is only a relative word, and will end up being reduced to zero!"[16]

Ardan takes the American ballistics engineers' unspoken dream to its logical extreme by declaring that he will travel in the projectile and that he plans not to return, either because he will colonize the moon or die trying. The novel's title then takes on a darkly humorous meaning, since his trip to the moon may be a one-way journey. His suicide mission reveals not just the lunacy behind the technological calculations, but also the beauty in exploring the unknown and the death drive of scientific progress. The novel's detailed maps of the moon and its precise calculations of the projectile's voyage cannot predict what will happen to the lunar explorers, leaving their adventures to the imagination. At the end of the novel, everyone on earth waits anxiously for the skies to clear in front of a mountain observatory only to find out that the capsule is stuck orbiting around the moon. The reading public had to wait six years for the novel's sequel to discover how the astronauts returned. Verne's novel map of the moon succeeded as a precursor both to the real manned missions to the moon and to such fictional fantasies of space and time as Italo Calvino's *Cosmicomiche*.

The Novel as Map

The monumental scale of the ambitious novels and novel series in the long nineteenth century generated the need for novel maps functioning as maps of the novels themselves. The ninety-odd works in Balzac's *La Comédie humaine*, the twenty novels of Zola's *Rougon-Macquart*, and the seven volumes and thousands of pages of Proust's *À la recherche du temps perdu* created vast textual spaces in elaborate semifictional worlds that required a system of transversal connections between texts that could articulate the interactions and movements of recurring characters. These spatial structures, distinct from the references to real places such as Paris or Venice, make the organic unity of each series possible since they fuse the separate texts together. But these same structures, as novel maps, disrupt any linear reading across texts, any absolute certainty about the identity of the characters, whose fictional lives develop amid the unwritten gaps of the texts beyond narrative.

As the model for all the grand novelistic projects in the nineteenth century, Honoré de Balzac's *La Comédie humaine* ingeniously deploys countless recurring characters to project the supreme illusion of his textual humanity inhabiting a textual France. In 1842, he wrote a foreword to the novel series thirteen years after the publication of his first signed novel, laying out the "plan" (both project and map) of his grand vision in order that his critics could understand the unity of the *Comédie humaine* even as it was being composed (like Proust, Balzac would continue adding to his work until his death). Balzac writes that he was inspired by how the natural sciences of his time had found the "unity of composition" in the animal kingdom, where "there is only one animal."[17] Loosely adopting naturalist Geoffroy Saint-Hilaire's protoevolutionary theories, Balzac extrapolates that what differentiates people, like animals, is their environmental pressures. His novels would then describe all the diversity of "Social Species," which necessarily develop in the large but fixed number of social milieus and situations possible in the modern world.

Balzac declares that he fills his novels with "men, women, and things, which is to say persons and the material representations they give of their thoughts."[18] The representation of "two or three thousand" figures or types that present themselves in a generation, according to Balzac, entailed finding the proper "cadres" or frameworks to organize them hierarchically into three main sections: "Studies of Manners" with its six subcategories (such as "Scenes from Private Life" and "Scenes from Political Life"), "Philosophical Studies," and "Analytical Studies."[19] These frameworks allow for a natural classification of the novel as well as the repetition of certain "phases" of life or typical situations that also create a national and a personal geography: "My work has its geography in the same way that it has its genealogy and its families, its places and its things, its persons and its facts"[20]

While the "natural divisions" mapped out in his foreword help the reader see the connections between the disparate texts, Balzac's human comedians resist as much as possible whatever frameworks are imposed on them. In his strange tale *Sarrasine* (famously analyzed in Roland Barthes's *S/Z*), Balzac presents a framed narrative in which the narrator attempts to seduce a woman by telling the story of a French sculptor, Sarrasine, who falls in love with an opera singer, la Zambinella, only to find out that this ideal woman is in fact a castrato. The anonymous narrator begins the tale straddling a windowsill at a party, half his body freezing from the snowy night outside, half his body warmed by the revelry in the palace. The narrator's numb leg foreshadows la Zambinella's castration and explains why he is able to recount Sarrasine's story, since he understands it both from the inside and the outside. Sarrasine's fatal inability to imagine la Zambinella as a castrato derives from his reduction of the singer to superficial signs that he mistakes for rigid gender categories. The final lines of the story have the narrator's interlocutor, Madame de Rochefide, mysteriously exclaim that "no one will have understood me!"[21]

The most protean of all of Balzac's characters, Vautrin, moves almost effortlessly between social categories and geographic borders. He first appears in *Le Père Goriot* as a gregarious middle-aged man who lives in a bourgeois pension house and who knows everything about everyone. At the end of the novel, the police capture him and reveal that he is the dangerous criminal Jacques Collin, known as "Trompe-la-mort" ("death cheater"). He shows up again years later as the Spanish priest Abbé Herrera, who saves a young man from drowning and uses him to swindle a rich banker out of a fortune. When he is finally caught, he manages to scheme his way to becoming a police informant and eventually chief of the Paris police. The character who most successfully defies classifications thus becomes the enforcer of social order. The diversity of Balzac's work is also the geographic diversity of Vautrin's Paris, where the criminal underworld exists superimposed on the seemingly respectable world of the social climber Rastignac or the honest judge Popinot. The novels divide themselves into mappable categories, but the texts share the same virtual space where characters meet while living incommensurable lives.

Later in the century, Émile Zola would take Balzac's formula to its logical extreme by narrating how a single, flawed family, the Rougon-Macquart, progressively came to occupy all the different social strata of its time. Just as Balzac implicitly integrates the novel's classification system into the narrative with Vautrin's promotion to chief of police, Zola's "genealogical tree" mapping the origins and destinies of each member of the family, and each novel in the series, becomes the diegetic focus of the last novel in the series, *Le Docteur Pascal*. Marcel Proust, likewise, organizes the volumes of his novel and his characters based on the spatial division between "Swann's Way" and "the Guermantes Way." *À la recherche du temps perdu* depicts how the changing technologies of progress and war reshaped the social and physical landscapes of France. Proust reveals that the two ways do in fact connect, just as the narrator of the novel we are reading decides in the end to write a book. Proust's massive novel, combining all elements of the nineteenth-century novel map (the maps of the self we saw in Staël's work, the maps of progress we saw in Verne, and Balzac's novel as map), bridged the way for the modernist novels of the twentieth century.

The Novel Map in the Twenty-First Century

After the bewildering array of spatial inventions in twentieth-century novels, after the "spatial turn" in French theory, and with the general malaise pervading contemporary French society, a nostalgic look back at the novel map of the nineteenth century was inevitable. Celebrated and controversial novelist Michel Houellebecq's novel, *Soumission* has as its narrator a professor of nineteenth-century French literature.[22] Yet Houellebecq's previous novel, *La Carte et le territoire* (winner of the 2010 Prix Goncourt), is arguably much more a nineteenth-century novel, not only because of its many intertextual references to Nerval, Stendhal, Verne, and Balzac, but because of its portrayal of what a novel map might look like in the twenty-first century.

La Carte et le territoire follows the long life of the visual artist Jed Martin, who gets his artistic break from an exhibition juxtaposing Google Maps satellite images with photographs of the same locations on Michelin regional maps. Martin emphasizes the nostalgic quality of outmoded maps (and photography as an outmoded medium), highlighting the artisanal detail of the maps, which is lost in GPS technology. The title of his exhibition (sponsored of course by Michelin) is "THE MAP IS MORE INTERESTING THAN THE TERRITORY."[23] Instead of the impersonal and pixelated view from space, Martin's photos (one assumes, since there are no images in the novel) reveal the materiality of the map as paper and his own perspective as photographer (figure 11.3).

The novel is also obsessed with the most recent reconfiguration of French space, which has abandoned its industrial heritage for the ravages of global tourism. France, the most visited country in the world, sells a hollowed-out vision of its past to rich foreign tourists, but cannot conjure up an image of what its future will be.

Two-thirds of the way into the novel, Houellebecq himself appears, the subject of an artwork by Martin. This strange *mise-en-abyme* of the author in his own work abruptly comes to an end when Houellebecq is found cut up into thousands of tiny pieces and the novel turns into a detective story in the vein of Balzac or Poe. Just as Staël and Proust had scattered portions of their identity across their texts, Houellebecq's gruesome murder leads the reader to question where exactly the author is to be found. The novelist was accused, correctly, of plagiarizing large sections of his novel from *Wikipedia*, but could face no charges because of the atomized nature of authorship on the website, where no one owns copyright. As his character Jed Martin says, "I think that I am more or less done with *the world as narration*—the world of novels and of films, the world of music too. I'm only interested in *the world as juxtaposition*—the world of poetry, of painting."[24] Houellebecq's genius may not be in his style, but in his ability to juxtapose, reframe, and remap the language that surrounds us in a way that allows us to see the inanity of contemporary discourse.

In an epilogue, the narrator describes Jed Martin's final monumental project to "account for the world" ("rendre compte du monde").[25] Taking over thirty years of his life, like Balzac or Proust's vast works, Martin creates dozens of "vidéogrammes" where he juxtaposes and superimposes ninety-six separate video tracks on a powerful computer. The description of his art here approaches a novel map, since he is said to film time-lapse video of computer motherboards (in French, "cartes-mères" or "mother maps") as they are gradually dissolved in acid. Martin layers onto these images video of his property in rural France and then photographs of people he has known. His project melds and modulates technology, nature, and personal memories that both look back to nineteenth-century industrialism and forward to a time in the future when, as he says, "The triumph of vegetation is total" ("Le triomphe de la végétation est totale").[26] Eerily reminiscent of Balzac's summary of Charles Bonnet's naturalist theory in the "Avant-propos" to the *Comédie humaine*, "The animal vegetates like the plant,"[27] Martin's (or Houellebecq's) assertion could either imply that human technological prowess will eventually be overtaken by an impersonal nature, or rather, following Balzac, that man's animal need to define a territory, to "vegetate," will take precedence over the radical deterritorialization that has defined modernity.[28]

Houellebecq's revisiting of the nineteenth-century novel deploys the concept of the novel map as a way to affirm the need for an artistic territory in the face of ever-accelerating spatial homogenization and technological change. While Houellebecq's nostalgia minimizes the social disruptions brought on by the political and industrial redistribution of space in the nineteenth century, his work proves that the novel map continues to be a powerful way to both represent and resist dominant spatial discourses. *La Carte et le territoire* demonstrates the ubiquity of the novel map in literary production from 1800 until the present, suggesting that an alternative to the literary history of movements and authorial genealogies would be one that studied the varied and often contradictory literary responses to the ever-changing configurations of space. Instead of looking for a supposed break caused by the modernist novel (perhaps Proust or Joyce), the new novel (Butor), or contemporary fiction (Houellebecq), the study of cartography in literature, of how writers invent novel maps, could find continuity in the ways literature apprehends, represents, and distorts space, allowing for a greater repertory of responses to our contemporary technological and spatial challenges.

Figure 11.3

I have tried to recreate the fictional artwork with my own photograph of the Grand Ballon de Guebwiller from a Michelin regional map above, and on the next page from Google Maps, courtesy Google Maps, 2011 TerraMetrics.

Figure 11.3 (continued)

Notes

1. Madame de Staël, *De la littérature*, ed. Gérard Gengembre and Jean Goldzink (Paris: GF Flammarion, 1991). All translations are my own.

2. "The inspiration [for the division of France into *départements*] was cartographic. The prime mover on the constitutional committee, Abbé Sieyès …, explained that he would begin 'by obtaining the great triangulated map of Cassini, which has without dispute the most exact positions; I would divide it first of all geometrically'" (Michael Biggs, "Putting the State on the Map: Cartography, Territory, and European State Formation," *Comparative Studies in Society and History* 44, no. 2 (April 1999): 389).

3. See David Harvey, *Paris, Capital of Modernity* (New York: Routledge, 2003).

4. See Henri Lefebvre, *La production de l'espace*, 4th ed. (Paris: Anthropos, 2000); Stephen Kern, *The Culture of Time and Space: 1880–1918* (Cambridge, MA: Harvard University Press, 1983); Eugen Weber, *Peasants into Frenchmen: The Modernization of Rural France, 1870–1914* (Stanford, CA: Stanford University Press, 1976).

5. Jacques Rancière, *La Parole muette: Essai sur les contradictions de la littérature* (Paris: Hachette Littératures, 1998), 13. While Staël's essay is often mistakenly credited with single-handedly inventing the modern notion of literature, her work certainly participated in the romantic aesthetic movement and her use of the word *literature* in its modern meaning contributed to the spread of the concept. See Jean-Luc Nancy and Philippe Lacoue-Labarthe, *L'absolu littéraire: Théorie de la littérature du romantisme allemande* (Paris: Seuil, 1978).

6. "[La carte] est un mélange problématique, où la transparence de l'illusion référentielle coexiste avec l'opacité d'un support qui matérialise cette image" (Christian Jacob, *L'Empire des Cartes: Approche théorique de la cartographie à travers l'histoire* (Paris: Albin Michel, 1992), 41).

7. Michel de Certeau, *L'invention du quotidien 1: Arts de faire* (Paris: Gallimard, 1990), 172–173.

8. Patrick M. Bray, *The Novel Map: Space and Subjectivity in Nineteenth-Century French Literature* (Evanston, IL: Northwestern University Press, 2013).

9. Louis Marin identifies the illusion of autobiography as a simulacrum unifying the I of the past (the object of writing) with the I that writes in the present (the subject) in "The Autobiographical Interruption: About Stendhal's Life of Henry Brulard," *Modern Language Notes* 93, no. 4 (1978): 597–617. See also E. S. Burt, *Regard for the Other: Autothanatography in Rousseau, De Quincey, Baudelaire, & Wilde* (New York: Fordham University Press, 2009).

10. Marie-Claire Vallois, "Voice as Fossil: Madame de Staël's *Corinne ou l'Italie*: An Archaeology of Feminine Discourse," *Tulsa Studies in Women's Literature* 6, no. 1 (Spring 1987): 51.

11. Madame de Staël, *Corinne ou l'Italie*, ed. Simone Balayé (Paris: Gallimard, 1985), 194.

12. David F. Bell, *Real Time: Accelerating Narrative from Balzac to Zola* (Urbana: University of Illinois Press, 2004).

13. "Il y a là [dans les sciences commençantes] une marge qui leur appartient [aux poètes], entre la vérité conquise, définitive, et l'inconnu, d'où l'on arrachera la vérité de demain" (Émile Zola, *Les Rougon-Macquart*, vol. 5, ed. Henri Mitterand (Paris: Gallimard, 1967), 1008).

14. Roland Barthes, *Mythologies: Édition illustrée*, ed. Jacqueline Guittard (Paris: Seuil, 2010), 102-3.

15. Jules Verne, *Vingt mille lieues sous les mers* (Paris: Librairie Générale Française, 1990), 553.

16. Jules Verne, *De la terre à la lune* (Paris: Livre de Poche, 2001), 237. Paul Virilio, too, has claimed that speed of communication and of military technology has nullified space, notably in Paul Virilio, *Vitesse et politique: Essai de dromologie* (Paris: Galilée, 1977), *L'Inertie polaire* (Paris: Christian Bourgois, 1990), and *Ce qui arrive* (Paris: Galilée, 2002).

17. Honoré de Balzac, *La Comédie humaine*, ed. Pierre-Georges Castex (Paris: Gallimard, 1976), vol. 1, 8.

18. Honoré de Balzac, *La Comédie humaine*, ed. Pierre-Georges Castex (Paris: Gallimard, 1976), vol. 1, 9.

19. Honoré de Balzac, *La Comédie humaine*, ed. Pierre-Georges Castex (Paris: Gallimard, 1976), vol. 1, 18.

20. Honoré de Balzac, *La Comédie humaine*, ed. Pierre-Georges Castex (Paris: Gallimard, 1976), vol. 1, 18-19.

21. Honoré de Balzac, *La Comédie humaine*, ed. Pierre-Georges Castex (Paris: Gallimard, 1977), vol. 6, 1076.

22. Michel Houellebecq, *Soumission* (Paris: Flammarion, 2015).

23. "LA CARTE EST PLUS INTÉRESSANTE QUE LE TERRITOIRE" (Michel Houellebecq, *La Carte et le territoire* (Paris: Flammarion, 2010), 82).

24. "Je crois que j'en ai à peu près fini avec *le monde comme narration*—le monde des romans et des films, le monde de la musique aussi. Je ne m'intéresse plus qu'au *monde comme juxtaposition*—celui de la poésie, de la peinture" (Houellebecq, *La Carte et le territoire*, 258–259).

25. Houellebecq, *La Carte et le territoire*, 420.

26. Houellebecq, *La Carte et le territoire*, 428.

27. Honoré de Balzac, *La Comédie humaine*, ed. Pierre-Georges Castex (Paris: Gallimard, 1976), vol. 1, 8.

28. As Gilles Deleuze and Félix Guattari have theorized, "Art begins perhaps with the animal, at least with the animal that carves out a territory and makes a house" ("L'art commence peut-être avec l'animal, du moins avec l'animal qui taille un territoire et fait une maison") (*Qu'est-ce que la philosophie?* (Paris: Éditions de Minuit, 1991), 174).

Bibliography

de Balzac, Honoré. *La Comédie humaine*. Vol. 1. Ed. Pierre-Georges Castex. Paris: Gallimard, 1976.

de Balzac, Honoré. *La Comédie humaine*. Vol. 6. Ed. Pierre-Georges Castex. Paris: Gallimard, 1977.

Barthes, Roland. *Mythologies: Édition illustrée*. Ed. Jacqueline Guittard. Paris: Seuil, 2010.

Bell, David F. *Real Time: Accelerating Narrative from Balzac to Zola*. Urbana: University of Illinois Press, 2004.

Biggs, Michael. "Putting the State on the Map: Cartography, Territory, and European State Formation." *Comparative Studies in Society and History* 44, no. 2 (April 1999): 374–405.

Bray, Patrick M. *The Novel Map: Space and Subjectivity in Nineteenth-Century French Literature*. Evanston, IL: Northwestern University Press, 2013.

Burt, E. S. *Regard for the Other: Autothanatography in Rousseau, De Quincey, Baudelaire, & Wilde*. New York: Fordham University Press, 2009.

de Certeau, Michel. *L'invention du quotidien 1: Arts de faire*. Paris: Gallimard, 1990.

Deleuze, Gilles, and Félix Guattari. *Qu'est-ce que la philosophie?* Paris: Éditions de Minuit, 1991.

Harvey, David. *Paris, Capital of Modernity*. New York: Routledge, 2003.

Houellebecq, Michel. *La Carte et le territoire*. Paris: Flammarion, 2010.

Houellebecq, Michel. *Soumission*. Paris: Flammarion, 2015.

Jacob, Christian. *L'Empire des Cartes: Approche théorique de la cartographie à travers l'histoire*. Paris: Albin Michel, 1992.

Kern, Stephen. *The Culture of Time and Space: 1880–1918*. Cambridge, MA: Harvard University Press, 1983.

Lefebvre, Henri. *La production de l'espace*. 4th ed. Paris: Anthropos, 2000.

Marin, Louis. "The Autobiographical Interruption: About Stendhal's Life of Henry Brulard." *Modern Language Notes* 93, no. 4 (1978): 597–617.

Nancy, Jean-Luc, and Philippe Lacoue-Labarthe. *L'absolu littéraire: Théorie de la littérature du romantisme allemande*. Paris: Seuil, 1978.

Rancière, Jacques. *La Parole muette: Essai sur les contradictions de la littérature*. Paris: Hachette Littératures, 1998.

de Staël, Madame. *Corinne ou l'Italie*. Ed. Simone Balayé. Paris: Gallimard, 1985.

de Staël, Madame. *De la littérature*. Ed. Gérard Gengembre and Jean Goldzink. Paris: GF Flammarion, 1991.

Vallois, Marie-Claire. "Voice as Fossil: Madame de Staël's *Corinne ou l'Italie*: An Archaeology of Feminine Discourse." *Tulsa Studies in Women's Literature* 6, no. 1 (Spring 1987): 47–60.

Verne, Jules. *De la terre à la lune*. Paris: Livre de Poche, 2001.

Verne, Jules. *Vingt mille lieues sous les mers*. Paris: Librairie Générale Française, 1990.

Virilio, Paul. *Ce qui arrive*. Paris: Galilée, 2002.

Virilio, Paul. *L'Inertie polaire*. Paris: Christian Bourgois, 1990.

Virilio, Paul. *Vitesse et politique: Essai de dromologie*. Paris: Galilée, 1977.

Weber, Eugen. *Peasants into Frenchmen: The Modernization of Rural France, 1870–1914*. Stanford, CA: Stanford University Press, 1976.

Zola, Émile. *Les Rougon-Macquart*. Vol. 5. Ed. Henri Mitterand. Paris: Gallimard, 1967.

12
African Cartographies in Motion

Dominic Thomas

As a child, I harbored a deep fascination for place names. I expanded my inventory of capital cities by combing through atlases, scrutinizing maps, collecting stamps, and poring over travel magazines. To this day, I continue to devote inordinate amounts of time when traveling to the available in-flight software that tracks where the flight is. My interest in location has not diminished, but one significant shift has occurred. Whereas the early focus was on *where places were*, my energies are now channeled toward improving my understanding as to *why they are there*. To this end, the entangled nature of African and European history and long-standing mobility between these continents, and in particular the ways these have shaped literary production, provide especially rich terrain for the process of exploring these questions. As Achille Mbembe has claimed, there is a growing "awareness of the imbrication of the here and the elsewhere and the presence of the elsewhere in the here and vice versa."[1] This newfound realization can be explained by "Africa's entry into a new era of dispersion and circulation, characterized by the intensification of migration and the implantation of new African diasporas around the world. With the emergence of these new diasporas, Africa no longer itself stands alone as a center. Today, it is made up of several centers between which there is constant *passage*, circulation, and facilitation."[2] Establishing the new coordinates of this relationship is therefore all the more challenging, precisely because of its multidimensionality. To explore a range of concrete and imagined cartographies that pertain to emerging African globalities, our attention will turn to a broad range of cultural, historical, literary, and political elements that have been graphically represented and physically mapped.

A significant component of nation building entails establishing, contesting, and defending national territories, only subsequently augmenting these mechanisms through the implementation of control and surveillance strategies that ultimately make it possible to monitor and safeguard these frontier spaces. The continent of Africa is no different when it comes to this process, although the legacy of instituting autonomous nation-states is inextricably linked to a broad range of cartographic mechanisms that contributed to varying degrees to the reorganizing of existing communities and to the tracing of national boundaries, the contours of which have for the most part survived these earlier exercises. Overwhelming evidence confirms the unprecedented nature of global interpenetration, underscoring the symbiotic (but also tenuous) connections between the *here* and the *elsewhere*. Indeed, if the Berlin Congress of 1884–1885 reaffirmed the centrality of Europe and accepted contours of the African continent, then more recent historical developments, including the Schengen Agreement, the abolishing of common border control, and militarized devices adopted to safeguard Fortress Europe, have been at odds with geopolitical realignments and shifts in planetary circulation both within continents *and* between regions.[3]

These realities are themselves anchored in much longer histories that include expeditions, European overseas conquest and expansion, anticolonial struggle, political independence, postcolonial nation building, and immigration. The underlying forces—curiosity, passion for adventure, greed, competition—fueled these expansionist drives whose realization was now conceivable, feasible, and even attainable as a result of momentous technological advances. Systematic remappings of the world gradually took place, achieved by stimulating interest in the elsewhere, enlisting support for the implementation of policy, and motivating individuals to accomplish these aspirations. Through a consideration of colonial games, literature, and museological practices, I propose to situate some of these historical antecedents to the globalized societies of the twenty-first century, considering how migration from formerly colonized territories to European metropolitan centers and new diasporic formations and networks have transformed the coordinates of the African literary landscape, thereby inducing observers to revisit and rethink previous assumptions pertaining to the *here* and the *elsewhere*.

These questions were of paramount importance in a manifesto published in France in 2007—the "Manifeste pour une 'littérature monde' en français" ("Manifesto for a 'World Literature' in French")—which was "an attempt to alter this dynamic, calling for a new transnational world-literature in the French language, open to the world ... with the centre placed on an equal plane with other centres [and] in which language [would be] freed from its exclusive pact with the nation."[4] This manifesto thus provided a much-needed opportunity to evaluate the terrain of publishing, drawing attention to the peripheral status and underrepresentation of certain authors and regions, while concurrently evaluating the magnitude of global circulation and distribution. But of course, power relations can be inscribed over a much longer transcolonial historical framework, introducing dynamics that in many ways now serve as precursors to discussions on globalization. For analogous reasons, these questions are central to the novel *Congo Inc.: Le testament de Bismarck* (2014) by In Koli Jean Bofane, a writer from the Democratic Republic of the Congo who lives in Belgium.[5] The subtitle *Le testament de Bismarck*—Bismarck's "last will" or "testament"—establishes a connection to the eponymous congress organized from November 15, 1884, to February 26, 1885, in response to Bismarck's initiative, a gathering that transformed the global cartography of power relations and proved to be a determining moment in African and European history.

Literature and cartography interact in both the colonial and postcolonial eras, and literature is especially noteworthy from a cartographic point of view in the African context. As one endeavors to enter the coordinates of African literary production today, questions of latitude, longitude, horizontality, and verticality remain very much at the forefront. Shifts in global migration patterns, forced population displacement, and disquieting images of an escalating humanitarian crisis as the outcome of lingering economic asymmetries have also had a dramatic impact on literary production. Globalization has resulted in what Helon Habila has described as a new generation of "post-nationalist" writers, engaged with twenty-first century questions.[6] African writing today is thus multisited and truly globalized, written and published on the African continent but also in diasporic frameworks, and African writers reside in *and* write about the cultural, political, and social realities of France, the United Kingdom, Belgium, Austria, Germany, and so on; a cursory look at recently published authors corroborates this, with for example In Koli Jean Bofane, Théo Annénisoh, Fatou Diome, Fiston Mwanza Mujila, Max Lobe, Chika Unigwe, Sami Tchak, Wilfried N'Sondé, Elizabeth Tchoungui, Léonora Miano, or Brian Chikwava.[7] A similar phenomenon can also be observed in the United States, where a growing number of African writers are living and writing (including Chris Abani, Ngũgĩ wa Thiong'o, Wole Soyinka, Teju Cole, Helon Habili, Zakes Mda, Alain Mabanckou, Abdourahman A. Waberi, Emmanuel Dongala, Patrice Nganang, and Dinaw Mengestu), or moving farther afield to Australia, Cuba, or Mexico, among other places of course.[8] Critical categories have burgeoned in an attempt to circumscribe these transitions and globalities. Among these, one finds *Afropeanism*,[9] which corresponds to "the recognition of a belonging to Europe, but above all to the Europe of tomorrow, that Europe whose history is being written right now. ... It is the unavoidable entry of the European component in the diasporic experience of peoples of sub-Saharan African ascendance."[10] But the notion of the *Afropolitan* is also gaining currency as a way of delimiting "the newest generation of African emigrants ... not citizens, but Africans of the world,"[11] who "belong to no single geography, but feel at home in many. ... Ultimately, the Afropolitan must form an identity along at least three dimensions: national, racial, cultural—with subtle tensions in between."[12] Before returning to these and other questions, I would like to first consider the role of circulation, movement, and travel in the process of inspiring cartographic imaginaries.

The Colonial Game

My interest in travel is of course fairly universal, and this is certainly not the place to delve further into the origins of my own particular attraction for the subject. However, this is an opportunity to explore the part travel played in colonial exploration and expansionism. In the French context, "Colonialism was seen as a mark of civilization, of national grandeur, of science and progress. The nation, which emerged out of the French Revolution, brought *liberty* and not *oppression, development* and not *exploitation,* to the peoples it was 'liberating.'"[13] However, the dual task of convincing the French population of the merits of overseas expansion and enlisting their support for the enterprise so as to transform France "from an exclusively hexagonal society (with the exception of a few colonial territories inherited from the Ancien Regime) to an imperial culture" was complex, and considerable propagandist resources were devoted to this.[14] School-age children were targeted, coupling *pedagogy* with *propaganda*, and thereby effectively shaping mentalities and mapping consciousness. These activities are confirmed in the following 1913 geography textbook:

> Let us insist right now upon the importance of emphasizing *our* colonial empire in your lessons on elementary geography. The colonies already play an important role in the economic life of our country; this shall only become more and more the case. It is thus essential that French youth be familiarized with the resources from the vast territory over which *our* flag waves. They must learn about the living conditions, their chances of success, and also the potential risks encountered by colonials in *our* overseas possessions.[15]

The molding of young minds in this indoctrination project was inseparable from broader nationalist imperatives, instilling a fervent sense of patriotic duty, such that "this blend of pedagogy, patriotism, and nationalism helped to cement the idea that colonialism was consubstantial to the Republic."[16] But the lessons to be learned were manifold, concerning as they did a recognition of the importance to the French economy *of*–and dependency *on*–colonies, implanting a deep familiarity with products and goods, while also stimulating the associated ingredient of adventure and danger. When German Chancellor Otto von Bismarck convened the *Kongokonferenz* (Berlin Congress), the partitioning and "scramble" for Africa that ensued corresponded to a veritable political game at which only the most adept, creative, inspired, motivated, and strategic at outsmarting and outmaneuvering the competition stood to benefit exponentially from various deals, negotiations, and treaties. This was much as in board games in which one has opponents, competition, and specific goals (victory and defeat), and in which the balance between strategy and chance is always fragile. In other words, all the elements–adventure, entertainment, and intrigue–necessary to maximize student interest in empire were thus deployed.

As Sandrine Lemaire has shown, the "Agence Générale des Colonies was developing treasures of ingenuity, made concrete by a multifaceted strategy of propaganda," with the consequence that

> from the spectacular but ephemeral expositions to the insertion of the colonial ideology into daily life through textbooks, food, games, and calendars, the empire was everywhere. At school, at home, or at work, the French were acquiring a specific

national and colonial culture. This was in part thanks to propaganda's manifold seductive charms. The best tactics were employed to convince the middle and working classes that the nation was not limited to metropolitan France—the Hexagon—but included "Greater France."[17]

With this context in mind, the role of maps becomes all the more important, especially given the extensive use of them in children's games. Maps provided people with information about the colonies, including representations of colonized peoples (many of which have survived into contemporary French society and are reflected in current mindsets),[18] while stimulating the imagination and appeal for remote spaces, concomitantly bolstering a sense of familiarity and appetite for adventure and travel in an era when such experiences were not widely available to the vast majority of people. The toys and games distributed during the interwar years "made the discovery and conquest of colonial lands fun," as Nicolas Bancel claims, while spreading "a clear political message on the empire and its people" and reaching into "the familiar and intimate universe of family, neighbors, and friend."[19] Furthermore, "This world introduced children to unheard-of adventure, through which they developed new relationships to space. … Conquest was a game, without any apparent consequences, a game that contributed to the construction of a mentality meant to prepare a young elite for the challenges of colonial life and the defense of the empire. These games of courage were preparing the youth to help reestablish France's power and conserve its empire."[20]

A vast array of games were produced, but I will restrict my focus to a selection from France and Germany. These include a two-sided board game featuring the *Jeu de l'Empire français* (The French Empire Game; see figure 12.1) and the *Course de l'Empire français* (The Race for the French Empire; see figure 12.2). These two games and a third, the *Jeu des échanges France–Colonies* (Trading Game; see figure 12.3), were all released in 1941. There was also a German colonial game from the 1910s, *Deutschland's Kolonien-Spiel* (German Colonial Game; see figure 12.4). These and other games were featured at the Getty Research Institute (GRI) in Los Angeles (December 7, 2013, to April 13, 2014) at the *Connecting Seas: A Visual History of Discoveries and Encounters Exhibit*, described as follows by Isota Poggi (GRI curator):

> Made in France at the outbreak of World War II, the game sought to educate children about the colonial world supporting the French economy. With tokens printed in vivid colors to represent places and natural resources in regions colonized by the French, from North Africa to Oceania to southeast Asia, this game encapsulated the mighty business opportunities that lay ahead for adventurous explorers willing to embark for faraway colonial lands. As described in the rules at the center of the board, the underlying purpose of the game was to admire, through play, the greatness of the French colonial undertaking. The colonization of a land was symbolically achieved first by hoisting the French flag on its soil, then by the establishment of a

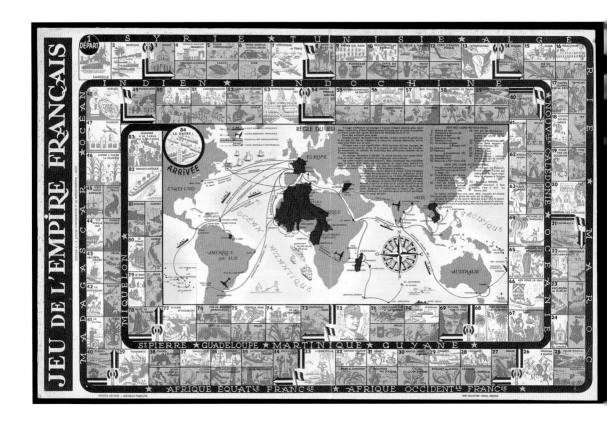

Figure 12.1

Raoul Auger, Imprimerie Delattre, 1941, *Jeu de l'Empire français* (The French Empire Game), two-sided game board, color lithograph, 32 × 50 cm, Getty Research Institute, Los Angeles

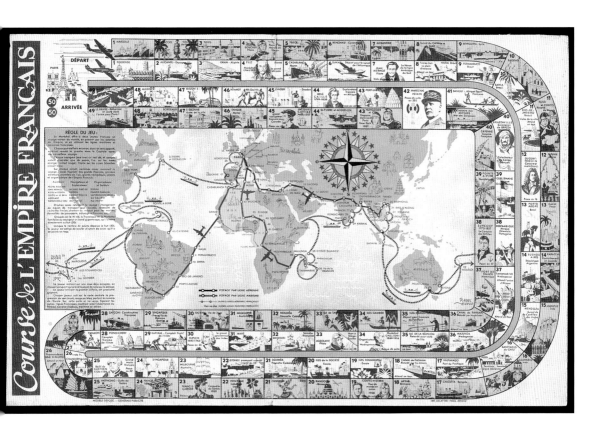

Figure 12.2

Raoul Auger, Imprimerie Delattre, 1941, *Course de l'Empire français* (The Race for the French Empire), two-sided game board, color lithograph, 32 × 50 cm, Getty Research Institute, Los Angeles

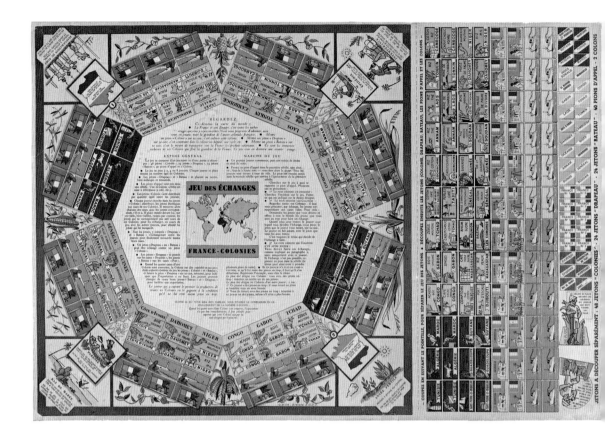

Figure 12.3

Jeu des échanges France–Colonies (Trading Game), 1941, O.P.I.M. (Office de publicité et d'impression), Breveté S.G.D.G. Lithograph on linen, 22 7/8 × 32 1/4 in., Getty Research Institute, Los Angeles.

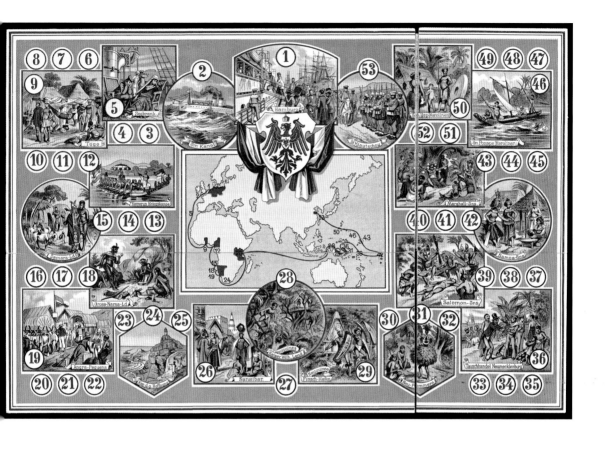

Figure 12.4

Deutschland's Kolonien-Spiel (Germany's Colonial Game), 1910s, Getty Research Institute, Los Angeles

hospital, a school, and ultimately a harbor. But the ultimate aim was to export the rich natural resources of the colonies back to France by boat. Images on the game provide a vivid picture of the vast variety of resources, including animals, plants, and minerals, that the colonies provided to France from all around the globe.[21]

Each game comprises two carefully demarcated circuits. In the *Jeu de l'Empire français*, red and white tiles slow down the player's progress while a privileged fast track, in blue, accelerates circuit completion. From the instructions, we learn that "each square is ascribed a particular meaning. In this way, as players make their way around the board, they will discover the beauties, resources, and attractions of imperial France, but also a series of obstacles." The journey begins in the French southern port city of Marseille, and eighty-four squares later, the victorious traveler arrives in the north of France in another port city, Le Havre, having covered an impressive physical distance over France's prodigious empire. Across the African continent, all the way to Indochina, across oceans, players familiarize themselves with the Temple of Angkor Wat, Tunis, Algiers, Dakar, Fez, Bamako, Saigon, Brazzaville, Djibouti, Pondicherry, and Tahiti, discover local products such as silk, rice, olive oil, iron, wine, tobacco, oranges, and lemons, while also being exposed to dramatic climatic variations, including torrential rainfall, high winds, and sandstorms, all the while overcoming a variety of challenges and travel delays (loss of identity papers, mechanical problems, missed connections).

In the *Course de l'Empire français*, young French people now living under Vichy-occupied France are generously offered by Marshal Pétain (to whom tile number 42 is devoted) the opportunity to travel around the world and visit the length and breadth of the French Empire, because in what corresponds to "a major, well-conceived, and organized marketing campaign directed at French youth, the Vichy regime created a 'kingdom' of distraction."[22] Two circuits (red or blue), including forty-nine tiles each, are available to players, each including "special" tiles that can speed up or slow down progress, tiles that feature examples of great French people, aviators, navigators, explorers, and soldiers of the French Empire. Players head out from Paris to either Marseille or Toulouse and end up in Marseille or Le Havre from where a plane will take them back to the capital (tile 50), having successfully navigated "France's impressive aerial and maritime network" (Rules of the Game). As with the *Jeu de l'Empire français*, players gain familiarity with French overseas possessions, "showcasing foreign locations, explorers and military figures, and products and businesses from the colonies. The two sides of the game show how the borders of the empire had, by the 1940s, come to be within close reach by sea and air,"[23] thereby narrowing the concrete and imagined gap between the *here* and the *elsewhere*.

In many ways, the individual tiles in these games resemble postage stamps, informed as they are by a similar representational logic, further strengthening the association with the metaphor of travel. The French colonial space is shaped by a conjunction of economic and political objectives, defined by aggressive extraction ambitions in the context of broader development goals. To this end, the games perfectly encapsulate official policies, encouraging identification and conscripting French youth to embrace these. Another game, the *Jeu des échanges France–Colonies* (Trading Game–French Colonies) is all the more convincing in this regard, conceptualized in such a way as to mainstream the ideals of the authorities as "players accumulate wealth by building a house, hoisting the French flag, and gradually expanding the larger infrastructure with hospitals, schools, and similar institutions to facilitate the export of products from the colonies. The game states: *It is the immense richness of the colonies that makes France great*" (GRI exhibition label).

In *Deutschland's Kolonien-Spiel* (Germany's Colonial Game), comprising fifty-three tiles, availability of the game is explained as a response to burgeoning interest in Germany's overseas activities: "The interest of German youth in our colonies grows every day and this game that offers exciting travel to our colonies is therefore most welcome." The port of Hamburg is the starting point for a riveting adventure, from where players depart on "a voyage from Germany via West Africa, around the Cape of Good Hope and across the Indian Ocean to be the first to arrive at Shanghai, where they are greeted triumphantly by a military parade" (GRI exhibition label). Much like the French context, it is the conjunction between adventure and patriotic ideals that heightens interest and inspires new generations to adhere to this vision, the upshot of which is German global dominance.

The French games we have considered were produced with the aim of promoting the grandeur of the French Empire, but when a German cigarette manufacturer in Dresden released a photomechanical print card collection album named *Deutsche Kolonien* in 1936, the images were instead commemorative, featuring "the German colonial empire lost as a consequence of World War I" and some 270 "color collecting cards [that are] pasted onto full-page black and white lithographed illustrations of colonial scenes, offering multiple glimpses of scenery, architecture, military action, indigenous peoples, artifacts, wildlife, and agricultural products of the various colonies."[24] Comparative analysis of the French and German contexts reveals the expediency of games in conditioning mindsets, disseminating information, and soliciting patronage and sponsorship for imperialist conceptions of territorial conquest as the basis of European supremacy.

Displaying Empire

Generations of young Europeans also responded enthusiastically to the adventures of Jean de Brunhoff's Babar the elephant, as indeed they did to the trials, tribulations, ordeals, and exploits of Belgian cartoonist Hergé's eponymous hero Tintin. These examples are best understood alongside a range of other practices that were instrumental in fostering greater identification *with* and interest *in* overseas expeditions, notably the International Colonial Exposition held in Paris in 1931. This was arguably "the century's primary showcase of Republican power"[25] and heralded "the advent of a resolutely modern and planned conception of colonial propaganda: an immense media campaign was launched a year before its opening, and week after week every newspaper related the progress that had been made. ... It was conceived with the specific goal of endearing French people to their colonial empire and of bringing together Western colonial powers through a shared glorification of the grandeur of the Occident."[26] Constructed for the International Colonial Exposition, the Palais de la Porte Dorée–located in the Bois de Vincennes on the eastern edge of Paris–has been the home of the Cité nationale de l'histoire de l'immigration (CNHI, National Center for the History of Immigration) since it opened in October 2007. The International Colonial Exposition, much like the global phenomenon of world's fairs, was a "larger-than-life metaphor for late twentieth-century writers trying to give shape and meaning to human experience as it unfolded in the twentieth century."[27] But the actual Palais de la Porte Dorée building, as Panivong Norindr writes, "was conceived as an architectonic colonial manifesto, a public and official display of French colonial policies, which determined its discourse, circumscribed its space, and revealed its ideology."[28] In many ways, the building itself, and in particular its remarkably elaborate façade, stand as precursors to the kind of visual interplay evidenced in the board-game designs we examined earlier. A plaque outside the CNHI offers some insight as to that history:

> At the request of the architect Albert Laprade, the sculptor Alfred Janniot (1889–1969) carved the colonial spirit out of stone and illustrated the Economic contribution of the colonies. Portrayed as Abundance, France sits on the throne at the top of the staircase. Colonial riches are spread at her feet, from Africa on her left and Asia on the right. For educational purposes, various inscriptions link each territory to the goods and raw materials it produces. ... The sculptural style of the 1920s and 1930s was used for imperial propaganda: muscular and imposing figures, wild and lush nature, ethnic faces that were recognizable and simplified according to the ethnographic codes of the time.

The irony of course of a former colonial palace, featuring colonized laborers hard at work in the process of enhancing France's grandeur has of course not been lost on critics, especially given that the space has been transformed into a museum of immigration, thereby seeking to encourage acknowledgment and recognition of the contributions made by both colonial subjects *and* postcolonial immigrants to the (re) building of France in the aftermath of two world wars.[29] Such stipulations are necessary in order to engage in the kind of reconditioning and deconstruction of mindsets that will pave the way for adequate reconsideration and reevaluation of the role of European societies in colonial history, in response to demographic changes in Europe's (postcolonial) population that have been accompanied by a growing desire to address colonial history, while also allowing for the systematic assessment of intricate legacies.[30]

Many of the devices at work in colonial-era modes of representation—games, international expositions, architecture, advertising, and so forth—relied heavily on the kind of iconography found today, whereby the "habits and spaces [of ethnic minorities] continue to be represented by fruits and scents in many advertisements, be they for tourist destinations or the display of cacao, bananas, coconut trees, vanilla, or sunny beaches. These representations send the nonwhite French back to the geographic, climatic, or cultural causes of their failure to integrate into the nation."[31] The disquieting tone of recent political rhetoric in the European Union and the persistent association of nonwhite populations with the *elsewhere* is indicative of the spectral presence of these earlier paradigms and perceptions of belonging.

Remapping the Traces and Remnants of Empire

The European landscape offers an overabundance of monuments to fallen soldiers in colonial wars and dedicated sites celebrating imperial expeditions and exploits analogous to those constructs staged in colonial-era games.[32] In Berlin, Carl Hagenbeck, one of the main organizers of animal and human exhibitions beginning in the 1870s,[33] wanted to establish a permanent exhibition in Berlin's Wedding-Rehberge district. Although this project to display Germany's African colonies did not come to fruition, an elaborate program of street naming did take place in the area. The neighborhood is known today as the Afrikanisches Viertel (African Quarter), and commemorates German colonies, explorers, and conquerors, thereby transforming the landscape of the colonial metropolis.[34] In the very city in which Chancellor Otto von Bismarck convened the historic *Kongokonferenz* that provided for the official partitioning of Africa, one can walk up and down Afrikanische Straße and adjacent and intersecting streets, including Transvaalstraße, Togostraße, Sansibarstraße, Senegalstraße, Ugandastraße, Sambesistraße, Kameruner Straße, Guineastraße, Ghanastraße, Kongo Straße, and Koloniestraße.

Various initiatives have been launched aimed at revisiting German colonial history, especially in the context of addressing a range of important issues pertinent to postcolonial Germany. A number of street-rededication campaigns have drawn attention to this history, an outcome of the activism of organizations such as *Berlin Postkolonial*.[35] Demands have been made for the authorities to engage in *Vergangenheitsbewältigung*—a process of "reckoning" aimed at coming to terms with the past and as a step toward "a 'decolonization' of Berlin street names, many of which are shameless references to the country's colonial past."[36] Of course, such practices are not limited to Berlin, and examples are also to be found on the African continent, where some cities have maintained colonial-era names (such as Brazzaville, Republic of the Congo, named after its founder, the explorer Pierre Savorgnan de Brazza), while others have abandoned these (such as Leopoldville, named in honor of the Belgian King Leopold II, renamed Kinshasa, Democratic Republic of the Congo) in a concerted effort to expunge those remnants of a colonial presence.

The contours that defined the *here* in opposition to some elusive *elsewhere* were of course premised on a unidirectional vertical relationship that relied on an intrinsic set of cultural and scientific discourses that helped shape perceptions of non-European spaces and peoples.[37] The dynamic only started to undergo structural adjustment with resistance and as the logical outcome of anticolonial struggle. The contours of the portrait of the elsewhere gradually emerged and were widely disseminated, and these depictions and perceptions were subsequently dismantled in such influential works as Edward Said's *Orientalism*.[38] Elsewhere, in literature, a range of remappings were also taking place, partially reversing colonial discourse, as African protagonists gradually began making the journey to European metropolitan centers. Ousmane Socé's 1937 novel, *Mirages de Paris*, is emblematic of this transition.[39] Fara, the main character, travels from Senegal to Paris to attend the International Colonial Exposition, discovering in the process those official mechanisms that had reconfigured the colonial space; in this construct, Africa has been relocated to Paris, and the "Africa" *in* "Paris" that Fara discovers is defined by a delicate if not precarious relationship since, as Christopher L. Miller has argued, "if the city of Paris has failed to conform to Fara's literary expectations, the exposition, in contrast, harmonizes perfectly with his old mirages."[40] Indeed, if "the pavilions of the exposition were designed to offer a mirage of the colonies, permitting a new and fantastic kind of trip"[41] for French visitors, as Fara himself navigates his way through the French Empire, he demystifies the foundations of this "state-sponsored hallucination,"[42] the premises of Republican ideals and values, and how these precepts have in turn shaped and justified expansionist motivations.

The bibliography of African texts addressing analogous issues and questions is considerable.[43] Socé's novel and "intercontinental" travel narrative constitute an important point on the literary map, to which numerous works would shortly thereafter be added,[44] most notably Laye Camara's *L'Enfant noir* (1954), Bernard Dadié's *Un nègre à Paris* (1959), and Cheikh Hamidou Kane's *L'aventure ambiguë* (1961), all of which contributed in significant ways to deconstructing the ambiguity of colonial myths and African perceptions of European colonial powers, exposing inconsistencies and inaccuracies in a discourse founded on an unquestioned occidental superiority.

Globalities: New African Identities and Locations

In the novel *Cruel City* (1954), Cameroonian author Mongo Beti maps out the imaginary colonial city of Tanga, structured around racialized categories and principles of exclusion. We thus learn that Tanga is subdivided along the following lines:

The other Tanga, the unspecialized part of the city, the Tanga to which the administrative buildings turned their backs—out of a lack of appreciation, no doubt—was the Tanga that belonged to the natives; this Tanga—of huts—fanned out over the Northern flank of the hill. This particular area of the city was divided into innumerable little neighborhoods, though these were actually just a series of little dips in the landscape, each of which had an evocative name. ... No one could say what the city would become, not the geographers, not the journalists, and even less the explorers.[45]

As it happens, these kinds of intentional cartographic alignments have been replicated in postcolonial urban contexts, as powerfully exemplified in numerous works focusing on the French *banlieues* (outlying urban housing projects), works in which the transcolonial connections are often rendered explicit. The main protagonist in Wilfried N'Sondé's *The Heart of the Leopard Children* (2007) remembers how his father, describing the city of Brazzaville during the colonial era, was "outraged" by "The separation between black and white neighborhoods. ... For example, he wondered why the white neighborhoods were off-limits to the black population after dark in what was a Black country."[46] Then, turning his attention to his current living conditions in France, he explains how "we spent all our holidays with the other youth and among the five buildings in the neighborhood, along the stretch of highway that ran past our project."[47] In fact, as Achille Mbembe has argued, "This treatment and these forms of humiliation, which were once tolerated only in the colonies, are now resurfacing in the metropolis itself; during sweeps and raids in the *banlieues,* they are applied not only to aliens—illegal immigrants and refugees—but increasingly to French citizens of African descent or the descendants of former African slaves."[48]

The novel *Congo Inc.: Le testament de Bismarck* by In Koli Jean Bofane mentioned earlier is thus especially relevant, because of its engagement with the intricate nature of postcolonial African history. Even though postcolonial works were often compelled to adopt imaginary topographic backdrops as a way of bypassing brutal repression and dictatorship, Bofane's novel relinquishes the fictitious framework and turns its attention instead to current political configurations (in the DRC, Rwanda, etc.) to tackle a broad range of issues that continue to impact the region as a result of civil conflict and political instability (human rights, gender inequities, genocide, ethnic conflict and cleansing, multinational corporate activity, environmental concerns, child soldiers, NGO activity). Indeed, if earlier "classics" of African literature published primarily during the 1950s and 1960s—such as *Things Fall Apart* (Chinua Achebe), *The Dark Child* (Camara Laye), and *The Ambiguous Adventure* (Cheikh Hamidou Kane)—examined the ways protagonists were forced to abandon traditional customs and practices identified primarily with rural existence and emigrate to colonial cities (and afterward to colonial metropolitan centers such as Paris, London, Lisbon, or Brussels) to gain social capital, then these "ambiguous" journeys yielded a newfound range of works during the

1980s and 1990s in which the focus of literature shifted to the postmigrant experiences associated with those attempts at achieving integration into European society. Bofane's novel certainly draws on this background, but his main protagonist, Isookanga, is instead lured away by advances in new technology as a result of the DRC's growing exposure to globalization from the setting of his rural upbringing to the centrifugal power of the capital, Kinshasa.

Colonial education and the French civilizing mission were the primary mechanisms deployed in the process of indoctrinating colonial subjects, but today Isookanga is able, from his remote rural village location, to "connect" to the "outside" world thanks to a large telecommunications tower installed by the Chinese. This makes it possible for him to devote his time to playing the online game *Raging Trade* under his avatar *Congo Bololo* (Bitter Congo). His adversaries—*Skulls and Bones*, *Mining Fields*, *Uranium and Security*, *American Diggers*—engage, to the soundtrack of American rapper and global icon Snoop Dogg, in relentless battles with the objective of amassing natural and territorial resources through rampant pillaging and senseless killing. When asked by his uncle "What is it exactly that you want to do," Isookanga responds: "Globalization, computer technology, Uncle."[49] Needless to say, drawing analogies between the logic informing the games we examined earlier and the imperatives of global multinational corporations and nation-states operating under newly established "good governance" clauses becomes a relatively straightforward step.

These (post)colonial globalized cartographies have produced new spatial and racial configurations as an outcome of such constitutive mutations, reconfiguring cultural and social environments, but also reshaping diasporic nexuses and systems in the process. Lingering disparities between the relatively prosperous areas of the world with those located in the global south persist and extend into the realm of literary appraisal, accessibility, and distribution.[50] Geostrategic features (waterways, climate, natural resources) explain *why* particular sites reached prominence as triumphant colonial powers set about exploitation. Corresponding factors can help elucidate the multifaceted ingredients of today's neocolonial configurations and globalized cartographies (access to cheap labor, tax incentives, emerging market yields, distribution networks, foreign direct investment determinants), at a historical moment when the *here* and the *elsewhere* have become imbricated in new and previously unanticipated ways. However, forms of situational apartheid, "territorial, social, and ethnic Apartheid" of the kind evoked by former French Prime Minister Manuel Valls in January 2015, continue to define human interaction. As Achille Mbembe has observed, "The foreigner is not only the citizen of another state. More significantly, he or she is different from us; his or her danger is real, for a genuine cultural distance separates us from him or her; in all these ways, the foreigner constitutes a mortal threat to our way of life."[51] Likewise, as Ursula Biemann's thought-provoking documentaries *Europlex* (2003) and *Sahara Chronicle* (2006–2009) confirm, political instability disproportionately impacts certain regions of the globe, migration patterns are unevenly distributed, and tremendous economic disparities also characterize twenty-first century existence.

These are the kinds of questions posed in Bofane's novel, where we learn that "The Chinese had gotten the point: overpopulated as they were, they had to find something to keep them busy. On the other hand, people on every continent wanted name brands so they could show off. Why not cater to them? Two billion arms could supply whomever they wanted, within whatever time they wanted, at the lowest possible cost. Nobody on earth could do better than that."[52]

Were one to design a twenty-first century version of the board games discussed earlier, new tiles would be needed, efforts to reconcile autonomous rule with development policies required, and imaginative display techniques implemented so as to accurately personify the kinds of infrastructure in place that define the transactions of multinational corporations and interactions with nation-states, how environmental priorities are addressed, while simultaneously accounting for the alarming assault on human rights. The type of committed literature the aspiring writer Lucien wishes to produce in Fiston Mwanza Mujila's pioneering novel *Tram 83* seeks to undertake this daunting task: "I think, unless I am mistaken, that literature deserves pride of place in the shaping of history. It is by way of literature that I can reestablish the truth. I intend to piece together the memory of a country that exists only on paper."[53]

Notes

1. Achille Mbembe, *Sortir de la grande nuit: Essai sur l'Afrique décolonisée* (Paris: Éditions La Découverte, 2010), 229.

2. Mbembe, *Sortir de la grande nuit*, 224.

3. See Dominic Thomas, "Fortress Europe: Identity, Race and Surveillance," in *Race, Violence, and Biopolitics*, ed. Alessandro Corio and Louise Hardwick, *International Journal of Francophone Studies* 17, nos. 3–4 (2014): 445–468.

4. "Pour une 'littérature monde' en français," March 15, 2007, http://www.lemonde.fr/livres/article/2007/03/15/des-ecrivains-plaident-pour-un-roman-en-francais-ouvert-sur-le-monde_883572_3260.html, cited here in "Toward a World Literature in French," trans. Daniel Simon, *World Literature Today* 83, no. 2 (March-April 2009), 54. See also Charles Forsdick, Alec G. Hargreaves, and David Murphy, eds., *Transnational French Studies: Postcolonialism and Littérature-monde* (Liverpool: Liverpool University Press, 2010).

5. In Koli Jean Bofane, *Congo Inc.: Le testament de Bismarck* (Arles: Actes Sud, 2014). Quotations are from *Congo Inc.–Bismarck's Testament*, trans. Marjolijn de Jager (Bloomington: Indiana University Press, forthcoming).

6. Helon Habila, "Introduction," in *The Granta Book of the African Short Story*, i–xvii (London: Granta, 2011).

7. See for example Paul Gilroy, *There Ain't No Black in the Union Jack: The Cultural Politics of Race and Nation* (Chicago: University of Chicago Press, 1987); Salman Rushdie, "The New

Empire within Britain," in *Imaginary Homelands: Essays and Criticism: 1981–1991*, 129–138 (London: Granta, 1991); Simon Gikandi, *Maps of Englishness: Writing Identity in the Culture of Colonialism* (New York: Columbia University Press, 1997); Mark Stein, *Black British Literature: Novels of Transformation* (Columbus: Ohio State University Press, 2004); Caryl Phillips, *The European Tribe* (London: Faber and Faber, 1987), and *Color Me English: Migration and Belonging Before and After 9/11* (New York: New Press, 2011); Ato Quayson, "Postcolonialism and the Diasporic Imaginary," in *A Companion to Diaspora and Transnationalism*, ed. Ato Quayson and Girish Daswani, 139–158 (Oxford: Blackwell, 2013).

8. See for example Dominic Thomas, "L'Afrique à l'université américaine," in *Penser et écrire l'Afrique aujourd'hui*, ed. Alain Mabanckou, 81-91 (Paris: Seuil, 2017).

9. See Sabrina Brancato, "Afro-European Literature(s): A New Discursive Category?," *Research in African Literatures* 39, no. 3 (2008): 1–13; Darlene Clark Hine, Trica Danielle Keaton, and Stephen Small, eds., *Black Europe and the African Diaspora* (Urbana: University of Illinois Press, 2009); Fatima El-Tayeb, *European Others: Queering Ethnicity in Postnational Europe* (Minneapolis: University of Minnesota Press, 2011); Nicki Hitchcott and Dominic Thomas, eds., *Francophone Afropean Literatures* (Liverpool: Liverpool University Press, 2014).

10. Léonora Miano, *Habiter la frontière* (Paris: Arche Éditeur, 2012), 86.

11. Taiye Selasi, "Bye-Bye Babar," March 3, 2005, http://thelip.robertsharp.co.uk/?p=76.

12. Selasi, "Bye-Bye Babar."

13. Pascal Blanchard, Sandrine Lemaire, Nicolas Bancel, and Dominic Thomas, "The Creation of a *Colonial Culture* in France, from the Colonial Era to the 'Memory Wars,'" in *Colonial Culture in France Since the Revolution*, ed. Pascal Blanchard, Sandrine Lemaire, Nicolas Bancel, and Dominic Thomas (Bloomington: Indiana University Press, 2014), 2.

14. Blanchard et al., "The Creation of a *Colonial Culture* in France," 3.

15. Joseph Fèvre and Henri Hauser, *Précis de géographie* (Paris: F. Alcan, 1913), 838, cited in Gilles Manceron, "School, Pedagogy, and the Colonies (1870–1914), in *Colonial Culture*, 124. Translation altered slightly.

16. Blanchard et al., "The Creation of a *Colonial Culture* in France," 16.

17. Sandrine Lemaire, "Spreading the Word: *The Agence Générale des Colonies (1920–1931)*," in *Colonial Culture*, 165–167.

18. See for example Adame Ba Konaré, ed., *Petit précis de remise à niveau sur l'histoire africaine à l'usage du président Sarkozy* (Paris: Éditions La Découverte, 2008); Makhily Gassama, ed., *L'Afrique répond à Sarkozy: Contre le discours de Dakar* (Paris: Éditions Philippe Rey, 2008).

19. Nicolas Bancel, "The Colonial Bath: *Colonial Culture* in Everyday Life," in *Colonial Culture*, 201–202.

20. Bancel, "The Colonial Bath: *Colonial Culture* in Everyday Life," in *Colonial Culture*, 207.

21. See Isotta Poggi, "Colorful Board Game Turns the French Colonies into Child's Play," February 24, 2014, http://blogs.getty.edu/iris/colorful-board-game-turns-the-french-colonies-into-childs-play.

22. Sandrine Lemaire, Catherine Hodeir, and Pascal Blanchard, "The Colonial Economy: Between Propaganda Myths and Economic Reality (1940–1955)," in *Colonial Culture*, 323.

23. Poggi, "Colorful Board Game Turns the French Colonies into Child's Play."

24. Collection inventory, Getty Research Institute, Los Angeles, http://archives2.getty.edu:8082/xtf/view?docId=ead/970031/970031.xml;chunk.id=ref1093;brand=default;query=deutsche%20kolonien.

25. Blanchard et al., "The Creation of a *Colonial Culture* in France," 16.

26. Nicolas Bancel, Pascal Blanchard, and Françoise Vergès, *La République Coloniale: Essai sur une utopie* (Paris: Albin Michel, 2003), 111–112.

27. Robert W. Rydell, *World of Fairs: The Century-of-Progress Expositions* (Chicago: University of Chicago Press, 1993), 3. See also Robert W. Rydell, "In Sight and Sound with All the Senses All Around: Racial Hierarchies at America's World's Fairs," in *The Invention of Race: Scientific and Popular Representations*, ed. Nicolas Bancel, Thomas David, and Dominic Thomas, 209–221 (New York: Routledge, 2014); Caroline A. Jones, *Sensorium: Embodied Experience, Technology, and Contemporary Art* (Cambridge, MA: MIT Press, 2006).

28. Panivong Norindr, "*La Plus Grande France*: French Cultural Identity and Nation Building under Mitterand," in *Identity Papers: Contested Nationhood in Twentieth-Century France*, ed. Steven Ungar and Tom Conley (Minneapolis: University of Minnesota Press, 1996), 233.

29. See Nicolas Bancel and Pascal Blanchard, "Incompatibilité: La CNHI dans le sanctuaire du colonialisme français," in *La Cité Nationale de l'Histoire de l'Immigration: Une collection en devenir, Hommes et Migrations* 1267 (2007): 112–127; Maureen Murphy, "Le CNHI au Palais de la Porte Dorée," in *La Cité Nationale de l'Histoire de l'Immigration: Une collection en devenir, Hommes et Migrations* 1267 (2007): 44–55, and *Un Palais pour une cité: Du musée des colonies à la Cité nationale de l'histoire de l'immigration* (Paris: Réunion des musées nationaux, 2007).

30. Some of this work has already taken place in substantive ways in such spaces as the International Slavery Museum (Liverpool, UK), the Royal Museum for Central Africa (Belgium), and the Tropenmuseum (Netherlands). See for example Herman Lebovics, *Bringing the Empire Back Home: France in the Global Age* (Durham, NC: Duke University Press, 2004); Catherine Coquio, ed., *Retour du colonial: Disculpation et réhabilitation de l'histoire coloniale* (Nantes: Librairie l'Atalante, 2008); Annie E. Coombes, "Museums and the Formation of National and Cultural Identities," in *Museum Studies: An Anthology of Contexts*, ed. Bettina Messias Carbonell, 231–246 (Oxford: Blackwell, 2004); Nancy E. Green, "A French Ellis Island? Museums, Memory and History in France and the United States," *History Workshop Journal* 63 (2007): 239–253; Dominic Thomas, ed., *Museums in Postcolonial Europe* (London: Routledge, 2010).

31. Achille Mbembe, "Provincializing France?," trans. Janet Roitman, *Public Culture* 23, no. 1 (2011): 110.

32. These include Lisbon's Padrao dos Descobrimentos (the Monument to the Discoveries) and the National Monument Slavernijverleden (National Monument to the legacy of slavery) in Amsterdam. See for example Robert Aldrich, *Vestiges of the Colonial Empire in France: Monuments, Museums and Colonial Memories* (New York: Palgrave Macmillan, 2005).

33. See for example Hilke Thode-Arora, "Hagenbeck's European Tours: The Development of the Human Zoo," in *Human Zoos: Science and Spectacle in the Age of Empire*, ed. Pascal Blanchard,

Nicolas Bancel, Gilles Boëtsch, Éric Deroo, Sandrine Lemaire, and Charles Forsdick (Liverpool: Liverpool University Press, 2008), 165–173.

34. See Christian Kopp and Marius Krohn, "Blues in Schwarzweiss: Die Black Community im Widerstand gegen kolonialrassistische Straßennamen in Berlin-Mitte," http://www.berlin-postkolonial.de/cms/index.php?option=com_content&view=article&id=78:afrikanisches-viertel&catid=10:mitte&Itemid=16.

35. See http://www.berlin-postkolonial.de.

36. Nadia Vancauwenberghe, "Africa in Berlin: Time for Some *Vergangenheitsbewältigung*?," *Exberliner* 108 (2012): 6.

37. See William B. Cohen, *The French Encounter with Africans* (Bloomington: Indiana University Press, 1984).

38. Edward W. Said, *Orientalism* (New York: Pantheon, 1978). See also Jean-Marc Moura, *La Littérature des lointains: Histoire de l'exotisme européen au xxe siècle* (Paris: Éditions Honoré Champion, 1998), and *L'Europe littéraire et l'ailleurs* (Paris: Presses Universitaires de France, 1998); Charles Forsdick, *Travel in Twentieth-Century French and Francophone Cultures: The Persistence of Diversity* (Oxford: Oxford University Press, 2005); Tim Youngs and Charles Forsdick, eds., *Travel Writing: Critical Concepts in Literary and Cultural Studies* (New York: Routledge, 2012).

39. Ousmane Socé, *Mirages de Paris* (Paris: Nouvelles Editions Latines, 1937).

40. Christopher L. Miller, *Nationalists and Nomads: Essays on Francophone African Literature and Culture* (Chicago: University of Chicago Press, 1998), 81.

41. Miller, *Nationalists and Nomads*, 67.

42. Miller, *Nationalists and Nomads*, 65.

43. See for example Bernard Mouralis, *Littérature et développement: Essai sur le statut, la fonction et la représentation de la littérature négro-africaine d'expression française* (Paris: Silex, 1984); Bennetta Jules-Rosette, *Black Paris: The African Writers' Landscape* (Urbana: University of Illinois Press, 1998); Abiola F. Irele, *The African Imagination: Literature in Africa and the Black Diaspora* (Oxford: Oxford University Press, 2001); Dominic Thomas, *Black France: Colonialism, Immigration, and Translationalism* (Bloomington: Indiana University Press, 2007).

44. See Aedín dí Loingsigh, *Postcolonial Eyes: Intercontinental Travel in Francophone African Literature* (Liverpool: Liverpool University Press, 2009).

45. Mongo Beti, *Cruel City*, trans. Pim Higginson (Bloomington: Indiana University Press, 2013), 11–15.

46. Wilfried N'Sondé, *The Heart of the Leopard Children*, trans. Karen Lindo (Bloomington: Indiana University Press, 2016), 48-49.

47. Ibid, 26.

48. Achille Mbembe, "The Republic and Its Beast: On the Riots in the French *Banlieues*," trans. Jean Marie Todd, in *Frenchness and the African Diaspora: Identity and Uprising in Contemporary France,* ed. Charles Tshimanga, Didier Gondola, and Peter J. Bloom (Bloomington: Indiana University Press, 2009), 51–52.

49. Bofane, *Congo Inc.*

50. On the sociology of publishing, reception, translation, and institutional recognition, see Claire Ducournau: "From One Place to Another: The Transnational Mobility of Contemporary Francophone sub-Saharan African Writers," *Yale French Studies* 120 (2011): 49–61.

51. Mbembe, "Provincializing France?," 103.

52. Bofane, *Congo Inc.*

53. Fiston Mwanza Mujila, *Tram 83*, trans. Roland Glasser (Dallas, TX: Deep Velum Publishing, 2015), 40–41.

Bibliography

Achebe, Chinua. *Things Fall Apart*. London: Heinemann, 1958.

Aldrich, Robert. *Vestiges of the Colonial Empire in France: Monuments, Museums and Colonial Memories*. New York: Palgrave Macmillan, 2005.

Bancel, Nicolas. "The Colonial Bath: *Colonial Culture* in Everyday Life." In *Colonial Culture in France Since the Revolution*, ed. Pascal Blanchard, Sandrine Lemaire, Nicolas Bancel, and Dominic Thomas, 200–208. Bloomington: Indiana University Press, 2014.

Bancel, Nicolas, and Pascal Blanchard. "Incompatibilité: La CNHI dans le sanctuaire du colonialisme français." In *La Cité Nationale de l'Histoire de l'Immigration: Une collection en devenir. Hommes & Migrations* 1267 (2007): 112–127.

Bancel, Nicolas, Pascal Blanchard, and Françoise Vergès. *La République Coloniale: Essai sur une utopie*. Paris: Albin Michel, 2003.

Beti, Mongo. *Cruel City*. Trans. Pim Higginson. Bloomington: Indiana University Press, 2013.

Blanchard, Pascal, Sandrine Lemaire, Nicolas Bancel, and Dominic Thomas. "The Creation of a *Colonial Culture* in France, from the Colonial Era to the 'Memory Wars.'" In *Colonial Culture in France Since the Revolution*, ed. Pascal Blanchard, Sandrine Lemaire, Nicolas Bancel, and Dominic Thomas, 1–47. Bloomington: Indiana University Press, 2014.

Bofane, In Koli Jean. *Congo Inc.: Le testament de Bismarck*. Arles: Actes Sud, 2014.

Bofane, In Koli Jean. *Congo Inc. – Bismarck's Testament*. Trans. Marjolijn de Jager. Bloomington: Indiana University Press, forthcoming.

Brancato, Sabrina. "Afro-European Literature(s): A New Discursive Category?" *Research in African Literatures* 39, no. 3 (2008): 1–13.

Cohen, William B. *The French Encounter with Africans*. Bloomington: Indiana University Press, 1984.

Coombes, Annie E. "Museums and the Formation of National and Cultural Identities." In *Museum Studies: An Anthology of Contexts,* ed. Bettina Messias Carbonell, 231–246. Oxford: Blackwell, 2004.

Coquio, Catherine, ed. *Retour du colonial: Disculpation et réhabilitation de l'histoire coloniale*. Nantes: Librairie l'Atalante, 2008.

Dadié, Bernard. *Un nègre à Paris*. Paris: Présence Africaine, 1959.

Ducournau, Claire. "From One Place to Another: The Transnational Mobility of Contemporary Francophone sub-Saharan African Writers." *Yale French Studies* 120 (2011): 49–61.

El-Tayeb, Fatima. *European Others: Queering Ethnicity in Postnational Europe.* Minneapolis: University of Minnesota Press, 2011.

Fèvre, Joseph, and Henri Hauser. *Précis de géographie.* Paris: F. Alcan, 1913.

Forsdick, Charles. *Travel in Twentieth-Century French and Francophone Cultures: The Persistence of Diversity.* Oxford: Oxford University Press, 2005.

Forsdick, Charles, Alec G. Hargreaves, and David Murphy, eds. *Transnational French Studies: Postcolonialism and Littérature-monde.* Liverpool: Liverpool University Press, 2010.

Gassama, Makhily, ed. *L'Afrique répond à Sarkozy: Contre le discours de Dakar.* Paris: Éditions Philippe Rey, 2008.

Gikandi, Simon. *Maps of Englishness: Writing Identity in the Culture of Colonialism.* New York: Columbia University Press, 1997.

Gilroy, Paul. *There Ain't No Black in the Union Jack: The Cultural Politics of Race and Nation.* Chicago: University of Chicago Press, 1987.

Green, Nancy E. "A French Ellis Island? Museums, Memory and History in France and the United States." *History Workshop Journal* 63 (2007): 239–253.

Habila, Helon. "Introduction." In *The Granta Book of the African Short Story*, i–xvii. London: Granta, 2011.

Hine, Darlene Clark, Trica Danielle Keaton, and Stephen Small, eds. *Black Europe and the African Diaspora.* Urbana: University of Illinois Press, 2009.

Hitchcott, Nicki, and Dominic Thomas, eds. *Francophone Afropean Literatures.* Liverpool: Liverpool University Press, 2014.

Irele, Abiola F. *The African Imagination: Literature in Africa and the Black Diaspora.* Oxford: Oxford University Press, 2001.

Jones, Caroline A. *Sensorium: Embodied Experience, Technology, and Contemporary Art.* Cambridge, MA: MIT Press, 2006.

Jules-Rosette, Bennetta. *Black Paris: The African Writers' Landscape.* Urbana: University of Illinois Press, 1998.

Kane, Cheikh Hamidou. *L'aventure ambiguë.* Paris: Julliard, 1961.

Konaré, Adame Ba, ed. *Petit précis de remise à niveau sur l'histoire africaine à l'usage du président Sarkozy.* Paris: Éditions La Découverte, 2008.

Kopp, Christian, and Marius Krohn. "Blues in Schwarzweiss: Die Black Community im Widerstand gegen kolonialrassistische Straßennamen in Berlin-Mitte." http://www.berlin-postkolonial.de/cms/index.php?option=com_content&view=article&id=78:afrikanisches-viertel&catid=10:mitte&Itemid=16.

Laye, Camara. *L'Enfant noir.* Paris: Plon, 1954.

Lebovics, Herman. *Bringing the Empire Back Home: France in the Global Age.* Durham, NC: Duke University Press, 2004.

Lemaire, Sandrine. "Manipulation: Conquering Taste (1931–1939)." In *Colonial Culture in France Since the Revolution*, ed. Pascal Blanchard, Sandrine Lemaire, Nicolas Bancel, and Dominic Thomas, 285–295. Bloomington: Indiana University Press, 2014.

Lemaire, Sandrine. "Spreading the Word: *The Agence Générale des Colonies* (1920–1931)." In *Colonial Culture in France Since the Revolution*, ed. Pascal Blanchard, Sandrine Lemaire, Nicolas Bancel, and Dominic Thomas, 162–170. Bloomington: Indiana University Press, 2014.

Lemaire, Sandrine, Catherine Hodeir, and Pascal Blanchard. "The Colonial Economy: Between Propaganda Myths and Economic Reality (1940–1955)." In *Colonial Culture in France Since the Revolution*, ed. Pascal Blanchard, Sandrine Lemaire, Nicolas Bancel, and Dominic Thomas, 320–332. Bloomington: Indiana University Press, 2014.

dí Loingsigh, Aedín. *Postcolonial Eyes: Intercontinental Travel in Francophone African Literature*. Liverpool: Liverpool University Press, 2009.

Manceron, Gilles. "School, Pedagogy, and the Colonies (1870–1914)." In *Colonial Culture in France Since the Revolution*, ed. Pascal Blanchard, Sandrine Lemaire, Nicolas Bancel, and Dominic Thomas, 124–131. Bloomington: Indiana University Press, 2014.

Mbembe, Achille. "Provincializing France?" Trans. Janet Roitman. *Public Culture* 23, no. 1 (2011): 85–119.

Mbembe, Achille. "The Republic and Its Beast: On the Riots in the French Banlieues." In *Frenchness and the African Diaspora: Identity and Uprising in Contemporary France*, ed. Charles Tshimanga, Didier Gondola, and Peter J. Bloom, trans. Jean Marie Todd, 47–69. Bloomington: Indiana University Press, 2009.

Mbembe, Achille. *Sortir de la grande nuit: Essai sur l'Afrique décolonisée*. Paris: Éditions La Découverte, 2010.

Miano, Léonora. *Habiter la frontière*. Paris: Arche Éditeur, 2012.

Miller, Christopher L. *Nationalists and Nomads: Essays on Francophone African Literature and Culture*. Chicago: University of Chicago Press, 1998.

Moura, Jean-Marc. *La Littérature des lointains: Histoire de l'exotisme européen au xxe siècle*. Paris: Éditions Honoré Champion, 1998.

Moura, Jean-Marc. *L'Europe littéraire et l'ailleurs*. Paris: Presses Universitaires de France, 1998.

Mouralis, Bernard. *Littérature et développement: Essai sur le statut, la fonction et la représentation de la littérature négro-africaine d'expression française*. Paris: Silex, 1984.

Mujila, Fiston Mwanza. *Tram 83*. Trans. R. Glasser. Dallas, TX: Deep Velum Publishing, 2015.

Murphy, Maureen. "Le CNHI au Palais de la Porte Dorée." In *La Cité Nationale de l'Histoire de l'Immigration: Une collection en devenir. Hommes & Migrations* 1267 (2007): 44–55.

Murphy, Maureen. *Un Palais pour une cité: Du musée des colonies à la Cité nationale de l'histoire de l'immigration*. Paris: Réunion des musées nationaux, 2007.

Norindr, Panivong. "*La Plus Grande France*: French Cultural Identity and Nation Building under Mitterand." In *Identity Papers: Contested Nationhood in Twentieth-Century France*, ed. Steven Ungar and Tom Conley, 233–258. Minneapolis: University of Minnesota Press, 1996.

N'Sondé, Wilfried. *The Heart of the Leopard Children*. Trans. Karen Lindo. Bloomington: Indiana University Press, 2016.

Phillips, Caryl. *Color Me English: Migration and Belonging Before and After 9/11*. New York: New Press, 2011.

Phillips, Caryl. *The European Tribe*. London: Faber and Faber, 1987.

Poggi, Isotta. "Colorful Board Game Turns the French Colonies into Child's Play." February 24, 2014. http://blogs.getty.edu/iris/colorful-board-game-turns-the-french-colonies-into-childs-play.

"Pour une 'littérature monde' en français." March 15, 2007. http://www.lemonde.fr/livres/article/2007/03/15/des-ecrivains-plaident-pour-un-roman-en-francais-ouvert-sur-le-monde_883572_3260.html.

Quayson, Ato. "Postcolonialism and the Diasporic Imaginary." In *A Companion to Diaspora and Transnationalism*, ed. Ato Quayson and Girish Daswani, 139–158. Oxford: Blackwell, 2013.

Rushdie, Salman. "The New Empire within Britain." In *Imaginary Homelands: Essays and Criticism: 1981–1991*, 129–138. London: Granta, 1991.

Rydell, Robert W. "In Sight and Sound with All the Senses All Around: Racial Hierarchies at America's World's Fairs." In *The Invention of Race: Scientific and Popular Representations*, ed. Nicolas Bancel, Thomas David, and Dominic Thomas, 209–221. New York: Routledge, 2014.

Rydell, Robert W. *World of Fairs: The Century-of-Progress Expositions*. Chicago: University of Chicago Press, 1993.

Said, Edward W. *Orientalism*. New York: Pantheon, 1978.

Selasi, Taiye. "Bye-Bye Babar." March 3, 2005. http://thelip.robertsharp.co.uk/?p=76.

Socé, Ousmane. *Mirages de Paris*. Paris: Nouvelles Editions Latines, 1937.

Stein, Mark. *Black British Literature: Novels of Transformation*. Columbus: Ohio State University Press, 2004.

Thode-Arora, Hilke. "Hagenbeck's European Tours: The Development of the Human Zoo." In *Human Zoos: Science and Spectacle in the Age of Empire*, ed. Pascal Blanchard, Nicolas Bancel, Gilles Boëtsch, Éric Deroo, Sandrine Lemaire, and Charles Forsdick, 165–173. Liverpool: Liverpool University Press, 2008.

Thomas, Dominic. *Black France: Colonialism, Immigration, and Translationalism*. Bloomington: Indiana University Press, 2007.

Thomas, Dominic. "Fortress Europe: Identity, Race and Surveillance." In *Race, Violence, and Biopolitics,* ed. Alessandro Corio and Louise Hardwick. *International Journal of Francophone Studies* 17, nos. 3–4 (2014): 445–468.

Thomas, Dominic. "L'Afrique à l'université américaine." In *Penser et écrire l'Afrique aujourd'hui*, ed. Alain Mabanckou, 81-91. Paris: Seuil, 2017.

Thomas, Dominic, ed. *Museums in Postcolonial Europe*. London: Routledge, 2010.

Vancauwenberghe, Nadia. "Africa in Berlin: Time for Some *Vergangenheitsbewältigung?*" *Exberliner* 108 (2012): 6.

"Toward a World Literature in French." Trans. Daniel Simon. *World Literature Today* 83, no. 2 (March–April 2009): 54–56.

Youngs, Tim, and Charles Forsdick, eds. *Travel Writing: Critical Concepts in Literary and Cultural Studies*. New York: Routledge, 2012.

III

GENRES AND THEMES

13
Popular Map Genres in American Literature

Martin Brückner

Introduction

In American literature, the rhetoric of mapping is a central trope in the texts of canonical as well as less familiar works. From Christopher Columbus's first travel reports to Mark Twain's *Adventures of Huckleberry Finn* (1884) to road novels like William Least Heat-Moon's *Blue Highways* (1982), U.S. authors have created elaborate word maps when describing their personal experiences, characters, settings, or spatial plots. To make sense of such literary mappings, critical studies have long pursued microanalytic approaches delving into word use, etymology, and poetics, or they have applied macroanalytic techniques exploring the history of narrative form, literary genre, and the role of the literary marketplace. From studies addressing matters of place and environment, movement and material culture in American literature, we have learned over time to think about American poems, short stories, plays, and novels as cartographically inflected repositories and proving grounds. U.S writers have explored literary themes and conventions associated with the heroic quest, the sentimental journey, the utopian/dystopian frontier tale, and the poetics as well as the politics of spatial feeling. All this has informed characters including hunters, emigrants, and runaway slaves next to men or women of leisure, working fathers and mothers, and modern representations of children.[1]

Regarding these approaches, scholars have often pointed out that actual maps, from the rare manuscript to the mass-produced printed artifact, have over the centuries informed the style of American literature. From Stephen Greenblatt to Ricardo Padrón, it has often been noted how in early modern texts maps were a physical part of contact reports and colonial histories, such as Captain John Smith's *The Generall Historie of Virginia* (1624) or Cotton Mather's ecclesiastical tome, *Magnalia Christi Americana* (1702). It has also been noted how they provided both the literal and the figurative plots for travel accounts ranging from *The Discoveries of John Lederer* (1672) to William Byrd's the *History of the Dividing Line* (c. 1730), or from Nicolas Biddle's famous edition of the *History of the Expedition under the Command of Captains Lewis and Clark* (1814) to Timothy Flint's *Recollections of the Last Ten Years, Passed in Occasional Residences and Journeyings in the Valley of the Mississippi* (1826).[2]

It has also frequently been observed that between the eighteenth and twentieth centuries many American authors seem to have written their fictional works with a map in hand: James Fenimore Cooper's *The Pioneers* (1823) and *The Chainbearer* (1845) abound with references to land surveyors' plats; Herman Melville's novels *Redburn* (1849) and *Moby-Dick* (1851) were composed while the author consulted atlas maps and geography handbooks published by Jesse Olney; Edgar Allan Poe's *The Narrative of Arthur Gordon Pym* (1838) provides geographic coordinates for navigating the South Pole; and Henry David Thoreau's *Walden* (1854) offers an intricate metaphor of national identity by meshing language, literary form, and a sketch map of Walden Pond. Similarly, if we fast-forward to more recent literature, maps are appended to works as different as William Faulkner's psychological novel, *Absalom, Absalom!* (1936), and John Steinbeck's travelogue, *Travels with Charley* (1962). Maps have been a critical reference point in poetry ranging from Walt Whitman's *Leaves of Grass* (1855) to Elizabeth Bishop's *Trial Balances* (1935) or *Geography III* (1976). Last but not least, maps have a constant presence—as both insert and narrative device—in the pages and endpapers of American fantasy and action novels from Frank Baum's *The Wizard of Oz* (1900) to Dan Brown's illustrated edition of *Angels & Demons* (2006).

As new scholarship in the form of reader response studies, the history of the book, and GIS-supported textual analysis has shown, we have come to realize that maps have had a much more pervasive influence than previously thought on the production and reception of American literature.[3] On the one hand, since the 1980s maps have increasingly come to be considered a vital, paratextual component supplementing sweeping literary surveys; published in anthologies that consider American literature an inclusionary discourse, the occasional historical map is now reprinted along with personal letters, political pamphlets, poetry, drama, and novels.[4] On the other hand, as maps have been shown to play a constitutive role in the making of American literature, they have also been recognized for assuming literary agency; case studies that read American literature through maps illustrate how particular cartographic works directly influenced the form and content of a text or even a genre.[5]

Whether authors have used maps as graphic inserts, direct citation, or wordy approximation, these maps illustrate how through the process of literary adaptation cartography has helped convey the American experience, redefine literary conventions in place-specific and often nationalistic terms, and explore the meaning of America in modern times. Maps accompanying texts illustrate how authors have envisioned unknown lands, processed the colonization of a vast continental expanse and numerous peoples, or reconciled personal desire for property with communal rules over ownership (Pagden; Buell; Newman). In the process of literary adaptation, map references and verbal maps reveal patterns of connection, linking particular literary conventions with spatial imaginings that have prescribed, for example, collective actions, personal perspectives, and above all, real or imaginary settings.[6]

Drawing on scholarship from literary geography, historical cartography, and studies in spatial theory, this chapter's principal goal is to provide a limited survey of map genres in conversation with American literature. To explore the form and function of literary maps, the chapter mostly addresses texts published with maps or referencing maps concretely. While territoriality and spatial thinking ostensibly inform all literary maps, the chapter concentrates on the readerly aspect of mobility, on the intersection between the act of reading as a function of movement and the map as the tool and repository providing orientation and direction. In trying to show the range of maps and their literary adaptations, the chapter is organized into three sections designed to track select literary genres across different, at times overlapping, periods defined by popular anthologies of American literature in the English language. The three sections are: "Maps in the Literature of Colonization," "Maps in the Literature of Nation Formation," and "Maps in the Literature of Two Gilded Ages, 1900/2000."[7]

Maps in the Literature of Colonization

The history of joining maps and American literature begins as early as 1493 with the publication of the bestselling pamphlet *De Insulis nuper inuentis* (1493), containing Christopher Columbus's letter to his patrons, King Ferdinand and Queen Isabella, in which descriptions of his experience were supplemented with maplike woodcuts (figure 13.1). Within a decade printers were quick to associate maps with travel accounts—for example, with the dual publication of Martin Waldseemüller's giant wall map, *Universalis Cosmographia Secundum Ptholomaei Traditionem et Americi* (1507) and Fracanzano da Montalboddo's volume, *Paesi novamente retrovati* (1507). The Waldseemüller map is today celebrated for inventing "America" because it was the first map to name the continent.[8] Montalboddo's anthology, on the other hand, is credited with publishing the first collection of reports detailing journeys to the American continent by, for example, Columbus, Amerigo Vespucci, and Gaspar Corte-Real. Within a short time, the genre of the travel report not only took Europe by storm but, with enterprising publishers adding woodcut illustrations, it became associated with maps and the literary trope of mapped information, settings, and events.

The topics of early modern travel reports revolved around questions of how to make contact with America, including tales of reconnaissance, the presence of salable natural resources and commodities, the discovery of unfamiliar flora and fauna, and recommendations on how to launch "plantations" or how to negotiate the presence of Native Americans, then called "savages." The language of the early reports shifted back and forth between personal diary and encyclopedic histories. Authors frequently struggled for words that were capable of reporting unfamiliar plants and objects, thus revealing a linguistic and literary uncertainty on how to render the experience of what literary historians have dubbed the "marvels" or "wonders" of the New World.[9] By

num copia salubritate admixta hominū : quæ nisi quis viderit: credulitatem superat . Huius arbores pascua & fructus / multū ab illis Iohanę differūt . Hæc præterea Hispana diuerso aromatis genere / auro metallisq; abundat. cuius quidem & omnium aliarum quas ego vidi : & quarum cognitionem baheo incolę vtriusq; sexus: nudi semp incedunt:

queadmodū edunt in lucem. præter aliquas fęminas. quę folio fronde ue aliqua: aut bombicino velo: pudenda operiunt: quod ipsę sibi ad id negocii parant. Carent hi omnes (vt supra dixi) quocunq; genere ferri. carent & armis: vtpote sibi ignotis. nec ad ea sunt apti. non propter corporis deformitatē: (cum sint bene formati) sed quia sunt timidi ac pleni formidine. gestant tamen p armis arundines sole perustas: in quaz; radicibus: hastile quoddam ligneum siccum & in mucronem attenuatum figūt : neq; his audent iugiter vti: nam sæpe euenit cū miserim duos vel tris homines ex meis ad aliquas villas: vt cum earum loquerent incolis: exiisse agmen glomeratū ex Indis: & vbi nostros appropinquare videbant: fugam celeriter arripuisse: despretis a patre liberis / & econtra. & hoc non q; cuipiam eoz; damnum aliquod vel iniuria illata fuerit: immo ad quoscunq; appuli: & quibus cū verbum facere potui: quicquid habebam sum elargitus: pannū aliaq; pmulta: nulla mihi facta versura : sed sunt natura pauidi ac timidi . Ceterum vbi se cernunt tutos: omni metu repulso: sunt admodū simplices ac bonę fidei: & in omnibus quæ habent liberalissimi: roganti q; possidet inficiatur nemo : quin ipsi nos ad id poscendum inuitant. Maximum erga omnes amorē preseferūt: dant queq; magna p paruis. minima licet re / nihiloue cōtenti: ego attn phibui ne tā minima & nullius precii hisce darentur: vt sunt lan

e e ij

Figure 13.1

Christopher Columbus, *De Insulis nuper inuentis* (1493). Courtesy, Newberry Library.

contrast, the image of maps—even if they were rudimentary, highly inaccurate, and in many instances imaginary—offered a more stable and supposedly more correct view for readers interested in tracking travel routes and picturing the geography and biodiversity of the New World.

Illustrated travel reports that included maps and maplike illustrations—from Georg Stuchs's *Newe vnbekanthe Landte vnd ein newe Weldte* (1513) and Hans Staden's *Warhaftige Historia und beschreibung eyner Landtschafft der Wilden Nacketen, Grimmigen Menschfresser-Leuthen in der Newenwelt America gelegen* (1557), to the bestselling editions of Theodor de Bry's anthology, *Newe Welt vnd americanische Historien* (c. 1650), which included Thomas Hariot's now canonical work, *A Briefe and True Report of Virginia* (1590)—shaped the public imagination in two ways. On the one hand, they introduced the myths of American cannibalism, the ubiquity of precious metals (El Dorado), and geographic fantasies of gaining access to the Garden of Eden. On the other hand, they established the tradition that writing and reading about journeys to and inside America required maps as much for the purpose of navigation as for navigating the text itself. For example, the maps appended to Captain John Smith's *Generall Historie of Virginia* (1624) or John Lederer's *Discoveries* (1672) contained dotted lines or large inky crosses, offering readers graphic signposts that cross-referenced the volumes' table of contents with travel routes and encounters with Native populations in the Chesapeake Bay area and in the Carolina territory through the use of two textual landscapes: the graphic map and the printed word.[10]

Smith's *Map of New England* (1612; figure 13.2) followed a somewhat different narrative model. The map offered a unique palimpsest mixing authorial and literary ambitions, which are matched by the map's companion text, *Description of New England* (1612). Toward the end of the report, after numerous observations about American nature and indigenous cultures, Smith offered advice on how to colonize ("plant") New England by assigning himself controversial powers. In the narrative, he demanded the trust and authority usually reserved for sovereigns and aristocratic members of society occupying government positions. On the map, he not only rebaptized native places with English names—such as "New London," "The River Charles," or "Oxford"—but inserted an image of his own likeness in the map's margins and thus in the iconographic place usually reserved for portraits, heraldry, or other tokens of homage celebrating royalty or sponsors of risky Atlantic expeditions.

While in principle the genre of the travel report remained relatively unchanged during the sixteenth and seventeenth centuries, the accompanying maps suggested a shift in narrative direction. The earliest reports were preoccupied with finding America and focused on the American water/land divide—that is, on the relationship of the Atlantic Ocean to the American coast, including the coastal enclaves, corridors, and possible inroads into the continental hinterland.[11] At the same time, the travel reports' narrative styles experimented with different linguistic options and literary devices as a

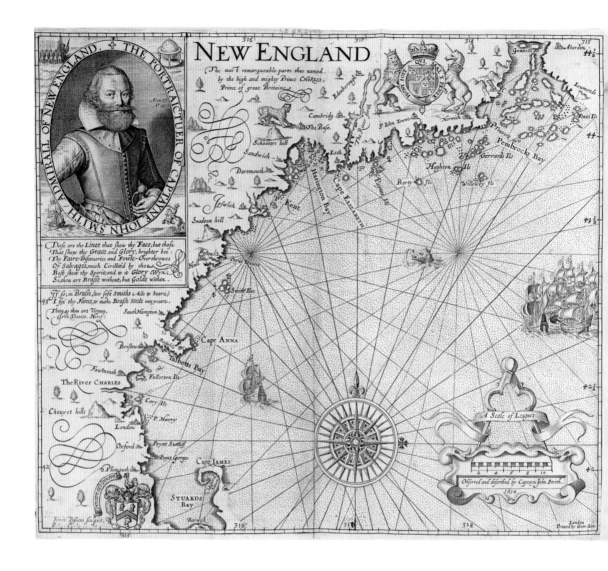

Figure 13.2

John Smith, "Map of New England." In *A Description of New England* (London, [1612] 1627). Rare Books Division, New York Public Library, Astor, Lenox and Tilden Foundations.

way of translating the unfamiliar into more familiar terms. While many early reports introduced to the English lexicon neologisms borrowed from European and Native American languages, most applied a Eurocentric approach to their literary craft by packing the American experience into established literary forms inflected by, say, the literary genres of historiography, the conversion narrative, or the emerging protocols of encyclopedic writing used in the natural sciences. By the mid- to late seventeenth century, and thus after nearly two centuries of accumulating information, travel writers began to locate their narratives inside a different hemispheric frame of reference, emphasizing (with many overlaps) the Atlantic world and the American continent as one continuous metasetting.

The representation of the Atlantic world as a literary setting was the product of accounts that conceptualized the New World not in continental terms but as a watery world defined by the Atlantic Ocean, waterways, and islands. Many geographic accounts, such as Richard Ligon's *A True and Exact History of the Island of Barbados* (1657), resembled in narrative structure and content nautical maps like those compiled by William Hack's *Waggoner Atlas* (1682) or William Blathwayt's *Atlas* (1675–1685), which merged sea charts with detailed sailing directions in an effort to facilitate both movement and settlement in the Atlantic or Pacific ocean. Ligon's history illustrates how the English-speaking world, driven by its growing focus on the Caribbean sugar and slave trade, conceived the experience of America as a succession of unrelated settings and a constant movement between the coasts of Europe, Africa, and American islands. Indeed, most of the early period's literary and even political imagination conceptualized America in uncontinental terms because the geological concept of continents was yet to be fully described and embraced.[12]

As a result, many European writers and their readers, including members of the British Parliament, imagined America as an archipelago of loosely connected islands linked by bodies of water, tropical storms, and fantasies of geographic autonomy or nightmares of slave rebellion. In his history, Ligon inserted the map *A Topographicall Description ... of Barbados* (1657; figure 13.3), in which toponyms and pictorial elements anticipate descriptions about land ownership, farming techniques, and slave culture. The trope of placing economically active fictional characters inside the setting of American islands in the Atlantic world was perhaps made most popular by Daniel Defoe's novel *Robinson Crusoe* (1719). Other works invested in geographic, economic, or spiritual mobility adapted the island theme, such as James Grainger's Georgic poem, *The Sugar-Cane* (1764); Thomas Paine's political essay, *Common Sense* (1776); Olaudah Equiano's autobiography, *The Interesting Narrative of the Life of Olaudah Equiano* (1789), describing his itinerant life as a slave during the 1750s and 1770s; the anonymous early American novel, *The History of Constantius and Pulchera* (1795); and J. Hector St. John de Crèvecoeur's *Letters from an American Farmer* (1782), in which rudimentary maps depicting the islands of Nantucket and Martha's Vineyard provided

Figure 13.3

Richard Ligon, *A Topographicall Description and Admeasurement of the Yland of Barbados in the West Indyaes: With the Mrs. Names of the Seuerall Plantacons* (1657). Courtesy, American Antiquarian Society.

visual plots for delineating the ideological concept of the "middle state" made popular in Crèvecoeur's now famous "Letter III: What Is an American?"

The focus on Oceanic and Caribbean geographies began to wane as literature devoted to traveling within the confines of the continent grew during the eighteenth century. Fueled by the territorial conflicts between France and England, these works and their maps included travelogues, natural histories, and geographies, which over the course of time expanded in geographic reach as well as literary purpose. While early tours like Sarah Kemble Knight's *Diary* (1704) of a journey from Boston to Connecticut or Alexander Hamilton's *Itinerarium* (1744) from Annapolis to Boston served to chart, albeit without the use of maps, the movement between different local cultures, later works that included maps, such as Peter Kalm's *Travels into North America ... Enriched with a Map* (1773) and Jonathan Carver's *Travels through the Interior Parts of North America* (1778), became international bestsellers.

The continental perspective emerged more fully in writings based on some of the most influential maps sponsored by government institutions. It was John Mitchell's *Map of the British and French Dominions in North America* (1755; figure 13.4) and his treatise *The Contest in America between Great Britain and France* (1757) that not only projected a continental reach for the British American colonies, but provided language reflecting imperial ambitions pursued first by the British Empire and then later by American politicians negotiating the territorial claims of the United States of America. A similar impact can be attributed to Lewis Evans's *A General Map of the Middle British Colonies* (1755), published along with his volume *Geographical, Historical, Political, Philosophical and Mechanical Essays* (1755). Both the Evans map and his literary analysis of the mid-Atlantic region contributed to the emerging Enlightenment narrative depicting America as a pastoral setting capable of accommodating European immigrants despite Native American land claims and the increasing dependence on slave-based economies during the late eighteenth and early nineteenth centuries.[13]

It is in the early texts accompanying maps that readers will find some of the best illustrations for understanding how maps informed literature of the late colonial period and beyond. Combining literary conventions pioneered by European travel narratives—emphasizing the experience of the traveler/narrator, the careful tally of natural resources, and even the aesthetic appreciation of natural phenomena—both the eloquent and the laconic accompaniment borrowed amply from cartography's vocabulary and visual orientation to better describe spaces, materials, and peoples. On occasion the authors' preoccupation with maps had its pitfalls—for example, when European travelers lost their way because they had naively followed Native American maps, which, unlike Western cartographic representations, marked places in spatiotemporal terms;[14] or when natural conditions, like shifting coastal sands, thwarted the authors' desire to establish literal and figurative boundaries, as described in William

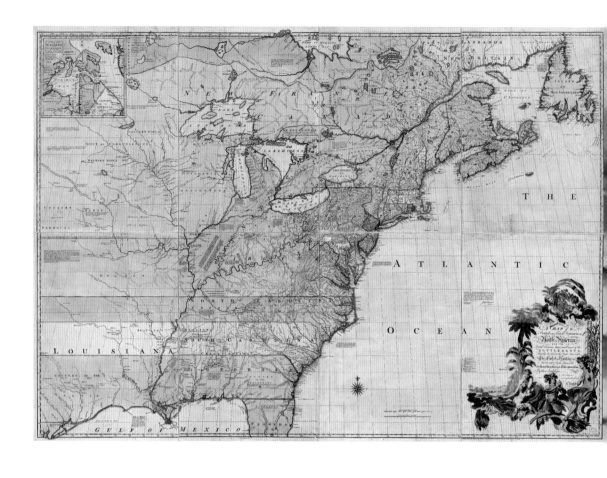

Figure 13.4

John Mitchell, *A Map of the British and French Dominions in North America* (London, 1755).
Library of Congress, Geography and Map Division.

Byrd's *History of the Dividing Line* (1730);[15] or when the map itself was deemed a tool superior to the English language, as suggested by Thomas Jefferson's first chapters in his *Notes on the State of Virginia* (1784).[16] In the examples where colonial authors purposefully prepared maps to accompany their writings, their textual representation borrowed from the map's graphic mode of communicating authority by referencing map titles and geodetic accuracy, the cartographic lexicon, and the supervisory "at a glance" method of visualization. Information was presented selectively or generalized, allowing writers to name, edit, or omit contested places or unpleasant experiences. While references to maps, coordinates, and measurements transformed geographic descriptions into literary maps, the same rhetorical ploy frequently silenced American voices, including Native Americans, European minorities, members of the working class, and the enslaved.

Maps in the Literature of Nation Formation

After the Revolution, map-illustrated travel writing evolved into a popular genre, allowing several generations of authors to explore the new nation in small- or large-scale accounts as well as to experiment with the genre's formal parameters.[17] Sketch maps accompanied the regional tour documented by William Bartram's *Travels through North and South Carolina, Georgia, East and West Florida, the Cherokee Country, etc.* (1791), which was quickly recognized for its elegiac tone and fantastical depiction of the American South. Similarly, Timothy Dwight's more ponderous *Travels in New England and New York* (1823) provided a new generation of writers with detailed descriptions and an early realist narrative model by meshing geography, statistics, and local history. By contrast, much more detailed maps accompanied the reports of transcontinental expeditions, such as William Clark's map supplementing Nicholas Biddle's *History of the Expedition under the Command of Captains Lewis and Clark* (1814), or Charles Preuss's maps published in John Fremont's polemical *Report of the Exploring Expedition to the Rocky Mountains in the Year 1842, and to Oregon and North California in the Years 1843–44* (1845).[18]

In these writings, even the most rudimentary map—like the one drawn by Bartram—emerged as a much-valued reading index for cross-referencing the journeys' geographic progress with literary musings about the narrator's psychological state, the region's cultural geography, or its natural environment. Indeed, it could be argued that in the course of the nineteenth-century, maps fostered literary strategies that turned map-sourced descriptions into poetic or philosophical masterpieces exploring questions unique to the American experience. In particular, these questions addressed identity concerns in the age of industrialization, as can be seen in the mappings and meditations about life in Concord, Massachusetts, and the United States in general in Henry David Thoreau's *Walden* (1854) or his essay "Walking" (1862).[19]

While travel writers used maps that sketched travel routes and traversed spaces by charting everything from small woods to the whole continent, perhaps the most popular form of vicariously experiencing a journey around the new nation was made available to ordinary people by a burgeoning schoolbook industry that incorporated local and national maps into geography lessons. Early slogans, like Noah Webster's 1788 exhortation that "a tour of the United States ought now to be considered as a necessary part of a liberal education," spurred many authors and readers to consult geography books and their companion atlases as a way of traveling the country without having to leave the comfort and safety of the home or classroom.[20]

Bestselling textbooks, such as Jedidiah Morse's *Geography Made Easy* (1784), shaped the geographic imagination of American writers for half a century by providing prototypes in literary mapping according to which foldout maps—like Amos Doolittle's *A Map of the United States of America* (1784; figure 13.5)—inventoried the geography, economy, and cultural customs of every region. A unique literary feature of subsequent editions published by Morse and a host of other geographers was the textbook journey to geographic landmarks, such as Niagara Falls in upstate New York, Natural Bridge in Virginia, Mammoth Cave in Kentucky, and so forth. At the same time, another narrative feature of early national textbooks was the literary construction of regionally inflected fictional character types, such as the thrifty New Englander or the lazy Virginia Gentleman. While biased and offensive even by contemporaneous standards, these place-based stereotypes informed geography lessons that required students to apply their textbook knowledge by touring the national map with both eyes and fingers in search of the places and people that represented the "American character."[21]

At this point in the survey it is important to note that after the heyday of travel narratives and geography books, not many genres in American literature emerged that contained actual maps documenting the act of inhabiting space, the culture or aesthetic of territoriality, or the emotional or psychological implications of moving across space or defining spatial relations in natural or human-made settings.[22] A bibliographical search of American literature tracing actual map inserts in printed works published between 1800 and 2000, especially in prose fiction, yields a surprisingly small number of examples. From the standpoint of the sociology and economics of literary production this may come as a surprise, because historians of book and print culture have demonstrated that in the course of the nineteenth century maps were rapidly becoming inexpensive artifacts, while publishers were increasingly willing to invest in book illustrations.[23] From the standpoint of literary studies, moreover, the lack of actual maps is almost baffling in light of the frequency with which American authors used map references in their creative efforts to convey the American experience in thematic terms, exploring the vagaries of westward migration, the journeys of ex-slaves, the dispossession of Native Americans, and the proprietary claims of new

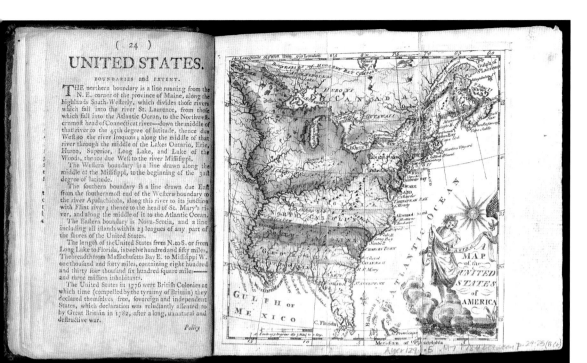

Figure 13.5

Amos Doolittle, *A Map of the United States of America* (1784). In Jedidiah Morse, *Geography Made Easy* (1784). Courtesy, Newberry Library.

settlements—emphases that correspond to many of the emergent thematic maps charting everything from diseases and crops to demography and history.[24]

Although the bulk of literary works, from eighteenth-century sentimental journeys to twentieth-century poetry, lacked the material presence of physical maps, three types of prose fiction displayed a strong affinity with popular map genres during the nineteenth century: picaresque narratives and small-scale overview maps; frontier stories and large-scale promotional maps; and sensational fiction and river maps.

Especially during the first decades after Independence, American readers would have found that small-scale maps, like the Doolittle map in Morse's school geography, not only depicted the nation, the continent, and the trans-Atlantic context, but provided a graphic aid for following the plot of many picaresque stories and novels. Like their European counterparts, American picaresque works chronicled the travels and travails of a single or sometimes multiple characters, usually of low social standing, whose movements ranged across the globe and tended to be episodic as well as highly unstructured. This was because their progress was randomly subjected to fate, storms, and pirates rather than stationary conditions, rational planning, or social contracts.[25]

In the anonymous novella, *The History of Constantius and Pulchera* (1795), the heroine or "picara" embarked on an improbable tour that took the reader from Philadelphia to various settings along the North Atlantic rim (Essex, Massachusetts; Bordeaux, France; Lisbon, Portugal; etc.), only to return in the end to the capital of the United States. Similarly, in James Butler's *Fortune's Football* (1798), the male protagonist, the "picaro," occupied in quick succession geographic locales that included London, Venice, Malta, Algiers, London, Quebec, Florence, Esfahān, Baghdad, Moscow, and again London. While the characters' rapid encounters with randomly placed settings were central to the story, picaresque tales like these required readers to consult small-scale maps—that is, maps showing large areas—in order to track the stories' plots. Indeed, the haphazard movement of characters in picaresque novels, such as Royal Tyler's *The Algerine Captive* (1797) or Charles Brockden Brown's *Memoirs of Carwin, the Biloquist* (1803–1805), loosely resembled the design of board games. Cartographic board games—from Thomas Jefferys's *Royal Geographical Pastime: Exhibiting a Complete Tour Round the World* (1770), to John Walker's *Walker's Geographical Pastime* (1834; figure 13.6), to Milton Bradley's *The Game of Round the World* (1873)—mimicked the journeys of novelistic characters, allowing readers to behave like players whose actions relied on chance rather than on personal intent or cultural reason.

A similarly calculated suspension of readerly protocol informed the mapping motifs in fiction grappling with life on the western frontier. A good example is Caroline Kirkland's fictionalized biography and novel of frontier manners, *A New Home—Who'll Follow? or Glimpses of Western Life?* (1839). It described the experience of a middle-class East Coast family that, after being displaced by the economic crash of 1837, traveled to southern Michigan following the paths laid out by large-scale promotional maps—that is, by maps showing a smaller area, like John Farmer's *An Improved Edition of a Map of the Surveyed Part of the Territory of Michigan* (1835; figure 13.7). Farmer's map and those designed by land speculators depicted a territory that was not only neatly subdivided according to the township system decreed by the U.S. Land Ordinance Act of 1785, but that contained the core elements of modern infrastructure, including roads, towns, river crossings, and so forth.

In the case of Kirkland's narrative, however, after enduring detours, impassible roads, and nonexistent settlements, the characters (and readers) quickly discovered that the map preceded the territory—that is, that the details on the map were imagined projections rather than factual representations of conditions on the ground. Or, as Kirkland wrote, according to an "emblazoned chart" advertising Michigan, "There were canals and railroads, with boats and cars at full speed. There was a steam-mill, a windmill or two; for even a land-shark did not dare to put a stream where there was scarce running water for the cattle; and a state-road, which had at least been talked of, and a courthouse and other county buildings." She reinforced this picture of advertising hype—more accurately, of the way land speculators tricked prospective settlers—by quipping that "besides all this, there was a large and elegantly decorated space for the name of the happy purchaser."[26] Similar tales commenting on the way maps bewitched sensible characters, or bemoaning the blind trust in the authority of maps, can be found in the sea novels by Herman Melville (*Mardi*, 1849; *Moby-Dick*, 1851), in the gothic tales of Edgar Allan Poe (*The Narrative of Arthur Gordon Pym*, 1837; "The Gold Bug," 1843), and in Charles Dickens's American novel, *The Life and Adventures of Martin Chuzzlewit* (1843–1844), which included a testy account of how promotional maps made false promises and lured unsuspecting emigrants to the American frontier.

Two classic American novels in particular intertwined the graphic logic of cartography with the delineation of territorial integrity, plot design, and the readerly habit of tracing stories with map in hand: Harriet Beecher Stowe's *Uncle Tom's Cabin* (1852) and Mark Twain's *The Adventures of Huckleberry Finn* (1885). In chapter 14 of Stowe's abolitionist novel, popular river maps—like *Norman's Chart of the Lower Mississippi River* (1858; figure 13.8)—were at the heart of one of the novel's more poignant scenes of literary mapping. Sitting on the moving deck of a riverboat, Tom surveyed the Southern plantation landscape from the perspective of the mapmaker, in the process revealing as much as hiding the conditions on land, here how the images of neatly surveyed plantations disguised the fact that their picturesque appearances were founded on the brutal regime of slavery. Throughout, the novel's plot presupposed a cartoliterate audience capable of following the fate of the principal characters, as they traveled southward on the river from state to state into the heart of American slavery

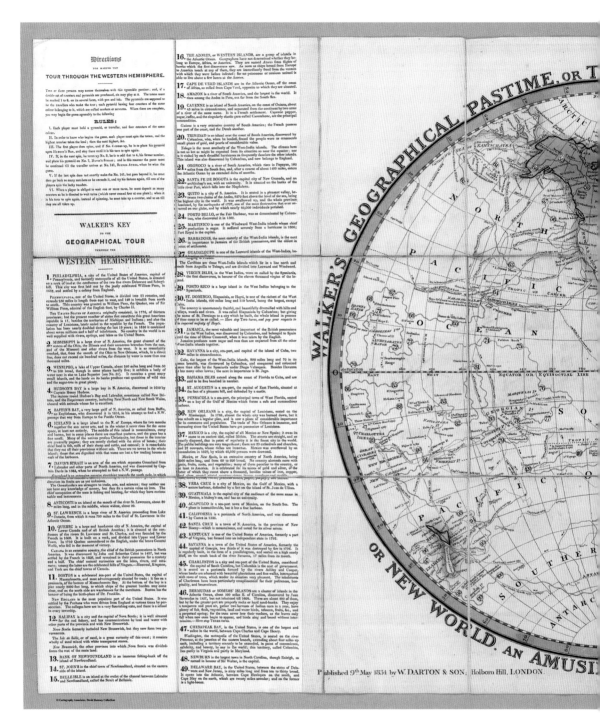

Figure 13.6

John Walker, *Walker's Geographical Pastime* (London, 1834). David Rumsey Map Collection, www.davidrumsey.com.

Figure 13.6 (continued)

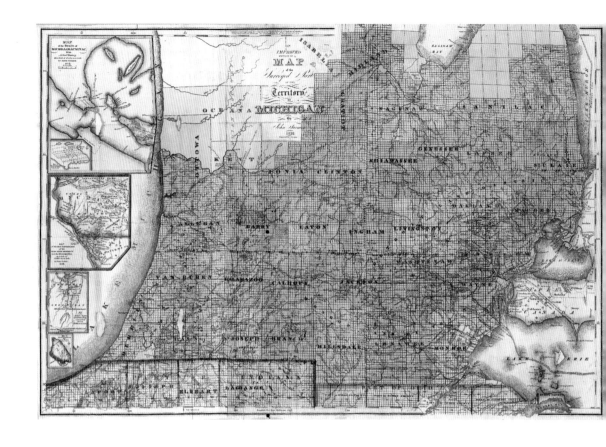

Figure 13.7

John Farmer, *An Improved Edition of a Map of the Surveyed Part of the Territory of Michigan* (New York, [1835] 1836). David Rumsey Map Collection, www.davidrumsey.com.

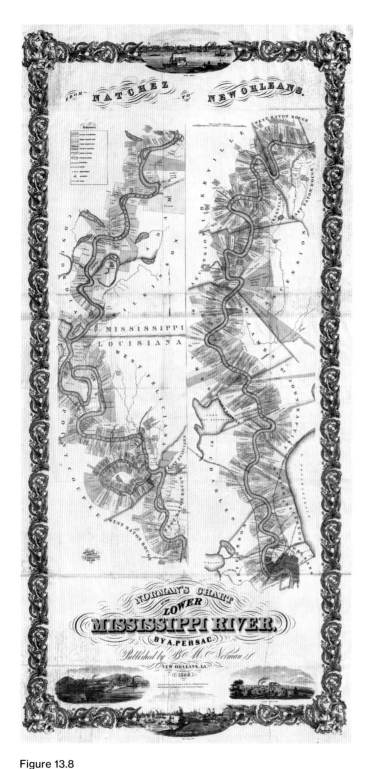

Figure 13.8

B. M. Norman, *Norman's Chart of the Lower Mississippi River by A. Persac* (New York, 1858). David Rumsey Map Collection, www.davidrumsey.com.

and northward along the treacherous route toward the Canadian border and freedom. While Stowe only implied the river map, she offered a more concrete map source in an episode reminiscent of the more academic textbook lessons discussed above: the New England character of Miss Ophelia (and thus Stowe's implied readership) studied the course of the river using "Morse's Atlas," linking the novel's plot either to Jedidiah Morse's *Modern Atlas Adapted to Morse's New School Geography* (1822) or to *Morse's North American Atlas* (1845), published by his son, Sidney E. Morse.

River maps loomed even larger in Twain's famous tale about childhood and rebellion, Reconstruction America, and the culture of racism. What is well known today is that as a former river pilot Twain used his intimate knowledge of the Mississippi River when creating verbal depictions of its meandering course, overgrown banks, and islands, thus adding descriptive authenticity to the otherwise fictional construction of the characters' vernacular speech and actions. What is perhaps less known is the fact that Twain was childhood friends with river pilots who not only helped create actual river maps, such as *Lloyd's Map of the Lower Mississippi River from St. Louis to the Gulf of Mexico* (1862; figure 13.9), but also served as models for Twain's fictional character of Tom Sawyer.[27]

More generally, the river map left its generic mark on the way both Stowe and Twain structured their novels' narrative arc. River maps tended to be designed in the tradition of the "strip" or "ribbon" road maps invented by John Ogilby, and thus would consist of narrow paper sections showing only the land located to the left and right of the river. At the same time, they were often assembled as scrolls that were housed in wooden boxes with a display window, allowing pilots (and also travelers) to manually turn the ribbon map to match the boat's position on the river. The narrative structures of the two novels behave similarly to these pilot maps: characters rarely stray from the river's course; compared to the detailed descriptions of the river, the hinterland remains mostly invisible; and the rapid movement of the novels' scenes, which critics today tend to consider a forerunner of a modern "filmic" technique and sensibility, resembles the mechanical motion of scrolling through pilot or ribbon maps (while anticipating the future reels of celluloid pictures).

Maps in the Literature of Two Gilded Ages, 1900/2000

The absence of maps representing places and providing orientation in the old and new canons of American fiction continued into the twentieth century. If we exclude nonfiction writings, a bibliographical subject search of paratextual metadata (using OCLC) reveals only a smattering of printed maps supplementing novels, poems, and plays. When maps were present, they had usually been added to anniversary editions or lesson plans for teaching classic works such as Melville's *Moby-Dick* or F. Scott Fitzgerald's *The Great Gatsby*.[28] Yet, the cartographic imagination abounds in

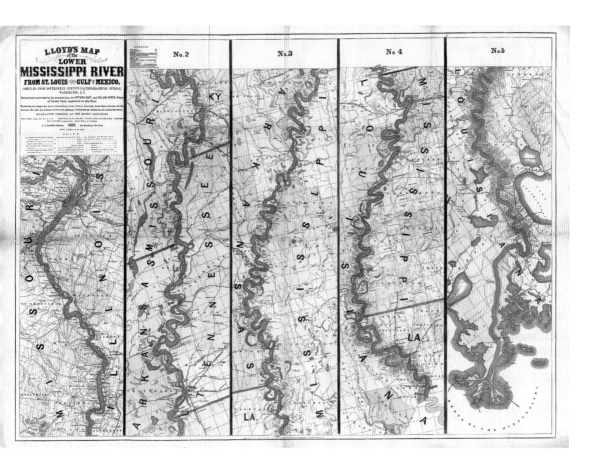

Figure 13.9

J. T. Lloyd, *Lloyd's Map of the Lower Mississippi River from St. Louis to the Gulf of Mexico* (New York, 1862). David Rumsey Map Collection, www.davidrumsey.com.

American literature, in particular in novels inspired by the rise of the modern city and suburbs, technological change, and transnational awareness. In particular, authors specializing in genres of "lowbrow" fiction, like the penny-press Western or detective mystery, emulated the graphic code of cartographic representation to give depth and structure to imaginary landscapes, characters, and actions. For example, Owen Wister's *The Virginian* (1902) and Zane Grey's *Riders of the Purple* Sage (1912) applied the language of cartography and the representational value of "maps" to better delineate places and perspectives, in particular when creating the illusion of utopian, islandlike settings located in the unruly territories of the western continent. Conversely, modernist mappings by authors like John Dos Passos playfully enlisted the reading experience of urban street maps. In the case of Faulkner's conception of "Yoknapatawpha County," a unique insert map accompanying *Absalom! Absalom!* offered graphic guidance to multiple plot strands and point of views, in short becoming a cartographic blueprint for tracking a multitude of objects, emotions, and voices. Through the invocation of the visual experience of the "map," twentieth-century literary mappings opened up fictive worlds that were multidimensional. Their spatiality was contingent on a dialectic pitting geographies against antegeographies, according to which imaginary places not only existed outside known spaces but through the act of mapping appeared similar to or contiguous with the physical geography of the here and now.[29]

Indeed, if we consider American literature a discourse network "1900/2000" (to borrow Friedrich Kittler's term), thus bracketing modern American literary history, we will discover three literary genres devoted explicitly to maps: the grand-tour account; journeys to fictional worlds; and the American road trip.

The theme of the mapped "grand tour" entered American fiction during the antebellum decades in the form of fictional or personal travel accounts describing visits to South America, the European continent, and the Holy Land for both children and adults. In works written for both audiences, rudimentary map inserts served the dual didactic purpose of teaching historical and national geography while providing blueprints for tracing the movement of fictionalized guides in works such as Samuel G. Goodrich's *Peter Parley's History of the Wanderings of Tom Starboard* (1834) or John L. Stephens's *Incidents of Travel in Egypt, Arabia Petraea, and the Holy Land* (1837).[30] After the Civil War, a burgeoning market for young adult fiction transformed the grand tour into gender-specific genres. Offering geographic authentication and lectoral guidance, map inserts were thus commonly found in Elizabeth W. Champney's "Vassar girls" adventures (see her *Three Vassar Girls in South America: A Holiday Trip of Three College Girls ...*, 1885) and in Hezekiah Butterworth's multibook "Zigzag" series for boys. For example, in Butterworth's *Zigzag Journeys in the Western States of America: The Atlantic to the Pacific* (1884), a foldout map was used to illustrate the size of the United States by fitting Europe into the space between California and the Mississippi River, while the volume's endpapers contained map fragments to illustrate diverse travel routes next to images showing various modes of transportation (figure 13.10). Each of these fictional travel accounts, wittingly or not, commented on the travelers' shifting perception of time and space, the impact of speed, or the experience of rapid relocation using novel forms of transportation. Inspired by Jules Verne's novel, *Around the World in Eighty Days* (1873), the American journalist Nellie Bly made headlines by embarking on the ultimate "grand tour"—later described with a map in *Nellie Bly's Book: Around the World in 72 Days* (1890)—in which she raced around the globe in order to break Verne's imaginary travel record, with newspaper audiences tracking her progress via maps and telegraphed reports.[31]

Changes in society and technology resonated strongly in the first American works of science fiction and later in the burgeoning genre of fantasy novels. Today, readers are perhaps most familiar with maps emphasizing the kind of stationary geographies, sweeping vistas of countries, continents, or parallel worlds exemplified by the extraterrestrial canals and waterways of Edgar Rice Burroughs's *A Princess on Mars* (1912) or the landscapes of Middle Earth in J. R. R. Tolkien's trilogy *Lord of the Rings* (1937–1949). Other imaginative settings occur in well-known works like Frank Baum's *Wizard of Oz* (1900), or in Christopher Paolini's bestselling children's book *Eragon* (2003). Less familiar are the fictional maps showing imaginary journeys to and territorial designs of unknown continents, the moon, and the deep sea that emerged parallel to scientific reconnaissance missions into the heart of Africa and Asia, reflecting innovations in engineering and military experiments that relied on steamships, submarines, balloons, and railroads.

Popular novels by Jules Verne and his imitators capitalized on what must have been an insatiable appetite for stories that mixed sensational plots and scientific theories. A story that caught the attention of American readers for nearly a century revolved around the "Hollow Earth" theory. In 1818, and thus long before the publication of Verne's *Journey to the Center of the Earth* (1864), the American Captain John Cleves Symmes introduced the theory of a Hollow Earth containing concentric spheres accessible via large openings (1,400 miles across) at both poles. Symmes's theory inspired the anonymous science fiction novel *Symzonia: Voyage of Discovery* (1820), in which the narrator, after gaining access to the inner surface of the earth at the South Pole, discovered a utopian society living deep inside the earth. The Hollow Earth fantasy of finding a parallel universe containing advanced albeit subterranean civilizations quickly became a staple of American popular fiction. Besides influencing Poe's *The Narrative of Arthur Gordon Pym* (1838) and purportedly also Melville's *Moby-Dick* (1851), the Hollow Earth story had perhaps its greatest following in John Uri Lloyd's science fiction and fantasy novel *Etidorhpa, or, The End of Earth* (1895), in which the narrator, on receiving instructions from an underground inhabitant and with map in hand, entered the earth's interior continents through a cave in Kentucky. American authors embraced Symmes's theory well into the twentieth century, but none offered as explicit

Figure 13.10

Hezekiah Butterworth, *Zigzag Journeys in the Western States of America: The Atlantic to the Pacific* (1884). Courtesy, Newberry Library.

a map as did William Reed's illustrated study of polar expeditions, *The Phantom of the Poles* (1906), showing the "Hollow Earth" (figure 13.11).

The final prominent genre containing the occasional map consists of "road" fiction and biographies that involved traveling long distances across the United States by car and more recently also by motorcycle, bicycle, and even on foot. Since the advent of the mass-produced car, the American media, and in particular car advertisements, have coined the "road trip" as a fantasy of unfettered geographic and social mobility. Fiction writers, many of whom were working freelance for movie and advertising companies, developed the theme of road travel into a productive metaphor, plot device, agent of action, even character—think Ian Fleming's *Chitty-Chitty-Bang-Bang: The Magical Car* (1964) or Stephen King's horror novel *Christine* (1983)—to convey life in a fast-paced America that was always on the move.

Classic twentieth-century narratives—from F. Scott Fitzgerald's *The Great Gatsby* (1924) and John Steinbeck's *The Grapes of Wrath* (1938) to Jack Kerouac's *On the Road* (1957) and Toni Morrison's *Song of Solomon* (1977)—created elaborate literary maps exploring the exigencies of modern travel in American society. But while these novels omitted physical maps, autobiographical writings of authors ranging from John Steinbeck to William Least Heat-Moon purposefully included maps as part of the literary experience. The endpapers of Steinbeck's travelogue, *Travels with Charley* (1962), depicted a road trip around the periphery of the United States with the goal of finding some answers to the question "What are Americans like today?" Heat-Moon's *Blue Highways* (1982), whose title came from the cartographic convention of drawing secondary roads in the color blue (figure 13.12), recounted the author's adventures and soul-searching meditations compiled during a 13,000-mile journey across America and through mostly forgotten small towns.

***

When seeking to reconcile the absence of map inserts with the widespread presence of verbal mappings in American literature, readers of fiction will quickly discover an impasse formulated best by Henri Lefebvre's assertion that "any search for space in literary texts will find it everywhere and in every guise."[32] But when Lefebvre asks the question "How many maps, in the descriptive or geographic sense, might be needed to deal exhaustively with a given space, to code and decode all its meanings and contents?," the answer for readers of American literature would have to be "many."[33] Inspired by spatial theories like Lefebvre's, recent approaches to American literature tend to be informed by critical methodologies whose discussion of maps and mapping has provided new perspectives and energy to literary criticism. Studies invested in "literary cartography" frequently call on terms and protocols developed by critics working on the margins of literary criticism—for example, by Michel de Certeau (maps are "a memorandum prescribing actions"), Gilles Deleuze and Felix Guattari ("the map does not reproduce an unconscious closed in upon itself; it constructs the unconscious"), or

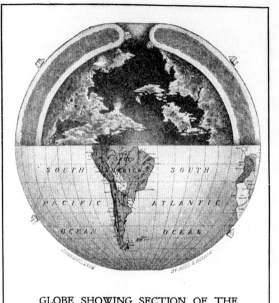

Figure 13.11
William Reed, *The Phantom of the Poles* (1906).
Courtesy, American Antiquarian Society.

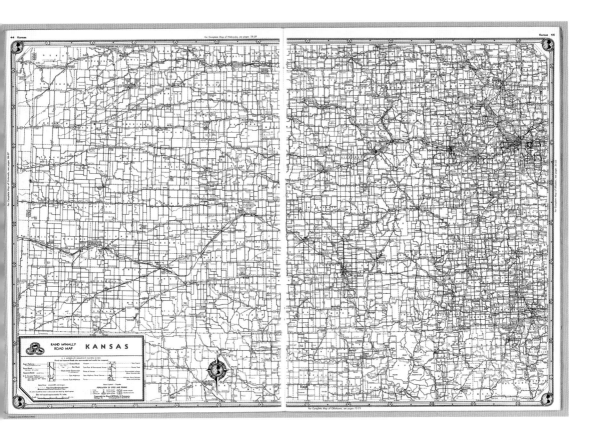

Figure 13.12

Rand McNally Road Map, Kansas (1940). David Rumsey Map Collection, www.davidrumsey.com.

Jean Baudrillard ("the map precedes the territory").[34] As the discourse on maps increasingly pervaded discussions of literary form and the tools of literary analysis, it was only a short leap to imbue established critical terms with a mapping function, from the classical rhetorical "topos," to the early modern neologism of "plot," to poststructural uses of "survey" and "metonymy." Moreover, after recognizing the representational power of maps, literary scholars now speculate about how maps have affected literary culture in America more broadly. If we consider Benedict Anderson's critique of nationalism in relation to the history of the novel and Edward Said's exploration of culture and imperialism, not to mention more recent discussions of transnational literature and identity, the discourse on maps and the language of mapping offer both a corrective lens and a constructive idiom for addressing a broad range of American literatures across time and space.[35]

That maps hold a unique sway over the way we conduct American literary studies can perhaps best be seen in debates over the configuration of American literary history in which maps have continually shaped our perspective on authors, genres, and the canon. Between the 1920s and 1950s (thus during the founding decades of American literary history as a scholarly field), wall posters titled the "Booklovers Map of America: A Chart of Certain Landmarks of Literary Geography" (1926) or "A Pictorial Map Depicting the Literary Development of the United States" (1952) offered snapshots of a map-based conception of literary production and consumption that plotted out book titles, iconic images, and author names across the surface of the national map (figure 13.13).

Recent editions of *The Norton Anthology of American Literature* or *The Heath Anthology of American Literature* still include maps in the endpapers—"North America to 1700" in front and "The United States: A Literary View" in back—thus suggestively turning the map into the graphic and material container of literary history.[36] Indeed, if we accept Franco Moretti's approaches to the novel, the conceptual tools of literary cartography emerge as a productive critical device because only by mapping the text (and Moretti really wants us to use graphic maps) will we "bring ... to light relations that would otherwise remain hidden."[37] With the aid of the computational powers of "literary GIS" applications, new methodological models are now able to recover map-based and map-producing approaches to and interpretations of American literature. Through digitization, both the rare map insert and the ubiquitous word map are easily transformed into mappable metadata, so that any given text now can be equipped with a map discourse function capable of bridging the gap separating textual and cartographic representations of spaces that over time we have come to consider as real as they were understood to be imaginary.

Figure 13.13

Amy Jones, *The Booklover's Map of the United States* (1949). Courtesy, Library of Congress. © 1949 R.R. Bowker, a ProQuest LLC affiliate. Amy Jones, Designer and Illustrator.

Notes

1. For sample approaches see Robert Lawson-Peebles, *Landscape and Written Expression in Revolutionary America* (Cambridge: Cambridge University Press, 1988); Myra Jehlen, "The Literature of Colonization," in *The Cambridge History of American Literature*, ed. Sacvan Bercovitch, vol. 1, 11–168 (Cambridge: Cambridge University Press, 1994); Lawrence Buell, *The Environmental Imagination: Thoreau, Nature Writing, and the Formation of American Culture* (Cambridge, MA: Harvard University Press, 1995); Hsuan L. Hsu, *Geography and the Production of Space in Nineteenth-Century American Literature* (Cambridge: Cambridge University Press, 2010).

2. Stephen Greenblatt, *Marvelous Possessions: The Wonder of the New World* (Chicago: University of Chicago Press, 1992); Ricardo Padrón, *The Spacious Word: Cartography, Literature, and Empire in Early Modern Spain* (Chicago: University of Chicago Press, 2004).

3. David Cooper and Ian N. Gregory, "Mapping the English Lake District: A Literary GIS," *Transactions of the Institute of British Geographers* 36 (2011): 89–108; Matthew Wilkens, "The Geographic Imagination of Civil War Era American Fiction," *American Literary History* 25, no. 4 (2013): 803–840; Franco Moretti, *Atlas of the European Novel* (London: Verso, 1998).

4. Sacvan Bercovitch, ed., *The Cambridge History of American Literature*, vol. 1 (Cambridge: Cambridge University Press, 1994); Myra Jehlen and Michael Warner, *English Literatures of America, 1500–1800* (New York: Routledge, 1996); Paul Lauter, ed., *The Heath Anthology of American Literature*, 7th ed. (New York: Cengage Learning, 2014).

5. Wayne Franklin, *Discoverers, Explorers, Settlers: The Diligent Writers of Early America* (Chicago: University of Chicago Press, 1979); Martin Brückner, *The Geographic Revolution in Early America: Maps, Literacy, and National Identity* (Chapel Hill: University of North Carolina Press, 2006); Martha Schoolman, *Abolitionist Geographies* (Minneapolis: University of Minnesota Press, 2014).

6. Phillip C. Muehrcke and Juliana O. Muehrcke, "Maps in Literature," *Geographical Review* 64, no. 3 (July 1974): 317–338; Peter Turchi, *Maps of the Imagination: The Writer as Cartographer* (San Antonio, TX: Trinity University Press, 2007); Ricardo Padrón, "Mapping Imaginary Worlds," in *Maps: Finding Our Place in the World*, ed. James R. Akerman and Robert W. Karrow Jr., 255–287 (Chicago: University of Chicago Press, 2007).

7. Nina Baym, ed., *The Norton Anthology of American Literature*, 8th ed. (New York: Norton, 2011); Lauter, *Heath Anthology*.

8. Edmundo O'Gorman, *The Invention of America: An Inquiry into the Historical Nature of the New World and the Meaning of Its History* (Bloomington: University of Indiana Press, 1961); Seymour L. Schwartz, *Putting "America" on the Map: The Story of the Most Important Graphic Document in the History of the United States* (Amherst, NY: Prometheus Books, 2007).

9. Greenblatt, *Marvelous Possessions*; Mary B. Campbell, *Wonder and Science: Imagining Worlds in Early Modern Europe* (Ithaca, NY: Cornell University Press, 1999); Ralph Bauer, *The Cultural Geography of Colonial American Literatures: Empire, Travel, Modernity* (Cambridge: Cambridge University Press, 2003).

10. Karen Ordahl Kupperman, *America in European Consciousness, 1493–1750* (Chapel Hill: University of North Carolina Press, 1995); Gavin Hollis, "The Wrong Side of the Map? The

Cartographic Encounters of John Lederer," in *Early American Cartographies*, ed. Martin Brückner, 145–168 (Chapel Hill: University of North Carolina Press, 2011).

11. Lauren Benton, *A Search for Sovereignty: Law and Geography in European Empires, 1400–1900* (Cambridge: Cambridge University Press, 2010).

12. Martin W. Lewis and Kären E. Wigen, *The Myth of Continents: A Critique of Metageography* (Berkeley: University of California Press, 1997); Brückner, *Geographic Revolution*, 76–95.

13. This depiction resonates strongly with map-based or map-supplemented texts, including Thomas Jefferson's *Notes on the State of Virginia* (1784) and John Filson's *Kentucke* (1784). See Leo Marx, *The Machine in the Garden* (London: Oxford University Press, 1964); Franklin, *Discoverers, Explorers, Settlers*; Thomas Hallock, *From the Fallen Tree: Frontier Narratives, Environmental Politics, and the Roots of a National Pastoral, 1749–1826* (Chapel Hill: University of North Carolina Press, 2003).

14. Mark Warhus, *Another America: Native American Maps and the History of Our Land* (New York: St. Martin's Press, 1997).

15. Franklin, *Discoverers, Explorers, Settlers*, 123–135.

16. Brückner, *Geographic Revolution*, 132–134.

17. Anette Kolodny, *The Land Before Her: Fantasy and Experience of the American Frontiers, 1630–1860* (Chapel Hill: University of North Carolina Press, 1984); Malcolm Lewis, "Rhetoric of the Western Interior: Modes of Environmental Description in American Promotional Literature of the Nineteenth Century," in *The Iconography of Landscape: Essays on the Symbolic Representation, Design and Use of Past Environments*, ed. Denis Cosgrove and Stephen Daniels, 179–193 (Cambridge: Cambridge University Press, 1988); Buell, *The Environmental Imagination*.

18. John Logan Allen, *Passage through the Garden: Lewis and Clark and the Image of the American Northwest* (Urbana: University of Illinois Press, 1975); Anne Baker, *Heartless Immensity: Literature, Culture, and Geography in Antebellum America* (Ann Arbor: University of Michigan Press, 2006).

19. Buell, *The Environmental Imagination*, 268–279.

20. Noah Webster, "On the Education of Youth in America" [Boston, 1788], in *Essays on Education in the Early Republic*, ed. Frederick Rudolph (Cambridge, MA: Harvard University Press, 1965), 77.

21. Brückner, *Geographic Revolution,* 169–172.

22. Martin Brückner, "The Cartographic Turn in American Literary Studies: Of Maps, Mappings, and the Limits of Metaphor," in *Turns of Events: American Literary Studies in Motion*, ed. Hester Blum, 44–72 (Philadelphia: University of Pennsylvania Press, 2016).

23. Martin Brückner, *The Social Life of Maps in America, 1750–1860* (Chapel Hill: University of North Carolina Press, 2017); Scott E. Caspar, Jeffrey D. Groves, Stephen W. Nissenbaum, and Michael Winship, eds., *A History of the Book in America*, Vol. 3: *The Industrial Book, 1840–1880*, 1–89 (Chapel Hill: University of North Carolina Press, 2007).

24. Susan Schulten, *Mapping the Nation: History and Cartography in Nineteenth-Century America* (Chicago: University of Chicago Press, 2012).

25. Cathy Davidson, *Revolution and the Word* (New York: Oxford University Press, 1986).

26. Caroline M. Kirkland, *A New Home–Who'll Follow? or Glimpses of Western Life?*, ed. Sandra Z. Zagarell (New Brunswick, NJ: Rutgers University Press, [1839] 1990), 83; see also 31.

27. Robert A. Holland, *The Mississippi River in Maps & Views: From Lake Itasca to the Gulf of Mexico* (New York: Rizzoli, 2008).

28. Martha Hopkins and Michael Buscher, *The Language of the Land: The Library of Congress Book of Literary Maps* (Washington, DC: Library of Congress, 1999).

29. Padrón, "Mapping Imaginary Worlds"; Turchi, *Maps of the Imagination*; Anthony Pavlik, "'A Special Kind of Reading Game': Maps in Children's Literature," *International Research in Children's Literature* 3, no. 1 (2010): 28–43.

30. Bruce E. Harvey, *American Geographics: U.S. National Narratives and the Representation of the Non-European World, 1830–1865* (Stanford, CA: Stanford University Press, 2001).

31. Edlie L. Wong, "Around the World and across the Board: Nellie Bly and the Geography of Games," in *American Literary Geographies*, ed. Martin Brückner and Hsuan L. Hsu, 296–324 (Newark: University of Delaware Press, 2007).

32. Henri Lefebvre, *The Production of Space*, trans. Donald Nicholson-Smith (Cambridge: Blackwell, 1991), 15.

33. Lefebvre, *Production of Space*, 85.

34. Michel de Certeau, *The Practice of Everyday Life*, trans. Stephen Rendall (Berkeley: University of California Press, 1984), 120; Gilles Deleuze and Félix Guattari, *A Thousand Plateaus: Capitalism and Schizophrenia*, trans. Brian Massumi (Minneapolis: University of Minnesota Press, 1987), 12; Jean Baudrillard, *Simulacra and Simulacrum*, trans. Sheila Faria Glaser (Ann Arbor: University of Michigan Press, 1994), 1.

35. Benedict Anderson, *Imagined Communities: Reflections on the Origin and Spread of Nationalism* (London: Verso, 1991); Edward W. Said, *Culture and Imperialism* (New York: Vintage, 1993).

36. Baym, *Norton Anthology of American Literature*.

37. Franco Moretti, *Atlas of the European Novel*, 3; see also his *Graphs, Trees, Maps* (London: Verso, 2005).

Bibliography

Adams, Percy G. *Travel Literature and the Evolution of the Novel*. Lexington: University of Kentucky Press, 1983.

Adler, Nanci J. "The Bicycle in Western Literature: Transformations on Two Wheels." 2012. Masters of Liberal Studies Theses, Paper 22. Rollins Scholarship Online.

Akerman, James R., and Robert W. Karrow Jr., eds. *Maps: Finding Our Place in the World*. Chicago: University of Chicago Press, 2007.

Allen, John Logan. *Passage through the Garden: Lewis and Clark and the Image of the American Northwest*. Urbana: University of Illinois Press, 1975.

Anderson, Benedict. *Imagined Communities: Reflections on the Origin and Spread of Nationalism*. London: Verso, 1991.

Baker, Anne. *Heartless Immensity: Literature, Culture, and Geography in Antebellum America*. Ann Arbor: University of Michigan Press, 2006.

Baudrillard, Jean. *Simulacra and Simulacrum*. Trans. Sheila Faria Glaser. Ann Arbor: University of Michigan Press, 1994.

Bauer, Ralph. *The Cultural Geography of Colonial American Literatures: Empire, Travel, Modernity*. Cambridge: Cambridge University Press, 2003.

Baym, Nina, et al., eds. *The Norton Anthology of American Literature*. New York: Norton, 1989–present.

Benton, Lauren. *A Search for Sovereignty: Law and Geography in European Empires, 1400–1900*. Cambridge: Cambridge University Press, 2010.

Bercovitch, Sacvan, ed. *The Cambridge History of American Literature*. Vol. 1. Cambridge: Cambridge University Press, 1994.

Boelhower, William. "Inventing America: A Model of Cartographic Semiosis." *Word & Image* IV (1988): 475–497.

Brückner, Martin. "The Cartographic Turn in American Literary Studies: Of Maps, Mappings, and the Limits of Metaphor." In *Turns of Events: American Literary Studies in Motion*, ed. Hester Blum, 44–72. Philadelphia: University of Pennsylvania Press, 2016.

Brückner, Martin. *The Geographic Revolution in Early America: Maps, Literacy, and National Identity*. Chapel Hill: University of North Carolina Press, 2006.

Brückner, Martin. *The Social Life of Maps in America, 1750–1860*. Chapel Hill: University of North Carolina Press, 2017.

Brückner, Martin, and Hsuan L. Hsu. *American Literary Geographies*. Newark: University of Delaware Press, 2007.

Buell, Lawrence. *The Environmental Imagination: Thoreau, Nature Writing, and the Formation of American Culture*. Cambridge, MA: Harvard University Press, 1995.

Bulson, Eric. *Novels, Maps, Modernity: The Spatial Imagination, 1850–2000*. London: Routledge, 2007.

Campbell, Mary B. *Wonder and Science: Imagining Worlds in Early Modern Europe*. Ithaca, NY: Cornell University Press, 1999.

Caspar, Scott E., Jeffrey D. Groves, Stephen W. Nissenbaum, and Michael Winship, eds. *A History of the Book in America*, Vol. 3: The Industrial Book, 1840–1880. Chapel Hill: University of North Carolina Press, 2007.

de Certeau, Michel. *The Practice of Everyday Life*. Trans. Stephen Rendall. Berkeley: University of California Press, 1984.

Cooper, David, and Ian N. Gregory. "Mapping the English Lake District: A Literary GIS." *Transactions of the Institute of British Geographers* 36 (2011): 89–108.

Davidson, Cathy. *Revolution and the Word*. New York: Oxford University Press, 1986.

Deleuze, Gilles, and Félix Guattari. *A Thousand Plateaus: Capitalism and Schizophrenia*. Trans. Brian Massumi. Minneapolis: University of Minnesota Press, 1987.

Dimock, Wai Chee. *Through Other Continents: American Literature across Deep Time.* Princeton, NJ: Princeton University Press, 2008.

Franklin, Wayne. *Discoverers, Explorers, Settlers: The Diligent Writers of Early America.* Chicago: University of Chicago Press, 1979.

Giles, Paul. *The Global Remapping of American Literature.* Princeton, NJ: Princeton University Press, 2011.

Gilroy, Paul. *The Black Atlantic: Modernity and Double-Consciousness.* Cambridge, MA: Harvard University Press, 1993.

Greenblatt, Stephen. *Marvelous Possessions: The Wonder of the New World.* Chicago: University of Chicago Press, 1992.

Gross, R. A., and M. Kelley, eds. A History of the Book in America, Vol. 2: *An Extensive Republic: Print, Culture, and Society in the New Nation, 1790–1840.* Chapel Hill: University of North Carolina Press, 2010.

Hallock, Thomas. *From the Fallen Tree: Frontier Narratives, Environmental Politics, and the Roots of a National Pastoral, 1749–1826.* Chapel Hill: University of North Carolina Press, 2003.

Harvey, Bruce E. *American Geographics: U.S. National Narratives and the Representation of the Non-European World, 1830–1865.* Stanford, CA: Stanford University Press, 2001.

Holland, Robert A. *The Mississippi River in Maps & Views: From Lake Itasca to the Gulf of Mexico.* New York: Rizzoli, 2008.

Hollis, Gavin. "The Wrong Side of the Map? The Cartographic Encounters of John Lederer." In *Early American Cartographies*, ed. Martin Brückner, 145–168. Chapel Hill: University of North Carolina Press, 2011.

Hopkins, Martha, and Michael Buscher. *The Language of the Land: The Library of Congress Book of Literary Maps.* Washington, DC: Library of Congress, 1999.

Howse, D., and J. W. Norman Thrower, eds. *A Buccaneer's Atlas: Basil Ringrose's South Sea Waggoner.* Berkeley: University of California Press, 1992.

Hsu, Hsuan L. *Geography and the Production of Space in Nineteenth-Century American Literature.* Cambridge: Cambridge University Press, 2010.

Huggan, G. Graham. *Territorial Disputes: Maps and Mapping Strategies in Contemporary Canadian and Australian Fiction.* Toronto: University of Toronto Press, 1994.

Jacob, Christian. *The Sovereign Map: Theoretical Approaches in Cartography throughout History.* Ed. Edward H. Dahl, trans. Tom Conley. Chicago: University of Chicago Press, 2006.

Jehlen, Myra. "The Literature of Colonization." In *The Cambridge History of American Literature*, vol. 1, ed. Sacvan Bercovitch. Cambridge: Cambridge University Press, 1994.

Jehlen, Myra, and Michael Warner. *English Literatures of America, 1500–1800.* New York: Routledge, 1996.

Kaveney, Roz. "Maps." In *The Encyclopedia of Fantasy*, ed. John Clute and John Grant. London: Orbit, 1999.

Kirkland, Caroline M. *A New Home—Who'll Follow? or Glimpses of Western Life?* Ed. Sandra Z. Zagarell. New Brunswick, NJ: Rutgers University Press, [1839] 1990.

Kolodny, Anette. *The Land Before Her: Fantasy and Experience of the American Frontiers, 1630–1860.* Chapel Hill: University of North Carolina Press, 1984.

Kupperman, Karen Ordahl. *America in European Consciousness, 1493–1750*. Chapel Hill: University of North Carolina Press, 1995.

Lackey, Kris. *RoadFrames: The American Highway Narrative*. Lincoln: University of Nebraska Press, 1999.

Lauter, Paul, ed. *The Heath Anthology of American Literature*. 7th ed. New York: Cengage Learning, 2014.

Lauter, Paul. *Reconstructing American Literature*. New York: Feminist Press, 1984.

Lawson-Peebles, Robert. *Landscape and Written Expression in Revolutionary America*. Cambridge: Cambridge University Press, 1988.

Lefebvre, Henri. *The Production of Space*. Trans. Donald Nicholson-Smith. Cambridge: Blackwell, 1991.

Lewis, Malcolm. "Rhetoric of the Western Interior: Modes of Environmental Description in American Promotional Literature of the Nineteenth Century." In *The Iconography of Landscape: Essays on the Symbolic Representation, Design and Use of Past Environments*, ed. Denis Cosgrove and Stephen Daniels. Cambridge: Cambridge University Press, 1988.

Lewis, Martin W., and Kären E. Wigen. *The Myth of Continents: A Critique of Metageography*. Berkeley: University of California Press, 1997.

Marx, Leo. *The Machine in the Garden*. London: Oxford University Press, 1964.

McMillin, T. S. *The Meaning of Rivers: Flow and Reflection in American Literature*. Iowa City: University of Iowa Press, 2011.

Moretti, Franco. *Atlas of the European Novel*. London: Verso, 1998.

Moretti, Franco. *Graphs, Trees, Maps*. London: Verso, 2005.

Muehrcke, Phillip C., and Juliana O. Muehrcke. "Maps in Literature." *Geographical Review* 64, no. 3 (July 1974): 317–338.

Newman, Andrew. *On Records: Delaware Indians, Colonists, and the Media of History and Memory*. Lincoln: University of Nebraska Press, 2012.

O'Gorman, Edmundo. *The Invention of America: An Inquiry into the Historical Nature of the New World and the Meaning of Its History*. Bloomington: University of Indiana Press, 1961.

Padrón, Ricardo. *The Spacious Word: Cartography, Literature, and Empire in Early Modern Spain*. Chicago: University of Chicago Press, 2004.

Padrón, Ricardo. "Mapping Imaginary Worlds." In *Maps: Finding Our Place in the World*, ed. James R. Akerman and Robert W. Karrow Jr., 255–287. Chicago: University of Chicago Press, 2007.

Pagden, Anthony. *European Encounters with the New World: From Renaissance to Romanticism*. New Haven, CT: Yale University Press, 1994.

Pavlik, Anthony. "'A Special Kind of Reading Game': Maps in Children's Literature." *International Research in Children's Literature* 3, no. 1 (2010): 28–43.

Post, J. B. *An Atlas of Fantasy*. Rev. ed. New York: Ballantine Books, 1979.

Said, Edward W. *Culture and Imperialism*. New York: Vintage, 1993.

Schoolman, Martha. *Abolitionist Geographies*. Minneapolis: University of Minnesota Press, 2014.

Schulten, Susan. *The Geographical Imagination in America, 1880–1950*. Chicago: University of Chicago Press, 2002.

Schulten, Susan. *Mapping the Nation: History and Cartography in Nineteenth-Century America*. Chicago: University of Chicago Press, 2012.

Schwartz, Seymour L. *Putting "America" on the Map: The Story of the Most Important Graphic Document in the History of the United States*. Amherst, NY: Prometheus Books, 2007.

Stafford, Barbara. *Voyage into Substance: Art, Science, Nature, and the Illustrated Travel Account, 1760–1840*. Cambridge, MA: MIT Press, 1984.

Sweet, Timothy. *American Georgics: Economy and Environment in American Literature, 1580–1864*. Philadelphia: University of Pennsylvania Press, 2001.

Tompkins, Jane. *Sensational Designs: The Cultural Work of American Fiction 1790–1860*. New York: Oxford University Press, 1985.

Turchi, Peter. *Maps of the Imagination: The Writer as Cartographer*. San Antonio, TX: Trinity University Press, 2007.

Warhus, Mark. *Another America: Native American Maps and the History of Our Land*. New York: St. Martin's Press, 1997.

Webster, Noah. "On the Education of Youth in America." [Boston, 1788.] In *Essays on Education in the Early Republic*, ed. Frederick Rudolph. Cambridge, MA: Harvard University Press, 1965.

Wilkens, Matthew. "The Geographic Imagination of Civil War Era American Fiction." *American Literary History* 25, no. 4 (2013): 803–840.

Wong, Edlie L. "Around the World and across the Board: Nellie Bly and the Geography of Games." In *American Literary Geographies: Spatial Practice and Cultural Production, 1500–1900*, ed. Martin Brückner and Hsuan L. Hsu, 296–324. Newark: University of Delaware Press, 2007.

14
Map Line Narratives

Jörg Dünne

Introduction

Modern maps are composite systems of signs. In the course of their historical development, these sign systems have become characterized by the increasingly standardized structure of coexisting texts, images, and numbers.[1] It is as a composite structure of texts, images, and geometric figures that the map becomes a central factor in terrestrial globalization and the colonization of the American continent, a shift particularly evident in Portuguese and Spanish maps around 1500.[2] Maps thus function as effective surfaces of operation[3] in which the diverse possibilities of their navigational applications are inextricably connected with the locatability of a specific territory through a given address.[4] As a result, the geopolitical correlation between colonial order and geographic positioning[5] is largely achieved through cartographic practices. And yet maps not only serve a specific form of the operationalization of political action, but also an operationalization of the imagination,[6] an exemplary depiction of which may be observed in the relationship between maps and fictional narrative texts.

The increasingly conventionalized coexistence of text, image, and number in the course of the early modern period is also significant for the central question of the following investigation, namely the question concerning the relationship between map and narrative. While medieval *mappaemundi* represent a direct inscription surface for elaborate texts and multiple images alike,[7] both text and image begin to detach somewhat from the surface of the map in the age of the mathematization[8] of those composite sign systems depicted on the map. What remains on maps as traces of a textual dimension are primarily the names of specific locations. The remnants of the image dimension, on the other hand, are conventional cartographic symbols[9] explained in the map's legend.

In terms of the relationship between map and narrative, it is necessary to clarify what becomes of those text passages excluded from the map, such as travel reports and miraculous tales, which the medieval map still seamlessly includes in the notation system of its heavily literary geography.[10] These texts do not simply disappear without a trace, rather they are transferred in part to the accompanying commentaries, as one can observe in the example of the *Carta marina* of the Swedish bishop Olaus Magnus from the year 1539,[11] which develops a differentiated letter-based reference system between map and commentary text (see figure 14.1). The result of such systems is the emergence of new and diverse forms of reciprocal relations. A central aim of the following reflections is to demonstrate how these reciprocal forms are instrumental in transforming maps into a "matrix" for the emergence of a new form of fictional situation in literary texts.

In this sense I would like to suggest an understanding of the early modern mathematization of cartographic spatiality as a sort of "historical a priori" for a mode of literary situation formation, specifically one that facilitates the formation of previously uncommon narrative chronotopes through a cartographic imagination. In the fictional texts of the early modern period, the "here" of the imagined, fictional space of a text[12] becomes increasingly associated with a mappable space[13] and thus contributes to the creation of what in studies of the history of the novel in particular has been called "factual fictions."[14] Maps participate in creating the lasting illusion of space as a given geographic frame in which the plot of a text can be established and described in detail as a form of eventful movement from one cartographically determinable point to another. Since the nineteenth century, however, yet another form of cartography may also be responsible for reevaluating the perception of geographic space, namely the imagination of geographic space as a static framework in which dynamization occurs through a form of movement largely conceived of as anthropocentric—such a form of cartography raises a new set of questions about the relationship between map and narrative.

In the following reflections I would like to return to the relationship between map and narrative in the early modern period to examine the way this relationship is transformed with lasting effects around 1800. More precisely, my aim is to examine how a dynamic depiction of movement comes to find its way into modern cartography as an alternative to the static "here" of a determined position. Much like the historical cartography of the early modern period, this dynamic form appears closely linked to contemporary modes of narration. What I regard as essential for transformations of this sort is the rise of so-called isometric cartography.

First, I differentiate three types of cartographic lines that may be relevant for narrative texts in different ways. I show how this is true by way of a discussion of Yuri M. Lotman's conception of narrative spaces, which can be described as based on a cartographic ordering of spaces since the early modern period. I then proceed to the novels of Jules Verne to illustrate the significance of a major cartographic-historical transformation around 1800 for literature. These novels, I argue, occupy a threshold between an early modern "static" depiction of space through the map and a fundamental dynamization of space through isometric cartography—a dynamization of space that asks for an alternative description of narrative events leading beyond Lotman's static approach to spatial order.

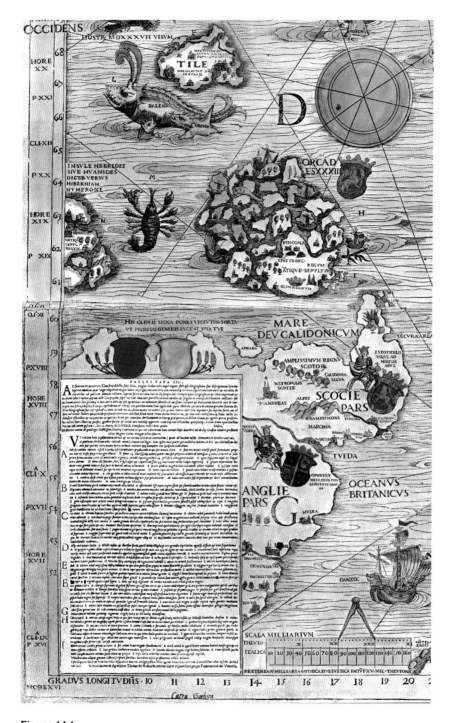

Figure 14.1

Olaus Magnus, *Carta marina* (colored version, 1572, sheets D and G), with large capital letters for each map sheet referring to the legend at the bottom left of the map, and small capital letters referring to a commentary printed separately (e.g., in Italian under the title "Opera breve"). *Source:* https://en.wikipedia.org/wiki/Image:Carta_Marina.jpeg.

Cartographic Lines

As a guiding principle for his specific form of "geopoetics," the writer Kenneth White speaks of the ambition "to read the lines of the world."[15] Yet while White appears to take as a point of departure lines inscribed in the biosphere, I would like to undertake a media and literary-historical application of this endeavor by shifting attention to the cultural techniques of worldmaking through the "practices of lines." In such a sense, cartography plays a crucial role as a practice with which the lines of the world are made visible, or rather with which the surface of the earth is first inscribed with lines.[16]

This occurs in various ways throughout early modern cartography. According to Christian Jacob,[17] a distinction can be made between several basic types of cartographic lines, which will be discussed in this section. The main object of this examination is restricted to cartographic lines in conjunction with the "smooth" space of the sea through which the development of new cartographic practices and methods of position determination are particularly conveyed in the early modern period.

The geometric construction lines of maps constitute the first type of cartographic lines. As early as the Middle Ages, an early type of map begins to take shape in contrast to the traditional *mappaemundi*, specifically utilizing such construction lines for a precise form of orientation or navigation. These are the so-called portolan charts, whose construction lines or rhumb lines follow the compass course of a ship or the particular wind direction (see figure 14.2).

Charted as a point on the map, a vessel's respective position on a portolan chart is determined by tracing a connecting line to another previously determined point, with each connecting line between two positions running parallel to a rhumb line. On the other hand, a second category in the typology of construction lines becomes increasingly important in the sixteenth century, namely the cartographic grid with the coordinates of longitude and latitude first conceived by Ptolemy. As constitutive of the formation of a spatial whole, the graticule makes it possible to determine positions not only in relation to previously set points, but also independently of each charted point on the map by way of two given coordinates in the grid. Construction lines on portolan charts correspond to concrete practices of motion—that is, to the resulting parcours lines, which will be examined in detail in the section to come. But in grid maps, the practices of navigation—the same practices that, as Michel de Certeau notes, mark the first step in the creation of a mappable space—are excluded from the map.[18] Geographic grid maps create the illusion of a given spatial whole detached from all concretely set routes. In creating such an illusion, they utilize a specific form of projection in which the curvature of the earth's surface is taken into account in its "flat" representation on a map. Among the projections developed in the early modern period, the best known is indeed the so-called Mercator projection (see figure 14.3) with its

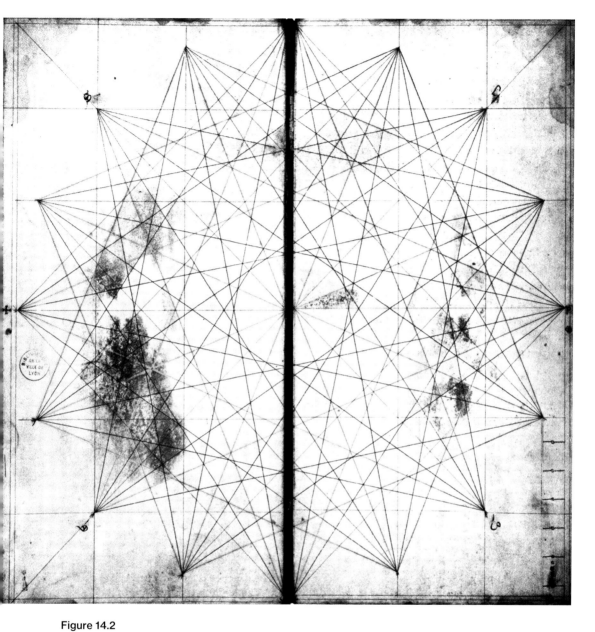

Figure 14.2
"Marteloio": Construction lines of a portolan chart (Bibliothèque de la Ville, Lyon). *Source:* Monique de la Roncière and Michel Mollat du Jourdin, *Portulane: Seekarten vom 13. bis 17. Jahrhundert* (Munich: Hirmer, 1984), 13.

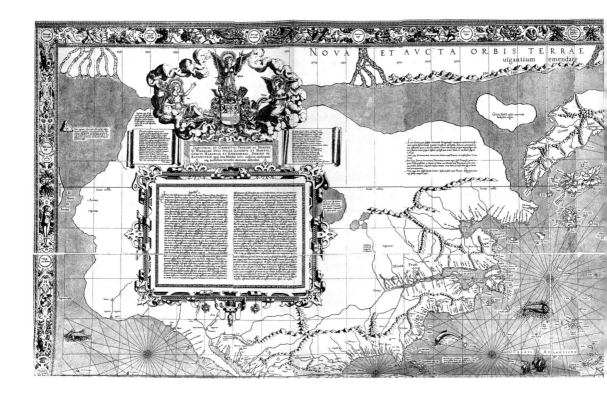

Figure 14.3

Gerardus Mercator, *Nova et Aucta Orbis Terrae Descriptio ad Usum Navigantium Emendate Accommodata* (1569, sheets 1–3, partial view of sheets 7–9). *Source:* Gerhard Mercator, *Weltkarte ad usum navigantium*, ed. Wilhelm Krücken and Joseph Milz (Duisburg: Mercator-Verlag, [1569] 1994).

conformal projection allowing lines of latitude and longitude to intersect consistently in a ninety-degree angle.

Parcours lines, which in most cases are not permanently drawn on maps, can be distinguished from construction lines in that they allow maps to be utilized as instruments of proper navigation practices. Parcours lines play a specifically important role on portolan charts. As discussed, these lines correspond closely with the construction lines of this particular type of map: on the surface of portolan charts, rhumb lines form "virtual" parcours lines[19] that connect one navigation point to the other while adhering to a determined compass course. These line formations function in conjunction with the so-called rutters, or route books, in which the given course from one navigation point to the other is charted in the form of a list. This allows a relational determination of position between points of navigation by specifying the compass course and distance—the portolan chart itself emerges primarily as an archive of the places reached through different parcours lines. The corresponding methods of navigation were developed in the Mediterranean and expanded on in the fifteenth century by Portuguese sailors while navigating the Atlantic coast of Africa in close proximity to the shoreline. While sufficiently accurate for shorter distances, this form of navigation encounters increasingly arduous difficulties[20] when practiced on the open seas, since it does not allow for accurate consideration of the curvature of the earth.

The relational determination of position from one point of navigation to the other appears in contrast to the use of grid maps for navigation practices. These practices require astronomical methods to determine position, a function independent from the relative position of a ship in relation to a previously documented position on the chart. In early modern navigation practices, however, only latitude can be determined with a sufficient degree of accuracy by astronomical bearing, while the exact determination of longitude remains an unsolved problem until the eighteenth century.[21] In most cases, navigation practices thus have to rely basically on both compass directions and the astronomical determination of latitude. On conformal maps developed for the use of sailors ("ad usum navigantium")—as in the previously mentioned Mercator projection—the constant planned course of a ship also appears as a straight line, albeit at the cost of increasingly declining accuracy with each advancing move toward the pole. Insofar as parcours lines can be found on still extant maps at all,[22] the real course of a ship can deviate drastically from an ideal "spatial script"[23] and require constant measures of assessment to determine discrepancies in the course and the resulting actions to correct them.

A third type of line appears chiefly in a specific and exemplary form on the navigation charts at the current center of attention, namely in the form of the so-called contour line. This line type refers to "terrain" as it is conceived on the map. The first two types of lines are constructed or generated respectively as a geometric grid or as the intended route of a ship. By contrast, this third category constitutes an irregular type of line. The contour lines on navigation charts are partially the result of mathematically determined points of orientation and may to some degree be assigned by observing the coast with the naked eye. As a result, navigation maps generally do not contain the topographic details of inland areas. Instead, they place a greater value on the relevant navigation information connected to the coastal line, such as headlands, navigable bays, shallows, and so on. Because it is the sea itself from which these contour lines emerge[24]–a persistent aspect of geographic knowledge since Strabo in antiquity–contour lines correspond most directly to the "lines of the world" in Kenneth White's geopoetics. As the subject of a specific form of geography, the sea itself possesses medial agency as an entity that makes its own lines of demarcation visible and that relinquishes to the human geographer the task of transferring this form of graph onto the paper or parchment of the navigation map.[25]

A fourth line type emerges from a generalization of the determination principles of contour lines. But before going directly into detail about this fourth type, the importance of which is growing steadily in modern cartography, I would like to first discuss the relevance of these line types for narrative forms, based as they are on a cartographic imagination. It is necessary to distinguish the cartographic imagination not only from cognitive cartography,[26] but also from the abstract topological forms of visualization of narrative events.[27] The cartographic imagination is based on a historical-epistemological interplay between cartography and literary history. This interplay is evident in both historical maps and contemporary texts regardless of whether the maps constitute a direct component of those texts in question.[28] I would like to draw on the connection between map and narrative primarily in the context of Lotman's well-known spatial-semiotic description of narrative spaces.

In the chapter devoted to space in his *Structure of Artistic Texts*,[29] Lotman conceives of narrative plots as the dynamization of a static spatial order occurring with the crossing of a usually impenetrable spatial border. Following this assumption, the transgression of a spatial border is constitutive of a transition to a topologically or topographically determined space, which generally exhibits semantic features other than those of the origin space. In a strictly functional application of this distinction, Lotman declares the hero to be the particular agent who succeeds in crossing the border. However, this refers in most of the examples he includes to an actual human actor. Lotman's structural theory of literary space is relevant for the question concerning the relationship between map and narrative insofar as it is based on a cartographic matrix: it is in this sense that he draws the comparison between a map and a "plotless" text–that is, a text without the border crossing that provides the condition of passage into another space.[30] For Lotman, a text that possesses a plot is first constructed through the movement of the actor on the map. Going one step further than Lotman, I would like to suggest that the connection between map and narrative is not simply a descriptive comparison, but also an insight into an actual historical-epistemological precondition for a specific form of literary event formation. Perhaps such a form of eventfulness, as based on Lotman's theory, cannot be regarded as independent from the stable, addressable space of the map as a *conditio sine qua non*, or as a necessary condition of possibility for a depiction of the event or eventful as a form of movement.[31]

I would therefore like to argue that Lotman's spatial plot model should not only be understood as an abstract topological model in its usual sense,[32] but also that topological spaces, as Lotman describes them in his *Structure of Artistic Texts*, should be considered inseparable from topographic space. The foundation of Lotman's spatial model may be understood in terms of its geographic conditions of possibility in the early modern period, or more specifically, in terms of its cultural-technical "spatialization" of the world through early modern cartography. Yet such an understanding certainly requires an imagination of the map type underlying Lotman's theory of the map as a "plotless" text in the form of a grid map. Because of its matrix of construction lines, the grid map implies a self-contained and thus "static" spatial whole as a given, a spatial whole that appears to precede the inscription of a concretely set route. As such it exists in contrast to the portolan chart. The plot is created within this matrix through the parcours line, or rather through the graph depicting the movement from a point of departure to a point of destination as the crossing of determined borders of significance. These borders can take a variety of forms—a notably common plot is certainly that of the border line between land and sea, particularly in the context of narrative texts that concern voyages at sea,[33] such as in the literature of shipwrecks and island fiction. It is in this context that the third of the outlined line types comes into play for Lotman's theory of space, namely the contour line. Yet in terms of Lotman's theory, it should be noted that not every parcours line of a cartographic matrix is already constitutive of a plot structure, but that a parcours line only comes to represent an event with a plot in the experience of an unexpected disturbance or interruption, such as in the example of the fraught border between land and sea—a paradigmatic illustration of such a disturbance, for example, would be a shipwreck on the coast of a lonely island.

Before I turn to a critique of the relationship between cartographic lines and narrative plots according to Lotman with regard to the underlying static spatial frame posited with this relationship, it is necessary to address a fourth type of cartographic line. This type, which has evolved historically from the contour line, gaining in importance in modern cartography throughout the nineteenth century, is the so-called isoline (also isarithm). Generally, lines of constant value belong to this category. In its historical trajectory, isoline cartography has been practiced since the seventeenth century. Perhaps the oldest known example of isoline cartography is a seventeenth-century map created by Edmond Halley, which displays the magnetic declination of specific locations. Lines of elevation or isohypses constitute a better-known and more widespread form of the isoline on maps, a form that emerges from the practice of measuring water depths, or bathymetry.[34] In this respect the third line type in question, the contour line of an island or a continent, already constitutes in essence an isoline, or more precisely, the line as the meeting point of both bathymetry and hypsometry.[35] Perhaps the best-known isolines in the history of geography are, however, isotherms introduced by Alexander von Humboldt in 1817, or those lines of the same mean temperature that have exerted lasting influence on the formation of modern climatology (see figure 14.4).

Throughout the nineteenth century isolines become increasingly widespread in cartography, expanding in the form of so-called pseudoisolines to the realm of statistical values such as population density, among others. As a result, the mapping of isolines becomes the basis of what has come to be known as "thematic" cartography.[36] At the beginning of the twentieth century, the extent of the popularity of isoline cartography is such that the German cartographer Max Eckert is able to speak of an infectious "iso-nosos," a rampant isoline disease.[37] Ultimately, each isoline map also facilitates the visualization of field lines or lines of force that illustrate the way gradients between the different values seek balance—such field lines always run perpendicular to isolines.

Thus isolines and field lines are important for the history of cartography and possibly for the history of a map-based literary imagination or narratology. Their importance stems largely from the fact that these lines are active in reintroducing a dynamization into the cartographic spaces that appeared to be immobilized since the end of portolan cartography and since ship routes vanished from cartography. Isoline maps can no longer be described simply as static frames for the geographic positioning of objects; instead they appear as the site of representation for a dynamic of movement. It is the representation of a dynamic that is not limited solely to the movement of vessels within the smooth space of the sea, but that also makes visible the forces that move the vessels themselves, such as meteorological dynamics. In this sense maps function more than ever as the site of the relations of forces that cannot be strictly "positioned" at a specific location on the surface of the earth, but that are instead only discernible in their field character across multiple locations. At the same time they also highlight the agency of nonhuman actors. Among their other functions, maps serve the visibility of what, to borrow from the philosopher Timothy Morton, could be called global "hyperobjects"—that is, objects that are "massively distributed in time and space relative to humans," exceeding local manifestation as well as a human time scale.[38] This is particularly evident in the example of climatography. In climatography, modern isoline cartography allows maps of climate zones, which have been known since antiquity and that are oriented on set lines of latitude, to appear as structures of relational values with which both regional and historical shifts and transformations can be depicted.

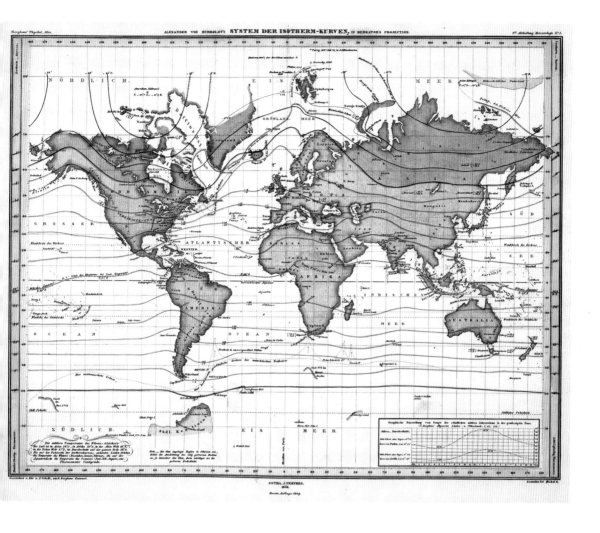

Figure 14.4

Hermann Berghaus, Alexander von Humboldt's System der Isotherm-Kurven (1949, first printed 1837 in his *Physikalischer Atlas*). *Source:* http://www.davidrumsey.com/maps5455.html.

Might this dynamization of space be useful beyond the context of the "address" of a body in space as it connects to a territory, beyond a dominant anthropomorphic understanding of a power of agency, and as the basis for an alternative description of narrative dynamics? If the narrative plot is defined by the characteristic movement across a border from an established point of departure to another equally determined destination by an almost exclusively human hero, the question is whether it is possible to conceive of another form of event dynamics in addition to this depiction. It is essentially a question that asks whether there is a form of event dynamics that emanates from a normal state of movement, rather than from stasis, and that expands the responsibility of movement beyond the sole province of human agency.

In the following section, I attempt to provide at least a rough sketch of the interplay between these two different logics of the event and their formation using the example of the work of Jules Verne, a nineteenth-century author who quite consciously utilizes maps in his narrative texts. Verne is very aware, I believe, that the function of maps in his novels not only serves an illustrative purpose and extends beyond the creation of a mere "reality effect"; he also recognizes their essential function for the formation of the narrative of travels, not only on land but specifically at sea. I would like, therefore, to turn to two of Verne's novels in which the journey at sea plays a central role.

Cartographic Lines and Narrative in the Work of Jules Verne

Many of Jules Verne's novels[39] include maps pertaining to the travels in the novel in which the first three of the relevant line types may be observed: the construction lines of the map, the parcours lines marking the travel routes of the heroes, and the contour lines connoting the countries and continents of the narrative's journeys. One of these novels is *Les Enfants du capitaine Grant* (1868), which relates the circumnavigation of the world, as in other books by Verne. The route of the protagonists is represented in a three-part map[40] corresponding to three original parts of the novel (see figures 14.5a–c).

The novel establishes its cartographic expertise largely in the form of the French geographer Jacques Paganel, who lands by accident on board the *Duncan*—the vessel of a group traveling in search of Captain Harry Grant, who goes missing after a shipwreck. As a character, Paganel is a notoriously scatterbrained scientist who mourns the "heroic" age of discovery in which cartographers were still the navigators themselves and in which, as Paganel wistfully recalls, the discovery of new territories corresponded to a form of geography as a sort of cartographic invention of the new. On the surface of the navigation map, the contour lines of the New World emerged little by little from the sea as the individual points of navigational orientation gradually yielded a closed coastal line:

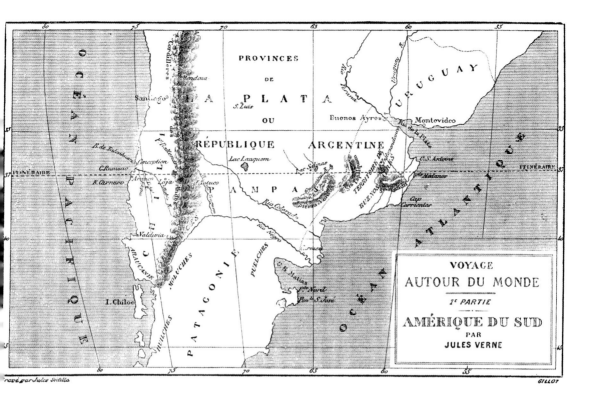

Figures 14.5a–c

Maps in Jules Verne's *Les Enfants du capitaine Grant*, containing the travelers' itinerary forming a (nearly) straight line along the 37th parallel of the Southern Hemisphere. *Source:* http://verne.garmtdevries.nl/en/maps/originals.html.

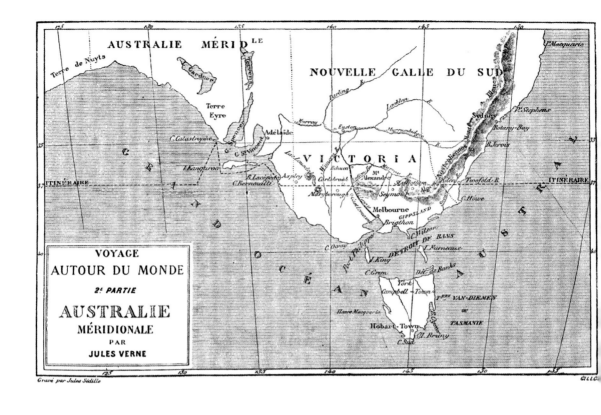

Figures 14.5a–c (continued)

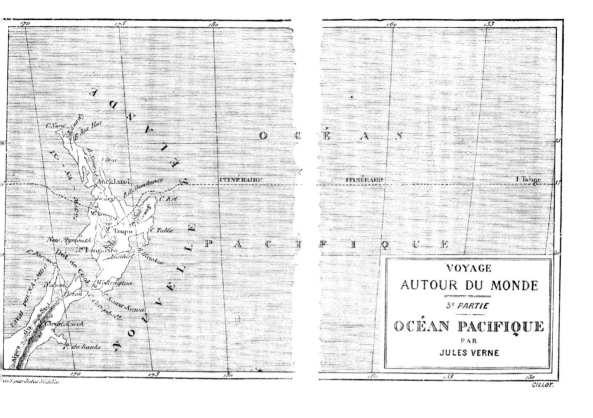

Figures 14.5a–c (continued)

> Is there indeed any more authentic satisfaction, any more real pleasure than when the navigator marks his discoveries on the chart he has on board? He sees how the land slowly takes shape before his eyes, island by island, headland by headland and, so to speak, emerges from the floods! First, the boundary lines are still vague, broken and interrupted! Here, a solitary cape, there an isolated bay, somewhat further a gulf lost in space. But then, the discoveries complement each other, the lines become joined, and the broken map line becomes transformed into a continuous line.[41]

Providing an account of an eventful plot of early modern travels, Paganel describes the gradual formation of coastal lines from a number of points and parcours lines—the portolan chart represents the surface of operation where the event of creation takes place as the emergence of America from the tides. His cartographic imagination appears remarkably stuck in an age characterized by practices of point-to-point localization rather than by a fixed location on a grid map.[42]

On the other hand, Paganel also describes the comprehensive "graticulation" of geographic space through the network of grid lines as the complete loss of the inventive power of cartography in the nineteenth century: "But now, this mine is almost exhausted! We have seen, recognized, and invented everything as far as new continents and new worlds are concerned, and it seems as if we ourselves, latecomers to geographic science, have nothing more to do!"[43] The complete graticulation of geographic space in cartographic grid lines means the comprehensive addressability of every last location on earth. For Paganel this also means the closing of the last of the famous "blank spaces" on the map, the last possibility now exhausted of that eventful border crossing into an island or an entire continent with inland secrets yet to be discovered.[44] Yet when he asks with weary resignation whether any purpose remains for a geographer in search of adventure, the captain of the *Duncan*, Lord Glenarvan, answers in disagreement:

> "Yes, there is something that remains, my dear Paganel, Glenarvan answered."
>
> "So what is it?"
>
> "What we are doing right now!"[45]

What the passengers on the *Duncan* practice is the invention of a new possibility of eventful travel in the age of the complete graticulation of the earth. It is an act of "traveling a line" along a single cartographic grid line. The aim of this journey is no longer to gain entrance to empty and unknown inland spaces, rather the primary objective is to traverse the earth and complete the circle of a parcours line that reflects the circumnavigation of the world. To travel a line is thus to question the traditional logic of the plot in the cartographic journey of discovery. At the same time, it also initiates a literary reflection on the relationship between maps and lines, as it shifts attention to a type of line that has until now functioned primarily as a means to the end of position determination.

The plot of the novel deals with the two underage children of Captain Grant as they search for their father after a message in a bottle lands in their hands by coincidence. Yet as a plot, the search for the lost father is more than a mere means to revisit and reenact the mythologizing intent of the Telemachy in the Homeric Odyssey.[46] Rather, within the narrative arrangement of the search for the father, the device of the only partially decipherable message in a bottle functions as the very prerequisite to the curious journey along a cartographic line. As this journey begins to occupy center stage, the search for the father fades increasingly into the background.

It is revealed in the fragmentary content of the bottle's message that the missing father suffered a shipwreck at the 37th parallel of the Southern Hemisphere. Neither the line of longitude nor the name of a location where the shipwreck occurred can be deciphered in any of the three languages in which the message is composed. The children endeavor to decode the message along with the help of their adult counterparts under the leadership of the owner of the search vessel, Lord Glenarvan, and Jacques Paganel. Yet on account of their faulty philological conjecturing,[47] they circle nearly the entire earth before finding the missing father on the island of Tabor,[48] an island that appears on numerous nineteenth-century maps. The characters in the narrative seem motivated in their travels by the search for the missing father against the background of ancient mythology and literature. However, I would like to suggest that the real interest of the text is to describe the circumnavigation of the world along a single line of latitude. This can be observed as a straight line on the novel's three maps, a line that runs around the world like a common thread from left to right, or from the beginning to the end of the search action.[49]

To read the text primarily on this cartographic matrix, and not solely in terms of the background of the contrasting search for the father, is to observe that the Vernian travel narrative ultimately does not follow the transgressive aim of crossing unknown borders; nor does the search for the father appear to be a sufficient explanation for its featured journey around the world. Instead, the text as a whole fulfills the geographer Paganel and his companions' newly discovered desire to travel a line. With as much accuracy as possible, the text thus ultimately aspires to draw a true horizontal line on the map as a narrative matrix of the text. In the delineation of this horizontal figuration, a simple line of latitude is transformed from the mere construction line of a grid map into the parcours line of a journey around the world.

To travel a line in the way Verne has demonstrated in literature[50]—that is, as a voyage with as little variation and as much accuracy as possible—is to depend on a set of specific technohistorical preconditions that are particularly evident in the voyage at sea. For it is not accidental that the idea of traveling a line around the world should emerge at this point in history. The sea voyages in Verne's novels of the 1860s and

1870s inhabit a period of transition between navigation as a form of travel by sailboat, in the tradition that had been well established since the early modern period, and that of the increasingly common form of travel by steamboat in the nineteenth century. As the ship used to complete the journey around the world, the *Duncan* is a combination of both sail and steamboat.[51] In this sense, the sailing vessel stands for the tradition of the individually determined route before the epoch of the established connections of regular world traffic. At the same time, the steam engine driving the *Duncan* makes it possible to follow a nearly straight line as the ship is freed from meteorological conditions, while sailing vessels remain forced time and again to deviate from the "straight line" of the ideal path and to pursue alternative paths by beating a course upwind.[52]

The previous textual analysis allows for a preliminary conclusion. Bypassing construction and parcours lines, Verne appears to reflect on the classic event logic of the mappable narrative, an event that emanates from static spaces. His circumnavigation of the world not only serves the didactic purpose of a literary lesson in the geography of the Southern Hemisphere, but the act of traveling a line also becomes a playful "constraint"[53] that makes a specific form of experience possible through its resulting construction lines—these construction lines facilitate the experience of the otherwise implicit static space of the map. The novel thus questions the "line" logic of the conventional transgression narrative, which is based on the first of the three line types described, yet it does so without completely departing from the framework of the traditional grid map. As something of a final point, I would like to provide the rough outline of an additional perspective that may be taken from Verne's texts; it is a view that extends beyond the steady course of movement along a graticule line on the earth's surface to isolines in particular as they continuously cross and intersect with this straight line. At the very least, these isolines might suggest the possibility of a narrative based on alternative forms of event logic. As such, it becomes possible to conceive of a narrative that goes beyond the scope of the framework of the map-based Lotmanian plot model, with its logic of addressable points of origin and destination in the movement of travel.

One element in particular transgresses the "static" logic of positioning that is both traceable on the grid map and indicative of human agency, an element already mentioned in the context of Verne's novel *Les Enfants du capitaine Grant*: the message in a bottle. The bottled message refers to a different logic of movement that exhibits the apparently static surface of the sea as a fluid and streaming space. Indeed, it is first through the movement of the ocean's currents that the lost Captain Grant is able to be located on a lonely island, as it is by way of the current's flowing motion that the captain's message in a bottle is brought from the island to its intended addressees.

The understanding of global sea currents is largely the result of the increasing research activity in the nineteenth century and the assistance of isolines, or field lines on maps (see figure 14.6).[54]

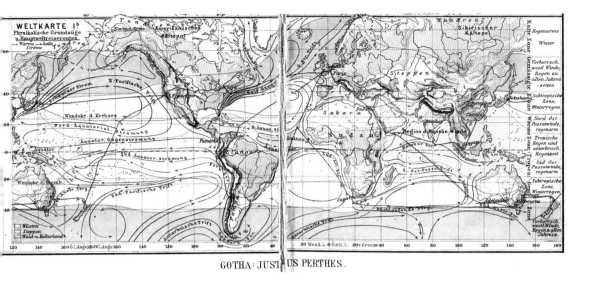

Figure 14.6

World map in Justus Perthes's *Taschen-Atlas* (1885), indicating sea currents and shipping routes. *Source:* Justus Perthes's *Taschen-Atlas*, ed. Hermann Habenicht (Gotha: Perthes, 1885, 21st ed.).

On the basis of such currents alone, it becomes possible to learn of the captain's fate. The search operation is set in motion by the reading of the bottled message. For a search to take place on the basis of a grid map, another sort of movement dynamics is initially required, a sort that appears to deviate from this form of the logic of the address. Yet, it is a recurring feature of Vernian narrative devices to treat the contingencies of the drifting currents as if they were ultimately compatible with the logic of locatability through an established address in a stricter sense. As early as in the balloon flight over the African continent in his debut novel *Cinq semaines en ballon* (*Five Weeks in a Balloon,* 1863),[55] Verne creates the impression that air streams can be controlled with such precision as to facilitate the exact flight of a balloon from point A to a determined point B—particularly as though it were as simple as a journey on foot or the movement of a steamboat at sea. In the similar case of *Tour du monde en quatre-vingts jours* (*Around the World in Eighty Days*, 1873),[56] the passing sailboats and ice yachts not only appear as the vehicles driving toward the complete closure of the remaining gaps in the world network. Their movements also fit so precisely in the planned itinerary and scheduled connecting stations of Phileas Fogg's circumnavigation of the world that something of a sense of fantastic hyperconnectivity begins to emerge.[57]

In *Les Enfants du capitaine Grant* a nonhuman agent becomes a "postal carrier" in the form of a shark that swallows the captain's message in a bottle from the island of Tabor,[58] an action that produces a twofold effect. On the one hand, the shark interferes with any possibility of locating the sender of the bottled message by tracing the sea's current from the approximate point the bottle was thrown into the water. On the other hand, the shark nonetheless succeeds in delivering the post with precision to the proper addressee, Lord Glenarvan, who is in the very position to begin the well-equipped search expedition with the help of the captain's two children. And yet the shark also sacrifices his life in his function as the postal carrier for Lord Glenarvan and the captain's children. Owing to the involuntary assistance of the shark, the free-floating movement of the message bottle in the constant motion of the current reverts fantastically to the precision of the logic of the established address, which corresponds to the concrete position determination of specific locations on the grid map. It is a persistent detail of Verne's work that the narrative conditions occur in conjunction with a cartographic framework as a concrete form of addressability. Such a feature appears in contrast to Herman Melville's *Moby-Dick* (1851) as a novel that is committed, to a much higher degree in its plot structure, to a smooth space beyond the cartographic logic of addressability.[59] Despite what at times appears to be their striking expansion to the fluidity of global currents and streams, Verne's plots ultimately undergo nonetheless a sort of reterritorialization in a Eurocentric sense. This is expressed in the fact that the European aristocracy, or rather the bourgeoisie, functions as both the point of departure and the final objective in the overwhelming majority of his novels.[60]

Yet, one of Verne's later and lesser-known novels also succeeds in creating at least the impression of what might be considered the necessary conditions of a book no longer structured along the lines of a cartographic matrix in the traditional sense, a book that instead functions solely by recording states of movement. A remarkable feature of Verne's later Tierra del Fuego novel *Le Phare au bout du monde* (*Lighthouse at the End of the World*, 1905) is the description of a "book within the book," a bundle of written records consisting not of fictional narratives, but purely of various weather reports and meteorological data from log records. This notebook kept by the lighthouse keeper at Cape Horn on the southern point of Tierra del Fuego is described as a "livre du phare," or lighthouse book:

> They had to keep the lighthouse logbook up to date, noting any incident which might occur, the passage of steam or sailing ships, their nationalities, their names when they sent them with their numbers, and finally the height of the tides, the direction and strength of the wind, the weather, the duration of the rain, the frequency of the storms, the rise and fall in the barometer, the temperature, and any other phenomena which would allow the meteorological chart of these areas to be drawn.[61]

In the context of such a book, a location is no longer something static; instead, it is the intersecting point of various forces. In such a location a body at rest occurs not as a general state but as an exception. To achieve such a state, the firmly established resting position of a lighthouse must first be secured through considerable technological effort before it can become the monitoring site for weather values and before it can serve as the point of orientation for ships at sea. The meteorological map referenced in the previous citation is ultimately a medial site of operation. It is the specific constellation of information whose mappability is achieved with the aid of isolines and not just by way of the positioning of lines on a grid map. In this sense, the static grid of construction lines becomes a palimpsest whose surface undergoes the repeated rewriting through isolines in constant transformation.[62]

Finally, the question clearly arises as to whether it is at all useful to consider the "livre du phare" as a narrative model of the novel, appearing as it does as a record of the dynamics of movement in this later of Verne's novels. Apart from the fact that the lighthouse book adheres to another mode of written recordkeeping altogether,[63] a translation of the dynamics of movement as recorded in the book into a literary event could hardly build on the depiction of the event as an act of the "setting into motion" of something once static in the sense of Lotman. On the contrary, it would be the halting of movement or immobilization that would constitute the event in such a dynamic model, such as in the case of a shipwreck or another accident in which the wind or other meteorological force rips the propelled body from its "general state" of constant movement. Such a form of plot formation remains to be fully examined from the perspective of literary criticism and in terms of narratology.[64] An investigation of this sort would likely reveal a series of examples for a sort of literary narrative based on an initial state of movement. These would not only include maritime plots, but also narratives in which the "meteorological" appears to be the overriding principle of action in another context—such as Guy de Maupassant's balloon-voyage chronicles[65] or the famous beginning of Robert Musil's *Mann ohne Eigenschaften*, in which an automobile accident is depicted as a force field modeled on the distribution of the meteorological lines of barometric pressure.[66] In these texts the anthropomorphic heroes of Lotmanian acts of transgression recede into the background of nonhuman actors forming the intersection of more complex forces.

In modern literature it is thus entirely possible to describe an alternative dynamic of the event in conjunction with the development of isoline and field cartography. Yet to accept this as such is also to embrace an alternative understanding of narrative and map alike, not merely as a sort of static matrix that provides the "physical" framework for geographic movement, but rather as the scene of the forces and relations of various bodies in motion. Such a shift in the understanding of maps may indeed correspond directly to the transformation of their current use. Current geomedia appear in opposition to the two-dimensional flat surfaces that depict imagined movements while at rest on a table or other flat base. With information processing in map form, these geomedia have become mobile devices with which spaces may be observed as fundamentally based on motion and usually not from a static point of view.[67]

Translated by Andrew Patten

Notes

1. For a general overview see Sybille Krämer and Horst Bredekamp, eds., *Bild, Schrift, Zahl* (Munich: Fink, 2003); for an overview with a particular emphasis on cartography see William Boelhower, "Inventing America: A Model of Cartographic Semiosis," *Word & Image* 4, no. 2 (1988): 475–497.

2. See Jürgen Osterhammel and Niels P. Petersson, *Globalization: A Short History*, trans. Dona Geyer (Princeton, NJ: Princeton University Press, 2005), 31–56.

3. For example, Latour views the map as a paradigm of an "immutable mobile," or a cultural technique with which knowledge is operationalized in becoming transportable while at the same time remaining stable (Bruno Latour, "Drawing Things Together," in *Representation in Scientific Practice*, ed. Michael Lynch and Steve Woolgar, 19–68 (Cambridge, MA: MIT Press, 1990)).

4. Yet, recent studies seeking to separate the "navigational" from what they regard as the "mimetic" tend to overlook the fact that historically the "navigational" use of maps appears initially as an outcome of the link between map and territory. See Valérie November, Eduardo Camacho-Hübner, and Bruno Latour, "Entering a Risky Territory: Space in the Age of Digital Navigation," *Environment and Planning D: Society and Space* 28, no. 4 (2010): 581–599. The critique of cartographic mimesis is valid to the extent that the relationship between map and territory is less a matter of "iconic" representation than of "indexical" relations. See Robert

Stockhammer, "'An dieser Stelle': Kartographie und die Literatur der Moderne," *Poetica* 33, nos. 3–4 (2001): 273–306.

5. See Carl Schmitt, *Der Nomos der Erde im Völkerrecht des Jus Publicum Europaeum* (Berlin: Duncker & Humblot, [1950] 1997).

6. See Jörg Dünne, *Die kartographische Imagination: Erinnern, Erzählen und Fingieren in der Frühen Neuzeit* (Munich: Fink, 2011), 17–44.

7. A similar relationship may be observed in globes of the late Middle Ages. See the example of Martin Behaim's globe in Robert Stockhammer, *Kartierung der Erde: Macht und Lust in Karten und Literatur* (Munich: Fink, 2007), 41.

8. See David Woodward, "Maps and the Rationalization of Geographic Space," in *Circa 1492: Art in the Age of Exploration*, ed. Jay A. Levinson, 83–87 (New Haven, CT: Yale University Press, 1991).

9. For an essential overview, see François de Dainville, *Le Langage des geographes: Termes, signes, couleurs* (Paris: Pivot, 1964).

10. On the survival of descriptive literary geography in early modern cosmography, see Frank Lestringant, *Mapping the Renaissance World: The Geographical Imagination in the Age of Discovery*, trans. David Fausett (Berkeley: University of California Press), 1994.

11. See Olaus Magnus, *Carta marina*, ed. Elena Balzamo (Paris: Corti, [1539] 2005). The text commentary to the *Carta marina* appears in an external commentary booklet with the title "Opera breve," which in turn becomes the core of the twenty-two books that make up the *Historia de gentibus septentrionalibus* of 1555 (*A Description of the Northern Peoples*, 3 vols., ed. Peter Foote, trans. Peter Fisher (London: Hakluyt Society, 1996–1998)).

12. For the fictional "Deixis am Phantasma," see Karl Bühler, *Sprachtheorie: Die Darstellungsfunktion der Sprache* (Stuttgart: Fischer UTB, [1934] 1982), 121–140; for the map as a "deictic surface," see Stockhammer, "'An dieser Stelle,'" 280.

13. On the "mappability" of literary spaces, see Stockhammer, *Kartierung der Erde*, 59–88.

14. See Lennard J. Davis, *Factual Fictions: The Origins of the English Novel* (New York: Columbia University Press, 1983).

15. See Kenneth White, "Presentation," Institute of Geopoetics, http://institut-geopoetique.org/en/presentation-of-the-institute(1989).

16. For a general introduction to the theory of lines, see Tim Ingold, *Lines: A Brief History* (London: Routledge, 2007); Sabine Mainberger, *Experiment Linie: Künste und ihre Wissenschaften um 1900* (Berlin: Kadmos, 2010).

17. See Christian Jacob, *The Sovereign Map: Theoretical Approaches in Cartography throughout History* (Chicago: University of Chicago Press, 2006), 118–129.

18. See Michel de Certeau, *Practice of Everyday Life*, trans. Steven Rendall (Berkeley: University of California Press, 2004), 121.

19. See Jacob, *The Sovereign Map*, 169–171.

20. See William Graham Lister Randles, "De la carte-portulan méditerranéenne à la carte marine du monde des grandes découvertes: La crise de la cartographie au XVIe siècle," in *Géographie*

du Monde au Moyen Age et à la Renaissance, ed. Monique Pelletier, 125–132 (Paris: Editions du C.T.H.S., 1989).

21. See Dava Sobel, *Longitude* (New York: Walker, 1995).

22. Representations of parcours lines on world maps are particularly evident in the cartography of travels around the entire world, such as Magellan's circumnavigation of the world in Battista Agnese's world map in his *Atlante nautico* from 1553, ed. Marica Milanesi (Venice: Marsilio, 1990).

23. On "spatial scripts," see Jörg Dünne, "Scénarios d'espace entre guide touristique et récit: Tentatives d'épuisement d'un lieu marocain," in *Scénarios d'espace: Littérature, cinéma et parcours urbains*, ed. Jörg Dünne and Wolfram Nitsch (Clermont-Ferrand: Presses universitaires Blaise Pascal, 2014), 177–181.

24. See Jacob, *The Sovereign Map*, 306. Jacob refers to Strabo's *Geography* II/5, 17, trans. Horace Leonard Jones (London: Heineman and Putnam, vol. 1, 1917), 462–463.

25. A clear example of this can be observed in the frequent appearance of early modern islands as geometric forms abstracted from "irregular" coastlines (see Jacob, *The Sovereign Map*, 143–153); it is also not without reason that coastlines appear additionally as a mathematical problem sparking debates concerning "fractal" geometry. See the exemplary case of Benoît Mandelbrot, "How Long Is the Coast of Britain? Statistical Self-Similarity and Fractional Dimension," *Science* 156, no. 3775 (1967): 636–638.

26. See Roger M. Downs and David Stea, *Maps in Minds: Reflections on Cognitive Mapping* (New York: Harper & Row, 1977).

27. See Marie-Laure Ryan, "Narrative Cartography: Toward a Visual Narratology," In *What Is Narratology? Questions and Answers Regarding the Status of a Theory*, ed. Tom Kindt and Hans-Harald Müller, 333–364 (Berlin: De Gruyter, 2003).

28. Ryan, "Narrative Cartography," 338, refers to this map type as the "internal maps of the textual world."

29. Yuri M. Lotman, *The Structure of the Artistic Text* (Ann Arbor: University of Michigan, 1977), 217–244.

30. Lotman, *The Structure of the Artistic Text*, 239.

31. In making this claim, however, I do not wish to assign the map a function as the definitive "media a priori" of narrative—on this notion, see Friedrich Kittler, *Gramophone, Film, Typewriter*, trans. Geoffrey Winthrop-Young and Michael Wutz (Stanford, CA: Stanford University Press, [1986] 1999), xxxix. I have equally little interest in turning the map into the a priori of occidental reason—see Franco Farinelli, "Von der Natur der Moderne: Eine Kritik der kartographischen Vernunft," in *Räumliches Denken*, ed. Dagmar Reichert, 267–302 (Zurich: vdf, 1996). Yet I would indeed argue that early modern cartography accounts for a significant share of the transformation and reconstruction of spaces of literary imagination.

32. See also the overview by Cornelia Ruhe, "Semiosphäre und Sujet," in *Handbuch Literatur & Raum*, ed. Jörg Dünne and Andreas Mahler, 170–177 (Berlin: De Gruyter, 2015).

33. See Margaret Cohen, *The Novel and the Sea* (Princeton, NJ: Princeton University Press, 2010).

34. See François de Dainville, "De la profondeur à l'altitude: Des origines marines de l'expression cartographique du relief terrestre par cotes et courbes de niveau," in *Le Navire et l'économie maritime du Moyen Age au XVIIIe siècle principalement en Méditerranée*, ed. Michel Mollat, 195–213 (Paris: S.E.V.P.E.N., 1958).

35. On the related phenomenon of "waterlines," see Bernhard Siegert, "Waterlines: Striated and Smooth Spaces as Techniques of Ship Design," in *Cultural Techniques: Grids, Filters, Doors, and Other Articulations of the Real*, trans. Geoffrey Winthrop-Young, 147–163 (New York: Fordham University Press, 2015).

36. Yet, see also Stockhammer, *Kartierung der Erde*, 44–45, for the problematic distinction between "physical" and "social."

37. See Werner Horn, "Die Geschichte der Isarithmenkarten," *Petermanns Geographische Mitteilungen* 103, no. 3 (1959): 225–232; this article basically lists different types of isolines.

38. Timothy Morton, *Hyperobjects: Philosophy and Ecology after the End of the World* (Minneapolis: University of Minnesota Press, 2013), 1.

39. See Terry Harpold, "Verne's Cartographies," *Science Fiction Studies* 32, no. 1 (2005): 18–42.

40. See Jules Verne, *Les Enfants du capitaine Grant* (Paris: Le Livre de poche, [1867–1868] 2004), 230, 569, 883.

41. In the only English translation available (*In Search of the Castaways*, trans. Charles Francis Horne (New York: Parke, 1911)), this and the following passages are missing. The original French quotation reads, "Est-il, en effet, une satisfaction plus vraie, un plaisir plus réel que celui du navigateur qui pointe ses découvertes sur la carte du bord? Il voit les terres se former peu à peu sous ses regards, île par île, promontoire par promontoire, et, pour ainsi dire, émerger du sein des flots! D'abord, les lignes terminales sont vagues, brisées, interrompues! Ici un cap solitaire, là une baie isolée, plus loin un golfe perdu dans l'espace. Puis les découvertes se complètent, les lignes se rejoignent, le pointillé des cartes fait place au trait" (Verne, *Les Enfants du capitaine Grant*, 90).

42. Particularly telling is a passage in the novel describing a mishap that Paganel experiences one day in his scholarly absentmindedness, namely the publication of a map of America on which he has mistakenly entered the name Japan (Verne, *Les Enfants du capitaine Grant*, 71). With this mistake he precisely reproduces the cartographic perception of Columbus as he travels to America and constantly projects his cartographically informed vision of Japan onto his perception of the Caribbean.

43. The original French quotation reads, "Mais maintenant cette mine est à peu près épuisée! On a tout vu, tout reconnu, tout inventé en fait de continents ou de nouveaux mondes, et nous autres, derniers venus dans la science géographique, nous n'avons plus rien à faire!" (Verne, *Les Enfants du capitaine Grant*, 90–91).

44. The only novel of Verne's that can claim to deal consistently with this sort of event logic is his debut novel *Cinq semaines en ballon* (Paris: Le Livre de poche, [1863] 2000).

45. Verne, *Les Enfants du capitaine Grant*, 91. The original French quotation reads, "–Si, mon cher Paganel, répondit Glenarvan.–Et quoi donc?–Ce que nous faisons!"

46. See on this topic Michel Serres, "Loxodromies des Voyages extraordinaires," in *Hermès I: La Communication*, 207–213 (Paris: Minuit, 1968).

47. Furthermore, Verne's novel also addresses the uncertain character created by the link between position determination on the map and in the narrative, using the example of the location name *Tabor*, which the geographer Paganel persistently misreads as the verb *aborder* ("to go on land"). Paganel thus embodies a failing cartographic knowledge of positioning, which allows the act of traveling a line to take the place of the successful practice of "localization."

48. On the character of the island as a phantom island, see Wolfgang Struck, "Die Zerstreutheit des Geographen: Jules Vernes andere Reise um die Welt," in *Weltnetzwerke–Weltspiele: Jules Vernes "In 80 Tagen um die Welt,"* ed. "Passepartout" (Konstanz: Konstanz University Press, 2013), 145.

49. Verne manipulates the legibility of the bottled message in his novel in a way that allows him to establish a known line of latitude but an unknown line of longitude. In doing so, he reproduces a staging of the critical transition phase of early modern navigation, shifting from compass navigation, in close proximity to the shores of the coast along rhumb lines, to astronomical navigation (see Randles, "De la carte-portulan"). Until the eighteenth century, only an adequately precise determination of lines of latitude was possible within the framework of astronomical navigation, while the determination of a reliable line of longitude remained impossible. As a result, ships were often forced to travel repeatedly to and from a line of latitude until happening on an intended goal.

50. Among similar examples from the late twentieth century is the noteworthy example of *Le Méridien de Greenwich* by Jean Echenoz (Paris: Minuit, 1979), which complements Verne's novel by ending with a journey along a meridian.

51. I cannot go into greater detail here about the way the network of world traffic in *Tour du monde* begins to detach from cartographic orientation and becomes content to depict the travel movements of the protagonist Phileas Fogg on account of connection diagrams in timetables. See Markus Krajewski, *Restlosigkeit: Weltprojekte um 1900* (Frankfurt am Main: Fischer, 2006), 23–63.

52. For more on the "meteorological" and the "uranological" in the context of Phileas Fogg, see Robert Stockhammer, "Der Globus und das Klobige: Uranologie und Meteorologie," in *Weltnetzwerke–Weltspiele*, 132–134.

53. For the significance of play and the game in the work of Verne, specifically in the context of *Tour du monde*, see Jörg Dünne and Kirsten Kramer, "Weltnetzwerke–Weltspiele–Welterzählungen," in *Weltnetzwerke–Weltspiele*, 15–22.

54. For the hypothesis concerning the predictability of sea currents due to experiments with sending bottled messages, see Struck, "Die Zerstreutheit des Geographen," 141–145.

55. Verne, *Cinq semaines en ballon*, 58.

56. Jules Verne, *Le Tour du monde en quatre-vingts jours*, ed. William Butcher (Paris: Gallimard, [1873] 2009), 188–199, 287–295.

57. See Jörg Dünne, "Ekstatische Konnektivität: Auf dem Segelschlitten," in *Weltnetzwerke–Weltspiele*, 300–304.

58. See Verne, *Les Enfants du capitaine Grant*, 7–16.

59. Herman Melville, *Moby-Dick, or, The Whale*, ed. Tom Quirk (New York: Penguin, [1851] 2003). On "whale charting" and Melville's novel, see Stockhammer, *Kartierung der Erde*, 187–209. While

Moby-Dick is also based on the attachment of plots to a determined meeting point between the white whale and Ahab, this meeting point, according to Stockhammer, would be located in an unmappable space.

60. The exception to this rule is of course Captain Nemo in Verne's *Vingt mille lieues sous les mers* (Paris: Le Livre de poche, [1869] 2006)—as a permanently deterritorialized character underwater, Captain Nemo never arrives at a "destination" and can never return to a place of origin.

61. Jules Verne, *Lighthouse at the End of the World*, ed. and trans. William Butcher (Lincoln: University of Nebraska Press, 2007), 19–20. The original French quotation reads, "Ils devaient tenir au courant le 'livre du phare,' y noter tous les incidents qui pourraient survenir, le passage des bâtiments à voile et à vapeur, leur nationalité, leur nom lorsqu'ils les enverraient avec leur numéro, la hauteur des marées, la direction et la force du vent, la relève du temps, la durée des pluies, la fréquence des orages, les hausses et baisses du baromètre, l'état de la température et autres phénomènes, ce qui permettrait d'établir la carte météorologique de ces parages" (Jules Verne, *Le Phare au bout du monde* (Paris: Gallimard, [1905] 1999), 52).

62. Here it is particularly important, however, to maintain at least the possibility of georeferencing as a distinctive characteristic of maps in contrast to the prevalent and vague use of the term *mapping* for the representation of all kinds of spatial relations; see the example of research by Robert T. Tally, *Spatiality* (New York: Routledge, 2013), 44–78.

63. On the list as an "operational" hybrid of alphabet and number instead of a purely alphabetic form of inscription, see Peter Koch, "Graphé: Ihre Entwicklung zur Schrift, zum Kalkül und zur Liste," in *Schrift, Medien, Kognition: Über die Exteriorität des Geistes*, ed. Peter Koch and Sybille Krämer, 43–82 (Tübingen: Stauffenburg, 1997).

64. In his book on film Gilles Deleuze provides a narratological model for this plot formation with his description of the "small" form of the action image: *L'Image-mouvement: Cinéma I* (Paris: Minuit, 1983), 220–242. This model may be relevant for the depiction of a form of the event or eventfulness as it emanates not from stasis but from movement. See Jörg Dünne, "Dynamisierungen: Bewegung und Situationsbildung," in *Handbuch Literatur & Raum*, 41–43.

65. See André Weber, *Wolkenkodierungen bei Hugo, Baudelaire und Maupassant im Spiegel des sich wandelnden Wissenshorizontes von der Aufklärung bis zur Chaostheorie* (Berlin: Frank & Timme, 2012), 313–330.

66. For a more detailed analysis see Michel Serres, *Hermes V: Le Passage du Nord-Ouest* (Paris: Minuit, 1980), 27–39.

67. See the contributions on geomedia in Jörg Döring and Tristan Thielmann, eds., *Mediengeographie: Theorie–Analyse–Diskussion* (Bielefeld: transcript Verlag, 2009); on the history of mobile, transportable media in general, see Martin Stingelin and Matthias Thiele, "Portable Media: Von der Schreibszene zur mobilen Aufzeichnungsszene," in *Portable Media: Schreibszenen in Bewegung zwischen Peripatetik und Mobiltelefon*, ed. Martin Stingelin and Matthias Thiele, 7–27 (Munich: Fink, 2010).

Bibliography

Agnese, Battista. *Atlante nautico*. Ed. Marica Milanesi. Venice: Marsilio, [1553] 1990.

Boelhower, William. "Inventing America: A Model of Cartographic Semiosis." *Word & Image* 4, no. 2 (1988): 475–497.

Bühler, Karl. *Sprachtheorie: Die Darstellungsfunktion der Sprache*. Stuttgart: Fischer UTB, [1934] 1982.

de Certeau, Michel. *Practice of Everyday Life*. Trans. S. Rendall. Berkeley: University of California Press, [1980] 2004.

Cohen, Margaret. *The Novel and the Sea*. Princeton, NJ: Princeton University Press, 2010.

de Dainville, François. "De la profondeur à l'altitude: Des origines marines de l'expression cartographique du relief terrestre par cotes et courbes de niveau." In *Le Navire et l'économie maritime du Moyen Age au XVIIIe siècle principalement en Méditerranée*, ed. Michel Mollat, 195–213. Paris: S.E.V.P.E.N., 1958.

de Dainville, François. *Le Langage des geographes: Termes, signes, couleurs*. Paris: Pivot, 1964.

Davis, Lennard J. *Factual Fictions: The Origins of the English Novel*. New York: Columbia University Press, 1983.

Deleuze, Gilles. *L'Image-mouvement: Cinéma I*. Paris: Minuit, 1983.

Döring, Jörg, and Tristan Thielmann, eds. *Mediengeographie: Theorie–Analyse–Diskussion*. Bielefeld: transcript Verlag, 2009.

Downs, Roger M., and David Stea. *Maps in Minds: Reflections on Cognitive Mapping*. New York: Harper & Row, 1977.

Dünne, Jörg. *Die kartographische Imagination: Erinnern, Erzählen und Fingieren in der Frühen Neuzeit*. Munich: Fink, 2011.

Dünne, Jörg. "Dynamisierungen: Bewegung und Situationsbildung." In *Handbuch Literatur & Raum*, ed. Jörg Dünne and Andreas Mahler, 41–54. Berlin: De Gruyter, 2015.

Dünne, Jörg. "Ekstatische Konnektivität: Auf dem Segelschlitten." In *Weltnetzwerke–Weltspiele: Jules Vernes "In 80 Tagen um die Welt,"* ed. "Passepartout," 300–304. Konstanz: Konstanz University Press, 2013.

Dünne, Jörg. "Scénarios d'espace entre guide touristique et récit: Tentatives d'épuisement d'un lieu marocain." In *Scénarios d'espace: Littérature, cinéma et parcours urbains*, ed. Jörg Dünne and Wolfram Nitsch, 177–196. Clermont-Ferrand: Presses universitaires Blaise Pascal, 2014.

Dünne, Jörg, and Kirsten Kramer. "Weltnetzwerke–Weltspiele–Welterzählungen." In *Weltnetzwerke–Weltspiele: Jules Vernes "In 80 Tagen um die Welt,"* ed. "Passepartout," 15–22. Konstanz: Konstanz University Press, 2013.

Echenoz, Jean. *Le Méridien de Greenwich*. Paris: Minuit, 1979.

Farinelli, Franco. "Von der Natur der Moderne: Eine Kritik der kartographischen Vernunft." In *Räumliches Denken*, ed. Dagmar Reichert, 267–302. Zurich: vdf, 1996.

Harpold, Terry. "Verne's Cartographies." *Science Fiction Studies* 32, no. 1 (2005): 18–42.

Horn, Werner. "Die Geschichte der Isarithmenkarten." *Petermanns Geographische Mitteilungen* 103, no. 3 (1959): 225–232.

Ingold, Tim. *Lines: A Brief History*. London: Routledge, 2007.

Jacob, Christian. *The Sovereign Map: Theoretical Approaches in Cartography throughout History*. Chicago: University of Chicago Press, [1992] 2006.

Kittler, Friedrich. *Gramophone, Film, Typewriter*. Trans. Geoffrey Winthrop-Young and Michael Wutz. Stanford, CA: Stanford University Press, [1986] 1999.

Koch, Peter. "Graphé: Ihre Entwicklung zur Schrift, zum Kalkül und zur Liste." In *Schrift, Medien, Kognition: Über die Exteriorität des Geistes*, ed. Peter Koch and Sybille Krämer, 43–82. Tübingen: Stauffenburg, 1997.

Krajewski, Markus. *Restlosigkeit: Weltprojekte um 1900*. Frankfurt am Main: Fischer, 2006.

Krämer, Sybille, and Horst Bredekamp, eds. *Bild, Schrift, Zahl*. Munich: Fink, 2003.

Latour, Bruno. "Drawing Things Together." In *Representation in Scientific Practice*, ed. Michael Lynch and Steve Woolgar, 19–68. Cambridge, MA: MIT Press, 1990.

Lestringant, Frank. *Mapping the Renaissance World: The Geographical Imagination in the Age of Discovery*. Trans. D. Fausett. Berkeley: University of California Press, 1994.

Lotman, Yuri M. *The Structure of the Artistic Text*. Ann Arbor: University of Michigan, 1977.

Magnus, Olaus. *Carta marina*. Ed. Elena Balzamo. Paris: Corti, [1539] 2005.

Magnus, Olaus. *Historia de gentibus septentrionalibus (A Description of the Northern Peoples)*. 3 vols. Ed. Peter Foote, trans. Peter Fisher. London: Hakluyt Society, [1555] 1996–1998.

Mainberger, Sabine. *Experiment Linie: Künste und ihre Wissenschaften um 1900*. Berlin: Kadmos, 2010.

Mandelbrot, Benoît. "How Long Is the Coast of Britain? Statistical Self-Similarity and Fractional Dimension." *Science* 156, no. 3775 (1967): 636–638.

Melville, Herman. *Moby-Dick, or, The Whale*. Ed. Tom Quirk. New York: Penguin, [1851] 2003.

Morton, Timothy. *Hyperobjects: Philosophy and Ecology after the End of the World*. Minneapolis: University of Minnesota Press, 2013.

November, Valérie, Eduardo Camacho-Hübner, and Bruno Latour. "Entering a Risky Territory: Space in the Age of Digital Navigation." *Environment and Planning D: Society and Space* 28, no. 4 (2010): 581–599.

Osterhammel, Jürgen, and Niels P. Petersson. *Globalization: A Short History*. Trans. D. Geyer. Princeton, NJ: Princeton University Press, 2005.

Randles, William Graham Lister. "De la carte-portulan méditerranéenne à la carte marine du monde des grandes découvertes: La crise de la cartographie au XVIe siècle." In *Géographie du Monde au Moyen Age et à la Renaissance*, ed. Monique Pelletier, 125–132. Paris: Editions du C.T.H.S., 1989.

Ruhe, Cornelia. "Semiosphäre und Sujet." In *Handbuch Literatur & Raum*, ed. Jörg Dünne and Andreas Mahler, 170–177. Berlin: De Gruyter, 2015.

Ryan, Marie-Laure. "Narrative Cartography: Toward a Visual Narratology." In *What Is Narratology? Questions and Answers Regarding the Status of a Theory*, ed. Tom Kindt and Hans-Harald Müller, 333–364. Berlin: De Gruyter, 2003.

Schmitt, Carl. *Der Nomos der Erde im Völkerrecht des Jus Publicum Europaeum*. Berlin: Duncker & Humblot, [1950] 1997.

Serres, Michel. *Hermes V: Le Passage du Nord-Ouest*. Paris: Minuit, 1980.

Serres, Michel. "Loxodromies des Voyages extraordinaires." In *Hermès I: La Communication*, 207–213. Paris: Minuit, 1968.

Siegert, Bernhard. "Waterlines: Striated and Smooth Spaces as Techniques of Ship Design." In *Cultural Techniques: Grids, Filters, Doors, and Other Articulations of the Real*, trans. G. Winthrop-Young, 147–163. New York: Fordham University Press, 2015.

Sobel, Dava. *Longitude*. New York: Walker, 1995.

Stingelin, Martin, and Matthias Thiele. "Portable Media: Von der Schreibszene zur mobilen Aufzeichnungsszene." In *Portable Media: Schreibszenen in Bewegung zwischen Peripatetik und Mobiltelefon*, ed. Martin Stingelin and Matthias Thiele, 7–27. Munich: Fink, 2010.

Stockhammer, Robert. "'An dieser Stelle': Kartographie und die Literatur der Moderne." *Poetica* 33, nos. 3–4 (2001): 273–306.

Stockhammer, Robert. "Der Globus und das Klobige: Uranologie und Meteorologie." In *Weltnetzwerke–Weltspiele: Jules Vernes "In 80 Tagen um die Welt,"* ed. "Passepartout," 132–134. Konstanz: Konstanz University Press, 2013.

Stockhammer, Robert. *Kartierung der Erde: Macht und Lust in Karten und Literatur*. Munich: Fink, 2007.

Strabo. *Geography*. 8 vols. Trans. Horace Leonard Jones. London and New York: Heineman and Putnam, 1917–1932.

Struck, Wolfgang. "Die Zerstreutheit des Geographen: Jules Vernes andere Reise um die Welt." In *Weltnetzwerke–Weltspiele: Jules Vernes "In 80 Tagen um die Welt,"* ed. "Passepartout," 141–145. Konstanz: Konstanz University Press, 2013.

Tally, Robert T. *Spatiality*. New York: Routledge, 2013.

Verne, Jules. *Cinq semaines en ballon*. Paris: Le Livre de poche, [1863] 2000.

Verne, Jules. *Le Phare au bout du monde*. Paris: Gallimard, [1905] 1999. English translation: *Lighthouse at the End of the World*. Ed. and trans. William Butcher. Lincoln: University of Nebraska Press, 2007.

Verne, Jules. *Les Enfants du capitaine Grant*. Paris: Le Livre de poche, [1867–1868] 2004. English translation: *In Search of the Castaways*. Trans. Charles Francis Horne. New York: Parke, 1911.

Verne, Jules. *Le Tour du monde en quatre-vingts jours*. Ed. William Butcher. Paris: Gallimard, [1873] 2009.

Verne, Jules. *Vingt mille lieues sous les mers*. Paris: Le Livre de poche, [1869] 2006.

Weber, André. *Wolkenkodierungen bei Hugo, Baudelaire und Maupassant im Spiegel des sich wandelnden Wissenshorizontes von der Aufklärung bis zur Chaostheorie*. Berlin: Frank & Timme, 2012.

White, Kenneth. "Presentation." Institute of Geopoetics, 1989. http://institut-geopoetique.org/en/presentation-of-the-institute.

Woodward, David. "Maps and the Rationalization of Geographic Space." In *Circa 1492: Art in the Age of Exploration*, ed. Jay A. Levinson, 83–87. New Haven, CT: Yale University Press, 1991.

15
Material Cartography: João Guimarães Rosa's Paratexts

Clara Rowland

> I have several such dried bits, which I use for marks in my whale-books. It is transparent, as I said before; and being laid upon the printed page, I have sometimes pleased myself with fancying it exerted a magnifying influence. At any rate, it is pleasant to read about whales through their own spectacles, as you may say.
>
> — Herman Melville, *Moby-Dick*

Mapping the Novel

In 1957, the Brazilian critic Antonio Candido published a piece titled "O Sertão e o Mundo" praising João Guimarães Rosa's major novel, *Grande Sertão: Veredas* (*The Devil to Pay in the Backlands*), which had come out the previous year.[1] In his essay, Candido develops an insightful reading of the novel's structure based on the interplay between three anthropological categories: the *land*, the *man*, and the *fight*. The entwinement of these main dimensions is the foundation for Candido's argument about the novel, in which men and action become a function of space: "An obsessive physical presence of the physical medium; a society whose rules and destiny depend on it; as a result, conflict among men."[2]

The importance of space in Candido's reading is paramount. Although Candido does not consider its implications at a theoretical level, it is through the essay's considerations of the spatial organization of the novel that its key hermeneutic problems may be posited, and it is also in the spatial and geographic projections of the narrator's discourse that the effects on the reader become visible for the critic. Projected visuality, Candido indirectly suggests, is central to this 600-page novel.

A detailed reading of some of the essay's passages may show us how cartographic metaphors structure this view. In one of the essay's key moments, Candido explicitly posits the relation of the novel to a map:

> Lowlands where horses gallop, sierras where horses drag themselves; grey fields, ..., holms, corrals and villages. At every step, the tangible reality of the Northern part of Minas, up to Piauí, where the man of the South is a stranger. Bent over the map, we are able to identify the majority of toponyms and the approximate route of horse rides. Guimarães Rosa's world seems to be limited to observation. But we need to

be careful. Pressed by curiosity, the map collapses and recedes. Here, a void; there, an impossible combination of places; further on, a mysterious route, unreal names. ... Let us unfold this map thoroughly. Like a large ox-skin, the Northern territory of Minas Gerais extends itself, cut along the loin by the River São Francisco—physical accident and magical reality, water flow and river God, the axis of the Sertão.[3]

This much-cited passage has been discussed throughout Guimarães Rosa's critical reception, of which mapping is an integral part. After extensive travels and geographic research, Alan Viggiano, in 1971, was proud to present a large number of identifiable toponyms to counter Candido's indeterminacy thesis, situating the novel's actions in accurate maps of the region.[4] More recently, the critic Willi Bolle, reenacting the protagonist's and Viggiano's travels along the landscape of Minas Gerais, concluded that obsessive geographic rigor and poetic indeterminacy balance each other through the intricate and self-canceling paths of the novel's characters.[5] The labyrinthine space of the novel hovers between referentiality and indeterminacy, critics seem to argue. In this, Guimarães Rosa's work appears to respond to one of the main tensions in the relations between literature and cartography.

Less noted, however, is the rhetorical shift that brings the figure of the map into Candido's reading; not at all obvious, the map appears as a referent that slowly slides into a discourse occupied with an immersion in the novel's landscape. What triggers the shift to the map as a possible figure of space is the recognition of the alien condition of the man from the South (the critic, the reader, Candido himself) in relation to the landscape described: "... where the man of the South is a stranger. Bent over the map, we are able to identify. ..." But if the map is initially posited as a possible *double* for the novel—both map and book are materially observed by the extraneous reader; the map mirrors and validates the novel, both isomorphically participating, "like a large ox-skin," in the landscape described—it will later be revealed as a metaphor for the novel itself. This may be seen through the deictics used ("*here*, a void; *there*, an impossible combination of places"), which seem to hesitate between the book and the map (*where* exactly is this void?), leading the way for the establishment of a relation between the materiality of the map and the materiality of the novel, while in the same gesture denying the very possibility of a map. One might think of Melville's famous description of Queequeg's birthplace in *Moby-Dick*: "Queequeg was a native of Kokovoko, an island far away to the West and South. It is not down in any map; true places never are."[6] *Grande Sertão: Veredas* seems to call into question its own referentiality, and to suggest a poetic cartography, a mapping of true places, one might say, where the text at the same time elicits and collapses cartographic representation.

To contemporary readers, Candido's description of the novel is entirely familiar. The two maps that make up the inside cover flaps of *Grande Sertão: Veredas* (figures 15.1 and 15.2) seem to match Candido's comments, expanding the meaning of the maps of Minas Gerais at their basis through a series of symbols, references, and illustrations suggestive of a symbolic and poetic relation to space. Furthermore, the maps' cartographic aspirations are established through the sole elements that could be superimposed on an actual map—the rivers that run through the territory. In fact, it has not escaped readers of the novel's illustrations that each of the flaps corresponds to one of the banks of the São Francisco River. Candido's reading seems to echo this division: "If we think of [the river's] function in the novel, we realize that it divides the world in two parts, different in quality: the left bank and the right bank, loaded with the magical-symbolical meaning that this division represents to the primitive mentality. The right bank is auspicious; the left, inauspicious."[7] Indeed, if we take a closer look at the book's materiality, we will find that this split is represented at all levels. "The São Francisco divided my life in two parts"[8] is the central sentence of the narrator's discourse, in the exact middle of the novel, textually dividing *Grande Sertão: Veredas* in two. Materially, the text enacts what the illustrations suggest; the map is embedded in the novel, as a frame and as a figure for its spatial structuring. Both a container and a contained, it seems to recall the ambiguities of the third term that Walter Moser relates to the encyclopedic project:

> In their complementarity, System and Dictionary would suffice to constitute the encyclopedia. One is entitled to wonder what is the contribution of the third element, the Map, also called "world map," "figured system," "table." At first sight it seems to have the status of an extra, an element that is not necessary, because it adds to a work that is already complete. It seems to be redundant information coming from a transsemiotic translation, which says the same thing in another semiotic register. The point is that as a synoptic table it belongs to another textuality. Making the transition to the pictorial, it situates itself on the margin of the verbal discourse that is common to the System and the Dictionary. The fact is also confirmed by the technical problem its inclusion in the book represents. The marginality of the map, however ... is contradicted by the fact that the ideal of a synoptic representation is an insistent fantasy throughout the history of the encyclopedic project whose center it apparently aims to occupy. Many citations could confirm this idea of Novalis: "The more difficult it is to put a book in a picture, the less good it is."[9]

It is therefore surprising to note that the famous maps of *Grande Sertão: Veredas* were only added to the second edition of the novel, published in 1958—one year, that is, after Candido's famous article. The first edition contained no illustrations aside from Poty's drawings for the front and back cover. Whatever the correlation of these two elements—the critic's trope and the author's decision—the episode is significant of the cartographic vocation, one could say, of Guimarães Rosa's novel, and of the importance of the map as a figure for its reading. As a threshold element, calling attention to a

Figures 15.1 and 15.2

Front and back cover of the 6th edition of *Grande Sertão: Veredas* (Rio de Janeiro: José Olympio, 1968).

Figures 15.1 and 15.2 (continued)

visual and spatial interpretation of the novel's thematic threads, the map functions, I argue, as an interface where an image of the book is projected and reading indications are visually set forth.

The purpose of this chapter is to examine this relation, taking *Grande Sertão: Veredas* as its departure point in questioning the interconnectedness of two sets of relations: the relation between literary text, obscure referentiality, and literary landscape in the building up of Guimarães Rosa's sertão (explored in the next section); and the relation between cartography and the questioning of the material book (discussed in the third and fourth sections of the chapter). Guimarães Rosa's work, I argue, is an extreme case of the enactment of the implications of mapping in literature—not only, as we have begun to see, in its poetic transfiguration of the elusive map of the landscape of the sertão heartland, but also in the overlapping of the map and its metaphors with the unstable cartography of a book in permanent reconfiguration. In this questioning of the textuality of the literary map, I believe, resides its main theoretical interest for the relations between literature and cartography.

The Building Up of Guimarães Rosa's Sertão

To better examine the building up of a literary landscape, let me turn to "Meu Tio o Iauaretê" ("The Jaguar," in David Treece's translation)—a short story completed by Guimarães Rosa at the time he wrote *Grande Sertão: Veredas*, but published only posthumously—that shares many themes and other features with his novel. At the beginning of the story, the narrator warns his visitor: "Aqui é muito lugaroso" (literally, "Here it is very *lugaroso*" or "This place is very *lugaroso* / has too many places").[10] Typical of Guimarães Rosa's neological derivations, the coinage *lugaroso* (an adjective formed from *lugar*, place) seems to convey the idea, at the same time, of this obsessiveness of place and of the multifarous nature of this complex geography. It is also a warning: in the tense context of the short story (in a lost cabin in the woods, a strange man tells a visitor the tale of his metamorphosis from jaguar hunter to hunter of men, suggesting he will be his next prey), *lugaroso* means excessive in the sense of dangerously unlimited or uncontrollable.

If we take into account the construction of this short story—as in Dostoevsky's *hidden dialogue*,[11] a supposed dialogue between two characters is rendered through the direct discourse of just one voice—the sentence may be read differently. One of the first intelligible sentences in the story is an invitation to enter an uncanny place (a cabin without walls), addressed, we come to understand, to some foreigner at the threshold: "Uh huh, you wanna come in, come on."[12] Necessarily, the reader entering the text identifies with this invitation, thus beginning to perceive an unsettling coincidence between the space of the narrative and the text itself, or between text and territory. The story will explore this tension dramatically, representing the death of the narrator through an abrupt textual interruption. *Aqui é muito lugaroso*, then, could also mean the cartographic quality of this literary work: haunted by a labyrinthine structure, obsessed with spatial demarcations, and determined by its own spatial materiality. Again, the geography in the text and the geography of the text seem to overlap, suggesting a dangerous excessiveness.

If, as I began to suggest, Guimarães Rosa's prose presents itself as the poetic cartography of a territory suspended between its referential nature and the rhetorical destabilization of such referentiality, the inescapable immensity of the territory of the sertão "as big as the world"[13] may form the basis of this idea of fiction, where book, map, and landscape continuously collapse on one another, and where the reader is continuously confronted with the paradoxical legibility of that which has no limit (geographically) or form (textually). We are not far from the totalizing ambition (*opere mondo*) Franco Moretti finds in works such as *Faust*, *Moby-Dick*, or *Ulysses*.[14] But in the consistent play on form and on the reading conditions of a blurred and yet insistent cartography, Guimarães Rosa's fiction explicitly takes the inscribed cartography as a representational and hermeneutic problem.

The title of *Grande Sertão: Veredas* points in that direction. In one of the rare moments in which the text incorporates its own title (embedded reflexivity is one of the most important features of this novel), the narrator Riobaldo describes a denied perspective: "Do you know the great sertão? The ones who know it are the vultures, hawks, kites, and birds like that: they are always high up there, feeling the air with lowered feet, sizing up at a glance all joys and sorrows."[15] The *veredas*—small rivers, streams, marshes, or valleys with the function of routes—are pathways to this unlimited and inapprehensible space, impervious to any kind of totalizing view: this is the narrator's (and the reader's) position. The landscape is thus described in tension with the impossibility of legibility or representation, its inaccessible totality being construed through the immersed perspective of its characters. The title is therefore an opposition between different modes of construing the same space. Indeed, the most operative definition of the sertão that the novel offers the reader may be found on its first page: "The sertão describes itself. It is where the grazing lands have no fences."[16] In the terminology of Deleuze and Guattari, this immense, deserted space may be characterized as a *smooth space*: the points are subordinated to the trajectory in an unstable errancy in which direction is often blurred by the instability of the toponyms.[17] Traversed by the veredas who seek their way through the immense space of the backlands, the landscape does not allow for fixity or distinctions, for borders between inside and outside, relating the impossibility of a map disconnected from its *tours*, in Michel de Certeau's terms,[18] with the impossibility of narrating an ever-open action: "The sertão has neither windows nor doors."[19] Describing the formless form of this world may be the challenging task of Guimarães Rosa's fiction: a true place, one could say, is a "moving world." I propose now to see how the material book becomes a fundamental tool in this cartography of the unlimited.

Cartography and the Material Book

For a walk and back again, said the fox. Will you come with me? I'll take you on my back. For a walk and back again.

This epigraph from *Sagarana*, Guimarães Rosa's first book, taken from "Grey Fox," a children's story, points to a double movement that runs through his entire oeuvre. On the threshold of the book, on the one hand, it introduces traveling as a common theme of the short stories that constitute it, reinforcing the unity of the whole; on the other hand, as a preliminary invitation, akin to the one we've seen in "Meu Tio o Iauaretê," it suggests the superimposition of the description of a double movement (back and forth) and the reading experience. The suggestion, however, is not developed in *Sagarana*. The association of the act of reading with a movement bent on itself will be construed through the five books published by Guimarães Rosa, finding its clearest design in *Tutaméia*, in which rereading is clearly stated as a condition of legibility of the book. Suzi Sperber has emphasized the presence, in Guimarães Rosa's fiction, of characters engaged in back-and-forth movements.[20] It is a feature that should be considered not only from a structural point of view but also in the context of the tension between mapping and reading I underlined above. And in the two books published in 1956– *Grande Sertão: Veredas* and *Corpo de Baile*–it is important to note how the semantic field of travel begins to function as a reflexive or metaliterary problem. On one level, this move takes the form of an increasingly precise metaphorization of the theme—in which terms such as *travel*, *traverse*, and *road* are combined in relatively fixed structures, aimed at a representation of the relationship between experience and knowledge. In the variations dictated by the temporality of the trip, space seems to translate a permanent instability in reading the world. As Peter Brooks puts it: "If there is a knowledge provided by narrative ... it is of a particular sort: not only knowledge that comes too late, but recognition of the perpetual belatedness of cognition in relation to action."[21]

We may think of the most important descriptions of error in *Grande Sertão: Veredas*:

> I go through an experience, and in the very midst of it I am blind. I can see only the beginning and the end. You know how it is: a person wants to swim across a river and does, but comes out on the other side at a point lower down, not at all where he expected. Isn't life really a dangerous business?[22]
>
> Had I but guessed then what I later came to know, after many bolts from the blue. ... One is always in the dark, only at the last minute the lights come on. I mean, the

truth is not in the setting out nor in the arriving: it comes to us in the middle of the journey.[23]

In terms now centered on its temporal implications, this is again the problem of form in Guimarães Rosa's fiction, represented by the invalidation of the extremes of the trip: the center is the place of blindness, and the metaphorization of movement continually exposes the late temporality of the constitution—and apprehension—of the way. The impossible bird's-eye view of the sertão I have commented on had the birds contemplate "at a glance all joys and sorrows," but no simultaneity is allowed on the ground. "We live on the road,"[24] says the narrator of "Estória do Homem do Pinguelo," when reaffirming the fundamental unreadability of this world's confusing revelations.

The epigraph from *Sagarana* allows the reader to realize that it is in the form of the book that the metaphorical dimension assumed by this topography in motion finds its performative value, temporalizing the mapping experience through the material limits—beginning and ending—of the book. *Corpo de Baile* and *Primeiras Estórias* already suggest this. Both books are organized around the two extremes of the trip. Between the final departure of Miguilim as a child at the end of the first story and his expected arrival at Buriti Bom at the close of the long final text, the book's structure is defined by the idea of movement. The same is true of *Primeiras Estórias*, in which two inverted travels by plane open and close the book for the same character in its first and last story ("And life was coming toward him" is the formula that concludes both the travels and the reading experience).[25] The inscription of maps—and of maps of reading, I want to suggest—in the paratextual margin reflects this thematic insistence.

A careful reading of the editions prepared by Guimarães Rosa clearly reveals an engagement with the book as a form and with its physical materiality. In Johanna Drucker's terms for describing experimental typography, what is at stake is "an investigation upon the book as an artist's medium, rather than an editor or publisher's domain."[26] Guimarães Rosa's control over his own books is confirmed both by his publishing house, José Olympio, and by his intense correspondence with translators and foreign publishers. The material edited by Paulo Rónai in the posthumous publication of the volume *Estas Estórias* clearly exemplifies this aspect. We can see a set of provisional tables of contents designed by the author as indications for the illustrator.[27] We now know that the books were scrupulously edited under the instructions of Guimarães Rosa, who had access to all stages of production. The promotional texts were written or suggested by the author, and anecdotes of Guimarães Rosa converting typos into neologisms in his careful review of the proofs became famous.[28]

Ironically, the editions of Guimarães Rosa's work published after his death have for various reasons sacrificed these elements to the point that the question of the material book has become, for more than thirty years, essentially unreadable. Only on the fiftieth anniversary of the books he published in 1956 was an effort made to recover materials sacrificed in earlier editions. This recovery is even more significant if we consider that the implications of Guimarães Rosa's close involvement with the production process of the book go beyond a simple manifestation of authorial control. These editions now allow us to understand the complexity of what we may call his book projects and to see how paratextual materials are placed at the service of a general indication of reading and of a mapping of the reader's path. In particular, they make visible the affinities and correlations between Guimarães Rosa's work on the illustrated maps and the inclusion of duplicated tables of contents (reading and rereading indexes) in many of his works. In both we can see the inscription of a reflection on the effects of rereading, of repetition, and of a double assessment of the same graphic material that involves the temporal and spatial questioning of the printed form.

Paratextual Doubling in Guimarães Rosa's Work

The material perspective I have elaborated here allows us to see a number of Guimarães Rosa's other works in a new light. Indeed, from the gigantic enterprise that is *Grande Sertão: Veredas* to the minimalism of his short story collection *Tutaméia*, all Guimarães Rosa's works after *Sagarana* have some kind of doubling in their margins that frames the text in two noncoincidental presentational elements and some kind of enactment of the index. *Corpo de Baile*, in its first edition of 1956, has two tables of contents, one at the beginning and one at the end of the seven stories, that literally tie the work together. The two versions offer different genre classifications for the same texts, belatedly identifying in the book's *parabasis* a space for theory and reflexivity. *Primeiras Estórias*, published in 1962, offers an illustrated table of contents at the end of the book that visually translates the short stories that integrate it. *Tutaméia*, published in 1967, takes this doubling further by clearly prescribing a revision of the book's first reading in a "Rereading Table of Contents," as we will see in this final section. Mapping the book's textual territory, these indexes both build up the work's unity and subject it to a process of reconfiguration through repetition and variation.

My purpose here is to draw attention to the similarities between different forms of paratextual doubling in Guimarães Rosa's work, of which the maps embracing *Grande Sertão: Veredas* and the tables of contents of *Tutaméia* will be key examples, and to the implications of this coincidence between geographic and bibliographical descriptions for an interrogation of the textuality of the literary map. My suggestion is that in all these cases—whether conveyed through illustration or through a creative use of indexes, tables of contents, or epigraphs—the book builds up, in its own extremes, a dissonant cartography conveying, in the same gesture, the spatialization and the temporalization I have hinted at as major formal devices in Guimarães Rosa's work. In this, it brings to the fore the temporal implications of his inscribed visible and readable cartography. Through the usage of maps as tables and of tables as maps,

Guimarães Rosa's books seem to establish rereading as a constitutive movement of the reading. They suggest a movement upon themselves that affects the mapping of the book, defining repetition as the impossibility of fixating their form and guiding the act of reading to the paradoxical readability of a device in perpetual reconfiguration. Indeed, the infinity symbol that the reader finds at the end of *Grande Sertão: Veredas* could be the visual sign for this recursive movement. We saw earlier that the theme of traveling functioned as a trope for a permanent deferral in the reading of the world. The inscription at the book's liminal extreme of reading as a *tour* through the book's shifting places is another figure of this paradoxical legibility.

As we have seen, the split map on the cover flaps of *Grande Sertão* enacted this tension with the book's completeness. As a graphic index of sorts, the double map indicates and constitutes, at the margin of the book, a visual and spatial whole—significantly split and dependent on the reunification of the book's extremes to build its picture—that in the same gesture suggests and collapses the reference to the sertão and the book's production of totality.[29] On the other hand, giving the reader the first unitarian image of the book while also breaking it down into its components, the table of contents seems to respond in dissonance to the hybrid function of the map as a figure that overlaps, in Paul Zumthor's terms, description, representation, and interpretation.[30] In the margins of Guimarães Rosa's books, duplicated maps and tables of contents thereby seem to share a synthetic and structural function that at the same time builds up and threatens the book, inscribing in its paratexts the cartography of a visual and temporal reconfiguration.

Let us then turn to the most extreme case of duplication in Guimarães Rosa's work: the "Rereading Table of Contents" at the end of *Tutaméia*, Guimarães Rosa's 1967 short story collection and his most sophisticated experiment with the possibilities of the material book. It will be helpful to take a closer look at its two extremes (figures 15.3a–b).

In the first table of contents, the stories are listed in the order in which they appear in the book, and one notices from the outset that this order is alphabetical, with the exception of two stories whose titles, beginning with G and R, spell the sequence JGR (the author's initials, a signature of sorts). The second table of contents mirrors the first, but with some significant changes: four stories, whose titles were printed in italics in the first table, are now grouped together at the beginning as "prefaces" to the other forty stories. Two different images of the same book are conveyed, building up the dissonant cartography we began to see with earlier examples. In fact, the book's identity changes from one table of contents to the other, as title and subtitle of the book are reversed from *Tutaméia: Terceiras Estórias* to *Terceiras Estórias: Tutaméia*. The same textual material projects two different structures and two different spatial and unitarian images of a scattered whole. Significantly, it is only in the second of these images—explicitly posited as a rereading table of contents, to be found after the

TUTAMÉIA
(TERCEIRAS ESTÓRIAS)

"Daí, pois, como já se disse, exigir a primeira leitura paciência, fundada em certeza de que, na segunda, muita coisa, ou tudo, se entenderá sob luz inteiramente outra."

SCHOPENHAUER.

Aletria e hermenêutica	3	Melim-Meloso	92
Antiperipléia	13	No prosseguir	97
Arroio-das-Antas	17	*Nós, os temulentos*	101
A vela ao diabo	21	O outro ou o outro	105
Azo de almirante	24	Orientação	108
Barra da Vaca	27	Os três homens e o boi	111
Como ataca a sucuri	31	Palhaço da bôca verde	115
Curtamão	34	Presepe	119
Desenrêdo	38	Quadrinho de estória	122
Droenha	41	Rebimba, o bom	126
Êsses Lopes	45	Retrato de cavalo	130
Estória n.º 3	49	Ripuária	134
Estoriinha	53	Se eu seria personagem	138
Faraó e a água do rio	57	Sinhá Secada	142
Hiato	61	*Sôbre a escôva e a dúvida*	146
Hipotrélico	64	Sota e barla	167
Intruge-se	70	Tapiiraiauara	171
João Porém, o criador de perus	74	Tresaventura	174
Grande Gedeão	77	— Uai, eu?	177
Reminisção	81	Umas formas	180
Lá, nas campinas	84	Vida ensinada	184
Mechéu	88	Zingarêsca	189

Figures 15.3a–b

The two tables of contents of *Tutaméia* (Rio de Janeiro: José Olympio, 1967)

TERCEIRAS ESTÓRIAS
(TUTAMÉIA)

Índice de releitura

"Já a construção, orgânica e não emendada, do conjunto, terá feito necessário por vêzes ler-se duas vêzes a mesma passagem."
SCHOPENHAUER.

PREFÁCIOS:

Aletria e hermenêutica .. 3	*Nós, os temulentos* 101
Hipotrélico 64	*Sôbre a escôva e a dúvida* 146

OS CONTOS:

Antiperipléia 13	Mechéu 88
Arroio-das-Antas 17	Melim-Meloso 92
A vela ao diabo 21	No prosseguir 97
Azo de almirante 24	O outro ou o outro 105
Barra da Vaca 27	Orientação 108
Como ataca a sucuri .. 31	Os três homens e o boi .. 111
Curtamão 34	Palhaço da bôca verde .. 115
Desenrêdo 38	Presepe 119
Droenha 41	Quadrinho de estória .. 122
Êsses Lopes 45	Rebimba, o bom 126
Estória n.º 3 49	Retrato de cavalo 130
Estoriinha 53	Ripuária 134
Faraó e a água do rio .. 57	Se eu seria personagem .. 138
Hiato 61	Sinhá Secada 142
Intruge-se 70	Sota e barla 167
João Porém, o criador de perus 74	Tapiiraiauara 171
	Tresaventura 174
	— Uai, eu? 177
Grande Gedeão 77	Umas formas 180
Reminisção 81	Vida ensinada 184
Lá, nas campinas 84	Zingarêsca 189

reading—that structural positions are revealed. Throughout the reading the prefaces may have functioned as prefaces, but they are declared so only in a second moment. The purpose of this belated identification of an opening function and position for some of the stories is to call into question the border between text and paratext: the prefaces see their paratextual nature blurred and their exteriority disturbed from the moment they are positioned in a series of recognized and recognizably fictional stories. Book and margin are intermingled from the outset.

The temporality of repetition is of course what defines the originality of this play on the literary map and its effects on the cartography of the book. The idea of rereading is introduced through two epigraphs from Schopenhauer's preface to the first edition of *The World as Will and Representation*. The source is indeed significant, for Schopenhauer's is an extreme case in the tradition of the philosophical preface, given its bitter reflections on the book as a medium and on the necessity of reading "the book twice" for the book's legibility. For Schopenhauer, the material book is an inadequate medium. From the outset, from the very margin, the preface tries to compensate for the discrepancy between a fixed form, which requires margins, and the organicity of the thought that rejects them. If the thought that the book brings forth is organic and unique, it will not endure the divisions and borders a book has to have. Schopenhauer writes: "A *single thought*, however comprehensive, must preserve the most perfect unity. … But a book must have a first and a last line, and to this extent will always remain very unlike an organism, however like one its contents may be. Consequently, form and matter will here be in contradiction."[31] This contradiction makes the preface and its demands necessary for an understanding of the book. The role of the marginal elements then becomes central: only the preface allows the book to be read, since it is the preface that states its rereading.

Both epigraphs in Guimarães Rosa's book articulate a critique of the limited form of the material book. The first mentions the relationship between a patient first reading and a revelatory second reading: "Therefore, as I have said, the first reading demands patience, derived from the confidence that with a second reading much, or all, will appear in quite a different light." In terms of reading protocols and expectations, the first table of contents thus has a destabilizing function: before even beginning a first reading, the reader is already projected toward a later revision, toward a subsequent teleological revelation of meaning. We are not far from one of the effects of the split in the maps of *Grande Sertão* I discussed earlier.

Completing this tentative first reading, the hypothetical reader then encounters a "Rereading Table of Contents" at the end of the book with a second quote from Schopenhauer: "The structure of the whole, which is organic and not like a chain, in itself made it necessary sometimes to read twice the same passage."[32] If the first epigraph brought the reader to the necessity of a suspension based on the knowledge that the second reading would change much or everything, at the end of this process the reader is projected back to her first reading, as she is told that a rereading has already occurred. Attention is called to the center, now defined as a space of difference, for the revelatory character of a second reading is subject to a temporal slip–it comes late, or belatedly, just as awareness (for instance, that *Tutaméia* is a book with four prefaces) comes at a later moment.[33]

It is possible to see in this movement an enactment of the impossibility of distinguishing between reading and rereading that Matei Calinescu discusses in his work on the subject. He calls attention to the fact that a first reading is often "double," intermingling linearity (typically characteristic of a first reading) with "structural attention" (usually related to second readings).[34] But what we find described in this interplay of anticipation and retrospection is not exclusively a matter of a more attentive or "spatial" reading, as is the case in the main discussions of rereading. In the building of a double mapping frame, what seems at stake is the interplay between difference, repetition, and deferral. In this, Guimarães Rosa's cartography of the material book marks its originality, for the idea of a literary map subjects the cartographic gesture to the temporality of writing and reading. Embedded in the book's own space, the map is necessarily contaminated by its paths, just as the bird's-eye view of the sertão was accessible exclusively through the narrator's *veredas* or pathways. Its relation to the book is metonymic, one could say, and not exclusively metaphorical. Inscribed geography and reflexive bibliography, margin and center, complement each other in the double movement through which the book as a space is constructed and mapped out while at the same time threatened by the instability of its form.

If the first table of contents of *Tutaméia* projects the reader toward the end, from the story to its map, one could say, the second version is there to make her go back to the center, to the "middle of the passage," to the book as a transformed space. Tables of contents and maps offer a paradoxical prescription that, as in Schopenhauer's case, deals with the impossibility of fixing the book in a settled form because of the contradiction between the spatial projection of the book and its inherent shifting movements that defy stable forms and fixed delimitations. Marginal elements, then, characterize Guimarães Rosa's books as the space of this transitory relationship between a fixed form–the text, the map, the printed matter–and its revision enacted through repetition in its temporality. In that sense, the margin can never be interpreted as an accessory. It is rather a way of establishing the book's unstable legibility, of taking its apparent fixed materiality as the point of departure for an organic transformation. Through the margin, Guimarães Rosa builds up the cartography of the book, in its material impossibility that depends on materiality, as a map in permanent reconfiguration.

Notes

1. The review was lately republished by Candido, with a different title, in one of his books: Antonio Candido, "O Homem dos Avessos," in *Tese e Antítese*, 119–140 (São Paulo: Companhia

Editora Nacional, 1964). The only existing English translation of *Grande Sertão: Veredas* (by James L. Taylor and Harriet de Onís) was published in 1963 under the title *The Devil to Pay in the Backlands*. The name *sertão* is often associated with the word *desertão* (big desert) and refers to a wide region in Central-Western and Northeastern Brazil.

2. Antonio Candido, "O Homem dos Avessos," 123.

3. My translation. The original quotation reads, "Dobrados sobre o mapa, somos capazes de identificar a maioria dos topônimos e o risco aproximado das cavalgadas. O mundo de Guimarães Rosa parece esgotar-se na observação. Cautela, todavia. Premido pela curiosidade o mapa se desarticula e foge. Aqui, um vazio; ali, uma impossível combinação de lugares; mais longe uma rota misteriosa, nomes irreais. ... Desdobremos bem o mapa. Como um largo couro de boi, o Norte de Minas se alastra, cortado no fio do lombo pelo São Francisco,–vacidente físico e realidade mágica, curso d'água e deus fluvial, eixo do Sertão" (Candido, "O Homem dos Avessos," 124).

4. Alan Viggiano, *Itinerário de Riobaldo Tatarana* (Belo Horizonte: Comunicação/MEC, 1971).

5. Willi Bolle, *grandesertão.br. O romance de formação do Brasil* (São Paulo: Duas Cidades; Editora 34, 2004), 65.

6. Herman Melville, *Moby-Dick* (New York: Norton, 2002), 59.

7. My translation. The original quotation reads, "Atentando para a sua função no livro, percebemos com efeito que ele divide o mundo em duas partes qualitativamente diversas: o lado direito e o lado esquerdo, carregados do sentido mágico-simbólico que esta divisão representa para a mentalidade primitiva. O direito é o fasto; nefasto o esquerdo" (Candido, "O Homem dos Avessos," 124).

8. João Guimarães Rosa, *Grande Sertão: Veredas* (Rio de Janeiro: José Olympio, 1968), 257.

9. My translation. The original quotation reads, "Dans leur complémentarité, Système et Dictionnaire suffiraient pour constituer l'encyclopédie. On est en droit de se demander ce que peut encore y apporter le troisième élément, la Carte, appelé aussi 'mappemonde,' 'système figuré,' 'tableau.' A première vue, il semble avoir le statut d'un supplément, d'un élément qui n'est pas nécessaire, parce-qu'il s'ajoute à un ouvrage qui est dejá complet. Il semble constituer une information redondante, issue d'une traduction transsémiotique, qui dirait encore une fois la même chose dans un autre registre sémiotique. C'est que, tableau synoptique, il relève d'une autre textualité. Faisant la transition au pictural, il se situe en marge du domaine discursif verbal qui est commun au Système et au Dictionnaire. Le fait est d'ailleurs confirmé par the problème technique que pose son insertion dans le livre. Cependant, la marginalité de la carte ... se trouve être contredite par le fait que l'idéal d'une représentation synoptique se maintien avec l'insistence d'un fantasme à travers toute l'histoire du projet encyclopédique dont il semble même vouloir occuper le centre. De nombreuses citations pourraient confirmer cette idée de Novalis: 'Plus il est difficile de mettre un livre dans un tableau, moins il est bon'" (Walter Moser, "L'Encyclopédie Romantique: Le *Brouillon* de Novalis," in *Le Genre / Die Gattung / Genre* (Strasbourg: Université de Strasbourg, 1979), 503–504).

10. In David Treece's translation, the sentence is literalized in one of its possible readings: "I listen, with my ear to the ground. Horse galloping, pa-ta-pa ... I know how to follow their trail. Phaw ... I can't now, no point, *too many trails around here*. They've gone way off" (João Guimarães Rosa, "The Jaguar," in *The Oxford Anthology of the Brazilian Short Story*, ed. K. David Jackson (Oxford: Oxford University Press, 2006), 329).

11. Cf. the Bakhtinian description of hidden dialogicality: "Imagine a dialogue of two persons in which the statements of the second speaker are omitted, but in such a way that the general sense is not at all violated. The second speaker is present invisibly, his words are not there, but deep traces left by these words have a determining influence on all the present and visible words of the first speaker. We sense that this is a conversation, although only one person is speaking, and it is a conversation of the most intense kind, for each present, uttered word responds and reacts with its every fiber to the invisible speaker, points to something outside itself, beyond its own limits, to the unspoken words of another person" (Mikhail Bakhtin, *Problems of Dostoevsky's Poetics*, ed. and trans. Caryl Emerson (Minneapolis: University of Minnesota Press, 1984), 197).

12. João Guimarães Rosa, *Ficção Completa* II (Rio de Janeiro: Nova Aguilar, 1994), 328.

13. João Guimarães Rosa, *The Devil to Pay in the Backlands*, trans. J. L. Taylor and H. de Onís (New York: Knopf, 1963), 60.

14. Franco Moretti, *Modern Epic: The World-System from Goethe to García Márquez* (London: Verso, 1996). The Italian critic Ettore Finazzi-Agrò draws this connection in *Um Lugar do Tamanho do Mundo: Tempos e Espaços da Ficção em João Guimarães Rosa* (Belo Horizonte: Editora da UFMG, 2001), 29.

15. Rosa, *The Devil to Pay*, 465; the original quotation reads, "Sei o grande sertão? Quem sabe dele é urubu, gavião, gaivota, esses pássaros: eles estão sempre no alto, apalpando ares com pendurado pé, com o olhar remedindo a alegria e as misérias todas" (Rosa, *Ficção Completa*, 364).

16. Rosa, *The Devil to Pay*, 4.

17. As they write: "In striated space, lines or trajectories tend to be subordinated to points: one goes from one point to another. In the smooth, it is the opposite: the points are subordinated to the trajectory. ... The dwelling is subordinated to the journey; inside space conforms to outside space" (Gilles Deleuze and Félix Guattari, *A Thousand Plateaus: Capitalism and Schizophrenia*, trans. Brian Massumi (Minneapolis: University of Minnesota Press, 1987), 478).

18. Michel de Certeau, "Parcours et cartes," in *L'invention du quotidian,* 1: *Arts de faire* (Paris: Gallimard, 1990), 176.

19. Rosa, *The Devil to Pay,* 402.

20. Suzi Sperber, *Guimarães Rosa: Signo e Sentimento* (São Paulo: Ática, 1982), 113.

21. Peter Brooks, *Reading for the Plot: Design and Intention in Narrative* (Cambridge, MA: Harvard University Press, 1992), 212.

22. Rosa, *The Devil to Pay*, 27–28. The original quotation reads, "Eu atravesso as coisas—e no meio da travessia não vejo!–só estava era entretido na idéia dos lugares de saída e de chegada. Assaz o senhor sabe: a gente quer passar um rio a nado, e passa; mas vai dar na outra banda é num ponto muito mais embaixo, bem diverso do em que primeiro se pensou. Viver nem não é muito perigoso?" (Rosa, *Ficção Completa* II, 28).

23. Rosa, *The Devil to Pay*, 52. The original quotation reads, "Acertasse eu com o que depois sabendo fiquei, para lá de tantos assombros. ... Um está sempre no escuro, só no último derradeiro é que clareiam a sala. Digo: o real não está na saída nem na chegada: ele se dispõe para a gente é no meio da travessia" (Rosa, *Ficção Completa* II, 46).

24. Rosa, *Ficção Completa* II, 824.

25. João Guimarães Rosa, "Treetops," in *The Oxford Anthology*, 324.

26. Johanna Drucker, *The Visible Word: Experimental Typography in Modern Art 1909–1923* (Chicago: University of Chicago Press, 1994), 227.

27. João Guimarães Rosa, *Estas Estórias* (Rio de Janeiro, José Olympio, 1969), xii; xx.

28. João Guimarães Rosa, *Corpo de Baile:* Edição Comemorativa 50 anos (Rio de Janeiro: Nova Fronteira, 2006), 5.

29. As Hans Blumenberg writes (my translation): "On the other hand, the fascination of the book's power as a producer of totality. Its strength in taking as a unity things disparate, distant, contradictory, extraneous or familiar, or at least its strength in suggesting it, is an essential element of the book, whatever the subject upon which this unity is projected" (Hans Blumenberg, *Die Lesbarkeit der Welt* (Berlin: Suhrkamp, 1983), 17–18).

30. Paul Zumthor, *La Mesure du Monde* (Paris: Seuil, 1983), 338.

31. Arthur Schopenhauer, *The World as Will and Representation*, trans. E. F. Payne (Indian Hills, CO: Falcon's Wing Press, 1958), xx–xxi.

32. Actually, Guimarães Rosa's translation of Schopenhauer's words introduces an interesting distortion: where the original says "die selbe Stelle zwei Mal zu berühren" (in Payne's translation, "to touch twice on the same point"), Guimarães Rosa writes "to *read* twice," displacing the agency from the writer to the reader, stressing the reader's role in the rereading/rebuilding of the book.

33. It is evident that in this projection we have the contradictions inherent in the temporality of the preface that Derrida questions when discussing the Hegel prefaces: the confusion between a moment before the text and after the text, between anticipation and retrospection, which establishes a belated and repetitive temporality in the act of reading (see Jacques Derrida, "Hors livre–prefaces," in *La Dissémination* (Paris: Seuil, 1972), 13–14).

34. Matei Calinescu, *Rereading* (New Haven, CT: Yale University Press, 1993), 18–19 and passim.

Bibliography

Bakhtin, Mikhail. *Problems of Dostoevsky's Poetics*. Ed. and trans. Caryl Emerson. Minneapolis: University of Minnesota Press, 1984.

Blumenberg, Hans. *Die Lesbarkeit der Welt*. Berlin: Suhrkamp, 1983.

Bolle, Willi. *grandesertão.br. O romance de formação do Brasil*. São Paulo: Duas Cidades; Editora 34, 2004.

Brooks, Peter. *Reading for the Plot: Design and Intention in Narrative*. Cambridge, MA: Harvard University Press, 1992.

Calinescu, Matei. *Rereading*. New Haven, CT: Yale University Press, 1993.

Candido, Antonio. "O Homem dos Avessos." *Tese e Antítese*, 119–140. São Paulo: Companhia Editora Nacional, 1964.

de Certeau, Michel. "Parcours et cartes." In *L'invention du quotidian,* 1: *Arts de faire*, 175–180. Paris: Gallimard, 1990.

Deleuze, Gilles, and Félix Guattari. *A Thousand Plateaus: Capitalism and Schizophrenia*. Trans. Brian Massumi. Minneapolis: University of Minnesota Press, 1987.

Derrida, Jacques. *La Dissémination*. Paris: Seuil, 1972.

Drucker, Johanna. *The Visible Word: Experimental Typography in Modern Art 1909–1923*. Chicago: University of Chicago Press, 1994.

Finazzi-Agrò, Ettore. *Um Lugar do Tamanho do Mundo: Tempos e Espaços da Ficção em João Guimarães Rosa*. Belo Horizonte: Editora da UFMG, 2001.

Melville, Herman. *Moby-Dick*. New York: Norton, 2002.

Moretti, Franco. *Modern Epic: The World-System from Goethe to García Márquez*. London: Verso, 1996.

Moser, Walter. "L'Encyclopédie Romantique: Le *Brouillon* de Novalis." In *Le Genre / Die Gattung / Genre*, 499–516. Strasbourg: Université de Strasbourg, 1979.

Rosa, João Guimarães. *Corpo de Baile: Edição Comemorativa 50 anos*. Rio de Janeiro: Nova Fronteira, 2006.

Rosa, João Guimarães. *The Devil to Pay in the Backlands*. Trans. J. L. Taylor and H. de Onís. New York: Knopf, 1963.

Rosa, João Guimarães. *Estas Estórias*. Rio de Janeiro: José Olympio, 1969.

Rosa, João Guimarães. *Ficção Completa II*. Rio de Janeiro: Nova Aguilar, 1994.

Rosa, João Guimarães. *Grande Sertão: Veredas*. 6th ed. Rio de Janeiro: José Olympio, 1968.

Rosa, João Guimarães. "The Jaguar." In *The Oxford Anthology of the Brazilian Short Story*, ed. K. David Jackson, 328–354. Oxford: Oxford University Press, 2006.

Rosa, João Guimarães. "Treetops." In *The Oxford Anthology of the Brazilian Short Story*, ed. K. David Jackson, 319–324. Oxford: Oxford University Press, 2006.

Rosa, João Guimarães. *Tutaméia*. Rio de Janeiro: José Olympio, 1967.

Schopenhauer, Arthur. *The World as Will and Representation*. Trans. E. F. Payne. Indian Hills, CO: Falcon's Wing Press, 1958.

Sperber, Suzi. *Guimarães Rosa: Signo e Sentimento*. São Paulo: Ática, 1982.

Viggiano, Alan. *Itinerário de Riobaldo Tatarana*. Belo Horizonte: Comunicação/MEC, 1971.

VV.AA. *Em memória de João Guimarães Rosa*. Rio de Janeiro: José Olympio, 1968.

Zumthor, Paul. *La Mesure du monde*. Paris: Seuil, 1983.

16

Cartographies of War: Star Charts, Topographic Maps, War Games

Anders Engberg-Pedersen

Introduction

Warfare has long been a driving force in the development of cartography. It is easy to understand why: orientational knowledge is indispensable in times of war where the specific topographic lay of the land, the details of the terrain, and the possible routes through it can be decisive for the outcome of the conflict. As Thucydides relates in his *History of the Peloponnesian War*, the sudden death of a native guide could have as a consequence the destruction of almost an entire Athenian army.[1] A good military map is therefore one of the most important tools for the military commander. An apocryphal but often quoted statement attributed to Napoleon summarizes it well: "A detailed map is a weapon of war."[2] As such, military maps were often surrounded by secrecy, and sharing maps with competitive nations was considered an act of treason.[3] On the other hand, national cartographic centers have often published incorrect maps of their own territory with intentionally misplaced cities and nonexistent rivers so as to confuse the enemy.[4]

Yet, for all the importance of topography, actual and fabricated, within military cartography, war maps also reveal an important linkage of spatial representations and temporal projections. While the map itself constitutes a projection of three-dimensional space onto a two-dimensional plane, one of its central functions is to generate a viable projection in the fourth dimension. As an instrument of tactics and strategy, the orientational knowledge of space serves as the basis for attempts to orient oneself in time. From the ancient divination of celestial constellations and star charts to the gradual development of an accurate topographic military cartography in the eighteenth century to the birth of the modern war game and its transformation in contemporary virtual simulations, cartography has formed the site where possible worlds, scenarios, and hypothetical events are imagined, projected, and tested. Above all else, the war map is a symbolic tool for the management of future events.[5]

In this respect, war maps give rise to a precarious epistemology. They produce knowledge, not of the actual world, but of possible worlds whose very existence is in the balance. They operate with conjectures, probabilities, and guesses, and serve as tools to govern the type of event that Aristotle termed "potentialities."[6] Via Boethius's Latin translation, "contingens," they have become known as "future contingents" and designate events that are neither impossible nor necessary.[7] Evidently, like all plans for the future, the projections of military strategy are highly uncertain, but they are based on a specific and carefully calculated transformation of spatial order into events, and this translation is authorized in different ways from one cartographic genre to the next. We might therefore regard war maps as *event maps*—that is, maps that in various ways serve the complex translation of space into time and virtual scenarios into actual deeds. But how, under what conditions, to what extent, and with what validity do war maps convert symbolic spatial arrangements into actions and events?[8]

In this chapter I would like to examine three main genres within the cartography of war: star charts, topographic maps, and war games. At the same time, in a parallel track, I examine a number of literary works. For the questions concerning the workings of event maps in military cartography have been posed explicitly in fictional accounts of war. The war map forms a central topos in literary history, not least because the representation of war and the management of its complex events have challenged authors just as much as they have challenged military commanders and topographers. While a driving force in the history of military cartography has been the utilitarian one of optimizing the map as a tool for the management of contingent futures, of tightening the link between spatial organization and projected events, literary history offers a reflection on the precariousness of the link. In a number of fictional narratives, the encounter with military maps sparks a metareflection on the very possibility of representing and managing war—both with narratives and with maps. A full-fledged history of war mapping and war writing is well beyond the scope of this chapter. Instead I sketch out a basic typology of three main genres in the history of military cartography, and I juxtapose them with a handful of literary scenes that reflect on the assumptions about the military event map inherent in each genre. Comparing astrological star charts and Friedrich Schiller's drama *Wallenstein*, the topographic map and the realist war novel, as well as war games and Roberto Bolaño's *The Third Reich*, the chapter explores some of the ways literature at once draws on, calls into question, and transforms military event maps.

The Celestial Event Map

In 1625 Johannes Kepler responded to a request he had received from Albrecht von Wallenstein, Duke of Friedland and soon to be Generalissimus of the Holy Roman Empire. Wallenstein had asked Kepler to update the horoscope he had made for him in 1608, for his life seemed out of sync with the original horoscope. Some of the predicted life events had occurred too early and some too late. In particular he wanted to know how long he would continue in military service and whether he would be blessed with luck in all his military endeavors.[9] The letter he received from the famous scientist, however, contained a clear rebuke and a lecture on the limits of the science of astrology: the stars did indeed influence events in the sublunary realm, but it was mere superstition to believe that it was possible "to predict particular matters and *futura contingentia* from the skies."[10] Nevertheless, Kepler updated the horoscope and predicted a series of events until the month of March in 1634, in which, as he wrote, "horrible disorder" threatened.[11]

As supreme commander of the Habsburg forces, Wallenstein followed a long tradition in politics and war of looking to the stars for guidance in terrestrial affairs. Since its emergence in ancient Mesopotamia, the science of astrology informed the decisions of rulers and generals seeking the most propitious moment of action for their political and military endeavors. The basic assumption was that celestial constellations carry an effective force, that spatial arrangements in the sky translate into terrestrial events. As Ptolemy puts it in *Tetrabiblos*, the standard reference for all things astrological for over a millennium, "The cause of both universal and of particular events is the motion of the planets, sun, and moon; and the prognostic art is the scientific observation of precisely the change in the subject natures which corresponds to parallel movements of the heavenly bodies through the surrounding heavens."[12]

The nature of the correspondence was the task of the science of astrology to determine and interpret. Over the centuries astrologers developed elaborate and competing systems of spatial notations that could determine the nature of the influence. The significance of a star event was, among other things, contingent on its position in the twelve houses of the zodiac and on the planetary aspect. The aspect concerned the planets' position relative to one another, such as a *conjunction* when two or more planets are lined up along the same longitude. While the rules and the exact layout of the spatial notations within the system were the topic of intense debate among astrologers, the variations all relied on the belief that the exact position of the stars within the spatial system at a given point in time was key to the influence they exerted.[13] The science of astrology is infused with a spatial metaphysics: the celestial configuration in itself carries an "effective power."[14] Star events are made possible because astrologers project a spatial field infused with metaphysical assumptions onto the planets and their movements. For the astrologer looking up at the skies, the heavens constitute a field of possible events at once celestial and terrestrial.[15]

Charts and Horoscopes

This double reference point transforms the heavens themselves into a vast map. The differences from a conventional geographic map are of course immediately evident. Against the background of infinite space the starry skies lack all topography, they do not contain any coordinate system, and with the planets in constant movement they are ever mutable. Infused with metaphysical assumptions, however, they function as spatial representations of earthly events. The stars could therefore be read and interpreted as maps that offered a temporal orientation in terrestrial affairs. Actual star charts such as the *Imagines coeli Septentrionales et Meridionales zodiaci*, the very first printed star chart made by Albrecht Dürer and two collaborators in 1515, often include the zodiac constellations and their astrological symbols (figure 16.1). But while such charts depict the basic elements of the astrological belief system, they had little use as maps of terrestrial events, since they represented the fixed stars but not the wandering stars (i.e., the planets).

The central astrological map was therefore the horoscope. On astrological nativities, the metaphysical spatial projection is particularly evident, and the horoscope Kepler made for Wallenstein in 1608 is exemplary in this regard (figure 16.2). Divided into twelve contiguous triangles, each representing the houses of the zodiac with their individual significance, the horoscope shows the exact position of the sun, the moon, and the planets within this spatial order at the time of Wallenstein's birth. Kepler first noted the "*Conjunctionem magnam Saturni et Jovis in domo prima*"—that is, the major conjunction or alignment of Saturn and Jupiter in the first house—giving those planets the greatest influence on Wallenstein's life.[16] Based on such observations, Kepler proceeded to predict a number of events in Wallenstein's life and concluded that the horoscope "was not a bad nativity, but contained exceptionally important signs."[17]

As in the case of Kepler and Wallenstein, horoscopes were made to reveal the personal characteristics of individuals, the course of their lives, or the propitious moment to undertake an action. In their function as event maps, horoscopes are therefore only indirectly concerned with the spatial relations of celestial bodies. As their etymology suggests (from the Greek *hōro-skopos*: *hōro* 'time' and *skopos* 'observer'), horoscopes constitute a snapshot of the movements of the stars at a particular time. They map a fleeting celestial event, and, extrapolating from the patterns of the constellations, astrologers transform the spatial order into knowledge of future events for a given individual.

Kepler's wry response to Wallenstein, however, marks a fundamental limitation in the epistemology of the astrological nativity map. He writes:

> I state this solely for the purpose of removing the illusion entertained by the subject of the nativity that all the *Particularia* can be predicted from the heavens. This much is true, that from the heavens follow heavenly *Particularia*, but not terrestrial ones, neither *specialia* nor *individua*, rather, all terrestrial *Eventus* take their form and shape from terrestrial causes, since every particular has its particular cause.[18]

According to Kepler, the celestial pattern of Wallenstein's horoscope can only be regarded as an indicator of general trends, but it is beyond the ken of astrology to predict particular events. It is not simply the case that the prediction of future events is precarious and uncertain; its spatial pattern *cannot* be translated into particular

Figure 16.1

The first printed star chart by Albrecht Dürer from 1515. In the corners, surrounding the zodiac constellations and their astrological symbols, four major astrologers—Aratus, Ptolemy, Marcus Manilius, and Al-Sufi—represent the combined wisdom of, respectively, Greek, Egyptian, Roman, and Islamic astrology (Albrecht Dürer, Konrad Heinfogel, and Johannes Stabius, *Imagines coeli Septentrionales et Meridionales zodiaci*, 1515). Courtesy of Daniel Crouch Rare Books, crouchrarebooks.com.

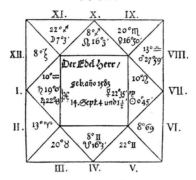

Figure 16.2

The original horoscope Kepler made for Albrecht von Wallenstein in 1608 (Johannes Kepler, *Die Astrologie des Johannes Kepler: Eine Auswahl aus seinen Schriften*, ed. Heinz Artur Strauss and Sigrid Strauss-Kloebe (Munich: Oldenbourg, 1926), 185).

"terrestrial *Eventus*." Too many terrestrial factors muddle the picture. As an event map, the horoscope does not contain a detailed topography of events, only larger tendencies and probabilities. While the metaphysics of the astrological system remains intact, Kepler goes out of his way to emphasize that his map cannot provide the information Wallenstein demands of it.

In the exchange between two of the leading figures of the time, one in astronomy, the other in warfare, we encounter the clash of opposing conceptions of the epistemology of the nativity map. On the one hand, the horoscope is an epistemically highly robust representation that accurately maps spatial constellations onto future contingents at all levels, if only for the trained beholder. On the other hand, the horoscope functions as a representation that accurately maps the positions of the stars onto supraindividual future events, but that generates no reliable knowledge at the level of the individual and the particular, only random fictions. The horoscope as a large-scale event map contrasts with the horoscope as a small-scale event map, so to speak. That Kepler nevertheless agrees to update the horoscope and offers a series of often fairly particular predictions is a good example of what Dan Edelstein has called the "Super-Enlightenment"—the curious blend of the hermetic and the occult with what we today recognize as well-reasoned science. With its amalgamation of astronomical observation and astrological metaphysics, the horoscope precisely maps the "epistemological no-man's land" in which many scientists operated at the time.[19]

Schiller's Poetic Stars

In fictionalized form the celestial event map plays a central role in Friedrich Schiller's *Wallenstein*. By the time Schiller conceived the drama in the late eighteenth century, astrology was no longer enmeshed with the scientific curriculum and had been banished to the field of the occult.[20] Written with hindsight, his drama reflects a late Enlightenment response to the epistemological follies and metaphysical superstitions of the past. But, as I will show, it also transforms the celestial event map into a productive poetic device that comes to organize the event structure of the fictional text. In the following, *Wallenstein* will therefore serve as a salient example of how literature has engaged with and retooled military cartography.

A key scene in the drama's engagement with astrology opens the third part of the play. Surrounded by maps, globes, and quadrants, Wallenstein is facing a large blackboard inspecting a *speculum astrologicum*—a diagram, in other words, of the planets' position relative to one another—when he suddenly exclaims: "Fortunate aspect!"[21] According to the diagram, three planets have entered into a significant conjunction: Jupiter, Mars, and Venus. In Wallenstein's interpretation their alignment is the outcome of a celestial warfare. Jupiter and Venus act as military commanders who have outmaneuvered the cause of the conflict, Mars, and have forced him into submission.

As he puts it: "Now they have vanquished the old enemy, / and bring him to me as a prisoner"[22]

Elated by the "Glücksgestalt," the figure of luck visible on the *speculum astrologicum*, Wallenstein is, in his own interpretation, looking at a diagram of his military future. Indeed, the astrological beliefs of Schiller's Wallenstein are such that he, like the historical Wallenstein, has transformed his palace into a vast map of the heavens, painting the zodiac signs on the walls and surrounding himself with pictures of the planets.[23] Immersed in his own star chart, he has become oblivious to its alleged referent, to what Kepler called the "terrestrial causes." When Sesina, one of Wallenstein's allies, is captured and Wallenstein's plans to betray the emperor are revealed, this unforeseen event therefore challenges the entire metaphysical order of his belief system. The auspicious constellation on his star chart did not translate into auspicious military events on the ground.

In an attempt to save the science of astrology, however, Wallenstein reconceptualizes the celestial event map. No longer a representation of the immediate future as it is about to unfold, the stars now represent what *ought* to unfold according to the natural course of things: "The stars don't lie, *that* however / took place against the course of the stars and against destiny. / The art is honest, but this false heart / brings lie and deceit into the truthful heavens. / Divination is based only on truth, / Where nature exceeds its bounds, all science errs."[24] The event map of the heavens is transformed from an ontological map into a normative map: from events that will be, to events that should have been. And in the end Wallenstein's celestial map loses all its metaphysical underpinnings. Paying no heed to the warnings of his astrologer, Seni, Wallenstein eventually abandons astrology as a guide in military affairs. As he puts it: "Such signs I do not fear."[25] Deprived of their signifying power, the heavens have ceased to be a map and now constitute only an infinite, meaningless territory. Where Kepler lectured Wallenstein on the limits of astrology, Schiller seemingly dismissed it entirely.

Readers and Astrologers

Seemingly. Even though *Wallenstein* recapitulates the end of military astrology and the metaphysical notions associated with it, Schiller at the same time endows the stars with a poetic function. While in early drafts the astrological material appeared only as a "ridiculous caricature," as he writes to Goethe, he became increasingly intrigued by the material and put it to use as a literary device in at least two ways.[26] At a fundamental level, the dramatic action is organized by the malfunction of the astrological map as a transformer of constellations into actions and events. While the heavens are filled with potentially significant movements, *Wallenstein* as a play is pervaded by an almost complete terrestrial eventlessness. The notorious hesitation of its protagonist as he ponders the heaven and the star charts generates a wealth of hypothetical scenarios, but none of them are ever actualized.[27] Eventually his virtual web of alternate worlds leads only to his demise—the one actual event of the play that in turn brings an end to the proliferation of possible scenarios. Instead of developing the drama through actions as the dramatic genre prescribes (*drama*, from the Greek *dran* 'to do, act'), Schiller organizes the drama around its absence, the central void of Wallenstein's nonaction.[28] At an impasse, the stalled dramatic action progresses only through distant events in the margins of the play such as Sesina's capture. Schiller thereby transforms conventional dramatic progression into a prolonged debate about the metachoice of choosing whether to choose or not to choose. And this peculiarly undramatic structure of the drama is a direct effect of the celestial map. Not of its incorrect interpretation, but of the failure to choose among its multiple latent possibilities.

At a different level, however, Schiller addresses the hermeneutics of the map. To see how, it is necessary to consider for a moment the interplay of maps and texts. Once astrological maps enter the play, they become part of a different signifying system that changes the way they operate. Yuri Lotman's statement about literary texts that they are "secondary, model-building systems" that construct possible worlds of signification, applies equally well to the signifying systems of astrological representations.[29] As the official master semiotician of the play, Seni explains the underlying hermetic principle in simple terms: "Nothing in the world is insignificant."[30] From the point of view of semiotics both astrology and literature share a hermetic ground insofar as they are representational models in which everything is potentially significant. Once one such model enters another one, however, it is subject to resignification by the operative principles in the new sphere of signs. While Schiller recapitulates the demise of the heavens as an astrological and military signifying system, he also refunctionalizes the stars and endows them with a different hermeneutic significance determined by the symbolic economy of the literary realm.

The resignification process is entwined with the permutations of the tragic irony that pervades the play. Watching or reading a play on a major historical character, the spectators and readers of *Wallenstein* knew full well the fate that awaited him. Moreover, in the prologue Schiller states that Wallenstein's treason can in large part be ascribed to his belief in the stars.[31] Readers are therefore not only fully aware that his interpretation of the *speculum astrologicum*, namely that the constellation of the stars forms a "figure of luck," is misguided; they are instructed to regard all astrological claims as pure superstition. Yet, just when Wallenstein's own belief in the stars begins to waver, they appear to carry an actual predictive force. Toward the end of the play Schiller intervenes in the natural celestial order and rearranges the stars. While astronomically impossible, Schiller's literary stars morph into an ominous celestial pattern a mere three days after the auspicious conjunction: "The signs are in a horrible position, close, close / Are the nets of doom. ... Come, read it yourself in the constellation of the planets, / That calamity threatens you from false friends."[32] While Seni incorrectly believes the false friends to be the Swedes—the ostensible enemy with whom Wallenstein has been negotiating clandestinely—and Wallenstein dismisses Seni's warning entirely, the readers, with their superior knowledge of the plot against Wallenstein among his allies, are invited to offer an interpretation of future events that confirms the astrological signs. False friends are indeed threatening his life. Astrological authority is transferred from Wallenstein and Seni to the readers, whose superior hermeneutics establishes the tragic irony of the play: just when the stars cease to signify to their main disciple, their astrological power is confirmed. At this level, astrology is not dismissed out of hand. But neither Wallenstein nor Seni are sufficiently skilled readers of the mutable map of heaven. For the reader-cum-astrologer the tragedy becomes one of a fatal error of cartographic misinterpretation rather than a superstitious belief in the occult power of the stars. Inserting one signifying system into another, Schiller transforms astrological maps into a literary device to capitalize on their suggestive predictive powers. In a double gesture, he recapitulates the demise of the astrological event map while, at the same time, he superimposes astrological and literary hermeneutics and makes of the heavens a truthful event map for the astute readers of literary astrology.

Topographies of War

From the metaphysics of the astrological map we move to the immanent workspace of the topographic military map. This genre became particularly important around 1800. By then the large-scale warfare of the Revolutionary and Napoleonic Wars had replaced the smaller cabinet wars that dominated the eighteenth century. As Carl von Clausewitz wrote, the French had unshackled "the horrible element of war from its old diplomatic and financial fetters: it now marched forth in its raw violence."[33] The scale and extensiveness of the new wars meant that accurate topographic military maps became a sine qua non of military operations. To manage their troops in the vast theater of war, commanders and officers became deeply reliant on the topographic knowledge that a detailed map could offer. (See figure 16.3.) As a result the period around 1800 witnessed an explosion in the production, dissemination, and use of military cartography.[34]

The effects of this media event were varied and profound. The Swiss military theorist Antoine-Henri Jomini developed a cartographically inflected theory of war, and in his *Précis de l'art de la guerre* he went so far as to claim that "strategy is the art of waging war upon the map."[35] In Prussia, topographer Johann Georg Lehmann, who also developed an influential method for depicting incline, sought to improve cartographic literacy among military topographers and planners for, as he wrote, "if even the best maps become dangerous objects in the hands of the strategist who does not know how to read and use them, how much more the bad ones that can lead even the most skilled officer astray if he does not know how to check them."[36] At around the same time his fellow Prussian, retired soldier and writer Georg Heinrich von Berenhorst, claimed that the officers should themselves become cartographic media. In the absence of maps

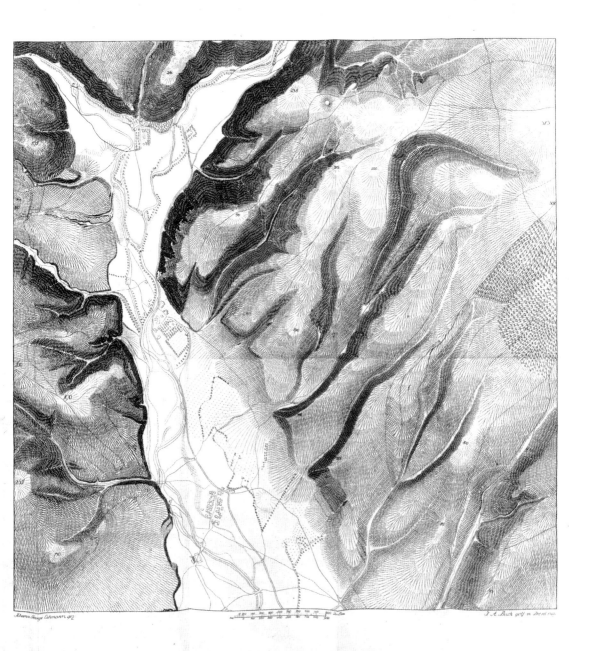

Figure 16.3

A topographic map by Johann Georg Lehmann from 1799 displaying hachures to mark the differentiations between various degrees of incline (Det Kongelige Bibliotek).

they should be able to project triangles and other geometric figures directly onto the terrain in order to transform it into a legible and manageable mental map. As he put it, the inside of an officer's skull must be "wallpapered with maps."[37]

As the ideas of these military thinkers indicate, the topographic military map was a tool that encompassed both spatial and temporal orientation. Commanders and officers needed detailed knowledge of the terrain, but their two-dimensional planes also served as the workspace on which various contingent futures were projected, simulated, and compared. This function is perhaps most clearly evidenced by the reports of some of the people who witnessed the emperor's work habits firsthand. Agathon-Jean-François Fain, Napoléon's personal secretary, describes the process:

> D'Albe was called whenever the Emperor wanted to read the dispatches on the map; with red and black pins d'Albe marked the sites occupied by our troops; he then highlighted the signs of the most important rivers, mountains, and borders with nuanced colors; finally he prepared the calculations of distance, underlined the scale and opened the compass next to the map. Once the dispatches had been applied to the map in this manner, the Emperor would begin to study it.[38]

In this setup, the map serves as the pivot of an information processing system. First, reports are procured from the actual world, which Bacler d'Albe, Napoleon's chief cartographer, proceeds to visualize with symbols on the map. Then Napoleon reaches a decision and dictates it to his secretary, who eventually transmits it back to the actual world. In this information processing system the two-dimensional plane of the map, along with the pins and the compass, functions as an experimental workspace in which various hypothetical scenarios can be tested. The military map, the instruments of inscription, and the pins together constitute a tool that can generate a plurality of worlds and a plurality of battles—all hypothetical, to be sure, but worlds and battles whose hypothetical management had immediate effects. As the aide-de-camp Louis François Lejeune writes in his memoirs: "This conjectural work prepared us for the more serious operations that we were about to undertake on the terrain."[39]

Compared to the star chart, the topographic military map constitutes a very different kind of event map. In the transition from celestial constellations to topographic maps of the terrain as the preferred cartographic tool for managing war, the mode of engagement with the maps shifts from hermeneutics to manipulation, from the teasing out of the future by way of astrological interpretation, to the invention of the best possible future by way of practical operations on a piece of paper.[40] The colored pencils, the pins, the compass are part of an operational praxis that seeks to realize the latent tactical and strategic knowledge that lies hidden within the map. Louis Marin has suggested that the printed, completed map constitutes a matrix of all possible movements.[41] Using it as a workspace for the projection of hypothetical military scenarios, however, the commanders sought to weigh the best possible movements against one another and choose the optimal virtual world with the greatest chance of being actualized in real events. According to the regulatory ideal of military cartography, the future is a map effect generated by the strategist.

Map Illusions

But was this regulatory ideal valid? To what extent might contingent futures invented on a map actually be realized out in the field? For a number of nineteenth-century war novels that was precisely the question. With different emphases, authors such as Stendhal and Tolstoy articulated a critique of the assumptions of the military event map. Tolstoy is the most explicit. In several scenes throughout *War and Peace* maps are at the center of attention. More than a mere literary prop that illustrates the pervasive use of cartography during the Napoleonic Wars, Tolstoy interrogates the military map as an object of knowledge.[42] He asks us to consider how information is ordered on maps, what semiotic models they rely on, what governing power they possess, and how virtual scenarios are transformed into actual events. At one point during a Russian war council the Prussian General Phull has had enough of the vocal critique of his dispositions:

> "Where's the difficulty? Nonsense—it's child's play!" He went over to the map and began poking at it with a desiccated finger, jabbering away as he demonstrated that the effectiveness of the Drissa camp was immune to all contingencies, every development had been foreseen, and if the enemy did try a pincer movement, then the enemy would inevitably be destroyed.[43]

The map spread out on the table is the emblem of Phull's theory of warfare. He excludes a consideration of time, of probability, and of the opponent's moves—that is, of the game-theoretical aspect of war. Such exclusions are highlighted when another general dismisses the objection that his disposition is useless, because it assumes knowledge of the enemy's situation though such knowledge is highly doubtful. "A certainty," the general rejoins, his elaborate battle plan will be as effectual on the terrain as it is on the map.[44] In Tolstoy's satire of military cartography, the map guarantees the actualization of virtual scenarios. But this is only because the map has lost all connection to the territory. Accordingly Tolstoy claims that not one but two Battles of Borodino took place, the planned one and the actual one: "On paper, every one of these columns arrived in position exactly on time and destroyed the enemy. As always in such dispositions everything had been superbly thought through, but as with all dispositions not a single column arrived anywhere on time."[45]

The split between the map and the territory reveals just how radical Tolstoy's claim really is. In war, symbols lose all referential power. The glue that connects signifiers to their signifieds is dissolved in the haze of chaotic events. As Tolstoy's own narrative depictions suggest, war is a stochastic phenomenon that comprises "[a] hundred million contingent factors" in which his characters continually get lost because "everything develops from the interplay of infinitely varied and arbitrary twists and turns!"[46] As such it can neither be managed nor comprehended. All cartographic attempts to govern contingent futures with scenarios and projections are doomed from the beginning. In Tolstoy's view, Phull's desiccated finger tapping the map during the war council touches a fantastic object: static, achronic, manageable, comprehensible, the map presents a negative image of war. Confined to the sphere of virtual events, the military event map is a grand illusion.

Sketching War

In his critique of the military event map, Tolstoy took his cue from Stendhal.[47] In the section of *The Charterhouse of Parma* that would win him accolades from both Balzac and Tolstoy, Stendhal constructs the Battle of Waterloo as a dizzying, fragmented literary topography. Immersing his protagonist Fabrice del Dongo in the swarm of contingent events of war, Stendhal erases even the faintest outline of any coherent plans in the process of being realized. As Fabrice is tossed around randomly on the battlefield he sees only the immediate surroundings "to the left" and "to the right," and experiences space in real time: "At this moment, the road disappeared into a thicket of trees."[48] Without a map or a vantage point from which he can survey the field and get his bearings, Fabrice's movements across the terrain chart a space that is continually in the making.

The topographic instability correlates with the unpredictability of military events. Around the single narrative line drawn by Fabrice as he moves around the battlefield, the scene constructs a nimbus of virtual plots. Surrounding Fabrice is a seemingly infinite number of potential events that only break into actuality when they enter the circumference of his experiential field. "Suddenly" musket shots bring down two soldiers next to Fabrice, and "suddenly" four soldiers appear out of the haze—soldiers that he first mistakes for the enemy.[49] The white smoke that veils everything is a figure of the plot cloud that surrounds the protagonist's every move. These invisible plots form a series of causal chains of pure effects. They are registered only on impact. Predictability therefore tends toward zero. Stendhal constructs a scene of military events composed solely of effects. Fabrice's one attempt at autonomous action, his one chance to link a cause to an effect, comes when he shoots a Prussian soldier. But later we learn that he missed the target and the soldier had been killed by someone else.

The protagonist of *The Charterhouse of Parma* serves the narrative function of a meeting point for a number of randomly intersecting events. One might regard Fabrice as a probe that has been submerged into the war matrix to make the logic of its event structure visible. And this logic is one in which there is no linear relationship between causes and effects. The two have been disjointed to such an extent that causes have no effects, while effects arise seemingly without causes. Events proliferate but they are not the strategic result of cartographic planning. Inside the war matrix the subject of war has ceased to be the author of his actions. He has become its object and effect.

We might say that in Stendhal's version of a grand Napoleonic battle, the regulatory ideal of the military event map meets its other in a literary topography of war that marks only the impossibility of all efforts to manage the complex interplay of events on the field. This applies to Stendhal's narrative version of the battle, but it is also curiously evident in his cartographic version of it. In the margins of a printed copy of *Charterhouse*, which he annotated for a revised second edition, Stendhal penciled a small sketch map (figure 16.4).

As we may glean from the text, the sketch depicts Fabrice's oblique movement on the battlefield. It is not oriented by any of the cardinal directions and the scale differs vastly. On the left the distance to Brussels suggests a small-scale map, but the section on the right-hand side is rendered in a very large scale. The dotted line marks Fabrice's movement as he crosses a ditch, veers to the right to join a contingent of French soldiers, and falls to the ground from fatigue. A soldier then tosses him a piece of bread, indicated by the French word *pain*. When Fabrice doesn't react, the soldier stuffs it into his mouth.

Without borders or a frame to create a sharp division between the representation and its outside,[50] the sketch indicates not just the experiential field of Fabrice, but the fact that this experiential field forms a small island surrounded by an epistemic void. As in the narrative, most of the battlefield in the sketch is a blank space, an invisible but constant threat. More than a question of orientation, of finding one's way, the sketch is an image of the complexity of events in the war matrix. Displaying the random incidents during battle and the limited point of view of a participant, Stendhal's simple sketch map forms the cartographic negative of the strategic military event map. In his laudatory review, Balzac labeled Stendhal's novelistic account of the battle a "military sketch" (*croquis militaire*) because it offered no overview of the events in the battle.[51] In like fashion Stendhal's actual sketch map depicts the stochastic operational logic of war. It displays war as the ungovernable *tout court*.

Playing at War

But might contingency be trained? Might the individual faced with numerous contingent futures somehow be taught to bring them under control? A third major genre in the cartography of war was developed as a response to these questions. Along with the advances in military cartography, the period around 1800 also witnessed the development of an elaborate tool designed to provide tactical war experience by proxy—*das Kriegsspiel*, or the modern war game.[52] Against the backdrop of large-scale Napoleonic warfare the game underwent two important changes. Earlier games were only slightly more elaborate versions of chess. Now, however, to the two-dimensional

LA GUERRE

sur son aide de camp, essayait de faire quelques pas ; il cherchait à s'éloigner de son cheval qui[a] se débattait renversé par terre et lançait des coups de pied furibonds.

Le maréchal des logis s'approcha de Fabrice. À ce moment notre héros entendit dire derrière lui et tout près de son oreille : C'est le seul qui puisse encore galoper. Il se sentit saisir les pieds ; on les élevait en même temps qu'on lui soutenait le corps par-dessous les bras ; on le fit passer par-dessus la croupe de son cheval, puis on le laissa glisser jusqu'à terre, où il tomba assis.

L'aide de camp prit le cheval de Fabrice par la bride ; le général, aidé par le maréchal des logis, monta et partit au galop ; il fut suivi rapidement par les six hommes qui restaient. Fabrice se releva furieux, et se mit à courir après eux en criant : *Ladri ! ladri !* (voleurs ! voleurs !) Il était plaisant[b] de courir après des voleurs au milieu d'un champ de bataille.

L'escorte et le général, comte d'A..., disparurent bientôt derrière une rangée de saules. Fabrice, ivre de colère, arriva aussi à cette ligne de saules ; il se trouva tout contre un canal fort profond[c] qu'il traversa. Puis, arrivé de l'autre côté, il se remit à jurer en apercevant de nouveau, mais à une très grande distance, le général et l'escorte qui se perdaient dans les arbres[d]. Voleurs ! voleurs ! criait-il maintenant en français. Désespéré, bien moins de la perte de son cheval[e] que de la trahison, il se laissa tomber au bord du fossé, fatigué et mourant de faim. Si son beau cheval lui eût été enlevé par l'ennemi, il n'y eût pas songé ; mais se voir trahir et voler par ce maréchal des logis qu'il aimait tant et par ces hussards qu'il regardait comme des frères ! c'est ce qui lui brisait le cœur[f]. Il ne pouvait se consoler de tant d'infamie, et, le dos appuyé contre un saule[g], il se mit à pleurer à chaudes larmes. Il défaisait un à un tous ces beaux rêves d'amitié chevaleresque[h] et sublime, comme celle des héros de la *Jérusalem délivrée*[10]. Voir arriver la mort n'était rien, entouré d'âmes héroïques[i] et tendres, de nobles amis qui vous serrent la main au moment du dernier soupir ! mais garder son enthousiasme

---

a. *son cheval qui, les deux jambes de devant coupées par un boulet, se débattait furieusement et lançait des coups de pieds furibonds* (R).
b. *assez plaisant* (Ch).
c. *un canal profond qu'il traversa <avec peine> péniblement* (R).
d. *son cheval et les voleurs qui se perdaient dans les arbres* (R).
e. *du cheval* (R).
f. *lui perçait le cœur* (R).
g. *comme Bayard mourant* (Ch).
h. *rêves de temps héroïques et d'amitiés chevaleresque. CV, 13 nov. 40,8 lodole,* first *parole* (?) (R).
i. *sublimes et tendres* (R).

Figure 16.4

Sketch map of Fabrice's oblique movement on the battlefield of Waterloo, penciled by Stendhal in the margins of the second edition of *The Charterhouse of Parma* (*La Chartreuse de Parme*, édition critique contenant les notes et additions de Stendhal, ed. Michel Crouzet (Orléans: Éditions Paradigme, 2007), 61). The map is also reprinted in the Di Maio edition, 712n.

LA GUERRE 61

entouré de vils fripons !!!a Fabrice exagérait comme tout homme indigné. Au bout d'un quart d'heure d'attendrissement, il remarqua que les boulets commençaient à arriver jusqu'à la rangée d'arbres à l'ombre desquelsb il méditait. Il se leva et chercha à s'orienter. Il regardait ces prairies bordées par un large canal et la rangée de saules touffusc : il crut se reconnaître. Il aperçut un corps d'infanterie qui passait le fossé et entrait dans les prairies, à un quart de lieue en avant de lui. J'allais m'endormird, se dit-il ; il s'agit de n'être pas prisonnier ; et il se mit à marcher très vite. En avançant il fut rassuré, il reconnut l'uniforme, les régiments par lesquels il craignait d'être coupé étaient français. Il obliqua à droite pour les rejoindre.

Après la douleur morale d'avoir été si indignement trahi et volé, il en était une autre qui, à chaque instant, se faisait sentir plus vivement : il mourait de faim. Ce fut donc avec une joie extrême qu'après avoir marché, ou plutôt couru pendant dix minutes, il s'aperçut que le corps d'infanterie, qui allait très vite aussi, s'arrêtait comme pour prendre position. Quelques minutes plus tard, il se trouvait au milieu des premiers soldatse.

– Camarades, pourriez-vous me vendre un morceau de pain ?
– Tiens ! cet autre qui nous prend pour des boulangers !

Ce mot dur et le ricanement général qui le suivit accablèrent Fabrice. La guerre n'était donc plus ce noble et commun élan d'âmes amantes de la gloire

a. *vils fripons et que l'on voit constamment des bas calculs du plus vil égoïsme, et les voir briller de la même bravoure que l'on reconnaît en soi, et pour laquelle on s'estime, voilà ce qui est plus fort qu'une âme de 17 ans\14 F[évri]er [1841] ; 7 alouettes et lapin CVa 24 octobre 1840* (Ch) ; *vils fripons !!! En pensant à la patrie, devoir faire attention à ce qu'on ne nous vole pas une étrille ! Fabrice exagérait comme tout homme indigné* (R).
b. *auprès desquels* (Ch) ; *auprès desquels. Ombre ? oui cela peut passer, le soleil est horizontal* (R).
c. *et débouchait dans les prairies* (R).
d. *et me laisser couper, il s'agit* (Ch).
e. *des derniers soldats* (R). ICI FIGURE DANS R CE CROQUIS AU CRAYON : BRUXELLES. - FOSSÉ. - PAIN. - FOSSÉ.

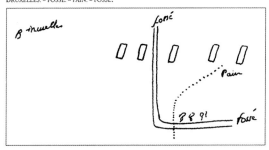

Figure 16.4 (continued)

board inventors added the illusion of a third dimension in the form of terrain differentiations. Johann Christian Ludwig Hellwig's game from 1803, *Das Kriegsspiel–ein Versuch die Wahrheit verschiedener Regeln der Kriegskunst in einem unterhaltenden Spiele anschaulich zu machen* (*The War Game–an Attempt to Illustrate the Truth of Various Rules of the Art of War in an Entertaining Game*), includes a board of 1,617 squares with two different obstacles represented by the colors green and red (figure 16.5).[53] To increase verisimilitude, other officers and inventors suggested playing directly on one of the topographic maps that were made and used for actual warfare.[54]

The second major change was the simulation of chance. In 1806 Johann Ferdinand Opiz brought a minor revolution to the war game. *Das Opiz'sche Kriegsspiel, ein Beitrag zur Bildung künftiger und zur Unterhaltung selbst der erfahrensten Taktiker* (*The Opiz War Game, a Contribution to the Education of Future Tacticians and to the Entertainment of even the most Experienced Tacticians*) included dice as a key component. Chance was thereby installed as the basic operative logic of the war game. As Clausewitz would later advocate in his treatise *On War*, commanders had to contend not just with the inscrutable will of the enemy but also with "friction"—the numerous unforeseen chance events that beset any concerted action in war. With Opiz's invention these unforeseen events were given a material correlate in the form of the dice.

As indicated by the various subtitles of the books just cited, the purpose of war games was and is to offer military training by proxy, to teach soldiers the basic rules of tactics and strategy in circumstances pervaded by contingency, and to do so in the comfort and safety of their homes or in the military academy. Georg Venturini, another inventor, wrote in 1797 that war games are the best tools with which "to teach young soldiers the often difficult doctrines of the art of war as if through experience."[55] While today computer simulation has become the preferred military training technology, board games still play a pedagogical role in the training of tactics and strategy. As Philip Sabin writes in his book *Simulated War*: "The most important function of wargames is to convey a vicarious understanding of some of the strategic and tactical dynamics associated with real military operations. Besides learning about the force, space and time relationships in the specific battle or campaign being simulated, players soon acquire an intuitive feel for more generic interactive dynamics associated with warfare as a whole."[56]

Compared to the two other cartographic genres, horoscopes and topographic maps, the war game constitutes a new kind of event map. Blending entertainment with serious purpose, war games create purely potential events separated from the immediate context that would allow them to be realized. Unlike the horoscope, whose power is supported by a metaphysical belief, and unlike the commander's military map, which forms part of a concrete political and military situation, the war game is not tethered to an immediate reality. Instead it allows for an almost infinite set of

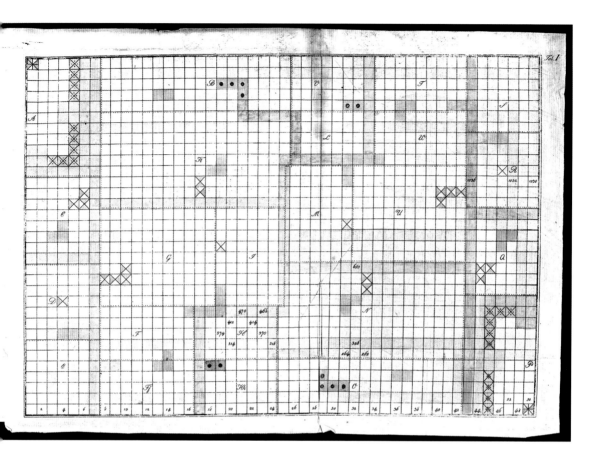

Figure 16.5

The board for Hellwig's war game features a space of 1,619 squares. The colors indicate different terrain types that complicate movement (Johann Christian Ludwig Hellwig, *Das Kriegsspiel—ein Versuch die Wahrheit verschiedener Regeln der Kriegskunst in einem unterhaltenden Spiele anschaulich zu machen* (Braunschweig: bei Karl Reichard, 1803)). (Forsvarets Bibliotek.)

situations and variations intended to teach the general logic of military operations in the form of a varied set of patterns. While hypothetical scenarios mushroom, the probability that they will ever be actualized—their chances of ontological success, so to speak—have diminished greatly. In the particular genre of historical war games that simulate famous wars and battles of the past, the inability of the game to effect a military event is obvious from the beginning. The simulation of historical events might change our understanding of them, but never the events themselves. Yet, as in the genre of counterfactual historiography, they lure with their nimbus of virtual alternate versions of the past.[57] As an event map, the war game in its different forms functions as a generator of a multiplicity of military fictions with, even in the best of cases, only a general and very slim chance of realization.

The Third Reich

How might a literary work interact with this third genre of military mapping and its tenuous link to actual events? Roberto Bolaño's *El Tercer Reich* (*The Third Reich*)[58] is named after a famous war game developed by John Prados in 1974: *The Rise and Fall of the Third Reich*, known to gamers simply as *The Third Reich*. Bolaño not only takes the specific game as the gravitational center of the plot, he models his text on a war game. Narrated by the protagonist himself, the novel follows the increasingly nervous and unreliable German war gaming champion Udo Berger during a vacation with his girlfriend Ingeborg in Spain. Udo sets up *The Third Reich* in his hotel room and spends most of his time trying to come up with a new "strategic variant" that will guarantee a German victory (figure 16.6).

The world outside his hotel room, however, slowly begins to take on the properties of the game. Not only do games of all sorts appear—local and international soccer games, card games, TV shows featuring simulations of accidents (i.e., various forms of sublimated warfare that all seem on the verge of breaking out of their sphere of virtuality)—but the world itself appears to be a series of variations of Udo Berger's life. The vacation itself is a repetition of his childhood vacations; Udo and Ingeborg befriend another couple of about the same age who also hail from Germany; the women are "almost the same size"; the man, Charly, mistakes their own town for the next one up the coast, which, incidentally, also has a hotel with the name Costa Brava, and so on. Just as the board and counters in *The Third Reich* are "like a stage set where thousands of beginnings and endings eternally unfold, a kaleidoscopic theater," the actual world appears to be "something unreal"—a proliferation of alternate versions of Udo, his game, and the Third Reich.[59] An avatar of the past, Udo himself appears to be living in a game, a ludic multiverse whose variations dissolve any sense of a solid, singular reality.

In a countermovement, the game gradually begins to leave the realm of fiction and break through the border between the virtual and the real. A game of entertainment,

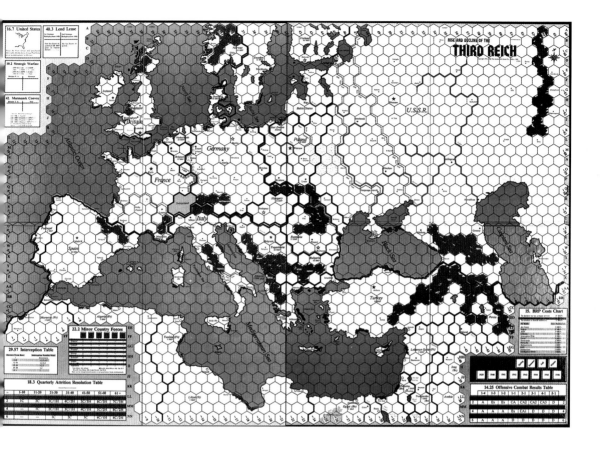

Figure 16.6

The board of John Prado's war game *The Rise and Decline of the Third Reich*, 3rd ed. (Baltimore: Avalon Hill Game Company, 1974). Private photo. Courtesy of John Prado.

The Third Reich is not designed to have immediate consequences. The board ought to produce only virtual events. Increasingly, however, the events in the actual world seem to be directly influenced by the action on the map. When Udo's local opponent, a man charred beyond recognition who goes by the name of El Quemado or The Burn Victim, slowly gains the upper hand over the German champion, the stakes seem to have risen to a new level where Udo's life is in the balance. People in town now believe that Udo is not just a commander of virtual Nazis on the board, but an actual Nazi, and the game a "trial of the war criminals."[60] Against the protestations of the increasingly unstrung Udo, who insists that a game is just a game, *The Third Reich* serves as a catalyst that transforms fiction into reality and the past into the present. It forms the basis not merely of a replay or a reenactment, but of a veritable repetition of World War II in miniature. Unlike conventional war games that produce only a nimbus of hypothetical worlds, Bolaño's literary war game becomes indistinguishable from actual war.

Blending actuality and games, realism and the fantastic, the past and the present, Bolaño fuses what the narrator tries to keep apart as "categorically opposed things."[61] The result is a novel made up of a haze of fictions, its events hovering in an undecidable ontological realm between the virtual and the real. Out of the entertaining or pedagogical virtual events of the traditional war game, Bolaño fashions a war delirium in which all distinctions collapse. Fictional war and actual war are merely variations of one another.

What is the point of these blends, one might ask? If we consider the literary war game as an event map, its curious temporality alerts us to the underlying purpose. Like Schiller's astrological event map in *Wallenstein*, *The Third Reich* serves the poetic function of generating and managing reader expectations. Everything gravitates toward a literary future in which a fatal event will bring the game to a conclusion. But at the same time Bolaño turns the event map around, so to speak. The game repeats a historical war, not a war to come. Udo's strategic variant turns out to be a perverse form of *Vergangenheitsbewältigung*. The variant is a *past contingent* with which he tries to rectify not the *actions* of the Nazis, but their *mistakes*. In his own words: "It's as if we want to know exactly how everything was done in order to change what was done wrong."[62] Not performing an act of retribution in a "game of atonement" as a compatriot sees it, Udo plays to lead the Nazis to victory.[63] Ostensibly pointing the readers' expectations toward a calamitous future, the map instead becomes a tool for dealing with the problematic events of the German past. The military mindset, Udo's latent belligerence, the insanity of the fanatical war delirium that made generals ponder their maps to conceive the best possible world of destruction, have been forgotten and repressed by "an amnesiac Europe," but now become manifest on the hexes of the board.[64] By means of the game, the novel makes explicit that war is anything but that. The entertaining virtual "events" produced by gamers with maps and dice are sublimations that repress their serious purpose as instruments of death and charred flesh.

In the end all reader expectations are thwarted. Udo loses the game, but only to return to Germany and withdraw from all gaming activities. The curiously anticlimactic nonevent of the game, however, precisely marks the central event of the game-cum-novel: the end of the obsession with war. As a response to the purpose of the traditional war game, *The Third Reich* inverts its raison d'être. Instead of teaching us to handle the potential events of future wars, it conjures from the past their horrific reality to make us desist from military activities altogether, be they actual or invented.

The Futures of War

Geography is the "eye of history" (*Historiæ oculus*), wrote Abraham Ortelius in 1570.[65] Surveying the main genres in the cartography of war, we are enticed to reorient his gaze. Within the military sphere, the map serves as a strategic tool that might offer a glimpse of the plural futures of war. Ortelius's statement is also pertinent in this context, however, for the futures of war have a history. While star charts and horoscopes, military topographic maps, and war games exist side by side as distinct cartographic genres, their genealogy marks three significant moments in the development of the military future. From the metaphysical grounding of astrological predictions in the movement of the celestial orbs, the future of the topographic map becomes an immanent product of strategic planning. Now grounded in the skill and creative genius of the commander rather than in a cosmological belief system, the future loses some of its givenness and emerges as an artifact that can, at least partially, be made. And with the emergence of war games and their recent instantiations in computer simulations, the future takes the form of a repetition, but a repetition of something that has never happened, of an imagined future that has already been played through multiple times. Accordingly, the primary mode of engagement with military maps shifts from hermeneutics, to manipulation, to play. Such shifts in cartographic habitus suggest a growing control over the futures of war, but it is accompanied by a widening of the military future and its increasing virtuality as possible scenarios burgeon and the nexus between map and event becomes ever more hypothetical.

In some of the main literary works that examine this nexus, the very notion of a "cartography of war" even appears to be a contradiction in terms. For authors such as Stendhal and Tolstoy conventional cartographic virtues like control, visibility, and power, which should ground the temporal projections and ensure the tractability of contingent futures, are deemed impotent in the context of war. In these texts, it seems that war is not simply a complex object that is difficult to govern, but one that evades the government of maps entirely. Texts and maps are linked not as supplemental media that enmesh literature with cartography, but as opposing ones in perpetual conflict. Yet, all the authors examined here at the same time co-opt military cartography, adapting and retooling it into productive poetic devices that structure the literary representation of war and the way we are instructed to read their texts. Whether it malfunctions as in Schiller, whether it serves as a foil for the narration of war as in Tolstoy, whether it is redrawn according to the precepts of a literary sketch as in Stendhal, or whether it is turned around and its purpose inverted as in Bolaño, the military event map has shaped the ways literature has represented and imagined the space and the events of war. And while these texts merely sample some of the literary refractions of military maps, they all suggest that it is from the friction between texts and maps that the silhouettes of the imagined futures of war are drawn.

Notes

1. Thucydides, *History of the Peloponnesian War* (Cambridge, MA: Harvard University Press, 2005), vol. 2, book 3, 173–175.

2. Unless otherwise indicated, all translations are my own. See Martin Rickenbacher, "Französische Kartierungen von Schweizer Gebieten zwischen 1760 und 1815," *Cartographica Helvetica* 41 (2010): 3–17, here 12. While the use of military maps reached new heights during the Napoleonic Wars, Napoléon's statement not only applies to the period around 1800. As John Hale writes about the Renaissance: "Maps and conquest seemed to go together" (John Hale, "Warfare and Cartography, ca. 1450 to ca. 1640," in *The History of Cartography*, Vol. 3.1: *Cartography in the European Renaissance*, ed. David Woodward, 719–737 (Chicago: University of Chicago Press, 2007), here 719).

3. See, for example, Ute Schneider, *Die Macht der Karten: Eine Geschichte der Kartographie vom Mittelalter bis heute* (Darmstadt: Primus Verlag, 2006), 116. For numerous examples during the Napoleonic Wars, see Henri Marie Auguste Berthaut, *Les Ingénieurs Géographes Militaires 1624–1831* (Paris: Impr. du Service géographique, 1902). In our time the geographic knowledge provided by satellites is also surrounded by secrecy. For years the U.S. military followed a policy of "selective availability" after they had launched GPS, considering it a "dual-use technology" that allowed only limited accuracy for civilians (they scrambled the signals), and retained its actual pinpoint capabilities for military purposes. See Laura Kurgan, *Close Up at a Distance: Mapping, Technology, and Politics* (Brooklyn, NY: Zone Books, 2013), 40.

4. See Mark Monmonier, *How to Lie with Maps* (Chicago: University of Chicago Press, 1996), 115–117; Christian Jacob, *The Sovereign Map: Theoretical Approaches in Cartography throughout History*, ed. Edward H. Dahl, trans. Tom Conley (Chicago: University of Chicago Press, 2006), 274.

5. See Anders Engberg-Pedersen, *Empire of Chance: The Napoleonic Wars and the Disorder of Things* (Cambridge, MA: Harvard University Press, 2015), 146–183.

6. Aristotle, *The Basic Works of Aristotle*, ed. Richard McKeon (New York: Modern Library, 2001), 47.

7. Boethius, *Commentarii in librum Aristotelis Peri hermeneias* (Lipsiae: In aedibus B. G. Teubneri, 1877–1880), 1:21.

8. As Benjamin Bühler and Stefan Willer note, all knowledge of the future relies on media—on language, models, and simulations. Examining the specificities of the formal makeup of event maps and their development over time is therefore necessary if we are to account for the changing nature of map events. See *Futurologien: Ordnungen des Zukunftswissens*, ed. Benjamin Bühler and Stefan Willer (Paderborn: Fink, 2016), 9.

9. Johannes Kepler, *Die Astrologie des Johannes Kepler: Eine Auswahl aus seinen Schriften*, ed. H. A. Strauss and S. Strauss-Kloebe (Fellbach: Verlag Adolf Bonz, 1981), 244–245.

10. Kepler, *Astrologie*, 233.

11. Kepler, *Astrologie*, 256.

12. Ptolemy, *Tetrabiblos*, ed. and trans. F. E. Robbins (Cambridge, MA: Harvard University Press, 1940), 221.

13. For an overview of these developments see Nicholas Campion, *A History of Western Astrology,* Vol. 1: *The Ancient World* (London: Continuum, 2008), and *A History of Western Astrology,* Vol. 2: *The Medieval and Modern Worlds* (London: Continuum, 2009).

14. Ptolemy, *Tetrabiblos*, 225.

15. The foundation that ensured the validity of the correspondence between the superlunary and the sublunary realms was a cosmological world picture in which the universe was considered a single entity with all elements connected from the celestial spheres to human beings, animals, and the botanical and mineral realms. For the classic explication of the "cosmic order" in the Elizabethan period, see E. M. W. Tillyard, *The Elizabethan World Picture* (New York: Vintage Books, 1959).

16. Kepler, *Astrologie*, 225.

17. Kepler, *Astrologie*, 225.

18. Kepler, *Astrologie*, 238–239. In the original German the quotation reads, "Dies melde ich allein zu dem End, auf daß ich dem Geborenen den Wahn nehme, als ob so ganz die *Particularia* aus dem Himmel vorzusagen seien. Einmal ist dies wahr, daß aus dem Himmel zwar wohl himmlische *Particularia* folgen, nicht aber irdische, weder *specialia* noch *individua*, sondern alle irdischen *Eventus* nehmen ihre Form und Gestalt aus irdischen Ursachen (terrestrial causes), allda ein jedes Particular seine Particular-Ursach hat."

19. See Dan Edelstein, "Introduction to the Super-Enlightenment," in *The Super-Enlightenment: Daring to Know Too Much*, ed. Dan Edelstein, 1–35 (Oxford: Voltaire Foundation, 2010), esp. 33. See also David Bates's essay "Super-Epistemology" in the same volume. The persistence in Kepler's thought of astrological themes such as the *harmonia mundi*, the analogy of the macro- and microcosmos, can be seen even in his mathematical calculations. See Hania Siebenpfeiffer, "Astrologie," in *Futurologien*, ed. Bühler and Willer, 379–392, esp. 385.

20. The changed status of astrology is unmistakable in a contemporary text by Johann Gottfried von Herder. In "On the Knowledge and Non-knowledge of the Future" from 1797 he delivers the following diatribe: "It wasn't the fact that astrology dealt with the science of the future that made it contemptible and ridiculous; but the fact that it dealt with it groundlessly, that it sought a science in combinations that did not contain any. It is the same with chiromancy, metoposcopy, with auspices and auguries. They sought omens where there could be none and deceived people's minds with a *false* science that they believed or pretended to be true" (Johann Gottfried von Herder, "Vom Wissen und Nichtwissen der Zukunft," in *Werke*, Vol. 8: *Schriften zu Literatur und Philosophie 1792–1800*, ed. Hans Dietrich Irmscher, 283–297 (Frankfurt am Main: Deutscher Klassiker Verlag, 1998), here 286). An early literary critique of the projection of the heavenly order onto the terrestrial one is already articulated by Cervantes in *Don Quixote*. See Jörg Dünne, *Die Kartographische Imagination*: *Erinnern, Erzählen und Fingieren in der Frühen Neuzeit* (Munich: Fink, 2011), 245–367.

21. Friedrich Schiller, Wallenstein (Frankfurt am Main: Deutscher Klassiker Verlag, 2005), 155. "Glückseliger Aspekt!"

22. Schiller, *Wallenstein*, 156. "Jetzt haben sie den alten Feind besiegt, / Und bringen ihn am Himmel mir gefangen."

23. Schiller, *Wallenstein*, 175, 155. In 1623, the historical Wallenstein commissioned Italian artists to decorate the 21-meter-long "Astrological Corridor" in his palace with frescoes that depict the seven known planets personified as ancient gods along with the signs of the zodiac. See Alena Hadravová and Petr Hadrava, "Science in Contact with Art: Astronomical Symbolics of the Wallenstein Palace in Prague," in *Science in Contact at the Beginning of the Scientific Revolution*, ed. Jitka Zamrzlová, Acta historiae rerum naturalium necnon technicarum, vol. 8, 173–210 (Prague: National Technical Museum, 2004), here 181–184.

24. Schiller, *Wallenstein*, 213. "Die Sterne lügen nicht, *das* aber ist / Geschehen wider Sternenlauf und Schicksal. / Die Kunst ist redlich, doch dies falsche Herz / Bringt Lug und Trug in den wahrhaft'gen Himmel. / Nur auf der Wahrheit ruht die Wahrsagung, / Wo die Natur aus ihren Grenzen wanket, da irret alle Wissenschaft."

25. Schiller, *Wallenstein*, 282. "Solche Zeichen fürcht' ich nicht."

26. Schiller writes to Goethe about his original plans for the astrological material: "Es macht auf die Einbildungskraft keine Wirkung und würde immer nur eine lächerliche Fratze bleiben" ("It has no effect on the imagination and would always remain merely a ridiculous caricature") (Schiller, *Wallenstein*, 672). See also Stefan Davies, *The Wallenstein Figure in German Literature and Historiography 1790–1920* (London: Maney Publishing for Modern Humanities Research Association, 2010), 42.

27. See also Joseph Vogl's analysis of the proliferation of virtual scenarios in *Über das Zaudern* (Zurich: Diaphanes, 2008).

28. Stefan Willer rightly calls him a "deficient dramatic subject" (Willer, "Stratege," in *Futurologien*, ed. Bühler and Willer, 245–256, here 247).

29. J. M. Lotman, *Die Struktur literarischer Texte* (Munich: Fink, 1972), 22.

30. Schiller, *Wallenstein*, 77. "Nichts in der Welt ist unbedeutend."

31. Schiller, *Wallenstein*, 16.

32. Schiller, *Wallenstein*, 282. "Die Zeichen stehen grausenhaft, nah, nahe / Umgeben dich die Netze des Verderbens. … Komm, lies es selbst in den Planetenstand, / Daß Unglück dir von falschen Freunden droht."

33. Carl von Clausewitz, "Über das Leben und den Charakter von Scharnhorst 1817," in *Verstreute kleine Schriften*, ed. Werner Hahlweg (Osnabrück: Biblio Verlag, 1979), 228.

34. A comprehensive account of Napoleonic war mapping is still to be written. I offer an initial sketch in *Empire of Chance: The Napoleonic Wars and the Disorder of Things*. See also Berthaut, *Ingénieurs*, as well as Max Eckart, *Die Kartenwissenschaft*, vol. 1 (Berlin: De Gruyter, 1921).

35. Antoine-Henri Jomini, *Précis de l'art de la guerre ou nouveau tableau analytique des principales combinaisons de la stratégie, de la grande tactique et de la politique militaire* (Osnabrück: Biblio Verlag, 1973), reprint of the Paris edition from 1855, vol. 1, 155. See also *Tableau analyt-*

ique des principales combinaisons de la guerre (St. Petersburg: Bellizard et Cie, Libraires de la Cour, 1830), 50.

36. Johann Georg Lehmann, *Die Lehre der Situation-Zeichnung oder Anweisung zum richtigen Erkennen und genauen Abbilden der Erd-Oberfläche in topographischen Charten und Situations-Planen*, pt. 1 (Dresden, 1816), 46. For a similar advocacy of a proper cartographic "critique" see Ernst Heinrich Michaelis, *Einige durch die Fortsetzung der Bohnenberger-Ammonschen Karte von Schwaben veranlaßte Bemerkungen über die topographische Kunst* (Bern, 1825), 130–131.

37. Georg Heinrich von Berenhorst, *Betrachtungen über die Kriegskunst, über ihre Fortschritte, ihre Widersprüche und ihre Zuverlässigkeit*, vol. 2 (Osnabrück: Biblio Verlag, 1978), 345. The book was originally published in 1797–1799.

38. Agathon-Jean-François Fain, *Mémoires du Baron Fain* (Paris: arléa, 2001), 29.

39. Louis François Lejeune, *Mémoires du général Lejeune*, vol. 1 (Paris, 1895), 259.

40. Discussing the early use of military maps in Holland, Wolfgang Schäffner labels such simultaneously practical and tactical engagement with maps "operational topography." See Wolfgang Schäffner, "Operationale Topographie: Repräsentationsräume in den Niederlanden um 1600," in *Räume des Wissens: Repräsentation, Codierung, Spur*, ed. Hans-Jörg Rheinberger, Michael Hagner, and Bettina Wahrig-Schmidt, 63–90 (Berlin: Akademie Verlag, 1997).

41. Louis Marin, "Les voies de la carte," in *Cartes et figures de la terre* (Paris: Centre Georges Pompidou, CCI, 1980), 50.

42. For a more comprehensive account of the role of maps in *War and Peace* see Anders Engberg-Pedersen, "Critique of Cartographic Reason: Tolstoj on the Media of War," *Russian Literature* 77, no. 3 (2015): 307–336, as well as *Empire of Chance*, chap. 6.

43. Leo Tolstoy, *Polnoe sobranie sochinenii*, vol. 11, ed. V. G. Chertkov (Moscow: Gosudarstvennoie Izdatelstvo, 1928–), 50.

44. Tolstoy, *Polnoe sobranie sochinenii*, vol. 9, 322.

45. Tolstoy, *Polnoe sobranie sochinenii*, vol. 12, 74.

46. Tolstoy, *Polnoe sobranie sochinenii*, vol. 11, 206, 131.

47. See Paul Boyer, *Chez Tolstoi, entretiens à Iasnaïa Poliana* (Paris: Institut d'études slaves de l'Université de Paris, 1950), 40.

48. Stendhal, *La Chartreuse de Parme*, ed. Mariella Di Maio (Paris: Éditions Gallimard, 2003), 90.

49. Stendhal, *Chartreuse*, 95–96. The poor visibility Stendhal describes is based in fact. Writing about their experiences, numerous officers later mentioned that thick smoke severely limited their field of vision. See H. T. Siborne, *Waterloo Letters: A Selection from Original and hitherto Unpublished Letters bearing on the Operations of the 16th, 17th, and 18th June, 1815, By Officers who served in the Campaign* (London: Cassel & Company, 1891).

50. The insertion of the rectangular box in the Crouzet edition changes the map fundamentally. Framed, the map no longer displays the fluidity of space and the surrounding field of potentiality that is key to Stendhal's conception of the state of war. See *La Chartreuse de Parme*, ed. Michel Crouzet (Orléans: Paradigme, 2007), 61.

51. Honoré de Balzac, *Écrits sur le roman* (Paris: Librarie Générale Française, 2000), 173.

52. See, above all, Philipp von Hilgers, *Kriegsspiele: Eine Geschichte der Ausnahmezustände und Unberechenbarkeiten* (Munich: Fink, 2008); Claus Pias, *Computer–Spiel–Welten* (Munich: sequenzia, 2002). See also Peter P. Perla, *The Art of Wargaming* (Annapolis, MD: United States Naval Institute, 1990); Hajo Holborn, "The Prusso-German School: Moltke and the Rise of the General Staff," in *Makers of Modern Strategy: From Machiavelli to the Nuclear Age*, ed. Peter Paret, 281–295 (Oxford: Oxford University Press, 1986).

53. For an in-depth analysis of the earlier version of Hellwig's game from 1780, see Pias, *Computer–Spiel–Welten*, 204–213.

54. See for example Georg Venturini, *Beschreibung und Regeln eines neuen Krieges-Spiels, zum Nutzen und Vergnügen, besonders aber zum Gebrauch in Militair-Schulen* (Schleswig: bey J. G. Röhß, 1797), xvi; Ludwig Freidrich Erck, *Anleitung zum zweckmässigen Studium der Kriegswissenschaft: Von einem norddeutschen Officiere* (Leipzig: Hahn, 1828), 35, 49, 57–58.

55. Venturini, *Beschreibung und Regeln*, xvi.

56. Philip Sabin, *Simulating War: Studying Conflict through Simulation Games* (London: Continuum, 2014), 31. Indeed, Christian Jacob suggests that the map is the "role-playing game par excellence" (Jacob, *The Sovereign Map*, 328).

57. See for example Niall Ferguson, *Virtual History: Alternatives and Counterfactuals* (New York: Basic Books, 1999).

58. The novel was written in 1989, but it was only published posthumously in 2010. The English translation followed in 2011.

59. Roberto Bolaño, *The Third Reich*, trans. Natasha Wimmer (New York: Farrar, Straus and Giroux, 2011), 49, 108, 200, 41.

60. Bolaño, *The Third Reich*, 250.

61. Bolaño, *The Third Reich*, 250.

62. Bolaño, *The Third Reich*, 206.

63. Bolaño, *The Third Reich*, 234.

64. Bolaño, *The Third Reich*, 83.

65. Abraham Ortelius, *Theatrum Orbis Terrarum* (Antverpiae: Apud Aegid. Coppenium Diesth, 1570), ("Benevolis lectoribus," no page number).

Bibliography

Aristotle. *The Basic Works of Aristotle*. Ed. Richard McKeon. New York: Modern Library, 2001.

Balzac, Honoré de. *Écrits sur le roman*. Paris: Librarie Générale Française, 2000.

Berenhorst, Georg Heinrich von. Betrachtungen über die Kriegskunst, über ihre Fortschritte, ihre Widersprüche und ihre Zuverlässigkeit. Vol. 2. Osnabrück: Biblio Verlag, 1978.

Berthaut, Henri Marie Auguste. *Les Ingénieurs Géographes Militaires 1624–1831*. Paris: Impr. du Service géographique, 1902.

Boethius. *Commentarii in librum Aristotelis Peri hermeneias*. Vol. 1. Lipsiae: In aedibus B. G. Teubneri, 1877–1880.

Bolaño, Roberto. *The Third Reich*. Trans. Natasha Wimmer. New York: Farrar, Straus and Giroux, 2011.

Boyer, Paul. *Chez Tolstoï, entretiens à Iasnaïa Poliana*. Paris: Institut d'études slaves de l'Université de Paris, 1950.

Bühler, Benjamin, and Stefan Willer, eds. *Futurologien: Ordnungen des Zukunftswissens*. Paderborn: Fink, 2016.

Campion, Nicholas. *A History of Western Astrology*, Vol. 1: The Ancient World. London: Continuum, 2008.

Campion, Nicholas. *A History of Western Astrology*, Vol. 2: The Medieval and Modern Worlds. London: Continuum, 2009.

Clausewitz, Carl von. "Über das Leben und den Charakter von Scharnhorst 1817." In *Verstreute kleine Schriften*, ed. Werner Hahlweg. Osnabrück: Biblio Verlag, 1979.

Davies, Stefan. *The Wallenstein Figure in German Literature and Historiography 1790–1920*. London: Maney Publishing for Modern Humanities Research Association, 2010.

Dünne, Jörg. *Die Kartographische Imagination: Erinnern, Erzählen und Fingieren in der Frühen Neuzeit*. Munich: Fink, 2011.

Dürer, Albrecht, Konrad Heinfogel, and Johannes Stabius. *Imagines coeli Septentrionales et Meridionales zodiaci*. 1515.

Eckart, Max. *Die Kartenwissenschaft*. Vol. 1. Berlin: De Gruyter, 1921.

Edelstein, Dan. "Introduction to the Super-Enlightenment." In *The Super-Enlightenment: Daring to Know Too Much*, ed. Dan Edelstein, 1–35. Oxford: Voltaire Foundation, 2010.

Engberg-Pedersen, Anders. "Critique of Cartographic Reason: Tolstoj on the Media of War." *Russian Literature* 77, no. 3 (2015): 307–336.

Engberg-Pedersen, Anders. *Empire of Chance: The Napoleonic Wars and the Disorder of Things*. Cambridge, MA: Harvard University Press, 2015.

Erck, Ludwig Freidrich. *Anleitung zum zweckmässigen Studium der Kriegswissenschaft: Von einem norddeutschen Officiere*. Leipzig: Hahn, 1828.

Fain, Agathon-Jean-François. *Mémoires du Baron Fain*. Paris: arléa, 2001.

Ferguson, Niall. *Virtual History: Alternatives and Counterfactuals*. New York: Basic Books, 1999.

Hadravová, Alena, and Petr Hadrava. "Science in Contact with Art: Astronomical Symbolics of the Wallenstein Palace in Prague." In *Science in Contact at the Beginning of the Scientific Revolution*, ed. Jitka Zamrzlová, Acta historiae rerum naturalium necnon technicarum, vol. 8, 173–210. Prague: National Technical Museum, 2004.

Hale, John. "Warfare and Cartography, ca. 1450 to ca. 1640." In *The History of Cartography*, Vol. 3.1: *Cartography in the European Renaissance*, ed. David Woodward, 719–737. Chicago: University of Chicago Press, 2007.

Hellwig, Johann Christian Ludwig. *Das Kriegsspiel–ein Versuch die Wahrheit verschiedener Regeln der Kriegskunst in einem unterhaltenden Spiele anschaulich zu machen*. Braunschweig: bei Karl Reichard 1803.

Herder, Johann Gottfried von. "Vom Wissen und Nichtwissen der Zukunft." In *Werke*, Vol. 8: *Schriften zu Literatur und Philosophie 1792–1800*, ed. Hans Dietrich Irmscher, 283–297. Frankfurt am Main: Deutscher Klassiker Verlag, 1998.

Hilgers, Philipp von. *Kriegsspiele: Eine Geschichte der Ausnahmezustände und Unberechenbarkeiten.* Munich: Fink, 2008.

Holborn, Hajo. "The Prusso-German School: Moltke and the Rise of the General Staff." In *Makers of Modern Strategy: From Machiavelli to the Nuclear Age,* ed. Peter Paret, 281–295. Oxford: Oxford University Press, 1986.

Jacob, Christian. *The Sovereign Map: Theoretical Approaches in Cartography throughout History.* Ed. Edward H. Dahl, trans. Tom Conley. Chicago: University of Chicago Press, 2006.

Jomini, Antoine-Henri. *Précis de l'art de la guerre ou nouveau tableau analytique des principales combinaisons de la stratégie, de la grande tactique et de la politique militaire.* Osnabrück: Biblio Verlag, 1973.

Jomini, Antoine-Henri. *Tableau analytique des principales combinaisons de la guerre.* St. Petersburg: Bellizard et Cie, Libraires de la Cour, 1830.

Kepler, Johannes. *Die Astrologie des Johannes Kepler: Eine Auswahl aus seinen Schriften.* Ed. H. A. Strauss and S. Strauss-Kloebe. Fellbach: Verlag Adolf Bonz, 1981.

Kurgan, Laura. *Close Up at a Distance: Mapping, Technology, and Politics.* Brooklyn, NY: Zone Books, 2013.

Lehmann, Johann Georg. *Darstellung einer neuen Theorie der Bezeichnung der schiefen Flächen im Grundriß oder der Situationszeichnung der Berge.* Leipzig: bei Johann Benjamin Georg Fleischer, 1799.

Lehmann, Johann Georg. *Die Lehre der Situation-Zeichnung oder Anweisung zum richtigen Erkennen und genauen Abbilden der Erd-Oberfläche in topographischen Charten und Situations-Planen.* Pt. 1. Dresden, 1816.

Lejeune, Louis François. *Mémoires du général Lejeune.* Vol. 1. Paris, 1895.

Lotman, J. M. *Die Struktur literarischer Texte.* Munich: Fink, 1972.

Marin, Louis. "Les voies de la carte." In *Cartes et figures de la terre.* Paris: Centre Georges Pompidou, CCI, 1980.

Michaelis, Ernst Heinrich. *Einige durch die Fortsetzung der Bohnenberger-Ammonschen Karte von Schwaben veranlaßte Bemerkungen über die topographische Kunst.* Bern, 1825.

Monmonier, Mark. *How to Lie with Maps.* Chicago: University of Chicago Press, 1996.

Ortelius, Abraham. *Theatrum Orbis Terrarum.* Antverpiae: Apud Aegid. Coppenium Diesth, 1570.

Perla, Peter P. *The Art of Wargaming.* Annapolis, MD: United States Naval Institute, 1990.

Pias, Claus. *Computer–Spiel–Welten.* Munich: sequenzia, 2002.

Ptolemy. *Tetrabiblos.* Ed. and trans. F. E. Robbins. Cambridge, MA: Harvard University Press, 1940.

Rickenbacher, Martin. "Französische Kartierungen von Schweizer Gebieten zwischen 1760 und 1815." *Cartographica Helvetica* 41 (2010): 3–17.

Sabin, Philip. *Simulating War: Studying Conflict through Simulation Games.* London: Continuum, 2014.

Schäffner, Wolfgang. "Operationale Topographie: Repräsentationsräume in den Niederlanden um 1600." In *Räume des Wissens: Repräsentation, Codierung, Spur*, ed. Hans-Jörg Rheinberger, Michael Hagner, and Bettina Wahrig-Schmidt, 63–90. Berlin: Akademie Verlag, 1997.

Schiller, Friedrich. *Wallenstein*. Frankfurt am Main: Deutscher Klassiker Verlag, 2005.

Schneider, Ute. *Die Macht der Karten: Eine Geschichte der Kartographie vom Mittelalter bis heute*. Darmstadt: Primus Verlag, 2006.

Siborne, H. T. *Waterloo Letters: A Selection from Original and hitherto Unpublished Letters bearing on the Operations of the 16th, 17th, and 18th June, 1815, By Officers who served in the Campaign*. London: Cassel & Company, 1891.

Siebenpfeiffer, Hania. "Astrologie." In *Futurologien: Ordnungen des Zukunftswissens*, ed. Benjamin Bühler and Stefan Willer, 379–392. Paderborn: Fink, 2016.

Stendhal. *La Chartreuse de Parme*. Ed. Mariella Di Maio. Paris: Éditions Gallimard, 2003.

Stendhal. *La Chartreuse de Parme*. Ed. Michel Crouzet. Orléans: Paradigme, 2007.

Thucydides. *History of the Peloponnesian War*. Vol. 2, book 3. Cambridge, MA: Harvard University Press, 2005.

Tillyard, E. M. W. *The Elizabethan World Picture*. New York: Vintage Books, 1959.

Tolstoy, Leo. *Polnoe sobranie sochinenii*. Vols. 9, 11, 12. Ed. V. G. Chertkov. Moscow: Gosudarstvennoie Izdatelstvo, 1928–.

Venturini, Georg. *Beschreibung und Regeln eines neuen Krieges-Spiels, zum Nutzen und Vergnügen, besonders aber zum Gebrauch in Militair-Schulen*. Schleswig: bey J. G. Röhß, 1797.

Vogl, Joseph. *Über das Zaudern*. Zurich: Diaphanes, 2008.

Willer, Stefan. "Stratege." In *Futurologien: Ordnungen des Zukunftswissens*, ed. Benjamin Bühler and Stefan Willer, 245–256. Paderborn: Fink, 2016.

Conclusion

Anders Engberg-Pedersen

The introduction to this book begins with a pair of quotations on maps from two writers with a profound interest in space. Robert Louis Stevenson found it hard to believe that there are people who do not care for maps; Herman Melville spoke of "true places" that transcend the parameters of maps.[1] As a frame for *Literature and Cartography*, the two statements establish the tension that runs through the individual chapters. Stevenson's fascination with maps expresses the desire with which numerous authors have engaged cartography productively down through the centuries. But Melville's pronouncement on the limits of maps—limits that he suggests literary texts may surpass—indicates how the disjunctions between texts and maps are equally important in the history and theory of literature and cartography. Given the scope of the volume, which examines the theoretical, historical, and generic character of these interactions between literature and cartography, it is appropriate to end with a brief summary of the book's three parts. The conclusion also offers some suggestions for the directions future studies might take.

Part I, on theory and methodology, has highlighted the fundamental role of fiction in both literature and cartography. Though we like to distinguish fiction from reality, the operations of literature and cartography reveal that this simple opposition is inadequate. The curious miniatures such as the *Swissminiatur* at Melide near Lugano, *Italia in miniatura* in Rimini, *Russie miniature* in Saint Petersburg, *Miniatürk* near Istanbul, and the Beijing *World Park* have captured the imagination of children and adults across the globe with their merger of representations and territories. Taking his starting point in such blends of the fictional and the real with which we have represented the world to ourselves *as* a world in itself, Jean-Marc Besse (chapter 1) rethinks the epistemology and ontology of literature and cartography. As he makes clear, maps operate a series of transitions and oscillations that merge the fictional and the real into "half-places" equally partaking of the world of the imagination and the world of things. On the basis of these theoretical statements, Besse offers an overview of cartographic effects in literary works—the ways maps set fiction in motion (Robert Louis Stevenson is a prime example), how they construct a frame around the plot (as in Émile Zola), how they can serve as a principle for the ordering of the narrative sequence (never more evident than in James Joyce), and how they at times clash with the literary text and call linearity into question (as evidenced by John Dos Passos or Alfred Döblin). The basic parameters of the various interactions of literature and cartography are thereby staked out.

One of the most dramatic changes in the study of both literature and cartography is the emergence in recent years of digital mapping. Big data and GIS have opened an array of new possibilities for scholars in the humanities. For one, literary scholars today are not just studying maps, they have also appropriated the cartographers' newfangled digital instruments and begun making maps. Under the banner of literary geography, Franco Moretti, Barbara Piatti, and many others have begun exploring everything that mapping has to offer as a methodology. In this approach, maps serve a number of functions. They illustrate a narrative, they provide the source of inspiration for new ideas, and they function as "analytic tools" that allow us to notice novel features of individual texts, of genres, or of whole corpora of texts that can only be read at a distance.[2] Picking up where Moretti left off, Barbara Piatti (chapter 2) guides us through the development of a detailed geographic information system (GIS) for literature and sketches out some of the ways literature can be mapped in interactive, digital atlases.

The quantification and visualization of unique literary works challenge basic theoretical and methodological assumptions in literary studies. The status of the individual work and the traditional temporal conception of literary history give way to data patterns and the geographic layering of texts, characters, and spatial figures from distant historical periods. With surveys of literary spatiality we might see the development of a new genre as an alternative to conventional literary history. But what happens to the space of fiction once it is superimposed on actual geographic locations? How does this blended object compare to "purely" fictional spaces? Is something essential lost or transformed in the transition from literary texts to literary maps? Robert Stockhammer (chapter 3) examines the unmappability of literature. Reminding us that both literature and cartography are media with distinct and highly complex constitutive elements, he cautions against using the conventions of the realist novel as a model for the signifying processes of literature on the one hand, and the Ptolemaic model of the map with its grid and coordinates as the model for cartography on the other. Not only do many modernist and so-called realist novels contain moments of unmappability that defy Euclidean conceptions of space and resist a straightforward correspondence with geographic space, but can even the strong referentiality effects produced by, for example, a toponym like Paris always be trusted? Does literature not in its most basic forms include "referential aberrations," as Paul de Man suggested, that challenge a straightforward, mimetic correspondence theory of literary space?[3] Such aberrations can take many shapes, ranging from a single point in the text as when Captain Ahab finally locates Moby-Dick in an uncertain space where mappability fails, to the mutual incompatibility of georeferences in one and the same fictional work as in Proust's *In*

Search of Lost Time, to the twisted, Möbius-like spatiality of Kafka's texts. Even if it is the attempts to conceive of literary space in cartographic terms that make such features of fictional texts clearly visible, these blank spots, charted only by the language of literature, mark one of the things that get lost on the digital maps currently at our disposal.

Maps, however, form an integral part of language itself. While cartographic metaphors have exploded in our map-saturated culture and so have deprived maps of some of their explanatory power as critical concepts, they were once used in a conscious attempt to hold metaphorical slippage at bay. The writings of two of the main thinkers of space, Kant and Foucault, abound in cartographic metaphors. Instead of diluting them further, however, Oliver Simons subjects them to a careful analysis (chapter 4) and shows the profound shift they undergo within critical thought. For Kant, the rigor of geometry makes it a model of philosophical writing. Striving for a language of exactitude that could illustrate the concepts of understanding and reason, his critical work is pervaded by cartographic tropes as he painstakingly maps the territory of the mind and draws the borders of knowledge. Kant's cartographic writing is thus a way of containing the seductive metaphors of pure invention and fantasy—the literary aspect of language—that always threaten philosophical discourse from within. With Foucault, however, the cartographic metaphor changes status. Geometry no longer serves as a bulwark against the dangerous inventions of language. It rather takes the fictions of literature as the model for a novel form of cartographic writing that invents and constructs new spaces of knowledge, as much as it describes and represents the territories of knowledge that already exist. The cases of Kant and Foucault demonstrate just how much the explicit meanings as well as the underlying assumptions of cartographic metaphors vary even among the master thinkers who have shaped the spatial turn over the past fifty years.

It is not only philosophers, however, who have wrestled with cartographic writing. Throughout his works, Jorge Luis Borges uses the language of mapmaking to test the functions and limits of language and representation in general. Considering language *as* cartography, Borges hovers between a critical and a utopian perspective framed by the scholastic debate between realism and nominalism. Is language, in other words, motivated by the things to which it refers—does it, in Borges's formulation, constitute "a map of the universe"—or is it merely an arbitrary collection of symbols? Bruno Bosteels (chapter 5) charts the linguistic, epistemological, and ontological consequences of this historical debate with its long philosophical heritage. And going beyond it, he pursues a third conception of language that Borges also connects to cartography, namely the pragmatist view that regards language as producer of new realities that grow the world rather than copy it. Teasing out the major and subtle differences of the language of cartography in both philosophical and literary texts is a productive undertaking that could reveal other divergences and affinities between the principal conceptions of space and language.

Turning from theoretical fundamentals to historical developments and contexts, part II opens with the birth of Western literature out of nautical guidebooks. Burkhardt Wolf (chapter 6) traces the links between literature and navigation from Homer to Joyce, as he plots Odysseus's course through literary history and the history of nautical instruments. Poetry and the tools of seafaring were closely intertwined from the beginning. In the reconstruction of Odysseus's actual travels, Phoenician *periploi* or sailing manuals came to be regarded as the basis of Homer's epic. As Odysseus travels into the Christian era in Dante's *Commedia*, however, the periploi give way to actual sea charts whose graticules mark an ordered space in which the cunning Odysseus no longer has a proper place. When modernity expanded the circumference of its known world, the instruments of navigation could not keep up. Compass deviations and map distortions undercut the explorations of the mind. Accordingly, when Odysseus travels through modern literature, he gets increasingly lost, suffering a transcendental homelessness ever more distant from his Homeric home. Charting the steady collapse of nautical and narrative orientation, the chapter shows the intimate correlation between literary orientation and the concrete instruments of navigation that either guide literature or make it suffer a shipwreck.

Sea charts are not the only maps that have served as narrative devices. In the Middle Ages, maps were part of a larger diagrammatic culture of thought that brought the operations of cartography and literature together. The famous *mappaemundi* of Beatus manuscripts, for example, trigger narratives of biblical stories or invite apocalyptic commentaries. As Simone Pinet has shown (chapter 7), the pervasive diagrammatic imagination in the Middle Ages made porous not just the divisions between the visual and the verbal, the cartographic and the literary, but also those between the geographic, the historical, and the spiritual. Indicating or generating texts in processes of accretion, as on the elaborate Ebstorf and Hereford maps, or through minimal symbolism as on T/O maps where the T denotes the Crucifixion, diagrammatic thought forged tight links between the cosmological and the human. In such figurae of medieval culture, between the two extremes of the diagrammatic and the encyclopedic, maps present a treasure trove of materials for narrative development, pedagogical instruction, mnemonic retrieval, silent meditation, or other functions, layering and merging theological, political, scientific, and personal meanings within clerical discourses. Inflected by recent theorizing, the diagram thereby opens up whole new perspectives on the richness of medieval visual culture, whose fluid exchanges are only beginning to be understood.

The Renaissance inaugurates the fetishization of the Ptolemaic model of the map—the planar coordinate grid of locatable positions. But as is often the case in the history of maps, cartographic dreams outstrip their realization. The maps produced in the early modern period suffered a number of limitations that cannot simply be attributed to the bias inherent in all maps. Turning solid theoretical foundations into practice was fraught with difficulties. Without any reliable means of measuring longitude, or compensating for magnetic declination when reading the compass, maps were often inaccurate, based on errors or estimates. When mapmakers prepared their maps they had to resort to a vast and heterogeneous archive of topographic texts and sketches of various types. In the Spanish imperial expansion in the sixteenth century, texts served as sources for making maps, but they also formed a mode of cartography in its own right. As Ricardo Padrón explains (chapter 8), the dominant cartographic mode in the spatial imagination of the time was still the linear geometry of the "itinerary" rather than the planar geometry of the gridded map. But often these modes of mapping imperial conquest coexisted side by side, forming hybrid spatialities that competed against one another. The fictional texts of the time not only embody such tensions in the transition from the European Middle Ages to the early modern period, they also reflect on the very nature and limitations of mapping and the consequences of its role in fulfilling imperial ambitions.

In the sixteenth and seventeenth centuries new technologies of woodcut and copperplate illustration give rise to an unprecedented production and dissemination of cartography. Circulating throughout society, maps frequently make their way into the writings of poets. In the period between 1550 and 1660 the aesthetic sensibility of the Baroque is particularly evident in the serpentine, meandering, ever-mobile forms of French literary hydrography. From Oronce Fine to Madeleine de Scudéry, Tom Conley (chapter 9) traces the interactions between fluvial literature and cartography and shows how rivers become the site of shifting functions and meanings, ranging from the elements of war and destruction, to the celebration of wealth and beauty and the development of a fluvial style of writing, to the representation of feminine virtue and the force of generation in Scudéry's famous *Carte de tendre*.

The exuberant Baroque sensibility reflected in these writings not only reveals the wide array of themes expressed by literary hydrography, it also raises the question of limit: What is suppressed or curtailed by cartographic representation? The frequent association of maps with science, control, and objectivity, particularly in the eighteenth century, should not make us forget the objects that seem to exceed or evade cartographic organization, or the will and the impulses behind them. Beneath the grid lurk exuberant natural phenomena and obscure desires often at odds with the product of mapping. At the turn of the nineteenth century, Goethe brilliantly reveals this internal contradiction of cartography in his novel *Elective Affinities*. As John K. Noyes makes clear (chapter 10), maps may simultaneously contain two opposing rationalities—an instrumental one of economic value, and an aesthetic one of pleasure and desire. Uniting such conflicting rationalities on the same flat surface, Goethe not only develops his novel from the friction of their incompatibility, he also shows how the apparent objectivity of cartography is an outgrowth of a hidden economy of desire that far surpasses the amorous entanglements of the novel. In the late eighteenth and early nineteenth centuries, questions of the taxonomy and cartography of various phenomena such as botany, zoology, ethnography, and language were debated by the leading scientists of the time. But when historical developments and natural processes are frozen in a table or on a map, what happens to the genealogies that explain them, what disappears in the a priori generalizations of the mapmakers' choice of forms, and which will to truth drives the mapping? As Goethe's novel and his reflections on contemporary science show, maps are steeped in desires and reveal more than they prefer to put on display.

The disjunction within the scientific discourse between cartographic order and the objects on which it is imposed has a counterpart in the reorganization of national space in nineteenth-century France. Following the tumult of the French Revolution and the beginning of heavy industrialization, the diverse, local territories, each with its own culture, were rationalized into the abstract, national unity of the French Republic. Inspired by the detailed Cassini maps of France based on accurate geometric triangulations, the borders of local regions were redrawn and replaced with the political *départements*. The culture and history were thereby subsumed under an isotropic expanse imbued with a distant, national ideology. This profound change in the perception of space was reflected in the literature of the nineteenth century. As Patrick M. Bray explains (chapter 11), French literature, in particular the novel, devised new ways to represent the lost spaces of the past and to resist the dominant spatial discourse. Reimagining the relationship between the self and lived spaces, novelists such as Stendhal, Victor Hugo, Gustave Flaubert, and Émile Zola sought to create alternative literary maps through the interaction of word and image, textual map and geographic space, striving to harmonize the rootedness of the autochthonous with the possibilities of movement and change opened up by the processes of modernization.

The national conception of space was not only one of rationalization. It was also deeply informed by colonialism. Since the Berlin Congress in 1884–1885 partitioned the African continent and set off the colonial race for Africa, the imbrication of the *here* and the *elsewhere* has been a condition of national self-consciousness as well as of lived experience in both Africa and Europe. Which spatial imaginations did this entanglement of the two continents result in? How did migration from formerly colonized territories to European metropolitan centers and new diasporic formations and networks transform the coordinates of the African literary landscape? Charting the history of the colonial and postcolonial spatial imaginary, Dominic Thomas (chapter 12) explores both how an imperial power persuaded its citizens of the legitimacy of the colonial enterprise through propagandistic board games, and the responses of colonial and postcolonial writers, who sought and seek to deconstruct the expansionist myths of the past and to invent relevant and accurate conceptions of belonging in today's multisited and truly globalized world.

As an ordering principle that cuts across distinct media, genre is a useful way to bring together literary texts and cartographic objects. Part III of the book explores some of the central genres within literature and cartography. As Martin Brückner shows (chapter 13), not all literary cartographies include actual, physical maps. Sometimes the maps precede the narrative, when authors write their works with a map in hand. At other times, certain textual genres allude to or invoke specific types of maps. In American literature, mapping has been a central trope during the past four hundred years in spite of the fact that there is a surprising scarcity of actual maps accompanying the texts. Conveying the changing American experience down through the centuries, literary texts instead use the rhetoric of cartography, referencing maps, measurements, and coordinates to transform geographic descriptions into literary maps. A number of key genres have emerged as a result. From early travel accounts to picaresque narratives, frontier stories, and didactic fiction to the grand-tour account, fantasy novels, and the American road trip, literary texts have displayed a strong affinity for popular map genres such as small-scale overview maps, large-scale promotional maps, river maps, or interstate highway maps. Surveying and correlating these genres, Brückner reveals the fundamental themes of territorial thinking and geographic mobility that lie at the heart of American literature.

But which types of movements in literary texts is it even possible to map? What would a basic typology of mappable movements look like, and might literary texts suggest models of mobility different from the one we usually operate with? Juxtaposing four types of map lines in the geometric construction of the map as they have appeared historically in different narratives, Jörg Dünne (chapter 14) suggests that a dynamization of space takes place around 1800 that allows for a new conception of space, one based fundamentally on movement. Plotting the literary narratives of Jules Verne along the isolines of the map, the traditional opposition between the static map and the mobile narrative gives way to a dynamic model in which the general state of space is one of movement rather than stasis, where agents are primarily a complex of forces and nonhuman actors such as storms and currents, and where literary "events" occur as the interruption of movement rather than their commencement. Inverting basic assumptions about maps and narratives, this highly suggestive alternate model has far-reaching implications for how we might conceive cartographic space and literary events in other contexts.

The dynamic model of space can arise from the particular construction of the narrative line, but other dynamisms emerge from the materiality of the book itself. Paratexts such as maps, book covers, tables of contents, and indexes offer concrete spaces for reflections on form, materiality, and movement. An author who, perhaps more than any other, has experimented with the relations between the center and the margins of the material text and used them to probe and challenge the traditional conception of cartography as a static, atemporal means of representation is João Guimarães Rosa. Throughout his oeuvre, as Clara Rowland shows (chapter 15), the vast geography of his native Brazil is at once constructed and destabilized by paratextual invitations to map the reader's path in an unlimited and unmappable space, or to reread the chapters in a different sequence. In the dissonance between temporalizing and spatializing gestures, Guimarães Rosa's material cartography seeks to delineate a literary space that is forever mutable. Such paratextual experiments remind us of the material layer not just of maps, but also of the literary medium. Guimarães Rosa's work may serve as a model in itself for further explorations of the materiality of paratexts—their inherent spatiality and their influence on narrative space.

The final chapter turns both to the past and to the future. It analyzes the three main genres of military mapping that all involve the invention of possible futures. Military maps not only offer a crucial means of orientation, they have become a sine qua non for tactical and strategic planning. As such, they constitute *event maps* that serve to bring about the future invented on their surface. This function endows them with a peculiar epistemology. Transforming spatial configurations into hypothetical scenarios, the military map is a tool for the management of possible futures. The nature of these futures, however, differs between the three main genres of military maps. From the star charts of astrology to the topographic map to the modern war game, the representation, character, and management of the futures of war undergo profound shifts. The typology of war mapping sketched by Anders Engberg-Pedersen (chapter 16) alerts us to its overlaps with literary history. Equally concerned with the plotting and management of future events, authors from Schiller to Bolaño have variously reflected, challenged, and transformed the assumptions and the epistemology of the military event map. Linking readers and astrologers, writers and commanders, entertainment and death, the interaction of war literature and military cartography has generated important metareflections on the very possibility of representing and managing the futures of war.

Far beyond the map's function as a tool of orientation in a fictional world, the chapters of this book have revealed just how complex the conjunctions and disjunctions of literature and cartography are. Some basic insights stand out. Thriving on the metaphorical fecundity of maps and mapping, scholars of culture at large have inadvertently weakened the explanatory power of cartographic concepts. To restore the power and usefulness of these terms, the book has underlined the need for greater awareness both of the history of cartography proper and of the history of philosophy and its changing use of spatial concepts. Where Kant could use cartographic metaphors in his quest for a language of exactitude, today these metaphors need to be subjected to scrutiny before they can be deployed as a productive means of thinking—whether as models to understand the poetics of literary texts or as tools in digital mapping projects. Thus the book takes a critical stance toward the cartographic turn, one that encourages greater self-reflection among literary cartographers to ensure the continued value of cartography and of its terminology for the study of literature.

On the other hand, historians of cartography would do well to broaden their focus and become more attuned to the larger cultural sphere in which questions of maps and mappings are key. As the preceding chapters have shown, literature in particular emerges as a prism that refracts and reflects on the nature and functioning of cartography, discussing its virtues and vices, its productive possibilities as well as its blind spots. And just as literature and cartography have colluded to produce hybrid forms of representation, they have also appeared as estranged cousins with a keen eye for their incompatibilities and for the shortcomings of the other.

What lies ahead? The chapters just summarized invite further explorations of the theoretical foundations, historical exchanges, and generic links between literature and cartography. For example, much more can be said about the role of maps in the writing process, the spatial organization of poetry, and the mappability of different literary genres. Some contributors have also hinted at larger projects beyond the scope of their individual chapters, namely new ways of writing literary history. Exchanging the traditional linear form for a spatial model, a cartographic literary historiography may involve using maps as a tool, as we have seen. It could also involve the examination of cartography as a topos that cuts across national literatures and allows for new comparative assessments of established literary periods. Further, technological developments in the mapping sciences will surely continue to transform the writing and analysis of literary texts. One innovation that deserves more attention in literary studies is the rise, in the twentieth century, of electronic navigation systems such as GPS. Where the map with its disembodied, distant view traditionally operates within a representational regime governed by questions of mimesis and truth, GPS offers an embedded experience of navigation by coordinates that shifts the focus to utility. Whether the geographic information is correct matters less than getting you to your destination.[4] But what are the implications for our understanding of literature, and of the spatial structure of literature in particular? Always placing the user at the center in a space designed for travel, GPS enables a spatial experience with many similarities to that of reading a fictional text. With more than a billion GPS receivers in use across the globe, it is worth examining how the spatial order of this already fairly old technology will inflect the literary organization of space. It might, for example, make us notice alternative literary cartographies whose spatial figuration calls for a rethinking of the object that has for so long served as a foil for the literary production of space—the map. This book has sketched a basic framework for literature and cartography and offered a range of models for how future work in the field may be undertaken, but it has only taken the initial steps toward these larger projects.

Notes

1. "I am told that there are people who do not care for maps, and find it hard to believe" (Robert Louis Stevenson, *Treasure Island*, ed. John Sutherland (Peterborough, Ontario: Broadview Press,

2011), 233); "It is not down in any map; true places never are" (Herman Melville, *Moby-Dick* (New York: Norton, 2002), 59).

2. See Franco Moretti, *Atlas of the European Novel, 1800–1900* (London: Verso, 2007), 3; see also Barbara Piatti, chapter 2, this volume.

3. See Robert Stockhammer, chapter 3, this volume; see also Paul de Man, *Allegories of Reading: Figural Language in Rousseau, Nietzsche, Rilke, and Proust* (New Haven, CT: Yale University Press, 1979), 10.

4. For an excellent history of the development of navigational systems in the twentieth century and their role in the conception of territory, see William Rankin, *After the Map: Cartography, Navigation, and the Transformation of Territory in the Twentieth Century* (Chicago: University of Chicago Press, 2016).

Bibliography

de Man, Paul. *Allegories of Reading: Figural Language in Rousseau, Nietzsche, Rilke, and Proust.* New Haven, CT: Yale University Press, 1979.

Melville, Herman. *Moby-Dick.* New York: Norton, 2002.

Moretti, Franco. *Atlas of the European Novel, 1800–1900.* London: Verso, 2007.

Rankin, William. *After the Map: Cartography, Navigation, and the Transformation of Territory in the Twentieth Century.* Chicago: University of Chicago Press, 2016.

Stevenson, Robert Louis. *Treasure Island.* Ed. John Sutherland. Peterborough, Ontario: Broadview Press, 2011.

Contributors

Jean-Marc Besse is senior researcher at the National Center for Scientific Research (CNRS) in Paris and director of the EHGO/UMR Géographie-cités team (CNRS / Paris I / Paris VII).

Selected publications: *Le voyage, le témoignage, l'amitié: Abraham Ortelius et Georg Hoefnagel en Italie (hiver 1577–1578)* (Paris: Large, 2012); coeditor with G. A. Tiberghien, *Opérations cartographiques* (Arles: Actes Sud, 2017). *La sombra de las cosas: Sobre paisaje y geografia* (Madrid: Biblioteca Nueva, 2010); *Le goût du monde: Exercices de paysage* (Arles: Actes Sud, 2009); *Face au monde: Atlas, jardins, géoramas* (Paris: Desclée de Brouwer, 2003); *Les grandeurs de la Terre: Aspects du savoir géographique à la Renaissance* (Lyon: ENS Editions, 2003); *Voir la terre: Six essais sur le paysage et la géographie* (Paris: Actes Sud, 2000).

Bruno Bosteels is Professor in the Department of Latin American and Iberian Cultures and the Institute for Comparative Literature and Society at Columbia University.

Selected publications: *Marx and Freud in Latin America: Politics, Religion, and Psychoanalysis in the Age of Terror* (London: Verso, 2011); *Alain Badiou, une trajectoire polémique* (Paris: La Fabrique, 2009); *The Actuality of Communism* (London: Verso, 2011); *Badiou and Politics* (Durham, NC: Duke University Press, 2011); "Siting the Event: Gego, Modernity, and the Cartographic Turn," in *Gego 1957–1988: Thinking the Line* (Ostfildern: Hatje Cantz, 2006); "From Text to Territory: Félix Guattari's Cartographies of the Unconscious," in *Deleuxe and Guattari: New Mappings in Politics, Philosophy, and Culture* (Minneapolis: University of Minnesota Press, 1998); "A Misreading of Maps: The Politics of Cartography in Marxism and Poststructuralism," in *Signs of Change: Premodern, Modern, Postmodern* (Albany: SUNY Press, 1996).

Patrick M. Bray is Associate Professor of French at Ohio State University.

Selected publications: Editor, *Understanding Rancière, Understanding Modernism* (New York: Bloomsbury Press. 2017); "Wandering in the Text: Spatial Approaches to *Indiana*," *Approaches to Teaching George Sand's* Indiana, Eds. David A. Powell and Pratima Prasad (New York: MLA, 2015); coeditor with Phillip John Usher, *Building the Louvre: Architectures of Politics and Art* (special issue of *L'Esprit Créateur* 54, no. 2

(Summer 2014)); *The Novel Map: Space and Subjectivity in Nineteenth-Century French Fiction* (Evanston, IL: Northwestern University Press, 2013); "Lost in the Fold: Space and Subjectivity in Gérard de Nerval's 'Généalogie' and Sylvie," *French Forum* 31, no. 2 (Spring 2006): 35–51.

Martin Brückner is Professor of English and American Literature and co-director of the Center for Material Culture Studies at the University of Delaware.

Selected publications: *The Social Life of Maps in America, 1750–1860* (Chapel Hill: University of North Carolina Press, 2017); *Early American Cartographies* (Chapel Hill: University of North Carolina Press, 2011); coeditor with Hsuan L. Hsu, *American Literary Geographies: Spatial Practice and Cultural Production, 1500–1900* (Newark: University of Delaware Press, 2007); *The Geographic Revolution in Early America: Maps, Literacy, and National Identity* (Chapel Hill: University of North Carolina Press, 2006).

Tom Conley is Abbott Lawrence Lowell Professor of Romance Languages and Literatures and of Visual and Environmental Studies at Harvard University.

Selected publications: *À fleur de page: Voir et lire le texte de la Renaissance* (Paris: Classiques Garnier, 2015). *An Errant Eye: Poetry and Topography in Early Modern France* (Minneapolis: University of Minnesota Press, 2011); *Cartographic Cinema* (Minneapolis: University of Minnesota Press, 2007); *The Self-Made Map: Cartographic Writing in Early Modern France* (Minneapolis: University of Minnesota Press, 1996); *The Graphic Unconscious in Early Modern French Writing* (Cambridge: Cambridge University Press, 1992).

Jörg Dünne is Professor of Romance Literature at the Universität Erfurt.

Selected publications: Coeditor with Wolfram Nitsch, *Scénarios d'espace: Littérature, cinéma et parcours urbains* (Clermont-Ferrand: Presses universitaires Blaise Pascal, 2014); *Die kartographische Imagination: Erinnern, Erzählen und Fingieren in der Frühen Neuzeit* (Munich: Fink, 2011); coeditor with Stephan Günzel, *Raumtheorie: Grundlagentexte aus Philosophie und Kulturwissenschaften* (Frankfurt am Main: Suhrkamp, 2006); *Asketisches Schreiben: Rousseau und Flaubert als Paradigmen literarischer Selbstpraxis in der Moderne* (Tübingen: Narr, 2003).

Anders Engberg-Pedersen is Associate Professor of Comparative Literature at the University of Southern Denmark.

Selected publications: Coeditor with Kathrin Maurer, *Visualizing War: Emotions, Technologies, Communities* (New York: Routledge, 2017); *Empire of Chance: The Napoleonic Wars and the Disorder of Things* (Cambridge, MA: Harvard University Press, 2015); "Critique of Cartographic Reason: Tolstoj on the Media of War," *Russian Literature* 77, no. 3 (2015): 307–336; "Sketching War: August von Larisch's Collection of Field Maps from the Russian Campaign of 1812," *Imago Mundi* 66 (2014): 70–81; "The Refraction of Geometry: Tristram Shandy and the Poetics of War, 1700–1800," *Representations* 123, no. 1 (Summer 2013): 23–52; "Det Symbolske Terræn–Kartografi og Krigens Poetik i det 19. Århundrede," *Kritik* 2 (2013); coeditor with Michael Huffmaster, Eric Nordhausen, and Vrääth Öhner, *Das Geständnis und seine Instanzen: Zur Bedeutungsverschiebung des Geständnisses im Prozess der Moderne* (Vienna-Berlin: Verlag Turia + Kant, 2011); "Die Verwaltung des Raumes: Kriegskartographische Praxis um 1800," in *Die Werkstatt des Kartographen: Materialien und Praktiken visueller Welterzeugung* (Munich: Fink, 2011).

John K. Noyes is Professor of German at the University of Toronto.

Selected publications: *Herder: Aesthetics against Imperialism* (Toronto: University of Toronto Press, 2015); coeditor with Hans Schulte and Pia Kleber, *Goethe's Faust: Theatre of Modernity* (Cambridge: Cambridge University Press, 2011); "The World Map and the World of Goethe's *Weltliteratur*," *Acta Germanica* 38 (2010): 128–145; *The Mastery of Submission: Inventions of Masochism* (Ithaca, NY: Cornell University Press, 1997); *Colonial Space: Spatiality in the Discourse of German South West Africa 1884–1915* (Chur/Philadelphia: Harwood Academic Publishers, 1992).

Ricardo Padrón is Associate Professor of Spanish at the University of Virginia.

Selected publications: "'The Indies of the West' or, the Tale of How an Imaginary Geography Circumnavigated the Globe," in *Western Visions of the Far East in a Transpacific Age, 1522–1657* (New York: Ashgate, 2012); "Mapping Imaginary Worlds," in *Maps: Finding Our Place in the World* (Chicago: University of Chicago Press, 2007); *The Spacious Word: Cartography, Literature, and Empire in Early Modern Spain* (Chicago: University of Chicago Press, 2004); "Mapping Plus Ultra: Cartography, Space, and Hispanic Modernity," *Representations* 79 (2002): 28–60; "Exiled Subjects, Paper Empires: Revisiting Love and Heroism in Herrera," *Hispanic Review* 70, no. 4 (2002): 497–520.

Barbara Piatti has been co-director of the trinational research project "A Literary Atlas of Europe" at the Institute of Cartography and Geoinformation, ETH Zurich (2006–2014). Since 2014 she has worked as an independent book author and researcher.

Selected publications: *Lake Lucerne & the Gotthard – Like You Have Never Seen This Landscape Before* (Meggen, Switzerland: Imaginary Wanderings Press, 2016); *Es lächelt der See: Literarische Wanderungen in der Zentralschweiz* (Zurich: Rotpunktverlag, 2013); *Die Geographie der Literatur: Schauplätze, Handlungsräume, Raumphantasien* (Göttingen: Wallstein, 2008).

Simone Pinet is Professor of Romance and Medieval Studies at Cornell University.

Selected publications: *The Task of the Cleric: Cartography, Translation, and Economics in Thirteenth-Century Iberia* (Toronto: University of Toronto Press, 2016); *Archipelagoes: Insular Fictions from Chivalric Romance to the Novel* (Minneapolis: University of Minnesota Press, 2011); coeditor with Cynthia Robinson, *Courting the Alhambra: Cross-Disciplinary Approaches to the Hall of Justice Ceilings* (Brill: Leiden, 2008); "Literature and Cartography in Spain: Etymologies and Conjectures," in *The History of Cartography,* Vol. 3.1: *Cartography in the European Renaissance* (Chicago: University of Chicago Press, 2007); "Where One Stands: Shipwrecks, Perspective, and Chivalric Fiction," special issue of *e-Humanista* 16 (2010): 381–394; "Para leer el espacio en el *Poema de Mio Cid*: breviario teórico," *La corónica*, Vol. 33.2 (Spring 2005): 195–208.

Clara Rowland is Professor of of Portuguese Studies at Universidade Nova de Lisboa.

Selected publications: Coeditor with Tom Conley, *Falso Movimento: ensaios sobre escrita e cinema* (Lisbon: Cotovia, 2016); "Deliveries of Absence: Epistolary Structures in Classical Cinema," in *The Writer on Screen: Screening Literary Authorship* (London: Palgrave Macmillan, 2013); "I Could Stop Here: Figures of Rereading in *Grande Sertão: Veredas*" in *Studies in the Literary Achievement of João Guimarães Rosa, the Foremost Brazilian Writer of the Twentieth Century* (Lewiston, ME: Edwin Mellen Press, 2011): 318–333; *A Forma do Meio: Livro e Narração na obra de João Guimarães Rosa* (Campinas: Editora da UNICAMP, 2011); "Revelações ópticas: visão e distorção nos contos de Clarice Lispector," *Românica* 19 (2010): 71–88; "Forms of Crossing: Book and Margin in the Work of Guimarães Rosa," *Variants: Journal of the European Society for Textual Scholarship* 6 (2009): 277–290.

Oliver Simons is Professor in the Department of Germanic Languages at Columbia University.

Selected publications: *Literaturtheorien zur Einführung* (Hamburg: Junius, 2009); *Raumgeschichten: Topographien der Moderne in Philosophie, Wissenschaft und Literatur* (Munich: Fink, 2007); coeditor with Elisabeth Wagner, *Bachmanns Medien* (Berlin: Vorwerk 8, 2008); coeditor with Arne Höcker, *Kafkas Institutionen* (Bielefeld: transcript Verlag, 2007); coeditor mit Alexander Honold, *Kolonialismus als Kultur: Literatur, Medien, Wissenschaften in der deutschen Gründerzeit des Fremden* (Tübingen: Francke Verlag, 2002); "Persuasive Maps and a Suggestive Novel—Hans Grimm's *Volk ohne Raum* and German Cartography in Southwest Africa," in *German Colonialism, Visual Culture, and Modern Memory* (London: Routledge, 2010); "Nach Euklid: Geometrie als Narrativ bei Husserl und Foucault," in *Wissen. Erzählen. Narrative der Humanwissenschaften* (Bielefeld: transcript Verlag, 2006); "Topographie und

Archäologie der Moderne: Alfred Döblins Amazonas-Roman," in *Wolfgang Koeppen & Alfred Döblin: Topographien der Moderne* (Munich: Iudicium Verlag, 2005).

Robert Stockhammer is Professor of Comparative Literature at the Ludwig-Maximilians-Universität Munich.

Selected publications: *1967: Pop, Grammatologie und Politik* (Paderborn: Fink, 2017); *Afrikanische Philologie* (Berlin: Suhrkamp, 2016); *Grammatik. Wissen und Macht in der Geschichte einer sprachlichen* Institution (Berlin: Suhrkamp, 2014); *Kartierung der Erde: Macht und Lust in Karten und Literatur* (Munich: Fink, 2007); editor, *TopoGraphien der Moderne: Medien zur Repräsentation und Konstruktion von Räumen* (Munich: Fink, 2005); *Ruanda: Über einen anderen Genozid schreiben* (Frankfurt am Main: Suhrkamp, 2005).

Dominic Thomas is Madeleine L. Letessier Professor of French and Francophone Studies at the University of California, Los Angeles.

Selected publications: *Africa and France: Postcolonial Cultures, Migration, and Racism* (Bloomington: Indiana University Press, 2013); *Noirs d'encre: Colonialisme, immigration et identité au cœur de la littérature afro-française* (Paris: Editions La Découverte, 2013); *Black France: Colonialism, Immigration, and Transnationalism* (Bloomington: Indiana University Press, 2007); *Nation-Building, Propaganda, and Literature in Francophone Africa* (Bloomington: Indiana University Press, 2002).

Burkhardt Wolf is Heisenberg Research Fellow at Humboldt University.

Selected publications: *Fortuna di mare: Literatur und Seefahrt* (Zurich/Berlin: Diaphanes, 2013); coeditor with Thomas Weitin, *Gewalt der Archive. Studien zur Kulturgeschichte der Wissensspeicherung* (Konstanz: Konstanz University Press 2012). coeditor with Elisabeth Wagner, *Odysseen* (Berlin: Vorwerk 8, 2008); *Die Sorge des Souveräns: Eine Diskursgeschichte des Opfers* (Zurich/Berlin: Diaphanes, 2004); coeditor with Anja K. Maier, *Wege des Kybernetes: Schreibpraktiken und Steuerungsmodelle von Politik, Reise, Migration* (Münster: Lit, 2004); *Von der Kunst kleiner Ereignisse: Zur Theorie einer "minoritären Literatur": Alexander Kluge und Gilles Deleuze* (Marburg: Tectum Verlag, 1998).

Index

Absalom! Absalom!, 346
Adorno, Theodor, 36
The Adventures of Huckleberry Finn, 339–344
Aeneid, 152
African cartographic literature, 299–315, 448
  board games, 303–309
  and colonialism, 299–309
  commemorative projects, 311–312
  post-colonial identity, 312–315
  texts/writers, 301, 312
Afropeanism, 301
Akbari, Suzanne Conklin, 175
Alexandreis, 189
Alfonso X, 184
L'Amadis de Gaule, 219, 226–227, 231, 246
American cartographic literature, 449
  colonization, 327–335
  gilded ages, 344–352
  histories and travel writing, 325, 327, 335
  literary mappings, 325
  and maps, 325–326
  nation formation, 335–344
Amodeo, Immacolata, 61, 64
"The Analytical Idiom of John Wilkins," 121
Anderson, Benedict, 352
Androuet du Cerceau, Jacques, 219, 227–231, 234, 246
Anne of Geierstein, 46–48
Apian, Peter, 206
Apollonius of Rhodes, 149
Araucana, 209–211
The Archaeology of Knowledge, 108–110, 113–114
Argonautika, 149

Aristotle, 27, 119–120
Around the World in 80 Days, 30, 380
"Art as Device," 7
Astrology, 413, 417, 419–420
Astronomical bearing, 367
Atlas of Remote Islands, 3
d'Aubigné, Agrippa, 220, 231–237, 246
L'aventure ambiguë, 312

Bachelard, Gaston, 1
Bachmann, Ingeborg, 65
Bacon, Francis, 162
Baïf, Jean-Antoine de, 224
Balzac, Honoré de, 282, 290–293
Bancel, Nicolas, 303
Barbar the elephant, 309
Baroque hydrography, 219–224, 447
  L'Amadis de Gaule, 219, 226–227, 231, 246
  Carte des rivières de France curieusement recherchée, 240–243
  hydrographic writing, 246
  Lesphere du monde, 220–224
  Les plus excellens bastiments de la France, 227–231, 234
  rivers, 219–224, 230–231, 234–238, 240
  Les Tragiques, 220, 231–236, 246
  "Voyage de Tours," 224–226
Barthes, Roland, 111, 286
Bartram, William, 335
Base maps, 90–93
"Le Bateau ivre," 286
Baudelaire, Charles, 285
Baudrillard, Jean, 352
Baum, Frank, 347

Bäumler, Andreas, 58
Beatus manuscripts, 177–180, 188
Becher, Ulrich, 58
Bell, David F., 285
Belleau, Remy, 225
Benjamin, Walter, 21–22, 32, 36–37
Bérard, Victor, 144–146, 148, 150–151, 166
Berenhorst, Georg Heinrich von, 420
Berghaus, Heinrich von, 81
Berlin Alexanderplatz, 36
"Berlin Chronicle," 37
Besse, Jean-Marc, 443
Beti, Mongo, 312
Biddle, Nicholas, 335
Biemann, Ursula, 314–315
Blue Highways, 349
Bly, Nellie, 347
Board games, 303–309, 338
Boesch, Hans, 58
Bofane, In Koli Jean, 300, 313–315
Bolaño, Roberto, 412, 430–432, 434
Bolle, Willi, 392
Book cover maps, 401
"Booklovers Map of America," 352
Books, 398–400, 404–405
Borges, Jorge Guillermo, 129
Borges, Jorge Luis, 22, 24, 119–132, 445
"Borges and I," 119
Borges's maps, 119–132
  cartographic metaphors, 127–132
  map of the universe, 119–120
  and nominalist tradition, 123–124
  "On Rigor in Science," 125–126
  "Partial Enchantments of the *Quixote,*" 126–127
  and Platonism, 124–125
  realism vs nominalism, 119–125, 131
Bosteels, Bruno, 4, 445
"Bötjer Basch," 55
Boyle, Nicholas, 263
Bray, Patrick M., 448
Brooks, Peter, 398
Brückner, Martin, 4, 6, 449
Brunhoff, Jean de, 309

Budgen, Frank, 36
Burroughs, Edgar Rice, 347
Bussola, 154
Butler, James, 338
Butor, Michel, 37, 110
Butterworth, Hezekiah, 346
Byrd, William, 335
"Byrhtferth's Diagram," 187

Calinescu, Matei, 405
Camara, Laye, 312
Candido, Antonio, 391–393
Caquard, Sébastien, 67
Carruthers, Mary, 175, 188
Carte des rivières de France curieusement recherchée, 240–243
"La Carte de Tendre," 243–246, 447
La Carte et le territoire, 292
Carter, Paul, 260
Cartes de Cassini, 279
Cartina marina, 361–362
Cartographic fiction, 21–39, 443–444. *See also* Maps
  factual fictions, 362
  fantasy literature, 28
  heterotopia, 28–30
  mimesis, 22–25
  narrative organization, 36–39
  reality effect of maps, 27–33
  reflexive classes, 24
  spatial plotting, 33–37
  and urban reality, 32–33
Cartographic present, 2–4
Cartographic turn, 4–5, 11, 450
Cartographic writing, 100–102, 166. *See also* Cartographic fiction
  functions of maps, 208–209
  metaphors, 101–102, 127–132, 445, 450
Cartography. *See also* Cartographic fiction; Literary cartography; Maps; Nature cartography; War cartographies
  cartographic lines, 364–372
  development of, 160–163

isometric, 362
and linguistic diversity, 267–270
and literature, 1–2, 6, 12–13, 199, 301, 396, 443, 450–451
and the material book, 398–400, 405
in medieval culture, 189
as mimesis, 127
and narration, 173
and philosophy, 100
political dimension of, 260–261, 263
and species categorization, 272
thematic, 370
ubiquity problem, 6–7
Cartography in the European Renaissance, 11
"Cartography Is Dead (Thank God)," 6
Cartoons, 309–310
The Castle, 84
Celestial event maps, 412–413, 417–418
Certeau, Michel de, 4, 261, 282, 349, 364, 397
The Chainbearer, 326
Champney, Elizabeth W., 346
Charpentier, Johann Friedrich Wilhelm, 255
The Charterhouse of Parma, 424–425
Chartier, Roger, 27–28
Chatwin, Bruce, 52
Chitty-Chitty-Bang-Bang, 349
Christine, 349
Cinq semaines en ballon, 380
Clark, William, 335
Clausewitz, Carl von, 420, 428
Clélie, 220, 243
Climate maps, 81
Climatography, 370
Cohen, Dan, 66
Coleridge, Samuel Taylor, 121, 123–124
Collateral space, 111
Colonialism, 299–312, 448
  and board games, 303–309
  and cartographic practices, 361
  and maps, 303
  post-colonial literature, 312–315
  propaganda, 302–303, 310
  revaluation of, 310–312

Columbus, Christopher, 327
La Comédie humaine, 290–292
Commedia, 152–153, 157–160
Commentary of the Apocalypse, 180
Compass, 153–157, 163
Compasso, 154
Compendium Historiale in Genealogia Christi, 187
Complementary space, 111
Congo Inc., 300, 313–315
Conley, Tom, 31, 447
The Contest in America between Great Britain and France, 333
Contour lines, 367–370
Conversations of German Refugees, 255
Cooper, James Fenimore, 326
Coordinate grid, 200, 207, 364
Corinne ou l'Italie, 283–285
Corpo de Baile, 398–400
Correa, Gaspar, 162
Correlative space, 111
Cortés, Hernando, 203
Cosgrove, Denis, 5
Cosmographia, 206
Cotgrave, Randle, 238
Coulon, Louis, 240, 243
Counterfactual narratives, 91
Countersciences, 108
Course de l'Empire français, 303, 308
Cratylism, 120
Critique of the Power of Judgment, 99, 101, 104–105
Cruel City, 312–313
Le Cygne, 285

Dadié, Bernard, 312
Dalché, Patrick Gautier, 189
Dante Alighieri, 152–153, 157–160
Das Jahr der Liebe (My Year of Love), 61
Das Kriegsspiel, 428
Data acquisition
  literary maps, 81
  thematic mapping, 79–83

Data mining, 65–66
La Débâcle, 34
"Deconstructing the Map," 5
Defoe, Daniel, 3, 30, 331
Deleuze, Gilles, 4, 113, 128–130, 166, 349, 397
Denkstil, 11
Derrida, Jacques, 5, 128
Description of New England, 329
Description of San Marco, 110
Deutsche Kolonien, 309
"Deutsches Requiem," 123
Deutschland, geognostisch-geologisch dargestellt, 255
Deutschland's Kolonien-Spiel, 303, 309
Diagrams, uses of, 173–174, 191. *See also* Medieval diagrams
Dickens, Charles, 339
Didascalicon, 177
Digital mapping, 444
Discipline and Punish, 113
The Discoveries of John Lederer, 329
Distance, 99
Döblin, Alfred, 36
Le Docteur Pascal, 291
Doctor Faustus, 36
Don Quixote, 24, 27–28, 126, 207
Doolittle, Amos, 336
Dordogne river, 237–238, 240
Döring, Jörg, 52, 64
Drawings, 191
Drucker, Johanna, 399
Dünne, Jörg, 7, 449
Dürer, Albrecht, 414
Dwight, Timothy, 335

L'Ébauche de Germinal, 34
Ebstorf mappamundi, 184
Eckert, Max, 370
Edelstein, Dan, 417
Edney, Matthew, 207–208
Elective Affinities, 253–254, 258–265, 272, 447
Elkins, James, 174
Encyclopedia, 393

L'Enfant noir, 312
Les Enfants du capitaine Grant, 372–380
Engberg-Pedersen, Anders, 450
Epigraphs, 404–405
Eragon, 347
Eratosthenes, 145
Ercilla, Alonso de, 199, 209–213
Escorial Atlas, 201
Espacio, 207
Esquivel, Pedro, 201
Essais, 220, 237–238, 246
Essarts, Nicolas Herberay des, 226
"Estória do Homem do Pinguelo," 399
Estrangement, literary, 7–11
Etidorhpa, or, The End of Earth, 347
Etymologies, 182
Evans, Lewis, 333
Evans, Michael, 176–177
Event maps, 412, 428–429, 432, 434

Factual fictions, 362
Fantasy literature, 28
Farmer, John, 339
Faulkner, William, 79, 346
Faust, 264
La Faute de l'abbé Mouret, 34
Fernández, Macedonio, 129
Fictionality. *See* Cartographic fiction
Figurae, 175–176
Fine, Oronce, 219–224, 246
First space, 46
Fitzgerald, F. Scott, 55, 58, 349
Flaubert, Gustave, 52
Fleming, Ian, 349
Focalizing, 87
La Fortune des Rougon, 34
Fortune's Football, 338
Foucault, Michel, 4–5, 28, 99, 106–114, 128, 445
Foucault's cartography, 106–114, 445
  archaeology, 112–113
  archive, 112–113
  cartographic images, 114
  poetics of space, 110–111

space and knowledge, 106–108
statements, 109–110
topological turn, 108–113
France
　board games, 303–308
　cartographic representations, 280, 448
　colonial propaganda, 302–303, 309–311
　literary cartography, 279, 283
　literary hydrography, 219–224, 234, 447
La Franciade, 230–231
"Frankfurter Poetik-Vorlesungen," 65
Fremont, John, 335
Frisch, Max, 52
"From Allegories to Novels," 119–122
Frontier stories, 338–339

Gaboriau, Émile, 30
Gallia, 236
Gama, Vasco da, 162
Game of Thrones, 28
Games, 303–309, 338, 425–430
Gastaldi, Giacomo, 203
The Great Gatsby, 349
Gaze tours, 206
"Gedankenreisen," 61
General estoria, 184
A General Map of the Middle British Colonies, 333
Generall Historie of Virginia, 329
Genette, Gérard, 87
Geoffrey of Vinsauf, 187–188
Geographic information systems (GIS), 444
　GIS visualizations, 50
　and maps, 73–74
Geographical, Historical, Political, Philosophical and Mechanical Essays, 333
Geography, 27, 46, 67, 221–224
Geography (Geographia), 31, 199–200
Geography Made Easy, 336
Geography of literature, 46
Geometry, 100, 114, 445
Geospace, 46
Germany, 303, 309, 311–312

Germinal, 34
Gerner, Alexander, 173
Gibbs, Fred, 66
GIS visualizations, 50
Gladstone, William E., 146
Global hyperobjects, 370
Global Positioning System (GPS), 451
Globalization, 300–301
Goethe, Johann Wolfgang von, 253–272, 447
Goethe's maps, 255–260, 267–269
Góngora, Luis de, 199, 209, 212–214
Gothic novels, 81
Gotthard region, 45–48
Grande Sertão: Veredas, 391–401, 404
Grand-tour accounts, 346–347
"Granite," 89
The Grapes of Wrath, 349
Graphical hybrid maps, 87
Grenier, Jean-Claude, 25
Grey, Zane, 346
Grid lines, 200, 207, 364, 376
Große Karte von Westphalen, 258
Guattari, Félix, 4, 349, 397
Gudrunlied, 153
Guillotière, François de la, 238
Gulliver's Travels, 74–79
Günzel, Stephan, 5

Habila, Helon, 301
Hagenbeck, Carl, 311
Half-thing, 30
Halley, Edmond, 369
Hamlet, 24
Hardy, Thomas, 79, 83
Harley, J. B., 5–6
Harvey, David, 4
Heart of Darkness, 260
The Heart of the Leopard Children, 313
Hegel, Georg, 145
Hegglund, Jon, 67
Hellwig, Johann Christian Ludwig, 428
Henry the Navigator, 160
Herder, Johann Gottfried, 270, 272

Hergé (Georges Remi), 309
Heterotopia, 28–30
Histoire universelle, 231
"History of Eternity," 122
History of the Dividing Line, 335
History of the Expedition under the Command of Captains Lewis and Clark, 335
The History of Constantius and Pulchera, 338
The History of Cartography, 11
Holcomb, Melanie, 176, 187–188, 191
Homer, 143–153, 166
Honorius of Autun, 182–184
Horoscopes, 414–417
Houellebecq, Michel, 3, 292–293
Hugh of St. Victor, 177
Hugo, Victor, 280
Humboldt, Alexander von, 370
Humboldt, Wilhelm von, 267–270
Hydrographic writing, 246
Hydrography, 219–224, 234, 447
Hypotyposes, 101–102

Ideas for a Philosophy of History, 270
Illustrated Guide to Paris, 22
Images
  maps as, 174
  as reminders/models, 174
Imaginary spaces, 64–65
"Imaginative-functional toponyms," 64
Imagines coeli Septentrionales et Meridionales zodiaci, 414
Imago mundi, 184
An Improved Edition of a Map of the Surveyed Part of the Territory of Michigan, 339
Indirect referencing, 50
"Instructions for Drafting a General Linguistic Map," 269
De Insulis nuper inuentis, 327
Internal mappability, 74, 79
International Colonial Exposition of 1931, 310
"Introduction to the Structural Analysis of Narrative," 111
Iser, Wolfgang, 55
Isidore of Seville, 182, 188
Island fiction, 3
Isolario, 3
Isolines, 369–370, 378, 381
Isometric cartography, 362
Itineraries, 203–209, 447

Jacob, Christian, 5, 280–282, 364
James, William, 119, 128–131
Jameson, Fredric, 4, 125
Jefferson, Thomas, 335
Jeu de l'Empire français, 303, 308
Jeu des échanges France–Colonies, 303, 309
Jomini, Antoine-Henri, 420
Joyce, James, 36, 79, 166

Kafka, Franz, 84
Kane, Cheikh Hamidou, 312
Kant, Immanuel, 99–107, 109, 111–114, 445
Kant's maps, 99–106
  cartographic images, 105–106, 114
  cartographic writing, 100–102, 445
  drawing a line, 102–104
  symbolic hypotyposes, 101–102
Keferstein, Christian, 255
Kehlmann, Daniel, 3
Kelly, Douglas, 187
Kepler, Johannes, 412–414, 417–418
Kerouac, Jack, 349
King, Stephen, 349
Kirkland, Caroline, 338–339
Kittler, Friedrich, 258, 346
Kitzinger, Ernst, 180
Korzybski, Alfred, 3
Krämer, Sybille, 175, 262
Der Kreis (The Circle), 58

Lake Lucerne and Gotthard, 45–48
Language
  arbitrariness of signs, 120
  cartographic metaphors, 101–102, 127–132, 445
  and geometry, 100
  as map of the universe, 119–120

of mapmaking, 119
mapping linguistic diversity, 267–270
performative dimension of, 86
referential effects of, 83
"The Language of Space," 110
Latini, Brunetto, 157
Latitude, determination of, 367
Le Coq, Carl von, 258
Least Heat-Moon, William, 349
Lederer, John, 329
Lefebvre, Henri, 5, 349
Lehmann, Johann Georg, 420
Lemaire, Sandrine, 302
Lenda da India, 162
Lesphere du monde, 220–224
Lichtenberg, Georg Christoph, 87
The Life and Adventures of Martin Chuzzlewit, 339
Ligon, Richard, 331
"A Literary Atlas of Europe," 49–50, 81
Literary cartography, 12–13, 45–68
 categories and terms, 64
 and culture, 60–61
 functions of, 58–60
 geospace/first space, 46
 GIS visualizations, 50
 hermeneutics and distant reading, 65–67
 imaginary spaces, 64–65
 indirect referencing, 50
 literary maps, 49–50, 60–63, 81
 maps as representations, 48–49
 projected spaces, 52–55
 and reading, 61
 strong landscape, 46–48
 studies in, 349–352
 weight of settings, 58
Literary geography, 46, 67
"A Literary Map of Manhattan," 49–50
"Literary Maps of Switzerland," 60–63
Literature. *See also* Literary cartography
 and astrology, 419–420
 and cartographic terms, 6
 cartographic turn, 4–5, 11

and cartography, 1–2, 6, 12–13, 199, 301, 396, 443, 450–451
as cartography, 199
and estrangement, 7–11
geography of, 46
mappability of, 73–74, 444–445
and projected places, 52
and representation, 280–282
De la littérature, 279–280
Ljungberg, Christina, 27
Lloyd, John Uri, 347
Lloyd's Map of the Lower Mississippi River from St. Louis to the Gulf of Mexico, 344
Longitude, determination of, 367
Lopez, Tomas, 27
Lord of the Rings, 28, 347
Lotman, Yuri M., 362, 368–369, 419
Lugaroso, 396–397
Lumbroso, Olivier, 34
Lyotard, Jean-François, 127

Macrobian maps, 182
Madame Bovary, 52
Magnetism, 153
Man, Paul de, 83
"Manifeste pour une 'littérature monde' en français," 300
Mann, Thomas, 36
Mannered mapping, 219, 226
Map genres, 325–352
 American literature, 325–326
 Atlantic as setting, 331
 continental perspective, 333
 frontier stories, 338–339
 geographic accounts, 331
 grand-tour accounts, 346–347
 island theme, 331–333
 lowbrow and modernist fiction, 346
 map inserts, 336–338, 344, 346
 picaresque narratives and board games, 338
 promotional maps, 339
 river maps, 339–344
 road trip writing, 349

Map genres (cont.)
    science fiction/fantasy, 347–349
    shortcomings of, 333–335
    textbooks, 336
    travel reports, 325, 327–331, 335–336
Map line narratives, 361–382, 449
    borders, 369
    bottled messages, 378–380
    cartographic lines, 364–372
    circumnavigation, 376–378
    contour lines, 367–370
    grid lines, 364, 376
    isolines, 369–370, 378, 381
    and literary space, 368–369
    parcours lines, 367, 369
    and plot, 368–369, 381–382
    portolan charts, 154, 364, 367, 369, 376
    sea currents, 378–380
    and static positioning, 378
Map of New England, 329
Map of the British and French Dominions in North America, 333
The Map and the Territory, 3
A Map of the United States of America, 336
Mappability, 73–85
    internal, 74, 79
    referential, 79, 90–91
    unmappability, 83–85, 91, 444–445
Mappaemundi, 180, 184, 209, 361, 446
"Mappe-Mond suivant la projection des cartes reduites," 272
Mapping, 6, 106
    alternative forms of, 85–86
    and archival materials, 202–203, 207–208
    base maps, 90–93
    development of, 201
    digital, 444
    imaginary spaces, 64–65
    linguistic diversity, 267–270
    and literary performativity, 86–87
    mannered, 219, 226
    as metaphor, 85
    modes of, 207–208

projections, 87–90
and surveying, 201–202
"Mapping Emotions in Victorian London," 65–66
"Mapping the Republic of Letters," 65
Maps. *See also* Literary cartography; Map genres; Map line narratives; Novel maps
    and authenticity, 27, 31
    base, 90–93
    blank spaces of, 260–261, 272
    book cover, 401
    climate, 81
    and colonialism, 303
    and constructivism, 5–6
    coordinate grid, 200, 207
    culture of, 2
    defined, 73–74, 85, 175
    as diagrams, 174
    estranging, 7–11
    event, 412, 428–429, 432, 434
    and geographic information systems (GIS), 73–74
    graphical hybrid, 87
    inserts, 336–338, 344, 346
    language of, 119
    limitations of, 7
    literary, 352
    and literary scholarship, 4
    Macrobian/zonal, 182
    marine, 154
    in medieval culture, 174–175
    and mimesis, 22–25
    mosaic/mural, 180–182
    and narrative, 1, 33–37, 173, 180, 184–187, 191, 282, 361–362, 368–369, 382
    navigational/mimetic use of, 90
    novels as, 289–292, 391–392
    origins of, 207
    and ownership, 264–265, 272
    and parks, 25
    and plot, 9
    reality effect of, 27–33
    refunctionalization of, 261

    and representation, 282
    as representations, 48–49
    river, 339–344
    and space, 24–25, 31–32
    as tables, 400–4001
    and territory, 3
    thematic, 79–83
    T/O, 182–187
Margins, 405
Marin, Louis, 422
Massey, Doreen, 4
Mathematics, 100
Maury, Matthew Fontaine, 163
Mbembe, Achille, 299, 313–314
McLuhan, Marshall, 3
Measuring the World, 3
Medieval diagrams, 173–191, 446
    accretion, 184
    Beatus maps, 177–180, 188
    clerical uses of, 177
    as devices, 188–191
    figurae, 175–176
    graphic strategies, 176
    historical format, 187–188
    lists, 188
    Macrobian/zonal maps, 182
    mappaemundi, 180, 184, 209, 361, 446
    mosaic/mural maps, 180–182
    T/O maps, 182–187
    types of, 177
    wind poems, 188–189
Melville, Herman, 84, 164, 326, 339, 347, 392, 443
Las Meniñas, 24
Mensch, Jennifer, 265
Mercator, Gerhard, 163, 201
Mercator projection, 201, 364–367
Meridiana: The Adventures of Three Englishmen and
Three Russians in South Africa, 30
Metaphors, 101–102, 127–132, 445, 450
"Meu Tio o Iauaretê," 396
Military event maps, 411–412, 423–424, 428–429, 432, 434, 450

Miller, Christopher L., 312
Miller, J. Hillis, 5, 83
Mimesis, 22–25, 127
"Mineralogische Geographie der Chursächsischen Lande," 255
Miniature parks, 25, 443
Mirages de Paris, 312
Mitchell, John, 333
Mithridates, 269
Mobile, 37
Moby-Dick, 84, 164–165, 326, 339, 347, 380, 392
Modern Atlas Adapted to Morse's New School Geography, 344
Moll, Herman, 74
Monsieur Lecoq, 30
Montage, 36
Montaigne, Michel de, 220, 237–238, 246
Montalboddo, Fracanzano da, 327
Montauk, 52
More, Thomas, 3, 28
Moretti, Franco, 52, 73, 79, 352, 397, 444
Morrison, Toni, 349
Morse, Jedidiah, 336, 344
Morse's North American Atlas, 344
Morton, Timothy, 370
Mosaic maps, 180–182
Moses, Walter, 393
Les mots et les choses, 106–107
"MS. Found in a Bottle," 164
Müffling, Friedrich von, 258
Mujila, Fiston Mwanza, 315
Mural maps, 180
Die Murmeljagd (The Woodchuck Hunt), 58
"My First Book," 33
The Mysterious Island, 3

Napoléon Bonaparte, 263, 422
Narrative
    counterfactual, 91
    and knowledge, 398
    and maps, 1, 33–37, 173, 180, 184–187, 191, 282, 361–362, 368–369, 382

Narrative (cont.)
  omniscient, 87
  organization, 36–39
  projections, 89–90
  and space, 31, 39, 67, 111, 368, 372
The Narrative of Arthur Gordon Pym, 326, 339, 347
Nation building, 299, 335–344
"Naturalism Revived," 125
Nature cartography
  and desire, 263, 272–275, 448
  mathematical/aesthetic dimensions, 255
  and organization of life forms, 265–266
  and ownership, 264–265, 272
  and untamed nature, 254
Navarra discovery, 173
"'Nautilus' et 'Bateau ivre,'" 286
Navigation, 364–368
Navigational charts, 154, 364, 367, 369, 376
Neckham, Alexander, 154
Un nègre à Paris, 312
Nellie Bly's Book: Around the World in 72 Days, 347
New & Correct Map of the Whole World, 74
A New Home– Who'll Follow?, 338–339
"The Nightingale of Keats," 121–122
Nizon, Paul, 61
Nominalism, 119–122
Norindr, Panivong, 310
Norman's Chart of the Lower Mississippi River, 339
Northern Frisia, 50–52, 55
"Nota preliminar," 131
Notebooks of Inquiry (Carnets d'enquête), 34
Notes on the State of Virginia, 335
Notre-Dame de Paris, 280
Novel maps, 279–293
  as concept, 279–283
  embedded, 393, 405
  mapped self, 283–285
  and national identity, 284
  novel as map, 289–292, 391–392
  and technological progress, 285–289

Noyes, John K., 447
N'Sondé, Wilfried, 313
Nunes, Pedro, 163, 201

Odyssey, 143–153, 166, 446
  cartographic representations of, 145–148
  history of, 143
  locations of, 150
  in modern literature, 164–166
  polytropy in, 151–153
  as recursion, 149–151
  and sailing manuals, 143, 146, 148, 150
  topography, 144–145
"Of Other Spaces," 4
Ogilby, John, 344
Olaus Magnus, 362
Omniscient narrator, 87
"On Rigor in Science," 125
On the Road, 349
Opiz, Johann Ferdinand, 428
Das Opiz'sche Kriegsspiel, 428
The Order of Things, 108–109
Orlando furioso, 27
Ortelius, Abraham, 27–30, 79, 146, 201, 211, 433
"Our Poor Individualism," 121–122

Padrón, Ricardo, 272, 447
Paesi novamente retrovati, 327
Paolini, Christopher, 347
Paratextual doubling, 400–405, 450
Parcours lines, 367, 369
Parergon, 27–28
Paris, 21–22
Parks, 25
"Partial Enchantments of the *Quixote,*" 126–127
Pauwels, Yves, 230
Perec, Georges, 37
Le Père Goriot, 291
Peter of Poitiers, 187
The Phantom of the Poles, 349
Le Phare au bout du monde, 381

Philosophy
  and cartographic writing, 102
  and geometry, 100
Physikalischer Atlas, 81
Piatti, Barbara, 79, 444
Picker, Marion, 6
Pinet, Simone, 446
The Pioneers, 326
Plato, 119–120, 145
Pliny the Elder, 153
Les plus excellens bastiments de la France, 227–231, 234
Poe, Edgar Allen, 164, 326, 339, 347
Polar regions, 163
Polytropos, 151
Portolan charts, 154, 364, 367, 369, 376
Postel, Guillaume, 236
Postmodern Geographies, 4
The Postmodern Condition, 127
Pound, Ezra, 166
Prados, John, 430
Pragmatism, 129, 131–132
Pratt, Mary Louise, 261
Précis de l'art de la guerre, 420
Preface, role of, 404
Preuss, Charles, 335
Primeiras Estórias, 399–400
A Princess on Mars, 347
The Production of Space, 5
Projected spaces, 52–55, 64
Projections, map, 87–90
Protocartographic operations, 143–144
Proust, Marcel, 84, 86, 284, 290–291
Psychology, 129
Ptolemaic maps, 199–203, 209, 272
Ptolemy, 31, 74, 153, 199–200, 203, 364, 413

Radical empiricism, 129–130
Ramayana, 24
Rancière, Jacques, 174, 280
Reading, 61, 400–401, 404–405
Realism, 119–122
À la recherche du temps perdu, 84–87, 290–293

Recursion, 143–144, 149–151
Redburn, 326
Reed, William, 349
Referentiality, 83
Referential mappability, 79, 90–91
Reflexive classes, 24
"Reframing the Victorians," 66
Report of the Exploring Expedition to the Rocky Mountains in the Year 1842 ..., 335
Representation
  in French literature, 280–282
  hypotyposis, 101
  and maps, 48–49, 282
Rereading, 400–401, 404–405
Rest, Jaime, 124
The Return of the Native, 83
Riders of the Purple Sage, 346
River maps, 339–344
Rivers, 219–224, 230–231, 234–238, 240
Les Rivières de France, 240
The Road to Botany Bay, 260
Robinson Crusoe, 3, 30, 331
Rónai, Paulo, 399
Ronsard, Pierre de, 219, 224–226, 230, 246
Rosa, João Guimarães, 391–401, 404–405, 449–450
Rotae, 188–189
Rougon-Macquart, 290
Route books, 367
Route tours, 206
Rowland, Clara, 449
Royce, Josiah, 22, 24
Russell, Bertrand, 24

Sabin, Philip, 428
Sagarana, 398–399
Said, Edward, 352
Sailing manuals, 143, 146, 148, 150
San Pedro de Rocas, 180
Sanson, Nicolas, 220, 240–243, 246
Sarrasine, 290–291
Schalansky, Judith, 3
Schiller, Friedrich, 55, 412, 417–420, 434
Schmitz, Hermann, 30

Schopenhauer, Arthur, 404
Scott, Walter, 46
Scudéry, Madeleine de, 220, 243
Sea charts, 154, 364, 367, 369, 376
Sea currents, 378–380
Semiotics, 100
Semi-thing, 30
Settings, weight of, 58
Ship pilots, 202
Shklovsky, Viktor, 7, 9
Signs
  arbitrariness of, 120
  maps as, 361
Simons, Oliver, 445
Simulated War, 428
Smith, John, 329
Smooth space, 397
Socé, Ousmane, 312
Soja, Edward, 4, 6, 46
Soledad primera, 209, 212–214
Song of Solomon, 349
Soumission, 292
Space, 4–5
  abstract, 200–201, 207
  collateral/correlative, 111
  dynamization of, 362, 449
  fictional, 50
  imaginary, 64–65
  and knowledge epochs, 106–107
  and maps, 24–25, 31–32
  and narrative, 31, 39, 67, 111, 368, 372
  and plot, 33–37
  projected, 52–55
  as static framework, 362
  and temporal projections, 411
Spanish cartographic culture, 199–214
  and abstract space, 200–201, 207
  and archival materials, 202, 207–208
  cartographic modes, 209–214
  coordinate grid, 200, 207
  itineraries, 207–209, 447
  Ptolemaic maps, 199–203, 209
  travel narratives, 202–209

Spatial turns, 4–6, 106, 143
Sperber, Suzi, 398
Spitteler, Carl, 45
Staël, Madame Germaine de, 279, 282–285
Star charts, 413–417
State, 122–123
Steinbeck, John, 349
Steinberg, Saul, 87
Stendhal (Marie-Henri Beyle), 283, 423–425, 434
Sterne, Laurence, 1, 7, 9
Stevenson, Robert Louis, 3, 33–34, 79, 443
Stifter, Adalbert, 89
Stjernfelt, Frederik, 173
Stockhammer, Robert, 444
Storm, Theodor, 52–55
Stowe, Harriet Beecher, 339, 344
Strabo, 149
Structure of Artistic Texts, 368–369
A Study of Metaphors, 131
Super-Enlightenment, 417
Surveying, 201–202
Swift, Jonathan, 74
Symbolic hypotyposes, 101–102
Symmes, John Cleves, 347
Symzonia: Voyage of Discovery, 347

Tables of contents, 400–405
Tabula Mundi Geographico Zoologica, 270–272
Tag- und Jahreshefte, 264, 267
Tender Is the Night, 55–58
Terrarum Orbis maps, 182–187
De la terre à la lune, 289
Theatrum orbis terrarum, 28
Thematic maps, 79–83, 370
Theory of Prose, 9
The Third Reich, 412, 430–433
Thomas, Dominic, 448
Thoreau, Henry David, 326
Thousand and One Nights, 24
"Three Versions of Judas," 124
Tintin, 309
Tolkien, J. R. R., 347
Tolstoy, Leo, 423, 434

A Topographicall Description ... of Barbados, 331
Topographic military maps, 420–424
Topographies, 5, 83
Topography, 5, 221
Topology, 5, 108
Toponyms, 86, 225
Les Tragiques, 220, 231–236, 246
Tram, 83, 315
Travel narratives, 202–209, 325, 327–331, 335–336
Travels in New England and New York, 335
Travels through North and South Carolina, Georgia, East and West Florida, the Cherokee Country, etc., 335
Travels with Charley, 349
Treasure Island, 3, 33, 79
Trihedron, 107
Tristram Shandy, 1–2, 7–9
A True and Exact History of the Island of Barbados, 331
Turin mosaic, 180–182
Tutaméia, 398, 400–405
Twain, Mark, 339

Ulysses, 36, 67, 79, 166
Uncle Tom's Cabin, 339, 344
Ungern-Sternberg, Arnim von, 64
Universalis Cosmographia Secundum Ptholomaei Traditionem et Americi, 327
Unmappability, 83–85
Utopia, 3, 28
Utz, 52

Varieties of Religious Experience, 129–130
Vaugondy, Robert de, 272
Venturini, Georg, 428
Verne, Jules, 3, 30, 282, 286–289, 362, 372, 377–378, 380–381, 449
"Verortungen: Literatur und Literaturwissenschaft" ("Locations: Literature and Literary Studies"), 61
Verse maps, 210–211

Vie de Henry Brulard, 283
La vie mode d'emploi, 37
Viggiano, Alan, 392
Vingt mille lieues sous les mers, 286–289
Virgil, 152
The Virginian, 346
Virilio, Paul, 4
Visualization, 177–182
Vlyssis Errores, 146
Voß, Johann Heinrich, 146
"Le Voyage de Tours," 224–226
Voyages extraordinaires, 286

Walden, 326
Waldseemüller, Martin, 327
Walker, John, 338
Walker's Geographical Pastime, 338
Wallenstein, 412, 417–420
Wallenstein, Albrecht von, 412–414
War and Peace, 423
War cartographies, 411–434
  astrology, 413, 417, 419–420
  Bolaño's *The Third Reich,* 412, 430–433
  celestial event maps, 412–413, 417–418
  futures of, 433–434
  military event maps, 411–412, 423–424, 428–429, 432, 434, 450
  Schiller's *Wallenstein,* 412, 417–420
  star charts and horoscopes, 413–417
  temporal projections, 411
  topographic military maps, 420–424
  war games, 425–430
Weber, Ann-Kathrin, 60
"When Fiction Lives in Fiction," 126
White, Kenneth, 364, 368
Whittington, Karl, 175, 191
Wilhelm Meister's Apprenticeship, 255
Wilhelm Tell, 48, 55
Willkür, 258–260
Wind poems, 188–189
Wister, Owen, 346
Wizard of Oz, 347
Wolf, Burkhardt, 446

Wood, Denis, 6
Woodward, David, 199
The World as Will and Representation, 404
Wright, Edward, 201

"Xaver Z'Gilgen," 45

"The Zahir," 22
Zigzag Journeys in the Western States of America: The Atlantic to the Pacific, 346–347
Zimmermann, Eberhard, 270
Zola, Émile, 34, 37, 286, 290–292
Zonal maps, 182
Zumthor, Paul, 401